M. V. Fedoryuk Asymptotic Analysis

Mikhail V. Fedoryuk

Asymptotic Analysis

Linear Ordinary Differential Equations

Translated from the Russian
by Andrew Rodick

With 26 Figures

Springer-Verlag Berlin Heidelberg GmbH

Mikhail V. Fedoryuk †

Title of the Russian edition:
Asimptoticheskie metody dlya linejnykh obyknovennykh differentsial'nykh uravnenij
Publisher Nauka, Moscow 1983

Mathematics Subject Classification (1991): 34Exx

ISBN 978-3-642-63435-2

Library of Congress Cataloging-in-Publication Data. Fedoriuk, Mikhail Vasil'evich. [Asimptoticheskie metody dlia lineĭnykh obyknovennykh differentsial' nykh uravneniĭ. English] Asymptotic analysis : linear ordinary differential equations / Mikhail V. Fedoryuk ; translated from the Russian by Andrew Rodick. p. cm. Translation of: Asimptoticheskie metody dlia lineĭnykh obyknovennykh differentsial' nykh uravneniĭ. Includes bibliographical references and index.
ISBN 978-3-642-63435-2 ISBN 978-3-642-58016-1 (eBook)

DOI 10.1007/978-3-642-58016-1

1. Differential equations–Asymptotic theory. I. Title.
QA371.F3413 1993 515'.352–dc20 92-5200

Originally published by Springer-Verlag Berlin Heidelberg New York in 1993
Softcover reprint of the hardcover 1st edition 1993
Typesetting: Springer T_EX in-house system
41/3140 - 5 4 3 2 1 0 - Printed on acid-free paper

Contents

Introduction

In this book we present the main results on the asymptotic theory of ordinary linear differential equations and systems where there is a small parameter in the higher derivatives. We are concerned with the behaviour of solutions with respect to the parameter and for large values of the independent variable. The literature on this question is considerable and widely dispersed, but the methods of proofs are sufficiently similar for this material to be put together as a reference book. We have restricted ourselves to homogeneous equations. The asymptotic behaviour of an inhomogeneous equation can be obtained from the asymptotic behaviour of the corresponding fundamental system of solutions by applying methods for deriving asymptotic bounds on the relevant integrals.

We systematically use the concept of an asymptotic expansion, details of which can if necessary be found in [Wasow 2, Olver 6]. By the "formal asymptotic solution" (F.A.S.) is understood a function which satisfies the equation to some degree of accuracy. Although this concept is not precisely defined, its meaning is always clear from the context. We also note that the term "Stokes line" used in the book is equivalent to the term "anti-Stokes line" employed in the physics literature.

In Chapter 1 we study briefly the basic facts of the analytic theory of differential equations. In § 2, paragraph 4, and § 3, paragraph 3, we give results obtained in recent years on the moving of the boundary condition from a singular point of the equation to a non-singular one.

In Chapter 2 we consider second-order equations on a finite interval and on the half line. We give the asymptotic formulae for the solutions of equations with a small parameter in the highest derivative for the case where the equation has no turning points. We give the asymptotic formulae for large values of the independent variable, and also formulae that are applicable for both small values of the parameter and also large values of the independent variable (dual asymptotic behaviour). In § 5 analogous results are given for systems of equations of arbitrary order that are close to diagonal form. In § 8 we give examples which show that the existence of formal asymptotic behaviour does not always imply the existence of solutions having such asymptotic behaviour.

In Chapter 3 we consider second-order equations in the complex plane which have a large parameter. This part of the asymptotic theory is poorly represented in the literature. Asymptotic formulae for solutions are given in domains not containing turning points or small neighbourhoods of them. We describe the maximal domains for which the asymptotic formulae for the solutions are valid, and we give asymptotic formulae for transition matrices, which allows us to construct the asymptotic behaviour of solutions in the large. A series of applications is considered: the asymptotic behaviour of eigenvalues of equations with analytic coefficients (including non-selfadjoint ones or those having regular singular points), the asymptotic behaviour of the scattering matrix in quasiclassical approximation, and the asymptotic behaviour of the width of the gaps in the spectrum of the Sturm-Liouville operator with periodic potential.

In Chapter 4 we give the asymptotic formulae for the solutions of second-order equations in a real or complex neighbourhood of a turning point. We consider the cases where there is the fusion of turning points or of turning points and singular points of the equation.

In Chapter 5 we give results of the same type as those in Chapters 2 to 4, but for equations and systems of order greater than two. In § 6 we formulate results obtained with the help of the Maslov canonical operator. In § 8 we consider the scattering problem for the Stueckelberg system.

In this reference book we do not discuss results relating to the Orr-Sommerfeld equation, nor the method of averaging for equations with rapidly oscillating coefficients. Despite this limitation, we hope that this book will be useful in mathematics, mechanics and physics when the asymptotic methods of the theory of ordinary linear differential equations are needed.

M.V. Fedoryuk

Acknowledgements

The translator would like to thank Dr. R.K. Thomas for mathematical advice, Mr. David Woolley for his technical assistance, and Ms. Terri Moss for her secretarial skills.

A. Rodick

Springer-Verlag would like to thank Professor Boris Vainberg and Professor Michael Eastham for their valuable help with the publication of this book.

Chapter 1. The Analytic Theory of Differential Equations

In this chapter we consider linear equations and systems with holomorphic or meromorphic coefficients. We introduce the basic facts of the analytical theory of differential equations concerning the local structure of solutions. The reader can find more detailed information in the references [Ince, Wasow 2, Golubev, Kamke, Coddington, Hille].

§ 1. Analyticity of the Solutions of a System of Ordinary Differential Equations

1. Cauchy's Theorem. We formulate some ideas from the theory of analytic functions. Let $\zeta = (\zeta_1, \ldots, \zeta_m)$, where ζ_j are complex variables. The function $\phi(\zeta)$ is called *holomorphic* at the point $\zeta^0 = (\zeta_1^0, \ldots, \zeta_m^0)$ if it can be expanded as a power series

$$\phi(\zeta) = \sum_{|\alpha|=0}^{\infty} \phi_\alpha (\zeta - \zeta^0)^\alpha ,$$

which converges in some neighbourhood of the point ζ^0. Here the notation used is: $\alpha = (\alpha_1, \ldots, \alpha_m)$ is a multi-index, $\alpha_j \geqslant 0$ are integers, $|\alpha| = \alpha_1 + \ldots + \alpha_m$, $(\zeta - \zeta^0)^\alpha = (\zeta_1 - \zeta_1^0)^{\alpha_1} \times \ldots \times (\zeta_m - \zeta_m^0)^{\alpha_m}$. Thus the series for the function $\phi(\zeta)$ has the form

$$\phi(\zeta_1, \ldots, \zeta_m) = \sum_{\alpha_1, \alpha_2, \ldots, \alpha_m = 0}^{\infty} \phi_{\alpha_1 \ldots \alpha_m} (\zeta_1 - \zeta_1^0)^{\alpha_1} \ldots (\zeta_m - \zeta_m^0)^{\alpha_m} .$$

A vector function (or matrix function) is called *holomorphic* at the point ζ^0 if each of its components (elements) is holomorphic at this point. A function holomorphic at the point ζ_0 is also holomorphic in some neighbourhood of this point.

We consider the non-linear system of ordinary differential equations

$$\frac{dw_j}{dz} = f_j(z, w_1, \ldots, w_n) , \quad 1 \leqslant j \leqslant n .$$

Let us introduce the vector functions $w(z) = (w_1(z), \dots, w_n(z))^T$ and $f = (f_1, \dots, f_n)^T$, where the superscript T indicates a column vector. Then the system has the form

$$\frac{dw}{dz} = f(z, w). \tag{1}$$

We state Cauchy's problem:

$$w(z_0) = w^0, \tag{2}$$

where $w^0 = (w_1^0, \dots, w_n^0)^T$ is a given vector. A classical result is Cauchy's theorem.

Cauchy's Theorem. *Let the vector function $f(z, w)$ be holomorphic at the point (z_0, w^0). Then there exists a unique solution to the Cauchy problem (1), (2) that is holomorphic at the point z_0.*

This theorem has a local character: a holomorphic solution exists, generally speaking, only in some neighbourhood U of the point z_0. By extending this solution analytically along all paths with initial point z_0 at which this extension is possible, we obtain the full analytic vector function $\tilde{w}(z)$. This function can turn out to be multivalued and to have singular points. For linear systems the singularities of $\tilde{w}(z)$ are considered in § § 2, 3.

Cauchy's theorem carries over to higher order equations without any changes:

$$w^{(n)} = f(z, w, w', \dots, w^{(n-1)}). \tag{3}$$

We state the Cauchy problem:

$$w(z_0) = w_0, \quad w'(z_0) = w_1, \ \dots \ , w^{(n-1)}(z_0) = w_{n-1}. \tag{4}$$

If the function $f(z, \zeta_0, \zeta_1, \dots, \zeta_{n-1})$ is holomorphic at the point

$$z = z_0, \quad \zeta_0 = w_0, \ \dots, \ \zeta_{n-1} = w_{n-1}$$

for the totality of the variables $(z, \zeta_0, \dots, \zeta_{n-1})$, then there exists a holomorphic solution $w(z)$ to Cauchy's problem (3), (4) in some neighbourhood of the point z_0, and this solution is unique.

2. Linear Equations and Systems. We consider the system of n linear equations

$$\frac{dw}{dz} = A(z)w + f(z). \tag{5}$$

Here $w(z)$ and $f(z)$ are vector functions, $w = (w_1, \dots, w_n)^T$, $f = (f_1, \dots, f_n)^T$ and $A(z)$ is an $n \times n$ matrix with elements $a_{jk}(z)$. In component form, system (5) has the form

$$\frac{dw_j}{dz} = \sum_{k=1}^{n} a_{jk}(z)w_k + f_j(z)\,, \quad 1 \leqslant j \leqslant n\,. \tag{5'}$$

Suppose that the matrix function $A(z)$ and the vector function $f(z)$ are holomorphic in a simply-connected bounded domain D. Then the solution of the Cauchy problem (5), (2) exists, and it is unique and holomorphic in the domain D.

If the domain D is not simply-connected then the solution $w(z)$ of the Cauchy problem (5), (2) is an analytic vector function in D which is not in general single-valued.

The solutions of the nth order linear equation

$$w^{(n)} + q_1(z)w^{(n-1)} + \ldots + q_n(z)w = f(z) \tag{6}$$

have precisely the same properties. If all the coefficients $q_1(z), \ldots, q_n(z)$ and $f(z)$ are holomorphic in a simply-connected domain D then the solution of the Cauchy problem (6), (4) is holomorphic in D. If this domain D is not simply-connected then any solution of the equation is a function which is analytic in D.

3. Singular Points of Linear Equations. We consider the linear homogeneous system

$$\frac{dw}{dz} = A(z)w \tag{7}$$

with matrix function $A(z)$, holomorphic in a punctured neighbourhood of the point $z = a$ (that is, for $z \in U\backslash a$ where U is a neighbourhood of a). If the point a is singular for even one of the elements of the matrix $A(z)$ then a is called a *singular point* of the matrix $A(z)$. In this case the point $z = a$ is also called a *singular point* of the system (7).

The singular point at infinity is defined differently. With the change of variable $z = 1/\zeta$, (7) becomes

$$\frac{d\tilde{w}}{d\zeta} = -\zeta^{-2}A(\zeta^{-1})\tilde{w}\,, \tag{8}$$

where $\tilde{w}(\zeta) = w(\zeta^{-1})$. If the point $\zeta = 0$ is singular for the system (8) then the point $z = \infty$ is called *singular* for the system (7).

The structure of the solutions of (7) in a neighbourhood of a singular point is as follows.

If $a(\neq \infty)$ is a singular point of the system (7) then there exists a fundamental matrix (F.M.) $W(z)$ of the system with the form

$$W(z) = \Phi(z)(z-a)^P\,. \tag{9}$$

Here P is a constant matrix and the matrix function $\Phi(z)$ has the Laurent expansion

$$\Phi(z) = \sum_{-\infty}^{\infty} \Phi_k (z-a)^k ,$$

which converges in some annulus $0 < |z-a| < \rho$, the Φ_k being constant $n \times n$ matrices.

If $z = \infty$ is a singular point of the system (7) then there exists a fundamental matrix of this system with the form

$$W(z) = \Phi(z) z^P . \tag{10}$$

Here again P is a constant matrix and the matrix function $\Phi(z)$ has the Laurent expansion

$$\Phi(z) = \sum_{-\infty}^{\infty} \Phi_k z^{-k} ,$$

converging in some annulus $|z| > R$.

Similar results are true for n^{th}-order linear homogeneous equations.

If the matrix P in (9) and (10) cannot be reduced to diagonal form then $W(z)$ can contain logarithmic terms. For example if J is the single Jordan block of size $k \times k$

$$J = \begin{bmatrix} \lambda & 1 & & \\ & \ddots & 0 & \\ & \lambda & & \\ & & \ddots & 1 \\ 0 & & & \lambda \end{bmatrix}, \quad \text{then } z^J = \begin{bmatrix} z^\lambda & z^\lambda \ln z & & \frac{z^\lambda (\ln z)^{k-1}}{(k-1)!} \\ & z^\lambda & \ddots & \\ 0 & & \ddots & z^\lambda \ln z \\ & & & z^\lambda \end{bmatrix} .$$

Formulas (9) and (10) give only a general idea of the structure of the fundamental matrix near a singular point, since the effective calculation of the matrices P and $\Phi(z)$ from the matrix $A(z)$ is not possible in general (see § 3).

In the analytic theory of differential equations one considers the problem of the structure of the fundamental matrix in a neighbourhood of a pole of the matrix function $A(z)$. We introduce the following classification of singular points. The point a is called a *regular singular point* of the system (7) if the matrix function $\Phi(z)$ in (9) and (10) has a pole at the point a (or is holomorphic at this point). Otherwise the point a is called *irregular*.

This classification is indirect; the definition does not allow us to establish the character of the singular points directly from the matrix $A(z)$. In a similar way one classifies singular points of nth-order linear homogeneous equations with meromorphic coefficients.

One of the fundamental problems in the analytic theory of linear differential equations is the study of the structure of a F.M. (or of a fundamental system of solutions (F.S.S.) in the case of scalar equations) given the coefficient matrix $A(z)$ (or the coefficients $q_k(z)$ of the equation). The basic results obtained in this direction are given in § § 2 and 3.

§ 2. Regular Singular Points

1. Scalar Equations. We consider the homogeneous n^{th}-order linear equation

$$lw \equiv w^{(n)} + q_1(z)w^{(n-1)} + \ldots + q_n(z)w = 0\,. \tag{1}$$

The coefficients in this equation either have a pole or are holomorphic at the point a.

1.1 Conditions for a Regular Singular Point. For the point $a(\neq \infty)$ to be a regular singular point of equation (1) it is necessary and sufficient that the coefficients $q_k(z)$ have poles of order not greater than k at a, $1 \leqslant k \leqslant n$. At least one function $q_k(z)$ must have a pole at a for otherwise this point would not be singular.

If a is a regular singular point then equation (1) has the form

$$lw \equiv w^{(n)} + (z-a)^{-1}p_1(z)w^{(n-1)} + \ldots + (z-a)^{-n}p_n(z)w = 0\,, \tag{2}$$

where the $p_k(z)$ are holomorphic at a.

For the point $z = \infty$ to be regular singular for equation (1) it is necessary and sufficient that each coefficient $q_k(z)$ has a zero of multiplicity not greater than k at the point $z = \infty$. In this case equation (1) has the form

$$lw \equiv w^{(n)} + z^{-1}p_1(z)w^{(n-1)} + \ldots + z^{-n}p_n(z)w = 0\,, \tag{3}$$

where the $p_k(z)$ are holomorphic at $z = \infty$.

1.2 Structure of the Solution in a Neigbourhood of a Regular Singular Point. Let us consider an example. Euler's equation

$$z^n w^{(n)} + a_1 z^{n-1} w^{(n-1)} + \ldots + a_n w = 0\,,$$

where $a_1, \ldots, a_n$ are constants, has two singular points 0 and ∞, both regular. This equation is integrable. We seek a solution in the form $w = z^\rho$. Substituting this solution into the equation and dividing by z^ρ we obtain the *indicial equation*

$$\begin{aligned} f(\rho) \equiv \rho(\rho-1)\ldots(\rho-n+1) \\ + a_1\rho(\rho-1)\ldots(\rho-n+2) + \ldots + a_n = 0\,. \end{aligned}$$

If ρ_0 is a root of multiplicity k of this equation then the functions z^{ρ_0}, $z^{\rho_0}\ln z$, $\ldots$, $z^{\rho_0}(\ln z)^{k-1}$ are solutions of the Euler equation. The totality of all such solutions form a F.S.S. Similarly we can construct the solutions of equation (1) in the neighbourhood of a singular point. Equation (2) has indicial equation

$$f(\rho) \equiv \rho(\rho-1)\dots(\rho-n+1) \\ + p_1(a)\rho(\rho-1)\dots(\rho-n+2)+\dots+p_n(a)=0\,. \qquad (4)$$

Let $\rho_1,\dots,\rho_n$ be the roots of this equation. Fix ρ_k. If all the differences $\rho_j-\rho_k$ $(1\leqslant j\leqslant n,\ j\neq k)$ are not non-negative integers then equation (2) has a solution of the form

$$w_k(z)=(z-a)^{\rho_k}\phi_k(z)\,, \qquad (5)$$

where $\phi_k(z)$ is holomorphic at $z=a$ and, further, $\phi_k(a)\neq 0$.

If all the differences $\rho_j-\rho_k$ $(j\neq k,\ 1\leqslant j,\ k\leqslant n)$ are not integers then equation (2) has a F.S.S. $\{w_1(z),\dots,w_n(z)\}$ consisting of solutions of the form (5).

Similarly we can form the solutions of equation (3) in a neighbourhood of the regular singular point $z=\infty$. The indicial equation has the form

$$f(\rho) \equiv \rho(\rho-1)\dots(\rho-n+1) \\ + p_1(\infty)\rho(\rho-1)\dots(\rho-n+2)+\dots+p_n(\infty)=0\,. \qquad (6)$$

Let $\rho_1,\dots,\rho_n$ be the roots of this equation. Fix ρ_k. If all the differences $\rho_j-\rho_k$ $(1\leqslant j\leqslant n,\ j\neq k)$ are not integers then equation (3) has a solution of the form

$$w_k(z)=z^{\rho_k}\phi_k(z)\,, \qquad (7)$$

where $\phi_k(z)$ is holomorphic at $z=\infty$ and $\phi_k(\infty)\neq 0$.

If there is an integer among the differences $\rho_j-\rho_k$, $j\neq k$, then the solutions of equation (2) can contain $\ln^m(z-a)$, where $m>0$ is an integer. Formulae for a F.S.S. are also known in this case [Kamke, Coddington] but are unwieldy. It is simpler and more convenient to give an algorithm for them.

1.3 Frobenius' Method. Let $z=0$ be a regular singular point of equation (2). Thus we take $a=0$ for simplicity. We seek a formal series of the form

$$w(z)=z^{\rho}\sum_{j=0}^{\infty}c_jz^j\,, \quad c_0=1\,, \qquad (8)$$

such that

$$lw=f(\rho)z^{\rho}\,. \qquad (9)$$

We have

$$lw=f(\rho)z^{\rho}+[c_1f(\rho+1)-g_1]z^{\rho+1}+\dots \\ \dots+[c_jf(\rho+j)-g_j]z^{\rho+j}+\dots\,,$$

where $g_j = \sum_{k=1}^{j-1} g_{jk}(\rho)c_k$ and $g_{jk}(\rho)$ are polynomials. From the recurrence system of equations

$$c_j f(\rho + j) = g_j \tag{10}$$

we can successively find the coefficients $c_1, c_2, \dots$ as functions of ρ. Each one will be a rational function of ρ. Fix $\rho = \rho_1$. The following cases are possible.

1) ρ_1 is a root of the indicial equation and none of the differences $\rho_1 - \rho_2, \dots, \rho_1 - \rho_n$ is a positive integer. Then the series (8) with $\rho = \rho_1$ is a solution of equation (2) since we can find successively $c_1, c_2, \dots$ from system (10).

2) Let the conditions of 1) be satisfied and let ρ_1 be a root of multiplicity $m > 1$. Then the functions

$$w(z,\rho), \frac{\partial w(z,\rho)}{\partial \rho}, \dots, \frac{\partial^{m-1} w(z,\rho)}{\partial \rho^{m-1}}, \quad \rho = \rho_1 ,$$

are linearly independent solutions of equation (2). These solutions are polynomials in $\ln z$ of degree $0, 1, \dots, m-1$.

3) Let equation (4) have a root ρ_2 such that $\rho_1 - \rho_2 = k$, where $k(> 0)$ is an integer, and let $f(\rho_j + j) \neq 0$ for $1 \leqslant j < k$ and for $j > k$. Let m be the multiplicity of the root ρ_1. We look for the solution of equation (2) in the form of the series (8) where $c_0 = (\rho - \rho_2)^m$. We have

$$lw = f(\rho)(\rho - \rho_2)^m z^\rho .$$

From system (10) we find that the coefficients $c_1, \dots, c_{k-1}$ contain the factor $(\rho - \rho_2)^m$, and from the equation

$$f(\rho + k)c_k = g_k$$

it follows that c_k does not have a pole at $\rho = \rho_2$, since $f(\rho + k)$ and g_k are divisible by $(\rho - \rho_2)^m$. Therefore the point $\rho = \rho_2$ is not a pole for any of the coefficients c_k.

The series (8), with $\rho = \rho_2$, is a solution of equation (2), proportional to that found in case 1. The solution corresponding to the root ρ_2 has the form $\partial^m w/\partial \rho^m|_{\rho=\rho_2}$. If ρ_2 is a multiple root then the resulting solution can be obtained by successively differentiating, as in case 2.

We consider the remaining case where several of the differences $\rho_j - \rho_k$ are integers in a similar way. F.S.S. of equation (2) consists of solutions of the form

$$w = z^\rho \sum_{k=0}^{m} \phi_k(z)(\ln z)^k ,$$

where ρ is a root of the indicial equation and the $\phi_k(z)$ are holomorphic at the point $z = 0$.

1.4 Second-Order Equations. We consider the equation

$$z^2w'' + zp(z)w' + q(z)w = 0, \tag{11}$$

where $p(z)$ and $q(z)$ are holomorphic in the disk $|z| < R$. The point $z = 0$ is a regular singular point of equation (11). Let us construct the F.S.S. of equation (11) inside the disk $|z| < R$. We have

$$p(z) = \sum_{k=0}^{\infty} p_k z^k, \quad q(z) = \sum_{k=0}^{\infty} q_k z^k.$$

We seek a solution in the form

$$w = z^{\rho} \sum_{k=0}^{\infty} c_k z^k, \quad c_0 = 1. \tag{12}$$

Substituting this into (11) we obtain the recurrence system of equations

$$\begin{aligned} c_0 f(\rho) &= 0, \\ c_1 f(\rho + 1) + c_0 f_1(\rho) &= 0, \\ \cdots\cdots\cdots\cdots\cdots\cdots\cdots\cdots\cdots \\ c_k f(\rho + k) + c_{k-1} f_1(\rho + k - 1) + \ldots + c_0 f_k(\rho) &= 0, \end{aligned} \tag{13}$$

where

$$f(\rho) = \rho(\rho - 1) + p_0\rho + q_0, \quad f_k(\rho) = \rho p_k + q_k.$$

The indicial equation has the form $f(\rho) = 0$. Let ρ_1, ρ_2 be the roots of this equation.

1) Suppose that $\rho_1 - \rho_2$ is not an integer. Then $f(\rho_1 + k) \neq 0$ for any integer $k > 0$ and, from equations (13) with $\rho = \rho_1$ and $\rho = \rho_2$, we can find successively $c_1, c_2, \ldots$. In this case equation (11) has a F.S.S. of the form

$$w_1 = z^{\rho_1}\phi_1(z), \quad w_2 = z^{\rho_2}\phi_2(z),$$

where the $\phi_j(z)$ are holomorphic in the disk $|z| < R$ and $\phi_j(0) \neq 0$, $j = 1, 2$.

2) Suppose that $\rho_1 - \rho_2 = n$, where $n \geqslant 0$ is an integer. Then equation (11) has a F.S.S. of the form

$$w_1 = z^{\rho_1}\phi_1(z), \quad w_2 = aw_1 \ln z + z^{\rho_2}\phi_2(z),$$

where a is a constant and the $\phi_j(z)$ are holomorphic in the disk $|z| < R$.

Remark. It is possible that none of the solutions of equation (11) has a singularity at $z = 0$; for example if $\rho_1 \geqslant 0$ and $\rho_2 \geqslant 0$ are integers and $a = 0$.

1.5 Equations of Fuchs Type. Equation (1) is called an equation of *Fuchs type* if it has only regular singular points on the Riemann sphere (the compactification of the complex plane). Equation (1) is an equation of Fuchs type if and only if its coefficients have the form

$$q_k(z) = b_k(z) \prod_{m=1}^{l} (z - a_m)^{-k} \,,$$

where $b_k(z)$ is a polynomial of degree not greater than $l(k-1)$. Here, the points $a_1, \ldots, a_l$, and ∞ are all regular singular for equation (1).

A second-order equation of Fuchs type with singular points $a_1, \ldots, a_l, \infty$ has the form

$$w'' + \sum_{m=1}^{l} \frac{1 - \rho_1^{(m)} - \rho_2^{(m)}}{z - a_m} w' + \left[\sum_{m=1}^{l} \frac{\rho_1^{(m)} \rho_2^{(m)} \prod_{j=1}^{*l} (a_m - a_j)}{z - a_m} + Q_{l-2}(z) \right] \frac{w}{\prod_{j=1}^{l} (z - a_j)} = 0 \,.$$

Here $\rho_1^{(m)}$ and $\rho_2^{(m)}$ are the characteristic exponents corresponding to the singular point $z = a_m$, the asterisk denotes that $j \neq m$, $Q_{l-2}(z)$ is a polynomial of degree $l-2$ if $z = \infty$ is a singular point, and of degree $l-4$ if $z = \infty$ is not a singular point.

For $l = 1, 2, 3$ (and only for these l) the coefficients of a second order equation of Fuchs type are expressed in terms of singular points and their characteristic exponents. For $l = 1$ the equation can be reduced to the form $w'' = 0$, for $l = 2$ to the Euler equation and for $l = 3$ to the Papperitz (or Riemann) equation.

The fundamental linear ordinary differential equations which arise in problems of mathematical physics can be obtained from a second order equation with five regular singular points [Ince]. In this equation the difference of the characteristic exponents equals 1/2 for any singular point. The fusion of two such singular points forms a regular singularity with arbitrary difference of the characteristic exponents. The fusion of three or more singular points leads to an irregular singularity.

2. A System of Equations. We consider the system

$$w' = (z - a)^{-m-1} A(z) w \,, \tag{14}$$

where $w = (w_1(z), \ldots, w_n(z))^T$, $A(z)$ is an $n \times n$ matrix which is holomorphic at the point a, $A(a) \neq 0$, and m is a non-negative integer.

2.1 Singular Points of the First Kind. The number m is called the *rank* of the singularity a. Also, a is called a *singular point of the first* or *second kind* according as $m = 0$ or $m \geqslant 1$.

A singular point of the first kind is regular (§ 1); that is, system (14) has a fundamental matrix of the form

$$W(z) = \Phi(z)(z - a)^P \,. \tag{15}$$

Here P is a constant matrix and $\Phi(z)$ is holomorphic or has a pole at $z = a$. In contrast to scalar equations, conditions for a regular singular point for the system are unknown. A singular point of the second kind can be both regular and irregular. For example, the rank of $z = 0$ for the system

$$w_1' = w_2\,, \quad w_2' = z^{-2}w_1$$

is equal to 1, so that $z = 0$ is a singular point of the second kind. This system is equivalent to the Euler equation $z^2w'' - w = 0$ for which (and therefore also for the system) $z = 0$ is a regular singular point. There is a series of papers where sufficient conditions are obtained for $z = a$ to be a regular singular point for system (14).

Let $m = 0$ and suppose that none of the differences of the eigenvalues of $A(a)$ is a negative integer. Then system (14) has fundmental matrix of the form

$$W(z) = \Phi(z)(z-a)^{A(a)}\,, \tag{16}$$

where $\Phi(z) = \sum_{k=0}^{\infty} \Phi_k(z-a)^k$ is holomorphic at $z = a$, and $\Phi_0 = I$. We introduce an algorithm for the construction of the solution. Let $a = 0$ for simplicity, so that $A(z) = \sum_{j=0}^{\infty} A_j z^j$. Substituting (16) into (14) and multiplying by z^{-A_0} on the right we obtain the system

$$z\Phi'(z) + \Phi(z)A_0 = A(z)\Phi(z)\,.$$

Equating the coefficients of powers of z in this system we arrive at the recurrence system of equations

$$k\Phi_k + [\Phi_k, A_0] = \sum_{j=1}^{k-1} A_j\Phi_{k-j}\,, \quad k = 1, 2, \ldots\,,$$

where $[\Phi_k, A_0] = \Phi_k A_0 - A_0\Phi_k$ (the commutator of Φ_k and A_0). We consider therefore the matrix equation

$$\mu X + [X, C] = B\,,$$

where μ is a number, X, C and B are square matrices of the same order. It is known from linear algebra that if $\mu \neq \lambda_j - \lambda_k$ for any j and k, where λ_j are the eigenvalues of C, then this equation is solvable for any matrix B. We can therefore find successively $\Phi_1, \Phi_2, \ldots$ from the recurrence system. If further there is even one negative integer among the differences of the eigenvalues then (14) still has a fundamental matrix of the form (15) where $\Phi(z)$ is holomorphic at $z = a$ and P is a constant matrix which has no negative integers among the differences of its eigenvalues.

The singular point $z = \infty$ is studied similary. We consider the system

$$w' = z^{m-1}A(z)w\,,$$

where $A(z)$ is holomorphic at $z = \infty$, $A(\infty) \neq 0$ and m is a non-negative integer. The number $m+1$ is called the *rank* of the system. Also, the point $z = \infty$ is called *a singular point of the first* or *second kind* according as $m = 0$ or $m \geqslant 1$. The above results for a finite singular point are also true for the point $z = \infty$; in (15) and (16) $z - a$ must be replaced by z.

2.2 Systems of Fuchs Type. The system

$$w' = A(z)w \tag{17}$$

is called a system of *Fuchs type* if it has only regular singular points on the Riemann sphere. If in addition all these singular points are of the first kind then the system has the form

$$w' = \sum_{k=1}^{m} \frac{A_k}{z - a_k} w \,.$$

Here, the A_k are non-zero constant matrices, and $a_1, \ldots, a_m$ are the regular singular points.

3. The Monodromy Group. Suppose that the coefficients of equation (1) are holomorphic in a simply-connected domain D on the Riemann sphere, excluding the points $a_1, \ldots, a_m$ which can be poles of the coefficients. The singular points can be regular or irregular. We fix a point a_0, different from those indicated, and a small simply-connected neighbourhood U of a_0. In U there is a holomorphic F.S.S. $w(z) = (w_1(z), \ldots, w_n(z))$ for (1). Let γ_j be a simple closed curve starting and finishing at a_0, going around a_j in the positive direction (that is, γ_j is oriented counter-clockwise, a_j lies inside γ_j and the other singular points lie outside γ_j). Continuing the F.S.S. $w(z)$ analytically along γ_j we arrive at the F.S.S.

$$\gamma_j w(z) \equiv w^j(z) = w(z)T_j \,,$$

where T_j is a constant non-singular $n \times n$ matrix. If we extend the F.S.S. $w(z)$ analytically along γ_j^{-1} (that is, along γ_j in the clockwise direction) we obtain the F.S.S.

$$\gamma_j^{-1} w(z) = w(z)T_j^{-1} \,.$$

If we extend the F.S.S. analytically first along γ_j and then along γ_k we obtain the F.S.S.

$$\gamma_k \gamma_j w(z) = w(z) T_j T_k \,.$$

The matrix group G generated by $T_1, \ldots, T_m$ is called the *monodromy group* (or simply the *equation group*) of (1) in D. Any element of the group has the form $T_{i_1}^{k_1} \ldots T_{i_s}^{k_s}$, where $i_\alpha \in \{1, 2, \ldots, m\}$ and $k_1, \ldots, k_s$ are integers.

The monodromy group depends on the choice of the point $a_0 : G = G(a_0)$. If $b_0 \in D$ is a non-singular point and $b_0 \neq a_0$, the groups $G(a_0)$ and $G(b_0)$ are similar: that is, there is a constant non-singular $n \times n$ matrix $T = T(a_0, b_0)$ such that

$$G(b_0) = T^{-1}G(a_0)T\,.$$

The monodromy group for a system of the form (17) is defined in exactly the same way.

The monodromy group can be constructed explicitly only for a small number of classes of differential equations: for the hypergeometric equation, the Pochhammer equation, Bessel's equation and some others. In Chap. 3, § 4, examples are given in which the equation has a large parameter and the asymptotic behaviour of the generators of the group G can be found with respect to the parameter.

4. The Manifold of Solutions. We consider the system of n equations

$$tw' = A(t)w\,, \tag{18}$$

and we introduce the conditions

$$A(t) \in C^\infty\,, \quad 0 \leqslant t \leqslant a\,, \quad A(0) \neq 0\,. \tag{18'}$$

By analogy with the analytical case we call $t = 0$ a *regular singular point.* All the results stated in paragraph 2.1 are also true in this case, the only difference being that all the series (for instance the series for $\Phi(t)$ in (16)) no longer converge, but are asymptotic as $t \to +0$.

4.1 The Manifold of Bounded Solutions. Suppose that the matrix $A(0)$ does not have purely imaginary eigenvalues. Then by means of a linear substitution of the form $w = T\tilde{w}$, where T is a constant non-singular matrix, system (18) can be reduced to a form with $A(0)$ block-diagonal:

$$A(0) = \begin{bmatrix} A_- & 0 \\ 0 & A_+ \end{bmatrix}\,. \tag{19}$$

Here A_- and A_+ are square matrices of orders k and $n-k$, and the eigenvalues of A_- (A_+) lie in the left (right) half-plane:

$$\operatorname{Re}\lambda_1 < 0, \ldots, \operatorname{Re}\lambda_k < 0, \operatorname{Re}\lambda_{k+1} > 0, \ldots, \operatorname{Re}\lambda_n > 0\,. \tag{20}$$

A F.S.S. of (18) consists of the k solutions $w_1(z), \ldots, w_k(z)$, which are unbounded for $t \in (0, a] = I$, and $n - k$ solutions $w_{k+1}, \ldots, w_n$, which are bounded for $t \in I$. We denote by M^{n-k} the set of all solutions which are bounded for $t \in I$. Then M^{n-k} is an $n - k$ dimensional linear space over the field of complex numbers. Its elements have the form $w = \sum_{j=1}^{n-k} c_j w_{j+k}$ where the c_j are arbitrary constants.

The asymptotic behaviour of the bounded solutions as $t \to 0$ is the same as in the analytical case. It is clear that it is quite complicated and, in addition, essentially depends both on the Jordan normal form of the matrix $A(0)$ and also on whether there is an integer among the differences $\lambda_j - \lambda_k$ or not. Therefore these formulae are highly unsuitable for the numerical calculation of problems connected with (18) or with the inhomogeneous system

$$tw' = A(t)w + f(t)\,, \tag{21}$$

if the boundary condition is imposed at the singular point $t = 0$.

Although the structure of individual bounded solutions is complicated, the structure of the manifold M^{n-k} of all bounded solutions is simple [Abramov 1, Abramov 2]. We put

$$w = \begin{bmatrix} w_- \\ w_+ \end{bmatrix}, \quad f(t) = \begin{bmatrix} f_- \\ f_+ \end{bmatrix}, \quad V(t) = A(t) - A(0) = \begin{bmatrix} V_{11} & V_{12} \\ V_{21} & V_{22} \end{bmatrix}.$$

The vector $w_-(w_+)$ has k $(n-k)$ components and the orders of the matrices V_{jk} are clear. We seek an equation determining M^{n-k} in the form

$$w_-(t) = \alpha(t)w_+(t)\,, \tag{22}$$

where $\alpha(t)$ is some unknown matrix of order $k \times (n-k)$. Substituting (22) into (18) we obtain

$$\begin{aligned} tw'_+ &= (A_+ + V_{21}\alpha + V_{22})w_+\,, \\ t\alpha w'_+ &= (A_-\alpha + V_{11}\alpha + V_{12} - t\alpha')w_+\,. \end{aligned}$$

From this we find the equation for α:

$$t\alpha' = A_-\alpha - \alpha A_+ + V_{11}\alpha - \alpha V_{22} - \alpha V_{21}\alpha + V_{12}\,, \tag{23}$$

which is a Riccati matrix equation. The boundary condition for $\alpha(t)$ is

$$\lim_{t\to 0} \alpha(t) = 0\,. \tag{24}$$

The solution of the Cauchy problem (23), (24) for small t exists, is unique and has the series expansion

$$\alpha(t) = \sum_{j=1}^{\infty} \alpha_j t^j\,. \tag{25}$$

This series converges if $A(t)$ is holomorphic at $t = 0$ and is asymptotic under condition (18′). The matrices $\alpha_1, \alpha_2, \ldots$ are found by substituting (25) into (23) and equating the coefficients of powers of t to zero. A recurrence system of linear algebraic equations is obtained for $\alpha_1, \alpha_2, \ldots$. Thus for small $t > 0$ any bounded solution of (18) is a solution of (22) and any solution of (22) is a solution of (18) that is bounded as $t \to 0$.

This method is also applicable to the inhomogeneous system (21). Suppose that $f(t) \in C^\infty$ for $0 \leqslant t \leqslant a$ and the conditions on $A(0)$ are as above. We look for solutions that are bounded as $t \to 0$ in the form

$$w_- = \alpha(t)w_+ + \beta(t),$$

where $\alpha(t)$ is the matrix constructed above and the vector function $\beta(t)$ is the unique solution of the Cauchy problem

$$t\beta' = A_-\beta + (V_{11} - \alpha V_{21})\beta + f_- - \alpha f_+, \quad \lim_{t\to 0} \beta(t) = -A_-^{-1} f_-(0).$$

For small $t > 0$ there is an expansion

$$\beta(t) = \sum_{j=0}^{\infty} \beta_j t^j, \tag{26}$$

having the same properties as (25). The vectors $\beta_0, \beta_1, \ldots$ are determined from a recurrence system of linear algebraic equations.

4.2 A Second-Order System. We consider the homogeneous system

$$t^2 w'' = A(t)w. \tag{27}$$

This system can be reduced to a first order system of the form (18) but it is convenient to study it directly. We look for a solution in the form

$$w(t) = t^\rho \phi(t), \quad \phi(t) = \sum_{k=0}^{\infty} \phi_k t^k,$$

where the series $\phi(t)$ has the same properties as those in paragraph 4.1. Substituting this solution into (27) and equating coefficients of powers of t we obtain a recurrence system of equations, the first of which has the form

$$[(\rho^2 - \rho)I - A_0]\phi_0 = 0.$$

Consequently ρ must be a root of the defining equation

$$\det[(\rho^2 - \rho)I - A_0] = 0,$$

and the vector ϕ_0 is the eigenvector of A_0 corresponding to the eigenvalue $\lambda = \rho^2 - \rho$. Suppose that all the eigenvalues $\lambda_1, \ldots, \lambda_n$ of A_0 lie in the exterior of the parabola Π: $\operatorname{Re}\lambda + (\operatorname{Im}\lambda)^2 = 0$ in the complex λ-plane. Then there are n roots ρ_j such that $\operatorname{Re}\rho_j > 0$ and n roots such that $\operatorname{Re}\rho_j < 0$. The F.S.S. of system (27) consists of n linearly independent solutions that are bounded as $t \to 0$ and n that are unbounded. The equation for the manifold M^n of bounded solutions can be obtained by factorizing the differential operator corresponding to (27). We seek $n \times n$ matrices $\alpha(t)$, $\beta(t)$ which satisfy the identity

$$t^2\frac{d^2}{dt^2} - A(t) = \left(t\frac{d}{dt} - \beta(t)\right)\left(t\frac{d}{dt} - \alpha(t)\right).$$

From this we find that $\beta(t) = I - \alpha(t)$ and obtain the Riccati matrix equation for $\alpha(t)$

$$t\alpha' + \alpha^2 - \alpha = A(t).$$

It is evident that if w is a solution of the system

$$tw' = \alpha(t)w(t), \tag{28}$$

then w is a solution of system (27). We select $\alpha(t)$ such that (28) describes the manifold M^n by imposing the Cauchy condition

$$\lim_{t\to 0} \alpha(t) = \frac{1}{2}(I + \sqrt{I + 4A_0}).$$

Here $\sqrt{I + 4A_0}$ is chosen so that all the eigenvalues of this matrix lie in the upper half-plane Re $\rho > 0$, which is possible because of the conditions on A_0. The solution to the resulting Cauchy problem for $\alpha(t)$ exists for small $t > 0$ and is unique. The expansion (25) is valid for $\alpha(t)$, where the α_j are obtained from recurrence relations.

Let us consider the inhomogeneous system

$$t^2 w'' = A(t)w + f(t). \tag{29}$$

where $A(t)$ satisfies the conditions given above and $f(t) \in C^\infty$ for $0 \leqslant t \leqslant \alpha$. Then there is an n-dimensional manifold M^n of solutions, bounded for $t \to 0$, which is described by the equation

$$tw' = \alpha(t)w + \beta(t). \tag{30}$$

Here the matrix $\alpha(t)$ is as above, the vector function $\beta(t)$ is the solution of the Cauchy problem

$$t\beta' + \alpha\beta - \beta = f(t), \quad \lim_{t\to 0} \beta(t) = [\alpha(0) - I]^{-1} f(0). \tag{31}$$

The vector-function $\beta(t)$ has the series expansion (26) for small $t \geqslant 0$ where the β_j are determined from recurrence relations.

This method is widely used in computational mathematics [Abramov 2]. We consider the boundary problem on the interval $[0, a]$ for system (29):

$$w(t) \text{ is bounded as } t \to 0,$$
$$w(a) = w_0.$$

The boundary condition for $t = 0$ is equivalent to the boundary condition

$$t_0 w'(t_0) = \alpha(t_0)w(t_0) + \beta(t_0)$$

for small t_0. Approximate values for $\alpha(t_0)$ and $\beta(t_0)$ can be found by taking the first few terms of the expansions (25) and (26), which are determined by solving the corresponding systems of linear algebraic equations. Then we arrive at a boundary problem for (2) on the interval $[t_0, a]$ that does not contain singular points. This method can also be developed for irregular singular points (§ 3).

§ 3. Irregular Singular Points

1. The Scalar Equation. We consider the equation

$$w^{(n)} + q_1(z)w^{(n-1)} + \ldots + q_n(z)w = 0\,, \tag{1}$$

with coefficients which are holomorphic or have a pole at $z = \infty$. We have

$$q_j(z) = \sum_{-\infty}^{n_j} q_{jm} z^{-m}\,, \quad 1 \leqslant j \leqslant n\,. \tag{2}$$

These series converge for $|z| > R$. Here q_{j,n_j} is the first non-zero coefficient of the series: if $q_j(z) \equiv 0$ then we put $n_j = -\infty$.

1.1 Criterion for an Irregular Singular Point. Let

$$k = \max_{1 \leqslant j \leqslant n} n_j/j\,, \quad r = k + 1\,. \tag{3}$$

The number r is called the *rank*, and the number k the *subrank*, of the singular point $z = \infty$. Also, $z = \infty$ is a regular or irregular singular point according as $r = 0$ or $r \geqslant 1$.

Let $s \geqslant 1$ be an integer,

$$Q(\zeta) = q_0\zeta^p + \ldots + q_{p-1}\zeta\,, \quad \Psi(\zeta) = \sum_{m=0}^{\infty} \psi_m \zeta^{-m}\,, \tag{4}$$

where $\Psi(\zeta)$ is a formal series. The series

$$w(z) = z^{\rho} e^{Q(z^{1/s})} \Psi(z^{1/q}) \tag{5}$$

is called *normal* of order p if $s = 1$, and *subnormal* of order p/s if $s \geqslant 2$.

The subrank k is an integer or a fraction, so that $k = p/q$, where p and q are coprime natural numbers. If r is an integer then (1) has at least one formal solution of the form (5) of order r and $s = 1$. If $q \geqslant 2$ then (1) has not less than q formal solutions of the form (5) of order r where $s = q$. Series of the form (5) terminate only in exceptional cases and usually diverge. Their asymptotic behaviour will be shown below. Moreover, there are other solutions to (1) besides series of the form (5).

The ideas of rank and subrank carry over in the obvious way to the case of a finite singular point.

1.2 Local Structure of the Solutions: A Special Case. If $z = \infty$ is a regular singular point of equation (1) then all the solutions have roughly the same structure in the neighbourhood of this point. That is, there is a F.S.S. $\{w_1, \dots, w_n\}$ such that

$$w_j(z) \sim z^{\rho_j} (\ln z)^{m_j}, \quad z \to \infty .$$

An irregular singular point is a complicated conglomeration of singularities. For instance, the equation

$$w'' - (q(z) + 1)w' + q(z)w = 0, \quad q(z) = \frac{\rho(\rho - z - 1)}{z(\rho - z)},$$

has solutions $w_1 = z^\rho$, $w_2 = e^z$. The first of these has a singularity of the same type as in the case of a regular singular point, for the second $z = \infty$ is an essential singular point.

Let us consider a case where the structure of the F.S.S. can be described fully. Suppose that in equation (1)

$$q_j(z) = z^{jk} p_j(z), \quad 1 \leqslant j \leqslant n,$$

where the functions $p_j(z)$ are holomorphic at $z = \infty$. Let the roots $\rho_1, \dots, \rho_n$ of the equation

$$\rho_n + p_1(\infty)\rho^{n-1} + \dots + p_n(\infty) = 0 \tag{6}$$

be distinct and non-zero. The rank of equation (1) in this case is $r = k + 1$. Then (1) has n formal solutions of the form (5):

$$w_j(z) = z_j^\rho \exp\left\{\sum_{m=1}^{k+1} a_{jm} z^m\right\} \sum_{m=0}^{\infty} w_{jm} z^{-m}, \tag{7}$$

where

$$a_{j0} = \rho_j/(k + 1), \quad w_{j0} = 1 .$$

These formal series are asymptotic in some sector of the complex z-plane. Fix j and l, and consider the equation

$$\text{Re}\,[(\lambda_j - \lambda_l)z^{k+1}] = 0 . \tag{8}$$

It defines a finite number of rays which are called *Stokes lines*.

If S: $\alpha \leqslant \arg z \leqslant \beta$, $|z| > R$ is a sector, such that for a given j, S does not contain any of the directions in (8), then there is a solution $w_j(z)$ of equation (1) for which (7) is true as $z \to \infty$, $z \in S$. Moreover, one of the boundary lines of the sector S can be one of the rays of (8).

If these conditions are satisfied for all j, then (1) has an F.S.S. in S for which (5) is true as $z \to \infty$, $z \in S$. The whole complex z-plane is covered by a finite number of such sectors $S_1, \dots, S_N$, and they can be chosen so that the intersection of two adjacent sectors is non-empty. In each sector there is a F.S.S. with asymptotic expansion of the form (7); let us denote them $w_1(z), \dots, w_N(z)$. We have

$$w_j(z) = C_{jk} w_k(z) ,$$

where C_{jk} is a constant matrix of order $n \times n$.

The matrices C_{jk} are called the *Stokes matrices* (or *Stokes multipliers*). If all the Stokes multipliers $C_{j,j+1}$ are known (for $j = N$ we take $C_{N,1}$) then we can find the asymptotic behaviour of some F.S.S. of equation (1) as $z \to \infty$ in an arbitrary direction. However, we can find all these multipliers only in a restricted number of cases. Determining them is essentially equivalent to solving the differential equation.

1.3 Local Structure of the Solutions: The General Case. We consider equation (1) with irregular singular point $z = \infty$. Let $h \geqslant 1$, $l_j \geqslant 1$, $m_j \geqslant 0$ be integers,

$$\begin{aligned} Q_j(z) &= \sum_{k=0}^{l_j} a_{jk} z^{k/h} , \quad \Phi_j(z) = \sum_{k=0}^{m_j} \Psi_{jk}(z)(\ln z)^k , \\ \Psi_{jk}(z) &= \sum_{l=0}^{\infty} \phi_{jkl} z^{-l} , \quad \phi_{jk0} = 1 , \end{aligned} \tag{9}$$

and put

$$w_j(z) = z^{\rho_j} e^{Q_j(z)} \Phi_j(z) . \tag{10}$$

Equation (1) has a formal F.S.S. $\{w_1(z), \dots, w_n(z)\}$ consisting of solutions of the form (10). In some sectors of the form S: $\alpha \leqslant \arg z \leqslant \beta$ there is a F.S.S. of equation (1) for which the series (10) is asymptotic as $z \to \infty$, $z \in S$ [Coddington]. The algorithm for the construction of formal solutions and the sectors S is extremely complex for the general case.

Note that if $Q_j(z) \equiv 0$ then the corresponding solution $w_j(z)$ has the same structure as the solution in a neighbourhood of the regular singular point $z = \infty$.

2. A System of Equations. We consider the system of n equations

$$w' = z^r A(z) w , \tag{11}$$

where $r \geqslant 0$ is an integer, $A(z)$ is holomorphic at $z = \infty$ and $A(\infty) \neq 0$. Here the singular point $z = \infty$ can be both regular and irregular (§ 2).

We have

$$A(z) = \sum_{m=0}^{\infty} A_m z^{-m}$$

the series converging for $|z| > R$. Suppose that the eigenvalues $\lambda_1, \dots, \lambda_n$ of the matrix $A_0 = A(\infty)$ are distinct and non-zero, and let S be a sector in the complex z-plane which does not contain the Stokes lines (8) with j fixed. Then (11) has solutions of the form (7); in this formula w_{jm} are constant vectors and w_{j0} is an eigenvector of A_0, that is $A_0 w_{j0} = \lambda_j w_{j0}$. Series (7) is asymptotic for the solution $w_j(z)$ as $z \to \infty$, $z \in S$. The F.S.S. consisting of such solutions exists in S under the same conditions as in paragraph 2.3. The fundamental matrix for (11) has the form

$$W(z) = \Phi(z) z^R e^{Q(z)} ,$$

where R is a constant $n \times n$ matrix, $Q(z)$ is a diagonal matrix whose diagonal elements are polynomials of degree $r + 1$, and $\Phi(z)$ is the asymptotic series

$$\Phi(z) = \sum_{m=0}^{\infty} \Phi_m z^{-m} .$$

In the general case system (11) has a formal F.S.S. consisting of solutions of the form (9), (10), where the ψ_{jkl} are vectors and these series are asymptotic in some sectors of the complex z-plane.

We point out that the construction of the asymptotic expansion for the solution in the form (7) is difficult even for the case $n = 2$ of the scalar equation (1). The asymptotic methods of Chap. 5 are more effective.

3. The Manifold of Solutions

3.1 A First-Order System. We consider the system of n equations

$$t^{-r} w' = A(t) w + f(t) \tag{12}$$

on the half-line $I = [a, \infty)$ $(a > 0)$, where $r \geqslant 0$ is an integer, and $A(t)$, $f(t) \in C^\infty(I)$. Also, as $t \to \infty$, there are asymptotic expansions

$$A(t) = \sum_{m=0}^{\infty} A_m t^{-m} , \quad f(t) = \sum_{m=0}^{\infty} f_m t^{-m}$$

and $A_0 \neq 0$. Suppose that A_0 does not have purely imaginary eigenvalues. We can assume that A_0 is block diagonal, the blocks being square matrices A_-, A_+ of orders m and $n-m$ respectively. Moreover $\operatorname{Re} \lambda_j < 0$ ($\operatorname{Re} \lambda_j > 0$) for all the eigenvalues of A_- (A_+). This can be achieved by using the transformation $w = T\tilde{w}$, where T is a non-singular constant matrix. We introduce the blocks

$$A_0 = \begin{bmatrix} A_- & 0 \\ 0 & A_+ \end{bmatrix}, \quad w = \begin{bmatrix} w_- \\ w_+ \end{bmatrix}, \quad f = \begin{bmatrix} f_- \\ f_+ \end{bmatrix}, \quad A(t) - A_0 = \begin{bmatrix} V_{11} & V_{12} \\ V_{21} & V_{22} \end{bmatrix}.$$

The set M^m of solutions of (11) that are bounded for $t \in I$ is a linear manifold of dimension m. The homogeneous system (12) with $f(t) \equiv 0$ has m linearly independent solutions which are exponentially decreasing as $t \to \infty$ and $n-m$ linearly independent solutions that are exponentially increasing as $t \to \infty$. Then M^m can be described by the equation [Abramov 2]

$$w_+ = \alpha(t)w_- + \beta(t) \tag{13}$$

for $t \geqslant T \gg 1$. Here $\alpha(t)$ is the solution of the Riccati-type matrix equation

$$t^{-r}\alpha' = A_+\alpha - \alpha A_- + V_{22}\alpha - \alpha V_{11} - \alpha V_{12}\alpha + V_{21}\,, \quad t \geqslant T\,, \tag{14}$$

with the Cauchy condition at infinity

$$\lim_{t\to\infty} \alpha(t) = 0\,. \tag{15}$$

For $T \gg 1$ such a solution exists and is unique. The matrix $\alpha(t)$ has an asymptotic series expansion as $t \to \infty$

$$\alpha(t) = \sum_{k=1}^{\infty} \frac{\alpha_k}{t^k}\,. \tag{16}$$

Substituting (16) into (13) and equating coefficients of powers of t we get a recurrence system of linear algebraic equations from which we can find successively $\alpha_1, \alpha_2, \ldots$. These equations have the form

$$A_+\alpha_1 - \alpha_1 A_- = -\lim_{t\to\infty} tV_{21}(t)\,,$$

$$A_+\alpha_j - \alpha_j A_- = \phi_j(\alpha_1, \ldots, \alpha_{j-1})\,, \quad j = 2, 3, \ldots$$

The vector function $\beta(t)$ is a solution of the Cauchy problem

$$t^{-r}\beta' = A_+\beta + (V_{22} - \alpha V_{12})\beta + f_+ - \alpha f_-\,,$$

$$\lim_{t\to\infty} \beta(t) = -A_+^{-1} f_+(0)\,,$$

which exists and is unique for $t \geqslant T \gg 1$. As $t \to \infty$ there is an asymptotic expansion

$$\beta(t) = \sum_{k=0}^{\infty} \frac{\beta_k}{t^k}\,.$$

This method is widely used in numerical mathematics in transferring a boundary condition from infinity to a finite point [Abramov 2, Konyukhova].

3.2 A Second-Order System. We consider the system of n equations

$$w'' = t^r A(t) w \tag{17}$$

on the half-line I where $r \geqslant 0$ is an integer and the matrix $A(t) \in C^\infty(I)$ has an asymptotic series expansion. Suppose that the eigenvalues $\lambda_1, \ldots, \lambda_n$ of $A_0 = A(\infty)$ do not lie on the half-line $(-\infty, 0]$ in the complex λ-plane. Then the set M^m of solutions to the system (12) that are bounded at infinity is an n-dimensional linear space. We will obtain the equation for it by factorizing the differential operator in (17). We seek matrix functions $\alpha(t)$ and $\beta(t)$ such that

$$\left(\frac{d}{dt} - \beta\right)\left(\frac{d}{dt} - \alpha\right) = \frac{d^2}{dt^2} - t^r A(t). \tag{18}$$

From this we find that $\beta = -\alpha$ and obtain the Riccati matrix equation

$$\alpha' + \alpha^2 = t^r A(t), \quad T \leqslant t < \infty. \tag{19}$$

The Cauchy condition is imposed at infinity:

$$\lim_{t \to \infty} t^{-r/2} \alpha(t) = -\sqrt{A_0}. \tag{20}$$

Here $\sqrt{A_0}$ is the matrix with eigenvalues which lie in the half-plane $\operatorname{Re} \lambda > 0$ and $(\sqrt{A_0})^2 = A_0$. The solution to the problem (19), (20) for $T \gg 1$ exists and is unique.

The matrix function $\alpha(t)$ has an asymptotic series expansion

$$\alpha(t) = t^{r/2} \sum_{k=0}^{\infty} \frac{\alpha_k}{t^{k/2}} \tag{21}$$

in half-integer powers of t^{-1}. If r is an even number, then $\alpha(t)$ can be expanded as a series in integer powers of t^{-1}. Here $\alpha_0 = -\sqrt{A_0}$ and the substitution of (21) into (17) leads to a recurrence system of equations from which we can find successively $\alpha_1, \alpha_2, \ldots$.

It follows from (18) that if

$$w'(t) = \alpha(t) w(t), \tag{22}$$

where $\alpha(t)$ is a solution of (19), then $w(t)$ is a solution of (17). If $\alpha(t)$ satisfies the Cauchy condition (20) then $w(t) \in M^n$.

Let us consider the inhomogeneous system

$$w'' = t^r [A(t) w + f(t)],$$

where $A(t)$ satisfies the same conditions as above and $f(t)$ is given by the conditions of paragraph 3.1. Then the set M^n of solutions of this system which are bounded as $t \to \infty$ is a linear manifold of dimension n.

Because of (18) this system can be written in the form

$$\left(\frac{d}{dt} + \alpha\right)\left(\frac{d}{dt} - \alpha\right) w - t^r f(t) = 0. \tag{23}$$

We look for the equation for M^n in the form

$$\left(\frac{d}{dt} - \alpha(t)\right) w = \beta(t). \tag{24}$$

Substituting this relationship into (23) we obtain the equation for β:

$$\beta' + \alpha(t)\beta = t^r f(t).$$

We impose a Cauchy condition at infinity:

$$\lim_{t\to\infty} t^r \beta(t) = (\sqrt{A_0})^{-1} f_0.$$

The solution of this Cauchy problem exists and is unique for $t \geqslant T \gg 1$ and $\beta(t)$ has an asymptotic series expansion

$$\beta(t) = t^{r/2} \sum_{k=0}^{\infty} \frac{\beta_k}{t^{k/2}}, \quad t \to \infty.$$

If r is an even number then this series contains only integer powers of t^{-1}. We can obtain a recurrence system of linear algebraic equations for the coefficients $\beta_1, \beta_2, \ldots$.

3.3 Non-linear Systems. The method of transferring the boundary condition from a singular point to a non-singular point discussed above for linear systems is developed in [Abramov 2] for non-linear systems. We mention one of these results. Let us consider the system of n equations

$$w'' = f(t, w) \tag{25}$$

on the half-line I. Let

$$f(t, w) = A(t)w + g(t, w),$$

where $A(t) \in C^\infty$ and can be expanded in an asymptotic series of the form (12), and the vector-function $g(t, w)$ is a polynomial in w. Thus

$$g(t, w) = \sum_{|\alpha|=2}^{N} g_\alpha(t) w^\alpha,$$

where the vector functions $g_\alpha(t)$ have asymptotic series expansions

$$g_\alpha(t) = \sum_{k=0}^{\infty} g_{\alpha k} t^{-k}$$

as $t \to \infty$ and $g_\alpha(t) \in C^\infty(I)$ for all α. Here

$$\alpha = (\alpha_1, \dots, \alpha_n), \quad |\alpha| = \alpha_1 + \dots + \alpha_n, \quad w^\alpha = w_1^{\alpha_1} \dots w_n^{\alpha_n},$$

Suppose that the eigenvalues of A_0 do not lie on $(-\infty, 0]$. We impose the boundary condition at infinity

$$\lim_{t\to\infty} w(t) = 0. \tag{26}$$

Under our assumptions the solutions of the system (25) which satisfy (26) for each fixed t fill out the non-linear manifold M^n of dimension n in the space w, w'. We seek an equation for M^n in the form

$$w' = \alpha(t, w). \tag{27}$$

A solution of this system will be a solution of (25) if α satisfies the non-linear partial differential equation

$$\frac{\partial \alpha}{\partial t} + \frac{\partial \alpha}{\partial w}\alpha = f(t, w), \quad t \geqslant T. \tag{28}$$

This equation is considered in the domain $T \leqslant t < \infty$, $|w| \leqslant \varepsilon$, where $T \gg 1$ and $\varepsilon \ll 1$. For α in (27) we take the solution of the Cauchy problem at infinity

$$\lim_{t\to\infty} \alpha(t, w) = \gamma(w), \quad |w| \leqslant \varepsilon, \tag{29}$$

which is holomorphic in w. Here $\gamma(w)$ is the solution of

$$\frac{\partial \gamma}{\partial w}\gamma = A_0 w + \sum_{|\beta|=2}^{N} g_{\beta 0} w^\beta$$

such that

$$\gamma(w) = -\sqrt{A_0}\, w + O(|w|^2), \quad |w| \to 0.$$

The value $\sqrt{A_0}$ is the same as in paragraph 3.2.

The solution of (28), (29) exists and is unique and can be expanded for $|w| \leqslant \varepsilon$, $t \geqslant T$ in the convergent series

$$\alpha(t, w) = -\sqrt{A_0}\, w + \sum_{|\beta|=1}^{\infty} \alpha_\beta(t) w^\beta,$$

where $\alpha_\beta(t)$ as $t \to \infty$ has asymptotic series expansion

$$\alpha_\beta(t) = \sum_{l \geqslant 0,\, |\beta|+l \geqslant 2} \frac{\alpha_\beta^{(l)}}{t^l}, \quad |\beta| \geqslant 1.$$

One obtains a recurrence system of algebraic equations for $\alpha_\beta^{(l)}$.

For a more detailed discussion of the results obtained in this direction see [Abramov 2].

Chapter 2. Second-Order Equations on the Real Line

In this chapter we consider equations of the form

$$y'' + p(x, \lambda)y' + q(x, \lambda)y = 0$$

and systems of two first-order equations for real x. We establish asymptotic formulae for the solutions under the assumption that the equations have no turning points. The coefficients of the equations are assumed to be sufficiently smooth. The asymptotic formulae presented in this chapter are variously called the WKB-approximation (or WKB-asymptotic behaviour), quasi-classical approximation, short-wave approximation, high-frequency approximation.

§ 1. Transformations of Second-Order Equations

We introduce some transformations which are used in the study of the equation

$$y'' + p(x)y' + q(x)y = 0\,. \tag{1}$$

1. The substitution

$$y = \exp\left\{-\frac{1}{2}\int_a^x p(t)dt\right\} z$$

reduces (1) to the form

$$z'' + (q - p^2/4 - p'/2)z = 0\,.$$

We consider next the two-term equation

$$y'' + Q(x)y = 0\,. \tag{2}$$

2. The substitution $y'/y = w$ reduces (2) to the Riccati equation

$$w' + w^2 + Q(x) = 0\,.$$

3. The substitution

$$y = [\phi'(\xi)]^{1/2} z\,, \quad x = \phi(\xi)$$

reduces (2) to the form

$$\frac{d^2 z}{d\xi^2} + \left[Q(\phi(\xi))[\phi'(\xi)]^2 + \frac{\phi'''(\xi)}{2\phi'(\xi)} - \frac{3}{4}\left(\frac{\phi''(\xi)}{\phi'(\xi)}\right)^2 \right] z = 0\,.$$

Under these transformations the equation is still two-term. The expression

$$\frac{\phi'''}{\phi'} - \frac{3}{2}\left(\frac{\phi''}{\phi'}\right)^2 = \{\phi, \xi\}$$

is called the *Schwarzian derivative*.

4. Let us consider the equation

$$(P(x)y')' \pm Q(x)y = 0\,, \tag{3}$$

where $P(x) > 0$, $Q(x) > 0$. The substitution

$$\xi = \int_a^x \sqrt{\frac{Q(t)}{P(t)}}\,dt\,, \quad y = (P(x)Q(x))^{-1/4} z \tag{4}$$

reduces (3) to the form

$$\frac{d^2 z}{d\xi^2} \pm z + q(\xi) z = 0\,,$$

where

$$q = P^{-1/4} Q^{-3/4} \frac{d}{dx} P \frac{d}{dx} (PQ)^{-1/4}\,.$$

The transformation (4) is called the *Liouville transform*.

In particular, the equation

$$y'' \pm Q(x)y = 0 \tag{5}$$

is reduced to the form

$$\frac{d^2 z}{d\xi^2} \pm z + qz = 0\,, \quad q = -\frac{1}{4}\frac{Q''}{Q^2} + \frac{5}{16}\frac{Q'^2}{Q^3}\,.$$

The equation

$$P(x)y'' + R(x)y' + Q(x)y = 0$$

with Liouville transform

$$z = \phi(x)y\,, \quad \xi = \int_a^x \sqrt{\frac{Q(t)}{P(t)}}dt\,,$$

$$\phi(x) = \left|\frac{Q(x)}{P(x)}\right|^{1/4} \exp\left\{\frac{1}{2}\int_a^x \frac{R(t)}{P(t)}dt\right\}$$

is reduced to the form

$$\frac{d^2z}{d\xi^2} + \left[\frac{1}{2}\left(\frac{Q}{P}\right)'\left(-\frac{P}{Q}\right)^2\frac{\phi'}{\phi} - \frac{P\phi''}{Q\phi}\right]z = 0\,.$$

The Liouville transform reduces an equation with coefficients which are tamely (in a well-defined sense) increasing at infinity to an equation with almost constant coefficients.

Example. Let $Q(x) \sim ax^\alpha$ as $x \to \infty$, where $\alpha > -2$ and $a > 0$, and suppose that this asymptotic behaviour can be twice differentiated, that is, $Q'(x) \sim \alpha a x^{\alpha-1}$ and $Q''(x) \sim \alpha(\alpha-1)ax^{\alpha-2}$. Then

$$\xi(\infty) = \infty\,, \quad q = O(\xi^{-1-2/(\alpha+2)})\,.$$

Thus $q(\xi) \to 0$ as $\xi \to +\infty$ and $q \in L_1(0,\infty)$.

5. The system

$$u' = a_{11}(x)u + a_{12}(x)v\,, \quad v' = a_{21}(x)u + a_{22}(x)v$$

with the substitution

$$u = \rho\cos\theta\,, \quad v = \rho\sin\theta \tag{6}$$

is reduced to the form

$$\theta' = \frac{1}{2}(a_{21} - a_{12}) + \frac{1}{2}r\cos(2\theta + \psi)\,,$$

$$\frac{\rho'}{\rho} = \frac{1}{2}(a_{11} + a_{22}) + \frac{1}{2}r\sin(2\theta + \psi)\,.$$

Here

$$r^2 = (a_{11} - a_{22})^2 + (a_{12} + a_{21})^2\,,$$

$$\cos\psi = (a_{21} + a_{12})/r\,, \quad \sin\psi = (a_{11} - a_{22})/r\,.$$

This transformation (6) is called the *Prüfer transform*. In particular, equation (3) with the + sign and the substitution

$$y = (PQ)^{-1/4}\rho\sin\theta\,, \quad Py' = \rho\cos\theta$$

is reduced to the system

$$\theta' = \sqrt{\frac{Q}{P}} + \frac{1}{4}\sin 2\theta \frac{d}{dx}\ln(PQ)\,,$$
$$\frac{\rho'}{\rho} = -\frac{1}{2}\sin 2\theta \frac{d}{dx}\ln(PQ)\,.$$

6. The subsitution

$$y = u_1 + u_2\,,$$
$$y' = \left(\sqrt{PQ} - \frac{(PQ)'}{4PQ}\right)u_1 - \left(\sqrt{PQ} + \frac{(PQ)'}{4PQ}\right)u_2 \tag{7}$$

reduces (3) (with the − sign) to the system

$$\begin{bmatrix} u_1' \\ u_2' \end{bmatrix} = \left[\sqrt{\frac{Q}{P}}\begin{bmatrix} 1 & 0 \\ 0 & -1 \end{bmatrix} - \frac{(PQ)'}{4PQ}\begin{bmatrix} 1 & 0 \\ 0 & 1 \end{bmatrix} + \alpha_1 \begin{bmatrix} 1 & 1 \\ -1 & -1 \end{bmatrix}\right]\begin{bmatrix} u_1 \\ u_2 \end{bmatrix}, \tag{8}$$

where

$$\alpha_1(x) = \frac{1}{8(PQ)^{5/2}}\left[(P(\sqrt{PQ})')'\sqrt{PQ} - \frac{5}{4}P((\sqrt{PQ})')^2\right]. \tag{9}$$

This transformation has the same properties as the Liouville transform. In particular, $\alpha_1 \in L_1(b,\infty)$ for some b, if $P \equiv 1$ and $Q(x)$ has the same form as in the above example. But it is more convenient because it does not require a change of the independent variable and therefore we can use it for equations with complex-valued coefficients.

7. We look for a solution of (2), where $Q(x) > 0$, in the form

$$y = Q^{-1/4}(x)[A(x)e^{iS(x)} + B(x)e^{-iS(x)}]\,, \quad S(x) = \int_a^x \sqrt{Q(t)}dt\,. \tag{10}$$

If we also choose A and B such that

$$y' = iQ^{1/4}(x)[A(x)e^{iS(x)} - B(x)e^{-iS(x)}]\,,$$

then, to differentiate the solution $y(x)$ of the form (10), only the exponents are differentiated. Then A and B must satisfy the system of equations

$$A' = \frac{1}{4}\frac{Q'}{Q}e^{-2iS}B\,, \quad B' = \frac{1}{4}\frac{Q'}{Q}e^{2iS}A\,. \tag{11}$$

If $Q(x)$ is changing slowly for $x \gg 1$ (for instance if $Q(x) \to Q_0 > 0$, $Q'(x) \to 0$) then the coefficients in the right-hand sides of (11) are small for $x \gg 1$. We now introduce the function $R(x) = B(x)/A(x)$ which plays the role of the reflection coefficient in problems of wave propagation. Then $R(x)$ satisfies the Riccati equation

$$R' = \frac{Q'}{4Q}(e^{2iS} - R^2 e^{-2iS}).$$

8. We consider the Dirac-type system

$$Jy' + \begin{bmatrix} q_{11}(x) & q_{12}(x) \\ q_{21}(x) & q_{22}(x) \end{bmatrix} y = \lambda y, \quad J = \begin{bmatrix} 0 & 1 \\ -1 & 0 \end{bmatrix}, \quad y = \begin{bmatrix} y_1 \\ y_2 \end{bmatrix}, \tag{12}$$

with real coefficients and $q_{12}(x) = q_{21}(x)$. The transformation $y = T(x)z$, where

$$T(x) = \begin{bmatrix} \cos\phi & -\sin\phi \\ \sin\phi & \cos\phi \end{bmatrix}, \quad \phi(x) = -\frac{1}{2}\int_a^x (q_{11}(t) + q_{22}(t))dt,$$

reduces the system (12) to the form

$$Jy' + \begin{bmatrix} p(x) & q(x) \\ q(x) & -p(x) \end{bmatrix} y = \lambda y,$$

where

$$q(x) = q_{12}\cos 2\phi + \frac{1}{2}(q_{22} - q_{11})\sin 2\phi,$$
$$p(x) = \frac{1}{2}(q_{11} - q_{22})\cos 2\phi + q_{12}\sin 2\phi.$$

§ 2. WKB-Bounds

We consider the equation

$$y'' - Q(x)y = 0 \tag{1}$$

on the interval $I : a < x < b$ which can be finite or infinite. The function $Q(x) \in C^2(I)$ is complex-valued and satisfies the conditions:

1) $Q(x)$ is nowhere zero in I;

2) there is a branch of $\sqrt{Q(x)}$ of class $C^2(I)$ such that Re $\sqrt{Q(x)} \geqslant 0$ in I.

In all formulae we take this branch $\sqrt{Q(x)}$.

Let us consider 2) more fully. If $Q(x)$ is real, the condition already follows from 1). If $Q(x)$ is complex-valued, then the equation $z = Q(x)$, $x \in I$, defines a curve γ in the complex z-plane. The branch of $w = \sqrt{z}$ such that $w(1) = 1$ is a one-to-one map of the z-plane, with a cut along the half-line $(-\infty, 0]$, to the right half-plane Re $w > 0$. Consequently γ cannot intersect the half-line $(-\infty, 0]$.

1. WKB-Bounds. We denote

$$S(x_0, x) = \int_{x_0}^{x} \sqrt{Q(t)}dt\,, \tag{2}$$

$$\rho(x_0, x) = \left|\int_{x_0}^{x} |\alpha_1(t)|dt\right|\,, \tag{3}$$

where $\alpha_1(x)$ has the form (9) of § 1 for $P(x) \equiv 1$.

1.1 The Solution y_1. We put

$$\tilde{y}_1(x) = Q^{-1/4}(x)\exp\{S(x_0, x)\}\,. \tag{4}$$

If

$$\rho(a,\, x) < \infty\,, \quad x \in I\,, \tag{5}$$

then equation (1) has a solution y_1 such that, for $x \in I$,

$$\left|\frac{y_1(x)}{\tilde{y}_1(x)} - 1\right| \leqslant 2(e^{2\rho(a,x)} - 1)\,. \tag{6}$$

Bounds of the forms (6), (8) and (11) are called *WKB-bounds*. When $Q(x) = (ax+b)^{-4}$ the bound in (6) is exact.

The solution y_1 satisfies the boundary condition

$$\lim_{x\to a} y_1'(x)\Big/\left[\left(\sqrt{Q(x)} - \frac{Q'(x)}{4Q(x)}\right) y_1(x)\right] = 1\,. \tag{7}$$

For the derivative $y_1'(x)$ we have

$$\left|\frac{y_1'(x)}{\sqrt{Q(x)}\tilde{y}_1(x)} - 1\right| \leqslant \frac{1}{4}\left|\frac{Q'(x)}{Q^{3/2}(x)}\right| + 4\left(1 + \frac{1}{4}\left|\frac{Q'(x)}{Q^{3/2}(x)}\right|\right)(e^{2\rho(a,\,x)} - 1)\,. \tag{8}$$

1.2 The Solution y_2. We put

$$\tilde{y}_2(x) = Q^{-1/4}(x)\exp\{-S(x_0, x)\} \tag{9}$$

and we suppose that

$$\rho(x, b) < \infty, x \in I\,. \tag{10}$$

Then equation (1) has a solution y_2 such that for $x \in I$ there are the bounds

$$\left|\frac{y_2(x)}{\tilde{y}_2(x)} - 1\right| \leqslant 2(e^{2\rho(x,\,b)} - 1)\,,$$

$$\left|\frac{y_2'(x)}{\sqrt{Q(x)}\tilde{y}_2(x)}+1\right| \leqslant \frac{1}{4}\left|\frac{Q'(x)}{Q^{3/2}(x)}\right| + \\ + 4\left(1+\frac{1}{4}\left|\frac{Q'(x)}{Q^{3/2}(x)}\right|\right)(e^{2\rho(x,b)}-1). \tag{11}$$

The solution y_2 satisfies the boundary condition

$$\lim_{x\to b} y_2'(x) \Big/ \left[\left(\sqrt{Q(x)}+\frac{Q'(x)}{4Q(x)}\right) y_2(x)\right] = -1. \tag{12}$$

Asymptotic formulae follow immediately from the WKB-bounds:

$$y_1(x) \sim \tilde{y}_1(x), \quad x \to a; \quad y_2(x) \sim \tilde{y}_2(x), \quad x \to b.$$

2. Integral Equations for the Solutions. Let us construct the solution y_1 by reducing (1) to a system of integral equations. The substitution

$$y = \tilde{y}_1(x)(u_1+u_2), \tag{13}$$

$$y' = \tilde{y}_1(x)\left[\left(\sqrt{Q}-\frac{Q'}{4Q}\right)u_1 - \left(\sqrt{Q}+\frac{Q'}{4Q}\right)u_2\right],$$

as shown in § 1, reduces (1) to the system

$$u_1' = \alpha_1(x)(u_1+u_2), \quad u_2' + 2\sqrt{Q(x)}u_2 = -\alpha_1(x)(u_1+u_2).$$

This system is equivalent to the system of integral equations

$$\begin{aligned} u_1(x) &= c_1 + \int_a^x \alpha_1(t)(u_1(t)+u_2(t))dt, \\ u_2(x) &= c_2 e^{-2S(x_0,x)} - \int_a^x e^{2S(x,t)}\alpha_1(t)(u_1(t)+u_2(t))dt, \end{aligned} \tag{14}$$

or, in operator form,

$$u = u^0 + Ku, \quad u = (u_1, u_2)^T,$$

where K is the integral operator from (14). Since $a \leqslant t \leqslant x$ it follows from condition 2) that Re $S(x,t) \leqslant 0$ and therefore

$$|\exp\{2S(x,t)\}| \leqslant 1. \tag{15}$$

This leads to the bounds

$$|(Ku)_j(x)| \leqslant \int_a^x |\alpha_1(t)|(|u_1(t)|+|u_2(t)|)dt, \quad j=1,2.$$

We put $c_1 = 1$, $c_2 = 0$ and apply the method of successive approximations to (14):

$$u^{n+1} = u^0 + Ku^n.$$

This method converges and gives a solution $u(x)$ such that

$$|u_j(x) - c_j| \leqslant \exp\{2\rho(a, x)\} - 1, \quad j = 1, 2.$$

The WKB-bounds (6), (8) follow from this and from (13). Moreover

$$\lim_{x \to a} u_1(x) = 1, \quad \lim_{x \to a} u_2(x) = 0,$$

which proves (7). The solution y_2 is constructed in a similar manner.

If $Q(x)$ is real the WKB-bounds can be obtained by applying the Liouville transform (§ 1) to equation (1) and then reducing the equation obtained to integral form.

3. The Equation in Self-Adjoint Form. We consider the equation

$$(P(x)y')' - Q(x)y = 0. \tag{16}$$

Let $P(x)$, $Q(x) \in C^2(I)$ and suppose that

1) $P(x)$ and $Q(x)$ are nowhere zero in I;

2) there is a branch of $\sqrt{P(x)/Q(x)}$ of class $C^2(I)$ such that $\operatorname{Re} \sqrt{P(x)/Q(x)} \geqslant 0$ in I.

We denote

$$S(x_0, x) = \int_{x_0}^{x} \sqrt{\frac{Q(t)}{P(t)}}\, dt,$$

$$\tilde{y}_1(x) = [P(x)Q(x)]^{-1/4} \exp\{S(x_0, x)\},$$

$$\rho(x_0, x) = \left| \int_{x_0}^{x} |\alpha_1(t)| dt \right|,$$

where $\alpha_1(x)$ has the form (9) of § 1. Equation (16) has a solution $y_1(x)$ with the bound (6) for $x \in I$. Bounds similar to (8) and (11) also hold.

§ 3. Asymptotic Behaviour of Solutions of a Second-Order Equation for Large Values of the Parameter

1. Two-Term Equations. We consider the equation

$$y'' - \lambda^2 q(x) y = 0 \tag{1}$$

on the interval $I = [a, b]$. Here $\lambda > 0$ is a large parameter, and $q(x) \in C^\infty(I)$ is a complex-valued function.

Equation (1) can be written with the small parameter $\varepsilon = \lambda^{-1}$ in the highest-order term,

$$\varepsilon^2 y'' - q(x)y = 0,$$

and the dependence of the coefficients on ε has a very simple form. In this paragraph all the results concerning the asymptotic behaviour of the solutions will be formulated at first for equation (1) and only then for the case when the dependence of the coefficients on the parameter is more complicated.

1.1 Formal Asymptotic Solution (F.A.S.). If $q \neq 0$ is constant then (1) has two linearly independent solutions $y_{1,2} = \exp\{\pm\lambda x\sqrt{q}\}$. If $q \not\equiv$ const then we seek a solution of (1) in the form of an asymptotic series

$$y = e^{\lambda S(x)} \sum_{k=0}^{\infty} a_k(x)\lambda^{-k}.$$

It is convenient to represent this series as

$$y = \exp\left\{\int^x \sum_{k=-1}^{\infty} \lambda^{-k}\alpha_k(t)dt\right\}. \tag{2}$$

The substitution $y'/y = w$ reduces (1) to the Riccati equation

$$w' + w^2 - \lambda^2 q(x) = 0.$$

Substituting the expression (2) into this equation, we obtain the identity

$$\sum_{k=-1}^{\infty} \lambda^{-k}\alpha_k'(x) + \left(\sum_{k=-1}^{\infty} \lambda^{-k}\alpha_k(x)\right)^2 - \lambda^2 q(x) = 0.$$

Equating the powers of λ^{-1} to zero, we obtain a recurrence system for the unknown functions $\alpha_{-1}(x)$, $\alpha_0(x), \dots$. All the computations are of a formal nature.

The first of these equations is $\alpha_{-1}^2 = q$, so that $\alpha_{-1} = \pm\sqrt{q}$. For $\alpha_{-1} = \sqrt{q}$ (choosing the branch of the root as shown in paragraph 2) we have

$$\begin{aligned} \alpha_{-1}(x) &= \sqrt{q(x)}, \quad \alpha_0(x) = -\frac{q'(x)}{4q(x)}, \\ \alpha_1(x) &= \frac{1}{8}\frac{q''(x)}{q^{3/2}(x)} - \frac{5}{32}\frac{q'^2(x)}{q^{5/2}(x)}. \end{aligned} \tag{3}$$

Observe that $\alpha_1(x)$ is the same as in § 1, (9). We obtain the recurrence relation for successive functions $\alpha_k(x)$

$$\alpha_{k+1}(x) = -\frac{1}{2\sqrt{q(x)}}\left(\alpha_k'(x) + \sum_{j=0}^{k} \alpha_j(x)\alpha_{k-j}(x)\right). \tag{4}$$

If however $\alpha_{-1}(x) = -\sqrt{q(x)}$, $\sqrt{q}$ must be replaced by $-\sqrt{q}$ everywhere in (2) and (3). Equation (1) has therefore two formal asymptotic solutions:

$$\begin{aligned} y_1(x,\,\lambda) &= q^{-1/4}(x)\exp\left\{\lambda\int^x \sqrt{q(t)}dt + \sum_{k=1}^{\infty}\lambda^{-k}\int^x \alpha_k(t)dt\right\}, \\ y_2(x,\,\lambda) &= q^{-1/4}(x)\exp\left\{-\lambda\int^x \sqrt{q(t)}dt + \sum_{k=1}^{\infty}(-\lambda)^{-k}\int^x \alpha_k(t)dt\right\}. \end{aligned} \tag{5}$$

These asymptotic series terminate in two cases.

A. If $q(x) = (ax+b)^{-4}$ then $\alpha_k(x) \equiv 0$ for $k \geqslant 1$. Then (5) gives exact solutions of (1):

$$y_{1,2} = q^{-1/4}(x)\exp\left\{\pm\lambda\int^x \sqrt{q(t)}dt\right\}.$$

B. If $q(x) = (ax^2+bx+c)^{-2}$, then $\alpha_k(x) \equiv 0$ for $k \geqslant 2$. Equation (1) can also be integrated by quadratures but (5) does not give explicit solutions if $a \neq 0$.

The formulae (5) are clearly of no use at points where $q(x)$ vanishes. In fact if $q(x_0) = 0$, then the right-hand sides in formula (5) go off to infinity for $x = x_0$ but all the solutions of equation (1) are smooth functions.

Zeros of $q(x)$ are called *turning points* (or *transition points*) of equation (1). At a turning point the roots of the characteristic equation $\rho^2 - q(x) = 0$ coalesce.

1.2 The Principal Asymptotic Term. We assume that $q(x) \in C^2(I)$ and introduce the conditions

1) equation (1) has no turning points; that is, $q(x) \neq 0$ for $x \in I$;

2) there exists a branch of $\sqrt{q(x)}$ of class $C^2(I)$ such that Re $\sqrt{q(x)} \geqslant 0$ for $x \in I$.

The F.A.S., as follows from paragraph 1.1, may also hold when condition (2) does not hold. This condition is not necessary for the existence of solutions having an asymptotic expansion of the form (5). However, when it is not true, it is possible for there to be no such solutions, even if $q(x)$ is analytic (§ 8).

We denote

$$S(x_0,x) = \int_{x_0}^x \sqrt{q(t)}dt. \tag{6}$$

Then equation (1) has two solutions y_1 and y_2 of the form

$$y_{1,2}(x,\,\lambda) = q^{-1/4}(x)\exp\{\pm\lambda S(x_0,\,x)\}[1+O(\lambda^{-1})], \quad \lambda\to\infty. \tag{7}$$

The bound for the remainder term is uniform in $x \in I$, that is, $|O(\lambda^{-1})| \leqslant C\lambda^{-1}$ for $\lambda \geqslant 1$, where C does not depend on x.

The asymptotic formula (7) can be twice differentiated in x, that is

$$\begin{aligned} y_{1,2}^{(j)}(x,\,\lambda) &= (\pm\lambda\sqrt{q(x)})^j q^{-1/4}(x)\exp\{\pm\lambda S(x_0,\,x)\}[1+O(\lambda^{-1})], \\ &\lambda\to\infty,\ j=1,2. \end{aligned} \tag{8}$$

The bound for the remainder term is uniform in $x \in I$. The solutions y_1 and y_2 are linearly independent for $\lambda \gg 1$. The asymptotic formulae (7) are variously called the *WKB approximation, quasi-classical approximation, short-wave approximation, high-frequency approximation.*

We point out that for y_1 and y_2 we can take solutions satisfying the Cauchy condition

$$\begin{aligned} y_1(a,\ \lambda) &= A\,, \quad y_1'(a,\ \lambda) = A[\lambda\sqrt{q(a)} - q'(a)/(4q(a))]\,, \\ y_2(b,\ \lambda) &= B\,, \quad y_1'(b,\ \lambda) = -B[\lambda\sqrt{q(b)} + q'(b)/(4q(b))]\,, \end{aligned} \tag{9}$$

where A and B are constants.

All the assertions of paragraphs 1.1 and 1.2 arise from WKB-bounds (§ 2). In fact, comparing the equations (1) from § § 2 and 3, we arrive at

$$Q(x) = \lambda^2 q(x)\,, \quad \alpha_1(x;\ Q) = \lambda^{-1}\alpha_1(x;\ q)\,,$$

$$\rho(x_0, x; Q) = \lambda^{-1}\rho(x_0, x; q)\,.$$

Here the function $\alpha_1(x;q)$ is calculated in terms of $q(x)$ (see (3)) and the function $\alpha_1(x;Q)$ in terms of $Q(x)$. From formula (6) of § 2 we have

$$\left|\frac{y_1(x,\ \lambda)}{\tilde{y}_1(x,\ \lambda;\ a)} - 1\right| \leqslant 2\left(\exp\left\{\frac{2}{\lambda}\rho(a,x;q)\right\} - 1\right).$$

The right-hand side of this has order $O(\lambda^{-1})$ as $\lambda \to \infty$. Formulae (7) – (9) follow from formulae (7), (8), (11) and (12) of § 2.

1.3 Stucture of the Solutions for Real $q(x)$. The behaviour of the solutions essentially depends on the sign of $q(x)$.

A. Non-oscillatory solutions: $q(x) > 0$. We take the WKB-approximations in the form

$$\begin{aligned} y_1(x,\ \lambda) &\sim q^{-1/4}(x)\exp\{\lambda S(a,\ x)\}\,, \\ y_2(x,\ \lambda) &\sim q^{-1/4}(x)\exp\{-\lambda S(x,\ b)\} \end{aligned}$$

and fix $\lambda \gg 1$. The solution y_1 is equal to $q^{-1/4}(a) + O(\lambda^{-1})$ at $x = a$, and it is strictly monotonic increasing and exponentially large at $x = b$. The solution y_2 has the same properties if we interchange a and b.

B. Oscillatory solutions: $q(x) < 0$. In place of (1) it is convenient to consider

$$y'' + \lambda^2 q(x) y = 0\,, \tag{10}$$

where $q(x) > 0$. This equation has a F.S.S. of the form

$$y_{1,2}(x,\ \lambda) \sim q^{-1/4}(x)\exp\{\pm i\lambda S(x_0,\ x)\}\,,$$

which can be taken to be complex conjugates: $y_2 = \overline{y}_1$. The real solutions $y_3 = \operatorname{Re} y_1$, $y_4 = \operatorname{Im} y_1$ form a F.S.S. and have the form

$$y_3(x, \lambda) = q^{-1/4}(x)\{\cos[\lambda S(x_0, x)] + O(\lambda^{-1})\}$$
$$y_4(x, \lambda) = q^{-1/4}(x)\{\sin[\lambda S(x_0, x)] + O(\lambda^{-1})\},$$

that is, they are strongly oscillatory functions.

1.4 Asymptotic Expansions. Suppose that $q(x) \in C^\infty(I)$ and let conditions 1) and 2) be satisfied. Then equation (1) has two solutions of the form

$$y_{1,2}(x, \lambda) = q^{-1/4}(x)\exp\{\pm\lambda S(x_0, x)\} \times [1 + \sum_{k=1}^{N-1} \lambda^{-k} a_k^{\pm}(x) + O(\lambda^{-N})], \quad \lambda \to \infty. \tag{11}$$

Here $N \geqslant 1$ is arbitrary and the bound for the remainder term is uniform in $x \in I$. The asymptotic formulae (11) can be differentiated in x and in λ an arbitrary number of times, preserving the uniformity in x of the bound for the remainder term.

The coefficients $a_k^{\pm}(x)$ are determined by the formal identity

$$\exp\left\{\sum_{k=1}^{\infty}(\pm\lambda)^{-k}\int_{x_0}^{x}\alpha_k(t)dt\right\} = 1 + \sum_{k=1}^{\infty}\lambda^{-k}a_k^{\pm}(x).$$

Expanding the exponents in the formal series in powers of λ^{-1}, we obtain the functions $a_k^{\pm}(x)$. Let us write down the asymptotic expansion to within $O(\lambda^{-2})$. Thus

$$y_{1,2}(x, \lambda) = q^{-1/4}(x)\exp\left\{\pm\lambda\int_{x_0}^{x}\sqrt{q(t)}dt\right\} \times \left[1 \pm \lambda^{-1}\int_{x_0}^{x}\left(\frac{1}{8}\frac{q''(t)}{q^{3/2}(t)} - \frac{5}{32}\frac{q'^2(t)}{q^{5/2}(t)}\right)dt + O(\lambda^{-2})\right].$$

For $y_{1,2}$ we can take solutions satisfying the Cauchy conditions

$$\begin{aligned} y_1(a) = A, \quad y_1'(a) &= A\left[\lambda\sqrt{q(a)} - \frac{q'(a)}{4q(a)} + \sum_{k=1}^{N-1}\lambda^{-k}\alpha_k(a)\right], \\ y_2(b) = B, \quad y_2'(b) &= B\left[-\lambda\sqrt{q(b)} - \frac{q'(b)}{4q(b)} + \sum_{k=1}^{N-1}(-\lambda)^{-k}\alpha_k(b)\right]. \end{aligned} \tag{12}$$

1.5 Additional Parameters and Complex λ. We formulate a theorem on the analytical dependence of the solutions of linear differential equations on parameters. Let I be a segment of the x-axis and D a domain in the complex plane. We consider the Cauchy problem for the system of n equations:

$$y' = A(x, \mu)y, \quad y(x_0, \mu) = y_0(\mu). \tag{13}$$

Theorem. *Suppose that the matrix function $A(x, \mu)$ is continuous for $(x, \mu) \in I \times D$ and is holomorphic in D for each fixed $x \in I$. Suppose that the vector function $y_0(\mu)$ is holomorphic in D. Then the solution $y(x, \mu)$ of the Cauchy problem (13) is holomorphic in D for each fixed $x \in I$.*

Let us consider the Cauchy problem for the nth-order linear differential equation

$$y^{(n)} + q_1(x,\ \mu)y^{(n-1)} + \ldots + q_n(x,\ \mu)y = 0$$

$$y(x_0,\ \mu) = y_0(\mu), \quad y'(x_0,\ \mu) = y_1(\mu), \ldots, \quad y^{(n-1)}(x_0,\ \mu) = y_{n-1}(\mu).$$

If the coefficients $q_j(x, \mu)$ and the initial conditions $y_j(\mu)$ have the same properties as $A(x, \mu)$, then the solution $y(x, \mu)$ of the Cauchy problem is holomorphic in D for each fixed $x \in I$.

In particular, the solutions $y_{1,2}(x, \lambda)$ satisfying the Cauchy condition (12) are entire functions of λ for each fixed $x \in I$ if $A = B = 1$.

Let us consider the equation

$$y'' - \lambda^2 q(x,\ \mu)y = 0 \tag{14}$$

in either of the two following situations A and B.

A. μ is a real parameter and D is an interval of the real line. In this case we assume that $q(x, \mu) \in C^\infty(I \times D)$.

B. μ is a complex parameter and D is a domain in the complex μ-plane. Here we assume additionally that $q(x, \mu)$ is holomorphic in D for each fixed $x \in I$.

All the above results for (1) carry over completely to (14) if all the necessary conditions are satisfied uniformly in μ. We introduce the conditions:

1) $|q(x, \mu)| \geqslant \delta > 0$ for $(x, \mu) \in I \times D$, where δ does not depend on x or μ;

2) Re $\sqrt{q(x, \mu)} \geqslant 0$ for $(x, \mu) \in I \times D$;

3) All the partial derivatives of $q(x, \mu)$ are uniformly bounded in x, μ; that is

$$\left| \frac{\partial^{m+n}}{\partial x^m \partial \mu^n} q(x, \mu) \right| \leqslant C_{mn}$$

for $(x, \mu) \in I \times D$, where the constants C_{mn} do not depend on x or μ.

Then there exist solutions $y_{1,2}(x, \lambda, \mu)$ of the form (11) where we replace $q(x)$ by $q(x, \mu)$ in all the formulae. Formulae (11) can be differentiated in x, λ and μ an arbitrary number of times, preserving the uniformity in x of the bound for the remainder term. If μ is a complex parameter then the solutions $y_{1,2}$ satisfying the Cauchy condition (12) are holomorphic functions of λ, μ for all λ and $\mu \in D$ for each fixed $x \in I$.

Let G be an unbounded domain in the complex λ-plane. The asymptotic formulae (11) for the solutions remain true as $\lambda \to \infty$, $\lambda \in G$ if condition 2) is replaced by

2^+) Re $\lambda(\sqrt{q(x,\mu)})$ does not change sign for $\lambda \in G$, $|\lambda| \geqslant \lambda_0 > 0$, $\mu \in D$.

Examples. 1. Let $q(x) > 0$ for $x \in I$ and let $y_1(x,\lambda)$ be a solution of (1) satisfying the Cauchy condition (12), where $A = 1$. Then Re $(\lambda\sqrt{q(x)}) \geqslant 0$ for $x \in I$ and Re $\lambda \geqslant 0$, and hence the asymptotic formula (11) is applicable as $|\lambda| \to \infty$, Re $\lambda \geqslant 0$. We will show that this formula is suitable for $|\lambda| \to \infty$ and for any arg λ. The function $y_1(x,\lambda)$ is real for real λ; therefore $y_1(x,\overline{\lambda}) = \overline{y_1,(x,\lambda)}$. All the coefficients of powers of λ^{-1} in the expansion (11) are also real for real λ which proves the above assertion. All of this is also true for $y_2(x,\lambda)$.

2. Let $q(x) > 0$ and let $y_1(x,\lambda)$ be a solution of (10) for which the asymptotic series expansion (11) is valid. Since Re $(i\lambda\sqrt{q(x)}) \geqslant 0$ for Im $\lambda \leqslant 0$ this asymptotic formula is true as $|\lambda| \to \infty$, Im $\lambda \leqslant 0$. The solution y_2 can be taken as $y_2(x,\lambda) = \overline{y_1(x,\lambda)}$. We denote this F.S.S. by $\{y_1^+(x,\lambda),\ y_2^+(x,\lambda)\}$. Similarly there is a F.S.S. $\{y_1^-(x,\lambda),\ y_2^-(x,\lambda)\}$ for (10) which has the asymptotic expansion (11) as $|\lambda| \to \infty$, Im $\lambda \geqslant 0$. These F.S.S. are not generally the same.

1.6 More Complicated Dependence on λ. The two-term equation, with the second derivative multiplied by a small parameter, has the form

$$\varepsilon^2 y'' - q(x,\ \varepsilon)y = 0\,, \tag{15}$$

where $\varepsilon = \lambda^{-1}$. The standard assumptions concerning the dependence of the function q on the parameter ε are as follows. Let S be a sector in the complex ε-plane of the form $0 < |\varepsilon| < \varepsilon_0, -\alpha < \arg\varepsilon < \beta$, where $0 \leqslant \alpha,\ \beta \leqslant \pi$.

I. $q(x,\varepsilon) \in C^\infty(I \times S)$.
II. $q(x,\varepsilon)$ is holomorphic in ε in the sector S for each fixed $x \in I$.
III. There is an asymptotic expansion

$$q(x,\varepsilon) \sim \sum_{k=0}^{\infty} q_k(x)\varepsilon^k\,, \quad \varepsilon \to 0\,, \quad \varepsilon \in S\,,$$

which is uniform in $x \in I$ in each proper subsector S' of S (that is, for $-\alpha < -\alpha' \leqslant \arg\varepsilon \leqslant \beta' < \beta$).

If ε takes only real values then in place of S we take the interval $J:\ 0 < \varepsilon < \varepsilon_0$ and condition II is discarded. Conditions 1 and 2 take the form:

1) $q_0(x) \neq 0$, $x \in J$;
2) Re $(\varepsilon^{-1}\sqrt{q_0(x)}) \geqslant 0$, $x \in I$, $\varepsilon \in S$.

Then equation (15) has F.S.S. $y_1(x,\varepsilon)$, $y_2(x,\varepsilon)$ for which there is an asymptotic expansion of the form (11) as $\varepsilon \to 0$, $\varepsilon \in S$. This asymptotic expansion is uniform in x as $\varepsilon \to 0$, $\varepsilon \in S'$. The principal asymptotic term has the form

$$y_{1,2}(x,\varepsilon) = q_0^{-1/4}(x)\exp\left\{\varepsilon^{-1}\int_{x_0}^{x}\sqrt{q_0(t)}dt + \frac{1}{2}\int_{x_0}^{x}\frac{q_1(t)}{\sqrt{q_0(t)}}dt\right\}[1+O(\varepsilon)].$$

The asymptotic behaviour for the solution of (15) can be calculated in three ways.

1. The F.A.S. is sought in the form (2), that is

$$y = \exp\left\{\sum_{k=-1}^{\infty}\varepsilon^k\int^{x}\beta_k(t)dt\right\}.$$

The functions $\beta_k(x)$ are determined from recurrence relations as in paragraph 1.

2. We can make use of formulae (11) and rearrange the functions $\alpha_j(x,\varepsilon)$ in them as asymptotic series in powers of ε.

Since the asymptotic expansion is unique, the formulae obtained by these methods must be the same.

If we do not rearrange them, these approaches lead to asymptotic series of the form

$$\sum_{k=0}^{\infty}\phi_k(x)\varepsilon^k, \quad \sum_{k=0}^{\infty}\psi_k(x,\varepsilon)\varepsilon^k,$$

respectively, that is, to asymptotic series in the sense of Poincaré and in the sense of Erdelyi. The second of these is generally to be preferred since the formulae for ψ_k are simpler than those for ϕ_k. For this reason it does not always make sense to obtain the asymptotic series for a solution in powers of ε.

The asymptotic expansions with respect to asymptotic sequences of the form $\{\varepsilon^k\psi_k(x,\varepsilon)\}$ are non-unique when the ψ_k are not specified in advance. We can improve the WKB-approximation with the help of a successful choice of the ψ_k.

3. The principal asymptotic term for the solution of (1) is given by (7) with remainder term $O(\lambda^{-1})$. If we replace $q(x)$ by a function of the form $\tilde{q}(x) = q(x)[1+\lambda^{-2}\beta(x)]$, where $\beta(x)$ is an arbitrary smooth function, then the principal asymptotic term does not change. We choose $\beta(x)$ so that

$$y_{1,2}(x,\lambda) = \tilde{q}^{-1/4}(x)\exp\left\{\pm\lambda\int_{x_0}^{x}\sqrt{\tilde{q}(t)}dt\right\}[1+O(\lambda^{-2})].$$

To do this, the relationships

$$\pm\lambda\sqrt{q} - \frac{q'}{4q} \pm \frac{\alpha_1}{\lambda} = \pm\lambda\sqrt{\tilde{q}} - \frac{\tilde{q}'}{4\tilde{q}'} + O\left(\frac{1}{\lambda^2}\right),$$

must hold, from which we find that $\beta = 2\alpha_1/\sqrt{q}$. Consequently, (1) has solutions of the form

$$y_{1,2}(x,\lambda) = [\tilde{q}(x,\lambda)]^{-1/4} \exp\left\{\pm\lambda \int_{x_0}^{x} \sqrt{\tilde{q}(t,\lambda)}dt\right\} [1 + O(\lambda^{-2})] .$$

The principal term of the asymptotic series (7) satisfies (1) only if $q(x) = (ax+b)^{-4}$. The principal term of the expansion obtained precisely satisfies (1) also for $q(x) = (ax^2+bx+c)^{-2}$. There $\beta(x) \equiv \text{const}$. We can obviously take $\tilde{q}$ so that the "principal" asymptotic term would be the asymptotic series expansion to within $O(\lambda^{-N})$ for any N.

2. The General Second-Order Equation

2.1 Asymptotic Behaviour of the Solutions. We consider the equation

$$y'' + \lambda p(x)y' + \lambda^2 q(x)y = 0 \tag{16}$$

on the interval $I = [a,b]$, where $p(x),\ q(x) \in C^\infty(I)$. The roots of the characteristic equation are

$$p_{1,2}(x) = \frac{1}{2}(-p(x) \pm \sqrt{D(x)}), \quad D(x) = p^2(x) - 4q(x) .$$

The point $x_0 \in I$ is called a *turning point* of (16) if the roots of the characteristic equation coincide for $x = x_0$. Consequently, the turning points are the roots of the equation

$$D(x) \equiv p^2(x) - 4q(x) = 0 . \tag{17}$$

We introduce the conditions:

1) Equation (16) has no turning points, that is $D(x) \neq 0$ for $x \in I$.
2) There is a branch of $\sqrt{D(x)}$ such that Re $(\sqrt{D(x)}) \geqslant 0$ for $x \in I$.

Then there are solutions $y_{1,2}(x,\lambda)$ of (16) such that

$$y_{1,2}(x,\lambda) = (D(x))^{-1/4} \exp\left\{\pm\frac{\lambda}{2} \int_{x_0}^{x} \sqrt{D(t)}dt - \frac{1}{2}\int_{x_0}^{x} \frac{p'(t)}{\sqrt{D(t)}}dt\right\}[1 + O(\lambda^{-1})], \quad \lambda \to \infty . \tag{18}$$

The properties of these asymptotic forms are the same as those in paragraph 1.4. There exist asymptotic expansions to within $O(\lambda^{-N})$, that is, in (18) the expression in the square brackets can be replaced by

$$1 + \sum_{k=1}^{N-1} a_{kj}(x)\lambda^{-j} + O(\lambda^{-N}), \quad j = 1,2 .$$

For $y_{1,2}$ we can take solutions satisfying a Cauchy condition of the form (12).

The second-order equation

$$y'' + 2\lambda a(x,\ \mu)y' + \lambda^2 b(x,\ \mu)y = 0 \tag{19}$$

with the substitution $y = \exp\{-\lambda \int^x a\,dt\}z$ is reduced to the two-term equation

$$z'' + \lambda^2[b(x,\ \mu) - a^2(x,\ \mu) - \lambda^{-1}a'(x,\ \mu)]z = 0\,, \tag{20}$$

to which we can apply all the results obtained in paragraph 1. We will not repeat them but mention only that the turning points of (16) and (20) do not coincide, but are close for $\lambda \gg 1$. Let us suppose for simplicity, that the coefficients a and b do not depend on μ. The turning points of (20) are the roots of the equation

$$D(x) - \lambda^{-1}a'(x) = 0\,.$$

If x_0 is a turning point of (1), then equation (20) has a turning point $x_0(\lambda)$ of the form $x_0(\lambda) = x_0 + O(\lambda^{-1})$. This is a general property: a turning point is not invariant under changes of variables, but it may be asymptotically invariant.

2.2 Equations in Self-adjoint Form. We consider the equation

$$(P(x)y')' - \lambda^2 Q(x)y = 0\,, \tag{21}$$

where $P, Q \in C^\infty(I)$ and $P(x) \neq 0$. The equation

$$P(x)p^2 - Q(x) = 0$$

is called the *characteristic equation*, so that the turning points of (21) are the zeros of $Q(x)$. Conditions 1) and 2) take the form

$$Q(x) \neq 0\,, \quad \operatorname{Re}\sqrt{Q(x)/P(x)} \geqslant 0\,, \quad x \in I\,.$$

We give the principal terms in the asymptotic formulae for the solutions:

$$y_{1,2}(x,\lambda) = [P(x)Q(x)]^{-1/4}\exp\left\{\pm\lambda\int_{x_0}^{x}\sqrt{\frac{Q(t)}{P(t)}}dt\right\}[1 + O(\lambda^{-1})]\,. \tag{22}$$

3. Remarks on Asymptotic Series. In (11) we gave a finite number of terms of the asymptotic expansion for the solution. In many works on asymptotic methods, asymptotic expansions for solutions are given. For equation (1), with the conditions of paragraph 1.4, these results are formulated in the following form:

There exists a solution $y_1(x,\lambda)$ of equation (1) which can be expanded in an asymptotic series (5) as $\lambda \to \infty$, uniformly in $x \in I$.

Unfortunately, such a solution is non-constructive. In the final analysis, the proof of the existence of such a solution is based on the following fact.

Theorem (Nörlund). *Suppose that there is given an arbitrary formal series* $\sum_{N=0}^{\infty} a_n z^n$ *and a sector* S *of the form* $0 < |z| < r$, $\alpha < \arg z < \beta$, $0 < \beta - \alpha < 2\pi$. *Then there exists a function* $f(z)$, *holomorphic in* S, *having this series as its asymptotic expansion:*

$$f(z) \sim \sum_{n=0}^{\infty} a_n z^n, \quad z \to 0, \quad z \in S.$$

The solution $y_1(x, \lambda)$ for which (11) is true, can be constructed: it is the solution of the Cauchy problem (12) for (1). We can obtain an integral equation for this solution. Let us make the substitution

$$y = u_1 + u_2,$$

$$y' = \left[\sum_{j=-1}^{N-1} \lambda^{-j} \alpha_j(x) u_1 + \sum_{j=-1}^{N-1} (-\lambda)^{-j} \alpha_j(x) u_2\right].$$

Then (1) becomes

$$\begin{bmatrix} u_1' \\ u_2' \end{bmatrix} = \left[\sum_{j=-1}^{N-1} \lambda^{-j} A_j(x) + \lambda^{-N} Q_N(x, \lambda^{-1})\right] \begin{bmatrix} u_1 \\ u_2 \end{bmatrix}.$$

Here $A_j(x) = \mathrm{diag}(\alpha_j(x), (-1)^j \alpha_j(x))$ and the elements of the matrix function Q_N are bounded for $x \in I$ and $|\lambda| \geqslant \lambda_0 > 0$. The system so obtained reduces to a system of integral equations in the same way as in § 2, paragraph 2. We point out that this solution $y_1(x, \lambda)$ depends on N, and we denote it by $y_{1N}(x, \lambda)$.

The solution $y_{1N}(x, \lambda)$ cannot be expanded in an asymptotic series of the form (5) on I. To be precise, such an asymptotic series exists but is not uniform in x: near to $x = a$ the asymptotic series behaves as a boundary layer.

We show this for $N = 1$. Let us consider the system of integral equations (14) of § 2 where $c_1 = 1$ and $c_2 = 0$. We take the zero term approximation as $u_1^0(x) \equiv 1$, $u_1^0(x) \equiv 0$. Then the first term approximation is

$$u_1^1(x) = 1 + \lambda^{-1} \int_a^x \alpha_1(t) dt,$$

$$u_2^1(x) = -\lambda^{-1} \int_a^x \alpha_1(t) \exp\{2\lambda S(x, t)\} dt.$$

Suppose that $q(x) > 0$ for simplicity. Then $S(x, t)$ (as a function of t) takes its largest value on the integration path at the point $t = a$. Integration by parts leads to an asymptotic expansion that is uniform in $x \in I$

$$u_2^1(x) = -\sum_{k=1}^{N} \lambda^{-k-1} L^{k-1}\left(\frac{\alpha_1(x)}{2\sqrt{q(x)}}\right) + \exp\{2\lambda S(x,\, a)\} \sum_{k=1}^{N} \lambda^{-k-1} \left[L^{k-1}\left(\frac{\alpha_1(x)}{2\sqrt{q(x)}}\right)\right]\Bigg|_{x=a} + O(\lambda^{-N-1}), \tag{23}$$

where

$$L = -\frac{1}{2\sqrt{q(x)}}\frac{d}{dx}.$$

The function $F = \exp\{2\lambda S(x,a)\}$ is a typical boundary layer function, since

$$F(a,\lambda) = 1; \quad \lim_{\lambda\to+\infty} F(x,\lambda) = 0, \quad x \neq a,$$

and is appreciably distinct from zero only in a small (of order λ^{-1}) neighbourhood of the point a.

We have restricted ourselves to the first approximation, but we can show that for solutions u_1, u_2 of the integral equations there are asymptotic expansions of the form (23).

§ 4. Systems of Two Equations Containing a Large Parameter

1. Formal Asymptotic Solutions. Let us consider the system

$$y_1' = \lambda[a_{11}(x)y_1 + a_{12}(x)y_2], \quad y_2' = \lambda[a_{21}(x)y_1 + a_{22}(x)y_2]$$

with complex-valued coefficients $a_{jk}(x)$. We write it in matrix form

$$y' = \lambda A(x)y, \tag{1}$$

where $A(x) = (a_{jk}(x))$. Let $I = [a,b]$ be a finite interval and $a_{jk}(x) \in C^\infty(I)$.

System (1) can be reduced to a second order equation by eliminating one of the unknown functions, but to construct the F.A.S. it is more convenient to operate directly with the system itself. We assume that the eigenvalues $p_1(x)$, $p_2(x)$ of $A(x)$ are distinct for all $x \in I$. Then $p_{1,2}(x) \in C^\infty(I)$ and there are linearly independent eigenvectors $e_1(x)$, $e_2(x)$ of $A(x)$ of class $C^\infty(I)$. The matrix $T = (e_1(x),\, e_2(x))$ reduces $A(x)$ to diagonal form, that is

$$T^{-1}(x)A(x)T(x) = \Lambda(x) = \operatorname{diag}(p_1(x),\; p_2(x)).$$

Let $e_1^\star(x)$, $e_2^\star(x)$ be the rows of $T^{-1}(x)$; then

$$e_j^*(x)e_k(x) = \delta_{jk}\,, \quad e_j^* A(x) = p_j(x)e_j^*(x)\,. \tag{2}$$

We seek a F.A.S. of (1) in the form of an asymptotic series

$$y = e^{\lambda S(x)} \sum_{k=0}^{\infty} \lambda^{-k} f_k(x)\,. \tag{3}$$

Substituting this into (1), we obtain the recurrence system of equations

$$(A(x) - S'(x)I)f_0(x) = 0\,, \tag{4}$$

$$(A(x) - S'(x)I)f_{k+1}(x) = -f_k'(x)\,, \quad k = 0,\ 1,\ \ldots$$

It follows from the first equation that $S'(x)$ is an eigenvalue and $f_0(x)$ is an eigenvector of $A(x)$. Let us put $S'(x) = p_1(x)$. Then

$$f_0(x) = \alpha(x)e_1(x)\,,$$

where $\alpha(x)$ is a scalar function, and $\alpha(x)$ is determined from the equation

$$(A(x) - p_1(x)I)f_1(x) = -f_0'(x) \tag{5}$$

in the system (4). Multiplying both sides of this equation on the left by $e_1^*(x)$ and using (2), we obtain

$$e_1^*(x)f_0'(x) = 0\,, \quad x \in I\,.$$

Consequently

$$\alpha(x) = \exp\left\{-\int^x e_1^*(t)e_1'(t)dt\right\}\,.$$

Next, we find the vector function $f_1(x)$. We have $f_1(x) = \alpha_1(x)e_1(x) + \alpha_2(x)e_2(x)$, and substituting this into (5) we obtain

$$(p_2(x) - p_1(x))\alpha_2(x)e_2(x) = -f_0'(x)\,.$$

Multiplying this identity on the left by $e_2^*(x)$, we find that

$$\alpha_2(x) = \frac{\alpha(x)e_2^*(x)e_1'(x)}{p_1(x) - p_2(x)}\,.$$

The function $\alpha_1(x)$ is still undetermined; it is found from the next equation of the system (4).

Continuing these constructions we can obtain all the terms of the expansion (3). Putting $S'(x) = p_2(x)$ we find the second F.A.S. similarly.

2. Sufficient Conditions for the Existence of Asymptotic Behaviour of the Solutions. The point $x_0 \in I$ is called a *turning point* of (1) if the eigenvalues of $A(x_0)$ coincide. We introduce the following conditions.

(1) System (1) has no turning points, that is, $p_1(x) \neq p_2(x)$ for $x \in I$.
(2) The function Re $[p_1(x) - p_2(x)]$ does not change sign for $x \in I$.

Condition (2) will be discussed in § 8. We denote

$$\tilde{y}_j(x,\lambda) = \exp\left\{\lambda \int_{x_0}^{x} p_j(t)dt - \int_{x_0}^{x} e_j^*(t)e_j'(t)dt\right\}. \tag{6}$$

Recall that $e_j(x)$ is an eigenvector of class $C^\infty(I)$ of $A(x)$: $A(x)e_j(x) = p_j(x)e_j(x)$, and the vectors $e_j^*(x)$ are the rows of $(e_1(x),\, e_2(x))^{-1}$ (see (2)).

System (1) has two solutions of the form

$$y_j(x,\lambda) = \tilde{y}_j(x,\lambda)\left[e_j(x) + \sum_{k=1}^{N-1} f_{jk}(x)\lambda^{-k} + O(\lambda^{-N})\right], \quad \lambda \to \infty. \tag{7}$$

Here $N \geqslant 1$ is arbitrary, the asymptotic behaviour (7) is uniform in $x \in I$, and (7) can be differentiated in x and λ an arbitrary number of times, preserving the uniformity in x of the bounds for the remainder term. The solutions $y_1(x,\lambda)$ and $y_2(x,\lambda)$ are linearly independent for $\lambda \gg 1$.

Example. We consider the system (1) with a Hermitian matrix

$$A(x) = \begin{bmatrix} a(x) & c(x) \\ \bar{c}(x) & b(x) \end{bmatrix}.$$

Here $a(x)$, $b(x)$ are real-valued functions, and $c(x)$ is a complex-valued function. The eigenvalues of $A(x)$ are

$$p_{1,2}(x) = \frac{1}{2}[a(x) + b(x) \pm \sqrt{D(x)}], \quad D = (a-b)^2 + 4|c|^2,$$

and (1) has no turning points if for each $x \in I$ either $a(x) \neq b(x)$ or $c(x) \neq 0$. We have

$$e_j(x) = \begin{bmatrix} 2c(x) \\ b(x) - a(x) - (-1)^j\sqrt{D(x)} \end{bmatrix}, \quad e_j^*(x) = \frac{\bar{e}_j^T(x)}{|e_j(x)|^2},$$

from which it follows that (1) has a solution of the form

$$y_1(x,\lambda) = \frac{1}{\sqrt{c(x)}}\frac{1}{\sqrt[4]{D(x)}}\exp\left\{\lambda \int_{x_0}^{x} p_1(t)dt \right.$$
$$\left. + \frac{1}{2}\int_{x_0}^{x} \frac{c(t)}{\sqrt{D(t)}}\left(\frac{a(t)-b(t)}{c(t)}\right)' dt\right\}[e_j(x) + O(\lambda^{-1})].$$

Replacing $\sqrt{D}$ by $-\sqrt{D}$ we obtain the solution $y_2(x,\lambda)$.

3. Additional Parameters and Complex λ. In the case of the system

$$y' = \lambda A(x, \ \mu) y \, , \tag{8}$$

all the statements given in § 3, paragraph 1.5 for a second order scalar equation are again true if $\sqrt{q(x,\mu)}$ is replaced by $p_1(x,\mu) - p_2(x,\mu)$ in conditions (1), (2).

§ 5. Systems of Equations Close to Diagonal Form

Let us consider the system of n equations

$$y' = (\Lambda(x) + B(x))y \tag{1}$$

on the half-line $\mathbb{R}^+ : x \geqslant 0$. Here $\Lambda(x)$ and $B(x)$ are square matrices of order n, y is an n-vector and $\Lambda(x)$ is a diagonal matrix with diagonal elements $p_1(x), \dots, p_n(x)$. We assume that $\Lambda(x)$, $B(x) \in C(\mathbb{R}^+)$.

The fundamental results on the asymptotic behaviour of solutions of a system of the form (1) as $x \to \infty$ are best formulated straightaway for arbitrary n and not just for $n = 2$.

If $B(x) \equiv 0$, then (1) splits and has F.S.S.

$$\tilde{y}_j(x) = \exp\left\{\int_{x_0}^{x} p_j(t)dt\right\} f_j \, , \quad j = 1, 2, \dots, n,$$

where the f_j are vectors with components δ_{jk}. It is natural to assume that if the elements of $B(x)$ are small for $x \gg 1$, then the solutions of (1) are close to $\tilde{y}_j(x)$. We will give the relevant sufficient conditions.

1. Almost Diagonal Systems. System (1) is called *almost diagonal* if

$$\lim_{x\to\infty} \|B(x)\| = 0 \, . \tag{2}$$

However this condition is not sufficient to ensure that the solutions of (1) are close to solutions of the diagonal system. We bring in the condition:

(1) for some j and for $k \neq j$

$$\operatorname{Re}\,(p_k(x) - p_j(x)) \geqslant c > 0 \, , \quad x \geqslant a \geqslant 0 \, . \tag{3}$$

Theorem (Perron). *The almost diagonal system (1) has a solution $y_j(x)$ such that*

$$\lim_{x\to\infty} \frac{y_{jk}(x)}{y_{jj}(x)} = 0 \, , \quad j \neq k \, , \quad \lim_{x\to\infty} \left(\frac{y'_{jj}(x)}{y_{jj}(x)} - p_j(x)\right) = 0 \, .$$

This solution can be written in the form

$$y_j(x) = \exp\left\{\int_{x_0}^{x} p_j(t)dt + \int_{x_0}^{x} p_j^{(1)}(t)dt\right\}[f_j + u_j(x)], \tag{4}$$

where

$$\lim_{x\to\infty} p_j^{(1)}(x) = \lim_{x\to\infty} \|u_j(x)\| = 0.$$

If Re $(p_{j+1}(x) - p_j(x)) \geqslant c > 0$ for all j, then (1) has a fundamental matrix of the form

$$Y(x) = [I + U(x)]\exp\left\{\int_{x_0}^{x} \Lambda(t)dt + \int_{x_0}^{x} \Lambda_1(t)dt\right\}. \tag{5}$$

Here $\Lambda_1(x)$ is a diagonal matrix and

$$\lim_{x\to\infty} \|\Lambda_1(x)\| = \lim_{x\to\infty} \|U(x)\| = 0. \tag{6}$$

The asymptotic formulae (4) and (5) are rather coarse. In fact,

$$y_{jj}(x)\exp\left\{-\int_{x_0}^{x} p_k(t)dt\right\} = \exp\{o(x)\}, \quad x \to \infty.$$

The right hand side of this formula can, for instance, go off to infinity as $x \to \infty$. One cannot improve this result without additional assumptions on the behaviour of $B(x)$ as $x \to \infty$, even for $n = 1$.

Let us state some results analogous to Perron's theorem. Suppose that for some j and for all $k \neq j$ we have

$$\|B(x)\| = o(\text{Re}\,(p_j(x) - p_k(x))), \quad x \to \infty. \tag{7}$$

Then (1) has a solution $y_j(x)$ with bounds as $x \geqslant x_0 \gg 1$:

$$\begin{aligned} 0 < c_1 \exp&\left\{\text{Re}\, S_j(x_0, x) - (1+\delta)\int_{x_0}^{x} \|B(t)\|dt\right\} \\ &\leqslant \|y_j(x)\| \leqslant c_2 \exp\left\{\text{Re}\, S_j(x_0, x) + (1+\delta)\int_{x_0}^{x} \|B(t)\|dt\right\}, \end{aligned} \tag{8}$$

where

$$S_j(x_0, x) = \int_{x_0}^{x} p_j(t)dt.$$

Here $\delta > 0$ can be chosen as small as we please by increasing x_0 and the constants c_k can be chosen so that they do not depend on x_0.

It follows from (8) that

$$\lim_{x\to\infty} \frac{\ln \|y_j(x)\|}{\text{Re}\, S_j(x_0, x)} = 1.$$

If condition (7) is satisfied for all j then the solutions $y_1(x), \dots, y_n(x)$ form a F.S.S.

Let $\Lambda(x)$ be constant and let $B(x) \to 0$ as $x \to \infty$. Then the bounds in (8) hold for all $j = 1, \dots, n$. In particular,

$$\lim_{x\to\infty} \frac{\ln \|y_j(x)\|}{x} = \operatorname{Re} \lambda_j .$$

2. L-Diagonal Systems. System (1) is called L-*diagonal* if $\|B(x)\| \in L^1$ $[0, \infty)$, that is, if

$$\int_0^\infty \|B(x)\| dx < \infty . \tag{9}$$

We bring in a condition analogous to condition 1):

2) for $x \gg 1$ the differences $\operatorname{Re}\,[p_j(x) - p_k(x)]$ do not change sign for fixed j and for all k.

Theorem (N. Levinson). *The* L-*diagonal system (1) has a solution* $y_j(x)$ *such that*

$$y_j(x) = \exp\left\{\int_{x_0}^x p_j(t)dt\right\}[f_j + u_j(x)] , \quad \lim_{x\to\infty} \|u_j(x)\| = 0 . \tag{10}$$

Recall that the f_j are vectors with components δ_{jk}. If condition 2) is satisfied for all j, then (1) has a fundamental matrix $Y(x)$ of the form

$$Y(x) = [I + U(x)] \exp\left\{\int_{x_0}^x \Lambda(t)dt\right\} , \quad \lim_{x\to\infty} \|U(x)\| = 0 . \tag{11}$$

Condition 2) will be discussed in §8.

Levinson's theorem admits the following refinement. We will say that k belongs to the class $H_1(j)$ if for $0 \leqslant t \leqslant x < \infty$ we have

$$\operatorname{Re} S_{kj}(t,\, x) \leqslant c_{kj} < \infty ,$$

where

$$S_{kj}(t,\, x) = \int_t^x [p_k(s) - p_j(s)]ds .$$

Further, $k \in H_2(j)$ if

$$\lim_{x\to\infty} \operatorname{Re} S_{kj}(0, x) = -\infty .$$

Now suppose that each $k = 1, \dots, n$ belongs to one of $H_1(j)$, $H_2(j)$. Then the L-diagonal system (1) has a solution $y_j(x)$ of the form (10). There is the bound for the remainder term

$$\|u_j(x)\| \leqslant C\left[\int_x^\infty \|B(t)\|dt\right] + \sum_{k\in H_2(j)} \int_x^\infty |\exp\{S_{kj}(t,x)\}|\, \|B(t)\|dt\,. \tag{12}$$

If more exact information is known about $\|B(x)\|$ and $p_k(x)$ then (12) can be improved by applying the Laplace method for the asymptotic bound of the integrals.

Examples. 1. Suppose that all the eigenvalues $p_1(x), \ldots, p_n(x)$ are purely imaginary. Then

$$\|U(x)\| \leqslant C \int_x^\infty \|B(t)\|dt\,,$$

where $U(x)$ is as in (11). Indeed, in this case condition (2) is satisfied for all j, and $k \in H_1(j)$ for all k and j.

2. Let j be such that

$$\operatorname{Re}\, p_j(x) \leqslant \operatorname{Re}\, p_k(x)$$

for all $x \gg 1$ and for all k. Then

$$\|u_j(x)\| \leqslant C \int_x^\infty \|B(x)\|dt\,.$$

Further results concerning the asymptotic behaviour of the solutions of a system of equations will be given in Chap. 5.

3. Dual Asymptotic Behaviour. Let us consider the system

$$y' = [\lambda\Lambda(x) + \Lambda_1(x) + \lambda^{-1}B(x,\ \lambda^{-1})]y\,. \tag{13}$$

Here $\Lambda(x)$ is the same as in system (1), $\Lambda_1(x) = \operatorname{diag}(p_1^{(1)}(x), \ldots, p_n^{(1)}(x))$ and

$$\|B(x,\lambda^{-1})\| \leqslant Cb(x)\,, \qquad \int^\infty b(t)dt < \infty \tag{14}$$

for $x \geqslant x_0$, $\lambda \geqslant \lambda_0$. In this case, with suitable conditions on Λ and Λ_1 we can obtain asymptotic formulae for the solutions which are applicable both as $x \to \infty$ with $\lambda > 0$ fixed, and for $\lambda \to \infty$ with x fixed. Namely, suppose that for some j and for all $k \neq j$ we have

$$\int^\infty |\operatorname{Re}\,(p_j(x) - p_k(x))|dx = \infty\,, \tag{15}$$

$$\operatorname{Re}\,(p_j^{(1)}(x) - p_k^{(1)}(x)) = o(\operatorname{Re}\,(p_j(x) - p_k(x)))\,, \quad x \to \infty\,.$$

Then (13) has a solution of the form

$$y_j(x, \lambda) = \exp\left\{\lambda \int_{x_0}^{x} p_j(t)dt + \int_{x_0}^{x} p_j^{(1)}(t)dt\right\} [1 + \lambda^{-1} u_j(x, \lambda^{-1})]. \quad (16)$$

In this case for arbitrary $\lambda_1 \geqslant \lambda_0 > 0$ there exists $x(\lambda_1) < \infty$ such that for $\lambda \geq \lambda_1$ and $x \geqslant x(\lambda_1)$ there is the bound

$$\|u_j(x, \lambda^{-1})\| \leqslant k_j(x), \quad \lim_{x\to\infty} k_j(x) = 0. \quad (17)$$

If these conditions are satisfied for all j, then (13) has a fundamental matrix of the form

$$Y(x, \lambda) = [I + \lambda^{-1} U(x, \lambda^{-1})] \exp\left\{\lambda \int_{x_0}^{x} \Lambda(t)dt + \int_{x_0}^{x} \Lambda_1(t)dt\right\},$$

where

$$\|U(x, \lambda^{-1})\| \leqslant k(x), \quad \lim_{x\to\infty} k(x) = 0.$$

4. Scalar Equations. Let us consider the equation

$$y^{(n)} + \sum_{k=1}^{n} q_k(x) y^{(n-k)} = 0 \quad (18)$$

on the half-line $\mathbb{R}^+$, where $q_k(x)$ is a complex-valued function and $q_k(x) \in C^\infty(\mathbb{R}^+)$. Let

$$q_k(x) = a_k + r_k(x), \quad \int^{\infty} |r_k(x)|dx < \infty$$

and let $p_1, \ldots, p_n$ be the distinct roots of the equation

$$p^n + \sum_{k=1}^{n} a_k p^{n-k} = 0.$$

By Levinson's theorem equation (18) has a F.S.S. of the form

$$y_j^{(k-1)}(x) = p_j^{k-1} e^{p_j x} [1 + o(1)|, \quad x \to \infty,$$

where $j, k = 1, \ldots, n$.

Example. The Schrödinger equation

$$-\frac{h^2}{2m}\psi'' + [V(x) - E]\psi = 0$$

with the conditions $E > 0$, $V(x) \in L_1(-\infty, \infty)$ has two F.S.S. such that

$$\psi_{1,2}^{+}(x) = e^{\pm ikx} + o(1), \quad x \to +\infty,$$
$$\psi_{1,2}^{-}(x) = e^{\pm ikx} + o(1), \quad x \to -\infty,$$

where $k = \sqrt{2mE}/h$.

§ 6. Asymptotic Behaviour of the Solutions for Large Values of the Argument

1. WKB-Approximation. We will consider the equation

$$y'' - q(x)y = 0 \tag{1}$$

on the half-line $\mathbb{R}^+$, where $q(x) \in C^\infty(\mathbb{R}^+)$. We assume that conditions (1), (2) from § 3, paragraph 1.2 are satisfied for large x, that is

$$q(x) \neq 0\,, \quad \operatorname{Re} \sqrt{q(x)} \geqslant 0\,, \quad x \gg 1\,.$$

We introduce the notation

$$S(x_0, x) = \int_{x_0}^{x} \sqrt{q(t)}dt\,, \quad \alpha_1(x) = \frac{1}{8}\frac{q''(x)}{q^{3/2}(x)} - \frac{5}{32}\frac{q'^2(x)}{q^{5/2}(x)}\,, \tag{2}$$

$$\rho(x) = \int_x^\infty |\alpha_1(t)|dt\,.$$

1.1 Principal Asymptotic Term. Suppose that $q(x) \in C^2(\mathbb{R}^+)$. If the integral

$$\int^\infty |\alpha_1(t)|dt < \infty\,, \tag{3}$$

converges, then (1) has a F.S.S. of the form

$$y_{1,2}(x) = q^{-1/4}(x)\exp\{\pm S(x_0,\ x)\}[1 + \varepsilon_{1,2}(x)]\,, \tag{4}$$

where

$$\lim_{x\to\infty} \varepsilon_{1,2}(x) = 0\,. \tag{5}$$

There is a bound for the remainder term $\varepsilon_2(x)$,

$$|\varepsilon_2(x)| \leqslant C\rho(x)\,. \tag{6}$$

All the bounds stated in this paragraph are satisfied for $x \gg 1$.

Formulae (4) – (6) are direct consequences of the WKB-bounds (11) of § 2 at $b = \infty$ for the solution y_2.

It is somewhat more complex to give a bound for the remainder term $\varepsilon_1(x)$. There are two possibilities:

A. $\operatorname{Re} S(0,\infty) < \infty$. Then (6) applies for $\varepsilon_1(x)$,

B. $\operatorname{Re} S(0,\infty) = \infty$. Then

$$|\varepsilon_1(x)| \leqslant C\left[|\rho(x)| + \int_{x_0}^{x} |e^{2S(t,x)}||\alpha_1(t)|dt\right]\,. \tag{7}$$

1.2 Equation (1) with Real $q(x)$. There are the following two possibilities A and B.

A. Non-oscillatory solutions: $q(x) > 0$ for $x \gg 1$. Then

$$y_{1,2}(x) \sim q^{-1/4}(x) \exp\{\pm S(x_0,\, x)\}\,.$$

If in addition

$$q'(x) q^{-3/2}(x) \to 0\,, \quad x \to \infty\,, \tag{8}$$

and

$$S(0, \infty) = \infty\,, \tag{9}$$

then as $x \to \infty$

$$y_1(x) \to \infty\,, \quad y_2(x) \to 0\,.$$

Here $y_2(x)$ decreases exponentially and $y_1(x)$ increases exponentially.

The solution $y_2(x)$ is uniquely determined by its asymptotic behaviour at infinity: if $y(x)$ is such that

$$y(x) \sim q^{-1/4}(x) \exp\{-S(x_0,\, x)\}\,, \quad x \to \infty\,,$$

then $y(x) \equiv y_2(x)$.

The solution $y_2(x)$ has the following properties:

$$\begin{gathered} \lim_{x \to \infty} y_2(x) = 0\,, \\ \int^{\infty} |y_2^p(x)| dx < \infty\,, \quad \int^{\infty} |q(x)|^r |y_2(x)|^p dx < \infty\,, \end{gathered} \tag{10}$$

where $p > 0$ and r is arbitrary. In fact, from conditions (8), (9) and L' Hôpital's rule, it follows that

$$\lim_{x \to \infty} \frac{\ln q(x)}{S(x_0, x)} = \lim_{x \to \infty} \frac{q'(x)}{q^{3/2}(x)} = 0\,,$$

and the first relation in (10) is proved, since

$$\lim_{x \to \infty} \exp\{-S(x_0, x) + m \ln q(x)\} = 0$$

for arbitrary m. Further, for $n > 0$ and $x \geqslant x_0 \gg 1$,

$$\begin{aligned} \int_{x_0}^{x} q^m(t) \exp\{-nS(x_0, t)\} dt &= \int_{x_0}^{x} \sqrt{q(t)} \exp\{-(n + o(1)) S(x_0, t)\} dt \\ &\leqslant \int_{x_0}^{x} \sqrt{q(t)} \exp\left\{-\frac{n}{2} S(x_0, t)\right\} dt \\ &= -\frac{2}{n} \exp\left\{-\frac{n}{2} S(x_0, x)\right\} \Big|_{x_0}^{x} \to \text{const}\,, \\ &\quad x \to \infty\,, \end{aligned}$$

and the latter two relations in (10) are proved. If $y(x)$ satisfies any of the relations in (10) then $y(x) = (\text{const.})y_2(x)$.

B. Oscillatory solutions: $q(x) < 0$. For convenience we consider the equation

$$y'' + q(x)y = 0, \tag{11}$$

and suppose $q(x) > 0$ for $x \gg 1$. This equation has a F.S.S. of the form

$$y_{1,2}(x) = q^{-1/4}(x)\exp\{\pm iS(x_0, x)\}[1 + \varepsilon_{1,2}(x)].$$

Here the bound (6) is valid both for $\varepsilon_1(x)$ and $\varepsilon_2(x)$.

Both solutions $y_1(x)$, $y_2(x)$ are uniquely determined by their asymptotic behaviour as $x \to \infty$. If (8) holds then these solutions are uniquely determined by boundary conditions of *Sommerfeld radiation* type:

$$\lim_{x\to\infty} (y'/y \mp i\sqrt{q(x)}) = 0.$$

If the integral $\int^{\infty} q^{-1/2}(x)dx$ converges (diverges), then all solutions of (11) belong (do not belong) to the space $L_2[0, \infty)$.

1.3 Analysis of Condition (3). If $q(x) = x^{\alpha}$ then condition (3) holds for $\alpha > -2$. For α an integer this means that $x = \infty$ is an irregular singular point of equation (1) (Chap. 1, § 2).

Let us indicate some classes of functions for which conditions (1), (2), (3) and (8) are satisfied

1. $q(x) = ax^{\alpha}$, $\alpha > -2$, $a \neq 0$.
2. $q(x) = a(\ln x)^{\alpha}$, $-\infty < \alpha < \infty$, $a \neq 0$.
3. $q(x) = a\exp\{bx^{\alpha}\}$, $\alpha > 0, b > 0, a \neq 0$.

In this way $q(x)$ can increase as $x \to \infty$ with arbitrary speed and can even decrease, but not slower than x^{-2}. Condition (1) signifies some regularity in the behaviour of $q(x)$ as $x \to \infty$. If $q(x)$ behaves as in the cases 1–3 as $x \to \infty$ (for instance, $q(x) \sim ax^{\alpha}$) and if this asymptotic behaviour can be twice differentiated, then all the above conditions are satisfied.

The asymptotic formulae (4) are valid when $q(x)$ has a finite singular point x_0 and $x \to x_0$. Let $I = [x_0\, x_0 + \delta]$, $\delta > 0$, $q(x) \in C^2(I)$, let conditions (1), (2) be satisfied for $x \in I$ and suppose that the integral

$$\int^{x_0} |\alpha_1(x)|dx \tag{3'}$$

converges. Then formulae (4) are true, where $\varepsilon_j(x) \to 0$ as $x \to x_0$.

If $q(x) = a(x - x_0)^{\alpha}$, $a \neq 0$, then (3′) holds for $\alpha < -2$ and does not hold for $\alpha \geqslant -2$. We point out that for $\alpha = 2$ the point x_0 is a regular singular point of (1). In this case the asymptotic behaviour of the solution has a different form.

Let

$$\lim_{x\to\infty}\frac{q'(x)}{q^{3/2}(x)}=4\gamma\,,\qquad \int^{\infty}\left|\frac{d}{dx}\left(\frac{q'(x)}{q^{3/2}(x)}\right)\right|<\infty\,,\qquad \gamma\neq\pm1\,,$$

and for $x \gg 1$ the following functions are bounded

$$\exp\{\pm i\tilde{S}(x)\}\,,\qquad \tilde{S}(x)=\int_{x_0}^{x}\sqrt{q(t)}\sqrt{1-\frac{q'^2(t)}{16q^3(t)}}\,dt\,.$$

Then equation (11) has a F.S.S. such that

$$\begin{aligned}&y_{1,2}(x)\sim q^{-1/4}(x)\exp\{\pm i\tilde{S}(x)\}\,,\\&y'_{1,2}(x)\sim(-\gamma\pm i\sqrt{1-\gamma^2})\sqrt[4]{q}(x)\exp\{\pm i\tilde{S}(x)\}\end{aligned}\tag{12}$$

as $x\to\infty$. In particular, these formulae are true if

$$q(x)\sim ax^{-2}\,,\qquad x\to\infty\,,\qquad a>0\,,\qquad a\neq1/4\,,$$

and if this asymptotic behaviour can be differentiated. Formulae (12) can then be written in the form

$$y_{1,2}(x)=x^{-1\pm i\sqrt{a-1/4}+o(1)}\,,\qquad x\to\infty\,.$$

1.4 Differentiation of the Asymptotic Behaviour. Suppose that the conditions of paragraph 1.1 are satisfied, together with condition (8). Then the asymptotic formulae (4) can be twice differentiated, that is

$$y_{1,2}^{(j)}(x)\sim(\pm\sqrt{q(x)})^jq^{-1/4}(x)\exp\{\pm S(x_0,\,x)\}\,,\qquad x\to\infty\,,\qquad j=1,2\,.$$

1.5 The Further Terms of the Asymptotic Series. Comparison of formulae (4) of § 5 and (7) of § 3 shows that the principal asymptotic term of the solutions of $y''-\lambda^2q(x)y=0$ has the same form both for finite x with $\lambda\to\infty$ and for $\lambda=1$ with $x\to\infty$. This is true also for further terms of the asymptotic series: the asymptotic behaviour of the solutions of (1) has the form (5) of § 3, where $\lambda=1$, under regularity type conditions on the behaviour of $q(x)$ and its derivatives as $x\to\infty$. Roughly speaking, this condition is that the sequence $\{\alpha_j(x)\}$ must be asymptotic as $x\to\infty$, where $\alpha_j(x)$ is as in § 3, (4).

Example. Let $q(x)=x^{\alpha}\phi(x)$, where $\alpha>-2$, $\phi(x)=\sum_{n=0}^{\infty}\phi_nx^{-n}$ and this series converges for $x\geqslant x_0>0$. Then as $x\to\infty$

$$\alpha_j(x)\sim c_jx^{\beta_j}\,,\qquad \beta_j=\alpha/2-(\alpha/2+1)(j+1)\,,$$

where the c_j are constants, so that the sequence $\{\alpha_j(x)\}$ is asymptotic. This is also true for any of the functions $q(x)$ considered in paragraph 1.3.

Let us next bring in the functions $a_N^{\pm} = \sum_{j=1}^{N}(\pm 1)^j \alpha_j(x)$ and make the transformation

$$\begin{bmatrix} y \\ y' \end{bmatrix} = \begin{bmatrix} 1 & 1 \\ a_N^+ & a_N^- \end{bmatrix} \begin{bmatrix} u_1 \\ u_2 \end{bmatrix}.$$

For $N = 0$ this transformation is that of § 1, paragraph 4. Then equation (1) is reduced to the system

$$\begin{bmatrix} u_1' \\ u_2' \end{bmatrix} = \frac{1}{a_N^+ - a_N^-} \begin{bmatrix} q - a_N^+ a_N^- + a_N^{+\prime} & q - a_N^{-2} - a_N^{-\prime} \\ -q + a_N^{+2} + a_N^{+\prime} & -q + a_N^+ a_N^- + a_N^{-\prime} \end{bmatrix} \begin{bmatrix} u_1 \\ u_2 \end{bmatrix}.$$

This system will be L-diagonal (§ 5) under the condition that the integrals $\rho_N^{\pm}(x_0, \infty)$ converge, where

$$\rho_N^{\pm}(x_0, x) = \int_{x_0}^{x} |q - a_N^{\pm 2} - a_N^{\pm\prime}||a_N^+ - a_N^-|^{-1} dt. \tag{13}$$

In this case (1) has two solutions of the form

$$y_{1,2}(x) = q^{-1/4}(x) \exp\left\{ \int_{x_0}^{x} \pm\sqrt{q(t)}dt + \sum_{j=1}^{N} (\pm 1)^j \int_{x_0}^{x} \alpha_j(t)dt \right\} \times [1 + \varepsilon_{1,2}(x)], \tag{14}$$

where $\varepsilon_{1,2}(x) \to 0$ as $x \to \infty$. There is the bound

$$|\varepsilon_2(x)| \leqslant C\rho_N^-(x, \infty). \tag{15}$$

If $q(x)$ has the same form as in the example, then

$$\varepsilon_j(x) = O(x^{-\delta_N}), \quad \delta_N = (\alpha/2 + 1)(N + 1)$$

as $x \to \infty$, $j = 1, 2$.

2. Additional Material. All the asymptotic formulae for solutions as $x \to \infty$ obtained in this pargraph are based essentially on Levinson's theorem on the asymptotic behaviour of solutions of L-diagonal systems (§ 4, paragraph 2). Let us state some of the stronger results. We consider the equations

$$y'' + (1 + \alpha(x))y = 0, \tag{16}$$

and

$$y'' - (1 + \alpha(x))y = 0, \tag{17}$$

where

$$\lim_{x \to \infty} \alpha(x) = 0. \tag{18}$$

Suppose also that

$$\int^{\infty} |\alpha(t)|^n dt < \infty\,, \tag{19}$$

where $n \geqslant 1$ is an integer. For $n = 1$ the asymptotic behaviour of the solution is stated in § 5. We consider the case where $n > 1$.

A. $n = 2$. Equation (17) has a F.S.S. of the form

$$y_{1,2}(x) \sim \exp\left\{\pm\left(x + \frac{1}{2}\int_{x_0}^{x} \alpha(t)dt\right)\right\}, \quad x \to \infty\,, \tag{20}$$

The structure of solutions of the equation (16) is more complicated (see [Harris 2]). Let the function $\alpha(x)$ be continuous for $x \geqslant x_0$. We introduce the integrals

$$\beta_1(x) = \int_x^{\infty} \alpha(t)\cos 2t\,dt\,, \quad \beta_2(x) = \int_x^{\infty} \alpha(t)\sin 2t\,dt\,.$$

B. Let the integrals $\beta_1(x)$ and $\beta_2(x)$ be convergent for $x \geqslant x_0$ and let $\alpha\beta_j \in L_2(x_0,\, \infty)$, $j = 1, 2$. Then as $x \to \infty$ the equation (16) has solutions

$$\begin{aligned} y_1(x) &= \cos(x + \sigma(x)) + o(1)\,, & y_2(x) &= \sin(x + \sigma(x)) + o(1)\,, \\ y_1'(x) &= -y_2(x) + o(1)\,, & y_2'(x) &= -y_1(x) + o(1)\,, \end{aligned}$$

where

$$\sigma(x) = \frac{1}{2}\int_{x_0}^{x} \alpha(t)dt\,.$$

The case B is nonresonant. Next we introduce the integral

$$\beta_3(t) = \int_x^{\infty} \alpha(t)dt\,.$$

C. Let the integrals $\beta_1(x)$ and $\beta_3(x)$ be convergent for $x \geqslant x_0$ and let $\alpha\beta_j \in L_1(x_0, \infty)$, $j = 1$, 3. Suppose also that the conditions

$$\overline{\lim_{x\to\infty}} \int_{x_0}^{x} \alpha(t)\sin 2t\,dt = +\infty$$

and

$$\underline{\lim_{x\to\infty}} \int_{x_0}^{x} \alpha(t)\sin 2t\,dt = -\infty$$

cannot be satisfied simultaneously. Then as $x \to \infty$ the equation (16) has solutions

$$\begin{aligned} y_1(x) &= \rho(x)[\cos x + o(1)]\,, & y_2(x) &= [\rho(x)]^{-1}[\sin x + o(1)]\,, \\ y_1'(x) &= \rho(x)[-\sin x + o(1)]\,, & y_2'(x) &= [\rho(x)]^{-1}[\cos x + o(1)]\,, \end{aligned}$$

where

$$\rho(x) = \exp\left\{\int_{x_0}^{x} \alpha(t)\sin 2t dt\right\}.$$

D. Let

$$\tilde{\alpha}_j(x) = \int_x^{\infty} \alpha(t)\beta_j(t)dt, \quad j = 1, 2.$$

Let the integrals β_1, β_2, $\tilde{\alpha}_1$, $\tilde{\alpha}_2$ be convergent for $x \geqslant x_0$ and let $\alpha\beta_i\beta_j$, $\alpha\tilde{\alpha}_j \in L_1(x_0, \infty)$, $1 \leqslant i, j \leqslant 2$. Then as $x \to \infty$ equation (16) has solutions

$$\begin{aligned} y_1(x) &= \rho(x)[\cos(x + \sigma(x)) + o(1)], & y_2 &= \rho(x)[\sin(x + \sigma(x)) + o(1)], \\ y_1'(x) &= \rho(x)[-\sin(x + \sigma(x)) + o(1)], & y_2' &= \rho(x)[\cos(x + \sigma(x)) + o(1)], \end{aligned}$$

where

$$\rho(x) = \exp\left\{-\frac{1}{2}\int_{x_0}^{x} \alpha(t)\int_t^{\infty} \alpha(\tau)\cos 2(t-\tau)dt\, d\tau\right\},$$

$$\sigma(x) = \frac{1}{2}\int_{x_0}^{x} \alpha(t)dt + \frac{1}{4}\int_{x_0}^{x} \alpha(t)\int_t^{\infty} \alpha(t)\sin 2(t-\tau)dt\, d\tau.$$

Examples. 1. The equation

$$y'' + [1 + (a + \sin \lambda x)x^{-1}]y = 0, \quad \lambda \neq \pm 2$$

has solutions

$$y_1(x) = \cos(x + a \ln x) + o(1), \quad y_2(x) = \sin(x + a \ln x) + o(1).$$

2. If $\alpha(x) = x^{-1}\sin 2x + x^{-3/4}\sin x$, Conditions C are fulfilled.
3. The equation

$$y'' + (1 + x^{-1/2}\sin \lambda x)y = 0, \quad \lambda \neq \pm 1, \pm 2$$

has solutions

$$y_1(x) = \cos(x + \alpha \ln x) + o(1), \quad y_2(x) = \sin(x + \alpha \ln x) + o(1),$$

where

$$\alpha = 4[\lambda^2 - 4]^{-1}.$$

Here, Conditions D are satisfied.

Note that we can formally obtain formula (20) from (4):

$$\sqrt{1 + \alpha(x)} = 1 + \frac{1}{2}\alpha(x) + O(\alpha^2(x)), \quad x \to \infty,$$

so

$$\int_{x_0}^{x} \sqrt{1+\alpha(t)}dt = x + \frac{1}{2}\int_{x_0}^{x} \alpha(t)dt + \int_{x_0}^{x} O(\alpha^2(t))dt$$
$$= x + \frac{1}{2}\int_{x_0}^{x} \alpha(t)dt + C + o(1)\,.$$

E. $n = 3$. Equation (17) has a F.S.S. of the form

$$y_{1,2}(x) \sim \exp\left\{\pm\left(x + \frac{1}{2}\int_{x_0}^{x} \alpha(t)dt - \frac{1}{4}\int_{x_0}^{x} \alpha(t)\int_{x_0}^{t} \alpha(\tau)e^{2(\tau-t)d\tau}dt\right)\right\},$$

as $x \to \infty$. These formulae cannot be simplified further. Similar formulae are true for arbitrary $n > 1$, but as n increases they become more complicated [Bellman].

§ 7. Dual Asymptotic Behaviour

1. WKB-Asymptotic Behaviour. We consider the equation

$$y'' - \lambda^2 q(x)y = 0 \tag{1}$$

on the half-line $\mathbb{R}^+$, where $q(x) \in C^\infty(\mathbb{R}^+)$. Suppose that the conditions of § 3, paragraph 1.2 and of § 6, paragraph 1.2 hold simultaneously, that is

$$q(x) \neq 0\,, \quad \operatorname{Re}\sqrt{q(x)} \geqslant 0\,, \quad x \in \mathbb{R}^+\,, \quad \int^{\infty} |\alpha_1(x)|dx < \infty\,. \tag{2}$$

We show that we can then obtain asymptotic formulae for the solutions of equation (1) which are applicable simultaneously both for x fixed with $\lambda \to +\infty$, and for fixed $\lambda > 0$ with $x \to \infty$. We introduce the standard notation

$$S(x_0, x) = \int_{x_0}^{x} \sqrt{q(t)}dt\,, \quad \rho(x) = \int_{x}^{\infty} |\alpha_1(t)|dt\,, \tag{3}$$

$$\tilde{y}_j(x,\ \lambda) = q^{-1/4}(x)\exp\{\pm\lambda S(x_0, x)\}\,,$$

where the plus (minus) sign is taken for $j = 1$ ($j = 2$). We also assume that

$$\lim_{x\to\infty} q'(x)/q^{3/2}(x) = 0\,. \tag{4}$$

1.1 Decreasing Solutions. Suppose that $\operatorname{Re} S(0,\infty) = +\infty$. Then for fixed $\lambda > 0$ equation (1) has a unique (up to multiples) solution y_2 such that $y_2(\infty) = 0$ (§ 6, paragraph 1.1). We normalize y_2 by the condition

$$y_2(x,\ \lambda) \sim \tilde{y}_2(x,\ \lambda)\,, \quad x \to \infty\,,$$

which leads to a unique solution for any $\lambda > 0$. Applying the WKB-bound (11) of § 2 for $b = \infty$ we obtain

$$\left|\frac{y_2(x,\ \lambda)}{\tilde{y}_2(x,\ \lambda)} - 1\right| \leqslant 2(e^{2\rho(x)/\lambda} - 1)\,. \tag{5}$$

This bound holds for $x \geqslant 0$ and $\lambda > 0$. Since $\rho(0) < \infty$ and $\rho(\infty) = 0$, the right-hand side of (5) goes to zero if

(a) $\lambda \to \infty$ uniformly in $x \geqslant 0$;

(b) $x \to \infty$ uniformly in $\lambda \in [\lambda_0, \infty)$, where $\lambda_0 > 0$ is fixed.

In this way the asymptotic behaviour of the form $y_2 \sim \tilde{y}_2$ is dual. Let us fix $\lambda_0 > 0$. Then from (5) we obtain

$$y_2(x,\lambda) = \tilde{y}_2(x,\lambda)\left[1 + \frac{1}{\lambda}\phi_2(x,\lambda)\right]\,, \tag{6}$$

where for $x \geqslant 0$ and $\lambda \geqslant \lambda_0$ we have the bound

$$|\phi_2(x,\lambda)| \leqslant C \int_x^\infty |\alpha_1(t)| dt\,. \tag{7}$$

If y_1 and y_2 are linearly independent solutions, then $|y_1| \to \infty$ as $x \to \infty$. However, solutions which increase to infinity are usually of no interest since they have no physical meaning.

1.2 Oscillatory Solutions. We assume that $\operatorname{Re} S(0,\infty) < \infty$, which is satisfied, in particular, if $q(x) < 0$ for $x \gg 1$. Then equation (1) has two solutions which oscillate for $x \gg 1$ and can be chosen uniquely if we impose boundary conditions at infinity:

$$y_j(x,\lambda) \sim \tilde{y}_j(x,\lambda)\,, \quad x \to \infty\,, \quad j = 1, 2\,.$$

In this case (6) and the bound (7) hold for $j = 1, 2$.

1.3 Higher Approximations. Suppose that $q(x)$ satisfies conditions (2), (4) and the conditions of § 6, paragraph 5. Then

$$y_j(x,\lambda) = \tilde{y}_j(x,\lambda) \exp\left\{\sum_{k=1}^{N} (\pm\lambda)^{-k} \int_{x_0}^{x} \alpha_k(t) dt\right\} [1 + \lambda^{-N-1}\phi_{jN}(x,\lambda)]\,, \tag{8}$$

where

$$|\phi_{jN}(x,\ \lambda)| \leqslant C$$

for $\lambda \geqslant \lambda_0$ and $x \geqslant 0$. With the conditions of paragraph 1.1 these formulae hold for y_2 and with the conditions of paragraph 1.2 they hold for both y_1 and y_2.

2. Second-Order Poles. Let

$$q(x) = x^{-2}p(x), \tag{9}$$

where $p(0) \neq 0$ and $p(x) \in C^\infty$ for small $x \geqslant 0$. Then

$$\alpha_1(x) \sim \frac{1}{8\sqrt{p(0)}x}, \quad x \to +0, \tag{10}$$

and the integral $\int^0 |\alpha_1(x)|dx$ diverges. In this case it turns out that we cannot use the WKB-approximation

$$y_{1,2}(x) \sim q^{-1/4}(x)\exp\{\pm S(x_0, x)\}, \quad x \to +0.$$

Example. Let $p(x) \equiv a$. Then (1) is the Euler equation and its solutions are

$$y_j = x^{\gamma_j}, \quad \gamma_j = \frac{1}{2}\left(1 \pm \lambda\sqrt{a + \frac{1}{4\lambda^2}}\right), \quad j = 1, 2,$$

whereas the WKB-approximation is

$$\tilde{y}_j \sim x^{\beta_j}, \quad x \to 0, \quad \beta_j = \frac{1}{2}(1 \pm \lambda\sqrt{a}).$$

The exponents γ_j and β_j are not the same, but for $\lambda \gg 1$ they differ by a small amount of order $O(\lambda^{-1})$.

It turns out that if instead of λ we bring in the parameter

$$\mu = \sqrt{\lambda^2 + \frac{1}{4p(0)}}, \tag{11}$$

then the WKB-approximation can again be used. Let us make a transformation of the same form as in paragraph 4 of § 1.

$$y = u_1 + u_2, \quad y' = \left(\mu\sqrt{q} - \frac{q'}{4q}\right)u_1 - \left(\mu\sqrt{q} + \frac{q'}{4q}\right)u_2. \tag{12}$$

Then equation (1) is reduced to the system

$$\begin{pmatrix} u_1' \\ u_2' \end{pmatrix} = \left[\mu\sqrt{q}\begin{pmatrix} 1 & 0 \\ 0 & -1 \end{pmatrix} - \frac{q'}{4q}\begin{pmatrix} 1 & 0 \\ 0 & 1 \end{pmatrix} + \frac{\tilde{\alpha}_1(x)}{\mu}\begin{pmatrix} 1 & 1 \\ -1 & -1 \end{pmatrix}\right]\begin{pmatrix} u_1 \\ u_2 \end{pmatrix}, \tag{13}$$

where

$$\tilde{\alpha}_1(x) = \alpha_1(x) - \frac{1}{8\sqrt{p(0)}x}.$$

Because of (10) the function $\tilde{\alpha}_1(x) \in C^\infty$ for small $x \geqslant 0$, and so we can apply the results of § 4 to (13). If λ is fixed and $4\lambda^2 p(0) \neq -1$ then (1) has solutions y_1 and y_2 such that

$$y_j(x, \lambda) \sim \tilde{y}_j(x, \mu), \quad x \to 0. \tag{14}$$

Observe that the $\tilde{y}_j(x, \mu)$ are exact solutions of (1) if $q(x) = (ax^2 + bx + c)^{-2}$.

Let us turn our attention to dual asymptotic behaviour. It follows from (14) that

$$y_{1,2}(x,\ \lambda) \sim x^{1/2 \pm \mu\sqrt{p(0)}}, \quad x \to +0. \tag{15}$$

There are two possibilities:

A. $p(0) \notin (-\infty, 0]$. We choose a branch of the root such that $\operatorname{Re} \sqrt{p(0)} > 0$ and let $\lambda \geqslant \lambda_0 \gg 1$. Then $y_1 \to \infty$, $y_2 \to 0$ as $x \to +0$, and so there is a unique (to within a constant multiple) solution y_2 such that $y_2(+0) = 0$. We normalize it by the condition

$$y_2(x,\ \lambda) \sim \tilde{y}_2(x,\ \mu), \quad x \to +0. \tag{16}$$

Such a solution is unique. Formulae (6) – (8) hold for this solution, with λ replaced by μ and α_j by $\tilde{\alpha}_j$. The function $\tilde{\alpha}_1(x)$ has the form (13) and the functions $\tilde{\alpha}_j(x)$, for $j > 1$, satisfy the same recurrence relations as $\alpha_j(x)$.

B. $p(0) < 0$. In this case $\sqrt{p(0)}$ is a purely imaginary number, so that $|y_{1,2}| \sim \sqrt{x}$ as $x \to +0$. The condition

$$y_j(x, \lambda) \sim \tilde{y}_j(x, \mu), \quad x \to +0,$$

uniquely determines the solutions y_1 and y_2, for which formulae (6) – (8) are valid with the same changes as for case A.

3. The Sturm-Liouville Equation. We consider the equation

$$y'' - (q(x) - \lambda)y = 0 \tag{17}$$

on the half-line $\mathbb{R}^+$, where $\lambda > 0$ is a parameter. Suppose that $q(x)$ is real, $q(x) \in C^2(\mathbb{R}^+)$, $q'(x) > 0$ for $x \gg 1$ and $q(\infty) = \infty$. Equation (17) for $\lambda \gg 1$ has the unique turning point $x_0(\lambda) = q^{-1}(\lambda)$, $x_0(\infty) = \infty$. We give asymptotic formulae for the solutions that are applicable outside some neighbourhood $U(\lambda)$ of the turning point. The size of $U(\lambda)$ depends on the growth speed of $q(x)$ as $x \to \infty$. Let

$$\begin{aligned} \tilde{y}_{1,2}(x, \lambda) &= (\lambda - q(x))^{-1/4} \exp\left\{ \pm i \int_{x_0(\lambda)}^{x} \sqrt{\lambda - q(t)}dt \right\}, \quad x < x_0(\lambda), \\ \tilde{y}_0(x, \lambda) &= (q(x) - \lambda)^{-1/4} \exp\left\{ - \int_{x_0(\lambda)}^{x} \sqrt{q(t) - \lambda}dt \right\}, \quad x > x_0(\lambda). \end{aligned} \tag{18}$$

All the roots in these formulae are positive. In the sequel, we take $\lambda \geqslant \lambda_0 \geqslant 1$.

3.1 $q(x)$ with Polynomial Growth. Let

$$q(x) \sim ax^\alpha, \quad x \to \infty, \quad a > 0, \quad \alpha > 0,$$

and suppose that this asymptotic behaviour can be twice differentiated. In this case $x_0(\lambda) \sim (\lambda/a)^{1/\alpha}$. We put

$$k(\lambda) = \lambda^{(-1+2\alpha)/3} N(\lambda), \quad x_\pm(\lambda) = \left[\frac{\lambda}{a}(1 \pm k(\lambda))\right]^{1/\alpha},$$

where $N(\lambda)$ is an arbitrary positive function such that $N(\infty) = \infty$. Equation (17) has solutions $y_j(x, \lambda)$, $j = 1, 2$, such that for $0 \leqslant x \leqslant x_-(\lambda)$ there are the bounds

$$y_j(x, \lambda) = \tilde{y}_j(x, \lambda)[1 + \varepsilon_j(x, \lambda)], \quad |\varepsilon_j(x, \lambda)| \leqslant CN^{-3/2}(\lambda). \tag{19}$$

Both of these solutions are highly oscillatory for $\lambda \gg 1$, and they can be chosen to be complex conjugates. If $0 \leqslant x \leqslant k_0(\lambda/a)^{1/\alpha}$, where k_0 is constant with $0 < k_0 < 1$, then from (19) we have

$$|\varepsilon_{1,2}(x, \lambda)| \leqslant C\lambda^{-1/2-1/\alpha}.$$

Equation (17) has a solution $y_0(x, \lambda)$ of the form (18) such that for $x_+(\lambda) \leqslant x < \infty$ we have

$$y_0(x, \lambda) = \tilde{y}_0(x, \lambda)[1 + \varepsilon_0(x, \lambda)],$$

$$|\varepsilon_0(x, \lambda)| \leqslant C\lambda^{-1/2-1/\alpha}\left[\left(\frac{ax^\alpha}{\lambda} - 1\right)^{-1/2} + \left(\frac{ax^\alpha}{\lambda}\right)^{-1/\alpha-1/2}\right].$$

The right-hand side attains its greatest value at $x = x_+(\lambda)$ so that

$$|\varepsilon_0(x, \lambda)| \leqslant C[N^{-3/2}(\lambda) + \lambda^{-1/\alpha-1/2}].$$

Therefore the asymptotic behaviour of $y_0(x, \lambda) \sim \tilde{y}_0(x, \lambda)$ is dual: the remainder term $\varepsilon_0(x, \lambda) \to 0$ both for λ fixed with $x \to \infty$, and for x fixed with $\lambda \to \infty$. The solution y_0 decreases for $x \to \infty$ and $y_0(\infty, \lambda) = 0$. The second linearly independent solution increases exponentially as $x \to \infty$. The asymptotic formulae obtained can be differentiated in x and λ an arbitrary number of times, maintaining the uniformity in x of the bound for the remainder term.

3.2 Rapidly Increasing $q(x)$. A function $l(x)$, positive for $x \geqslant 0$, is called *slowly increasing* if

$$\lim_{x \to \infty} \frac{xl'(x)}{l(x)} = 0.$$

For such a function

$$\lim_{x\to\infty} \frac{l(kx)}{l(x)} = 1 \tag{20}$$

uniformly in x lying in any finite interval. The function $1/l(x)$ is also slowly increasing. Examples of slowly increasing functions are

$$(\ln x)^\alpha\,, \quad -\infty < \alpha < \infty\,; \quad (\ln x)^\alpha (\ln\ln x)^\beta\,, \quad -\infty < \alpha, \beta < \infty\,;$$

$$\exp\{(\ln x)^\alpha\}\,, \quad \alpha < 1\,.$$

We introduce the conditions: the functions $q'(q^{-1}(x))/x, q''(q^{-1}(x))/x$ are slowly increasing, and

$$\lim_{x\to\infty} \frac{q''(x)}{q'(x)\sqrt{q(x)}} = 0\,.$$

These conditions are satisfied, for instance, for $q(x)$ of the form $A\exp(Bx^\alpha)$, $A, B, \alpha > 0$ and $P(x)\exp(Q(x))$, where P and Q are polynomials with $Q(\infty) = \infty$, and $q(x)$ itself is said to be ***rapidly increasing***.

As is clear from the WKB-bounds in § 2, the remainder terms in (19) do not exceed $C(I_1 + I_2)$, where

$$I_1 = \left| \int_a^x \frac{q'^2(t)}{|q(t) - \lambda|^{5/2}} dt \right|\,, \quad I_2 = \left| \int_a^x \frac{|q''(t)|}{|q(t) - \lambda|^{3/2}} dt \right|\,.$$

Here $a = 0$ for $j = 1, 2$ and $a = \infty$ for $j = 0$.

Let us find a bound for I_1 when $a = \infty$. Making the change of variable $t = q(x)/\lambda$, $x = \phi(\lambda t)$ and making use of (20), we obtain

$$\begin{aligned} I_1 &= \lambda^{-3/2} \int_{q(x)/\lambda}^\infty \frac{\lambda t q'(\phi(\lambda t))}{\lambda t (t-1)^{5/2}} dt \\ &\leqslant C\lambda^{-3/2} q'(q^{-1}(\lambda)) \int_{q(x)/\lambda}^\infty \frac{t}{(t-1)^{5/2}} dt \\ &\leqslant C'\lambda^{-3/2} q'(q^{-1}(\lambda)) \left[\left(\frac{q(x)}{\lambda} - 1 \right)^{-3/2} + \left(\frac{q(x)}{\lambda} \right)^{-1/2} \right]\,. \end{aligned}$$

Also, I_2 is bounded in a similar way.

Let $N(\lambda)$ be a positive function with $N(\infty) = \infty$, and let

$$a_1(\lambda) = \lambda^{-1}[q'(q^{-1}(\lambda))]^{2/3}\,, \quad a_2(\lambda) = \lambda^{-1} \left[\frac{q''(q^{-1}(\lambda))}{q'(q^{-1}(\lambda))} \right]^2\,.$$

It follows from the conditions on $q(x)$ that $a_j(\infty) = 0$. Put $k_j(\lambda) = N(\lambda) a_j(\lambda)$, $k_-(\lambda) = \max(k_1(\lambda), k_2(\lambda))$ and let $x_-(\lambda)$ be the point such that $1 - \lambda^{-1} q(x_-(\lambda)) = k_-(\lambda)$. Equation (17) has solutions y_1 and y_2 of the form (19) such that for $0 \leqslant x \leqslant x_-(\lambda)$ we have

$$|\varepsilon_{1,2}(x,\lambda)| \leqslant CN^{-3/2}(\lambda).$$

If $\tilde{x}_-(\lambda)$ is the point such that

$$\lambda^{-1}q(\tilde{x}_-(\lambda)) - 1 = k_0, \quad 0 < k_0 < 1,$$

then for $0 \leqslant x \leqslant \tilde{x}_-(\lambda)$ we have

$$|\varepsilon_{1,2}(x,\lambda)| \leqslant C(a_1^{3/2}(\lambda) + a_2^{1/2}(\lambda_1)).$$

Let $x_+(\lambda)$ be the point such that $\lambda^{-1}q(x_+(\lambda)) - 1 = k_-(\lambda)$. Then equation (17) has a solution y_0 of the form (19) such that for $x_+(\lambda) \leqslant x < \infty$ we have

$$|\varepsilon_0(x,\lambda)| \leqslant Ca_1^{3/2}(\lambda)\left[\left(\frac{q(x)}{\lambda} - 1\right)^{-3/2} + \left(\frac{q(x)}{\lambda}\right)^{-1/2}\right]$$
$$+ Ca_2^{1/2}(\lambda)\left(\frac{q(x)}{\lambda} - 1\right)^{-1/2}.$$

In particular, $|\varepsilon_0(x,\lambda)| \leqslant C'N^{-3/2}(\lambda)$ and $y_0(+\infty,\lambda) = 0$. The asymptotic behaviour $y_0(x,\lambda) \sim \tilde{y}_0(x,\lambda)$ is dual as in paragraph 3.1. If $q(x) = A\exp(Bx^\alpha)$, where A, B, $\alpha > 0$, then

$$a_1^{3/2}(\lambda) \sim C_0 b_1^{1/2}(\lambda) \sim C_1\lambda^{-1/2}(\ln\lambda)^{1-1/\alpha}, \quad \lambda \to +\infty.$$

§ 8. Counterexamples

1. Perron's Example. We consider the system

$$y_1' = -ay_1, \quad y_2' = (\sin\ln x + \cos\ln x - 2a)y_2 + be^{-ax}y_1, \tag{1}$$

where a and b are constants, and $0 < a < \sqrt{2}$. For $b = 0$ the system splits and has solutions

$$y_1^0(x) = e^{-ax}, \quad y_2^0(x) = \exp\{x\sin\ln x - 2ax\}.$$

For $b \neq 0$ its solutions have the form

$$y_1(x) = c_1y_1^0(x), \quad y_2(x) = y_2^0(x)\left(c_2 + bc_1\int_0^x \exp\{-t\sin\ln t\}dt\right).$$

System (1) has the form

$$y' = A(x)y + B(x)y, \quad B(x) = \begin{bmatrix} 0 & 0 \\ be^{-ax} & 0 \end{bmatrix},$$

where $A(x)$ is a diagonal matrix. It is clear that

$$\|B(x)\| \to 0, \quad x \to \infty; \qquad \int^{\infty} \|B(x)\| dt < \infty.$$

However condition 2) of § 3 is not satisfied, since the function

$$\mathrm{Re}\,(p_2(x) - p_1(x)) = \sin \ln x + \cos \ln x - a$$

has infinitely many zeros on the half-line $x \geqslant 0$. Let us put $c_1 = c_2 = 1$; then

$$\frac{y_1(x)}{y_1^0(x)} = 1, \quad \frac{y_2(x)}{y_2^0(x)} = 1 + b \int_0^x \exp\{-t \sin \ln t\} dt,$$

from which it follows that $\lim_{x \to +\infty} y_2(x)/y_2^0(x) = \infty$.

2. Systems with a Parameter. We consider the system of two equations

$$y' = \lambda A(x) y$$

on the interval $I = [-\delta, \delta]$, $\delta > 0$. Suppose that $p_1(x)$ and $p_2(x)$ are the eigenvalues of the matrix $A(x)$. In § 4, paragraph 2 asymptotic formulae were given for the solutions of the system under the following assumptions:

1) $p_1(x) \neq p_2(x)$, $x \in I$;
2) $\mathrm{Re}\,(p_1(x) - p_2(x))$ does not change sign for $x \in I$.

If only condition 1) is satisfied then the system still has two formal asymptotic solutions of the form (7) of § 4. But if 2) is not satisfied then no solutions of the system with such asymptotic behaviour exist.

Examples. 1. Let us consider the system

$$y_1' = 0, \quad y_2' = \lambda(x + i) y_2 + \lambda a(x) y_1 \tag{2}$$

on the interval I, where $a(x) \in C^2(I)$. Here $p_1(x) = 0$ and $p_2(x) = x + i$, so that there are no turning points. However, condition 2) is not satisfied since $\mathrm{Re}\,(p_2(x) - p_1(x))$ changes sign on I.

One of the F.A.S. of (2) has the form

$$y_1(x) = 1 + O(\lambda^{-1}), \quad y_2(x) = b(x) + O(\lambda^{-1}), \tag{3}$$

where $b(x) = a(x)\,(x + i)^{-1}$. Let us assume that (2) has solutions y_1 and y_2 with the asymptotic formulae (3) as $\lambda \to +\infty$ uniformly in $x \in I$. Then $a(x)$ can be continued analytically into a complex half-neighbourhood of $x = 0$ of the form $|x| \leqslant \rho$, $\mathrm{Im}\, x \geqslant 0$, where $\rho > 0$.

The general solution of (2) has the form

$$y_1 = c_1, \quad y_2 = e^{\lambda S(x)} \left[c_2 + c_1 \int_{-\delta}^{x} e^{-\lambda S(t)} a(t) dt \right], \quad S(x) = x^2/2 + ix.$$

Comparing this with (3) we find $c_1 = 0(\lambda^{-1})$. Putting $x = -\delta$ and $x = \delta$ in the indentity $y_2 = y_2^0$ and subtracting the first relation from the second, we obtain

$$\int_{-\delta}^{\delta} e^{-\lambda S(t)} a(t)dt = O(\lambda^{-1} e^{-\lambda\delta^2/2}), \quad \lambda \to \infty .$$

After the change of variable $z = S(t)$ this relation becomes

$$\int_{\gamma} e^{-\lambda z} f(z)dz = O(\lambda^{-1} e^{-b\lambda}), \quad \lambda \to \infty . \tag{4}$$

Here $b = \delta^2/2 > 0$, γ is the arc of the parabola $z = it + t^2/2$, $-\delta \leqslant t \leqslant \delta$, and $f(z) = a(z)(z+i)^{-1}$, $z = z(t)$.

We make use of the following conditions concerning γ. Let γ be a smooth simple finite convex curve in the complex z-plane and let $f(z)$ be a smooth function on γ. Let γ touch the imaginary axis at $z = 0$, with $\operatorname{Re} z > 0$ at all other points of the curve and $\operatorname{Re} z = c > 0$ at the ends of γ.

If relation (4) holds then $f(z)$ can be continued analytically from γ into the domain bounded by γ and by the line joining its ends. Consequently $a(x)$ can be continued analytically into a complex half-neighbourhood of $x = 0$. This is also true for the system

$$y_1' = 0, \quad y_2' = \lambda(x+i)y + \lambda^{-N} a(x)y$$

for arbitrary $N \geqslant -1$.

2. Let us consider the system

$$y_1' = 0, \quad y_2' = \lambda(x + i\varepsilon)y_2 + \lambda y_1 \tag{5}$$

on $I = [-\delta, \delta]$. This system also has a F.S.S. of the form (3). We show that if $\delta > \varepsilon > 0$ then there does not exist a solution of (5) which has asymptotic formula (3) as $x \to +\infty$ uniformly in $x \in I$.

The same arguments as for example 1 lead to the relation

$$F(\lambda) \equiv \int_{-\delta}^{\delta} \exp\left\{-\lambda\left(i\varepsilon x + \frac{x^2}{2}\right)\right\} dx = O\left(\exp\left\{-\frac{\lambda\delta^2}{2}\right\}\right), \quad \lambda \to \infty .$$

As $\lambda \to \infty$ we have $F(\lambda) \sim \text{const}\cdot\lambda^{-1/2}\exp\{-\lambda\varepsilon^2/2\}$ which is a contradiction since $\delta > \varepsilon$. The same assertion is true for the system

$$y_1' = 0 \quad y_2' = \lambda(x + i\delta)y_2 + \lambda^{-N} y_1 ,$$

if $N \geqslant -1$.

§ 9. Roots of Constant Multiplicity

1. Second-order Equations. We consider the equation

$$y'' - 2\lambda a(x,\ \lambda^{-1})y' + \lambda^2 b(x,\ \lambda^{-1})y = 0 \tag{1}$$

on $I = [a, b]$, where

$$a(x, \lambda^{-1}) \sim \sum_{j=0}^{\infty} a_j(x)\lambda^{-j}\,, \quad b(x, \lambda^{-1}) \sim \sum_{j=0}^{\infty} b_j(x)\lambda^{-j} \tag{2}$$

as $\lambda \to \infty$. The conditions on the smoothness of the coefficients are the same as in § 3, paragraph 2.1.

Suppose that the roots of the characteristic equation

$$p^2 - 2a_0(x)p + b_0(x) = 0$$

coincide for all $x \in I$, that is, we must have

$$a_0^2(x) = b_0(x)\,, \quad x \in I\,. \tag{3}$$

In this case the solutions can be expanded in asymptotic series not in integer powers of λ^{-1} but in fractional powers. In fact, the substitution

$$y = \exp\left\{\lambda \int^x a(t, \lambda^{-1})dt\right\} z$$

reduces (1) to the form

$$z'' + \lambda q(x,\ \lambda^{-1})z = 0\,, \tag{4}$$

where $q(x, \lambda^{-1})$ can be expanded in an asymptotic series in powers of λ^{-1}, further

$$q_0(x) = b_1(x) - 2a_0(x)a_1(x) - a_0'(x)\,.$$

Equation (4) has the form (15) of § 3, the only difference being that λ multiplies the coefficient q instead of λ^2. Therefore the solution will have an asymptotic expansion in powers of $\lambda^{-1/2}$ if $q_0(x)$ satisfies the conditions given in § 3, paragraph 1.6.

2. Systems of Equations. Let us consider the system of two equations

$$y' = \lambda A(x,\ \lambda^{-1})y \tag{5}$$

with the same conditions on A as in § 4, paragraph 1. We have

$$A(x, \lambda) \sim \sum_{j=0}^{\infty} A_j(x)\lambda^{-j}\,, \quad \lambda \to \infty\,.$$

Suppose that the eigenvalues of $A_0(x)$ coincide, so that $p_1(x) = p_2(x) = p(x)$, say, for $x \in I$. We consider two cases.

A. Suppose that $A_0(x)$ can be reduced to diagonal form, that is, there exists a matrix $T(x)$ of class $C^\infty(I)$, non-singular for $x \in I$ and such that

$$T^{-1}(x)A_0(x)T(x) = p(x)I\,.$$

After the subsitution

$$y = \exp\left\{\lambda \int^x p(t)dt\right\} T(x)z\,, \tag{6}$$

the system (5) becomes

$$z' = B(x,\ \lambda^{-1})z\,,$$

where B can be expanded as an asymptotic series in powers of λ^{-1}. This system does not contain a small parameter in the derivative, and the asymptotic behaviour of the solutions can be obtained only when we can integrate the system $z' = B_0(x)z$.

B. Suppose $A_0(x)$ can be reduced to Jordan normal form, that is, there exists a matrix $T(x)$ such that

$$T^{-1}(x)A(x)T(x) = \begin{bmatrix} p(x) & q(x) \\ 0 & p(x) \end{bmatrix}.$$

The substitution (6) reduces (5) to the form $z' = \lambda B(x,\lambda^{-1})z$, where $B(x,\lambda^{-1})$ has an asymptotic expansion in powers of λ^{-1}. We have

$$\begin{aligned} z_1' &= \lambda q(x)z_2 + b_{11}(x)z_1 + b_{12}(x)z_2 + \dots\,, \\ z_2' &= b_{21}(x)z_1 + b_{22}(x)z_2 + \dots\,. \end{aligned}$$

After the substitution

$$z_1 = \sqrt{\lambda}w_1\,, \quad z_2 = w_2$$

we arrive at the system

$$\begin{bmatrix} w_1' \\ w_2' \end{bmatrix} = \left[\sqrt{\lambda}\begin{bmatrix} 0 & q(x) \\ b_{21}(x) & 0 \end{bmatrix} + O\left(\frac{1}{\sqrt{\lambda}}\right)\right]\begin{bmatrix} w_1 \\ w_2 \end{bmatrix}. \tag{7}$$

This system has the form (8) of § 4, the only difference being that its matrix has an asymptotic expansion in powers of $\lambda^{-1/2}$, not in powers λ^{-1}. Therefore all the results of § 4, paragraph 3 are applicable to (7). Case B is very rarely encountered in applications.

§ 10. Problems on Eigenvalues

In this and the following paragraphs we mention some applications of the above results. However in problems of mechanics, physics and other applied problems, equations involving a large parameter generally have turning points. Therefore the main application of asymptotic methods is contained in the following sections.

1. The Sturm-Liouville Problem. We consider the eigenvalue problem on the interval $I = [a, b]$ given by

$$y'' + \lambda^2 q(x) y = 0\,, \quad x \in I\,, \tag{1}$$

$$y(a) = y(b) = 0\,. \tag{2}$$

As is known [Coddington], for $q(x) \not\equiv 0$ this problem has infinitely many eigenvalues λ_n^2. It is required to find the asymptotic behaviour of the eigenvalues and eigenfunctions $y_n(x)$ as $n \to \infty$.

Suppose that $q(x) > 0$ for $x \in I$ and $q(x) \in C^\infty(I)$. Then $\lambda_n^2 > 0$ for all n. Therefore we can assume that $\lambda_n > 0$. In this case the asymptotic behaviour of the spectrum and eigenfunctions is well known [Titchmarsh] and the example considered is of a purely illustrative nature.

Suppose that $y_1(x,\lambda)$ and $y_2(x,\lambda)$ are a F.S.S. Then any solution has the form $y = c_1 y_1 + c_2 y_2$. The boundary conditions (2) lead to a system of two linear homogeneous algebraic equations for c_1 and c_2. Since $(c_1, c_2) \neq (0,0)$ the determinant of the system is zero and the equation for the eigenvalues is

$$F(\lambda) \equiv \begin{vmatrix} y_1(a,\lambda) & y_2(a,\lambda) \\ y_1(b,\lambda) & y_2(b,\lambda) \end{vmatrix} = 0\,. \tag{3}$$

Equation (1) has two solutions, which have the asymptotic expansions

$$y_{1,2}(x,\lambda) = q^{-1/4}(x) \exp\{\pm i \tilde{S}(x,\lambda)\}\,,$$

where

$$\tilde{S} = \lambda \int_a^x \sqrt{q(t)}dt + \sum_{k=1}^{\infty} (\pm\lambda)^{-k} \beta_k(t) dt \tag{4}$$

as $\lambda \to \infty$ uniformly in $x \in I$. The functions $\beta_k(x)$ are determined from the recurrence relations

$$\beta_0(x) = -\frac{q'(x)}{4q(x)}\,, \quad \beta_{k+1}(x) = \frac{i}{2\sqrt{q(x)}} \left(\beta_k'(x) + \sum_{s=0}^{k} \beta_s(x) \beta_{k-s}(x) \right)\,. \tag{5}$$

It follows from this that the $\beta_{2k+1}(x)$ are purely imaginary functions and that the $\beta_{2k}(x)$ are real. Denote

$$\Phi(\lambda) = \sum_{k=-1}^{\infty} c_{2k+1}\lambda^{-2k-1}, \tag{6}$$

where

$$c_{-1} = \int_a^b \sqrt{q(t)}dt, \quad c_k = -i\int_a^b \beta_k(t)dt, \quad k \geqslant 1.$$

The series $\Phi(\lambda)$ and all the following series are asymptotic. Substituting (4) into (3) we obtain the equation for the eigenvalues: $\sin\Phi(\lambda) = 0$, so that $\Phi(\lambda) = n\pi$ where $n \geqslant 0$ is an integer. Consequently there is an asymptotic expansion for λ_n in powers of n^{-1} given by

$$\lambda_n = \sum_{k=-1}^{\infty} d_{2k+1}n^{-2k-1}.$$

The principal asymptotic term is

$$\lambda_n = \pi n\left(\int_a^b \sqrt{q(t)}dt\right)^{-1} + O(n^{-1}).$$

The coefficients d_k can be computed by the Bürmann-Lagrange formula

$$d_k = \frac{\pi}{k!}\left(\frac{d}{dz}\right)^{k-1}[z\Phi(z^{-1})]^k|_{z=0}.$$

We next find the asymptotic behaviour of the eigenfunctions $y_n(x)$. From the first of the conditions (2) we find that $y_n(x) = c(y_1(x,\lambda_n) - y_2(x,\lambda_n))$, or $y_n(x) = \text{Im}\, y_1(x,\lambda_n)$, since the solutions y_1 and y_2 can be chosen as complex conjugates for $x \in I$. We normalize $y_n(x)$ by the condition $\int_a^b y_n^2(x) = 1$. Then we obtain

$$y_n(x) = c_n q^{-1/4}(x)\sin\tilde{S}(x,\lambda_n).$$

The form of $\tilde{S}$ is given in (4) and c_n is the normalizing factor which has an asymptotic series

$$c_n = \sqrt{\frac{2}{J}}\left(1 + \sum_{k=0}^{\infty} f_k n^{-2k-1}\right), \quad J = \int_a^b q^{-1/2}dx.$$

In fact,

$$\int_a^b q^{-1/2}\sin 2\tilde{S}(x,\lambda_n)dx = \frac{1}{2}\left[J - \int_a^b q^{-1/2}\cos 2\tilde{S}(x,\lambda_n)dx\right]$$

and integration by parts leads to the formula for c_n.

This method extends to a wider class of eigenvalue problems, for instance, those of the form

$$\begin{aligned} y'' + \lambda^2 q(x,\, \lambda^{-1})y &= 0\,, \\ a_{10}(\lambda)y(a) + a_{11}(\lambda)y'(a) + a_{20}(\lambda)y(b) + a_{21}(\lambda)y'(b) &= 0\,, \\ b_{10}(\lambda)y(a) + b_{11}(\lambda)y'(a) + b_{20}(\lambda)y(b) + b_{21}(\lambda)y'(b) &= 0\,. \end{aligned}$$

Here $a_{jk}(\lambda)$ and $b_{jk}(\lambda)$ are polynomials in λ with complex coefficients and

$$q(x, \lambda) = \sum_{k=0}^{\infty} q_j(x)\lambda^{-1}\,, \quad \lambda \to \infty\,.$$

This asymptotic expansion is uniform in $x \in I$, $q_0(x) > 0$ and $q_k(x)$ for $k \geqslant 1$ can be complex-valued.

If $q(x)$ is complex-valued as well, then even the simplest problems (1), (2) remain almost uninvestigated. The examples given in § 8 show that in this case there is clearly no "universal formula" for the asymptotic behaviour of the eigenvalues (see [Kostynchenko], Chap. 4, § 7).

2. The Redge Problem. We consider the eigenvalue problem on $I = [0, a]$, $a > 0$, given by

$$\begin{aligned} &-y'' + q(x)y = k^2 y\,, \\ &y(0) = 0\,, \quad y'(a) + iky(a) = 0\,. \end{aligned} \tag{7}$$

Here k is the spectral parameter and

$$q(x) = (a-x)^{\alpha} r(x)\,, \quad r(a) \neq 0\,, \quad \alpha > -1\,. \tag{8}$$

where $r(x) \in C^{\infty}(I)$. Thus $q(x)$ or its derivatives can have a singularity at $x = a$. This problem is not self-adjoint even for real $q(x)$ and therefore has complex spectrum. In the half-plane Im $k \leqslant 0$ there can be only a finite number of spectral points and we investigate (7) for Im $k \geqslant 0$.

The asymptotic expansions obtained in the preceding paragraph are not applicable here. The point is that $Q(x, k) = k^2 - q(x)$, for noninteger $\alpha > 0$, has no singularity at $x = a$, but all its derivatives of a sufficiently high order have one. Therefore for the construction of the asymptotic expansion of the solution we make use of the integral equations

$$y_1 = e^{-ikx}u_1\,, \quad u_1(x) = 1 + \frac{i}{2k}\int_0^x (1 - e^{2ik(x-t)})q(t)u_1(t)dt\,.$$

Let $K(x, t, k)$ be the kernel of the integral operator. Then for Im $k \geqslant 0$, $|K(x, t, k)| \leqslant |k|^{-1}|q(t)|$ and, since $\int_0^a |q(t)|dt < \infty$, the method of successive approximations converges for $k \in G$: Im $k \geqslant 0$, $|k| \geqslant R \gg 1$. Therefore

$$y_1(x, k) = e^{-ikx}\left[1 + \frac{i}{2k}\int_0^x (1 - e^{-2ik(x-t)})q(t)dt + O\left(\frac{1}{k^2}\right)\right]$$

and this asymptotic formula can be differentiated.

Further, $y_1(0,k) = 1$ and $y_1'(0,k) = -ik$, so that the solution $y_1(x,k)$ is an entire function of k for each fixed $x \in I$.

The solution y_2 is determined from the integral equation

$$y_2 = e^{ik(x-a)}u_2, \quad u_2(x) = 1 + \frac{i}{2k}\int_a^x (e^{2ik(t-x)} - 1)q(t)u_2(t)dt.$$

This solution satisfies the Cauchy condition $y_2(a,k) = 1$, $y_2'(a,k) = ik$, and it is an entire function of k for each fixed $x \in I$. Further,

$$y_2(x,k) = e^{ik(x-a)}\left[1 + \frac{i}{2k}\int_a^x (e^{2ik(t-x)} - 1)q(t)dt + O\left(\frac{1}{k^2}\right)\right].$$

We take an eigenfunction in the form $y = c_1y_1 + c_2y_2$. Then from (7) we obtain the equation for the eigenvalues

$$y_2(0)(y_1'(a) + iky_1(a)) - 2ik = 0.$$

Making use of the asymptotic expansion of the solutions, we can reduce this equation to the form

$$F(k) \equiv \int_0^a e^{-2ikt}q(t)dt = 2ik[1 + O(k^{-1})]. \tag{9}$$

Applying the saddle point method we obtain

$$F(k) = c_0e^{-2ika}k^{-\alpha-1}[1 + O(k^{-1})]$$

for $\operatorname{Im} k \geqslant 0$ and $k \to \infty$, where

$$c_0 = \exp\left\{i\frac{\pi}{2}(\alpha+1)\right\}2^{-\alpha-1}\Gamma(\alpha+1)r(a).$$

Then (9) becomes

$$e^{-z}z^{\alpha+2} = -a^{\alpha+2}r(a)\Gamma(\alpha+1)[1 + O(k^{-1})], \quad z = -2ika,$$

from which we find the asymptotic behaviour of the spectrum to be

$$k_n = \frac{\pi n}{a} + \frac{i(\alpha+2)}{2a}\ln n + \frac{i}{2a}[(\alpha+2)\ln(-2\pi i) - (\alpha+2)\ln a$$
$$- \ln(-r(a)\Gamma(\alpha+1))] + O\left(\frac{\ln n}{n}\right), \quad n \to \infty.$$

For $|k| \gg 1$ the spectrum consists of two series, corresponding to the values $n > 0$ and $n < 0$. If $q(x)$ is real then the spectrum consists of the pairs $(k_n, -\overline{k}_n)$.

§ 11. A Problem on Scattering

1. The Scattering Matrix. We consider the scattering problem [Landau] for the equation

$$y'' + \lambda^2 q(x) y = 0\,, \tag{1}$$

where $\lambda > 0$ and $q(x)$ is real and continuous for all $x \in \mathbb{R}$.

1.1 The Scattering Matrix (S-Matrix) and its Properties. Let us introduce the following conditions:

1) There exist finite limits $\lim_{x\to\pm\infty} q(x) = q_\pm > 0$.
2) The integrals

$$\int_{-\infty}^{0} |\sqrt{q(x)} - \sqrt{q_-}|dx\,, \quad \int_{0}^{\infty} |\sqrt{q(x)} - \sqrt{q_+}|dx\,.$$

converge.

Then (1) has two F.S.S. with the following asymptotic behaviour

$$\begin{aligned} y_{1,2}^{+}(x) &= q_+^{-1/4} \exp\{\pm i\lambda\sqrt{q_+}x\}[1 + o(1)]\,, \quad x \to +\infty\,, \\ y_{1,2}^{-}(x) &= q_-^{-1/4} \exp\{\pm i\lambda\ \sqrt{q_-}x\}[1 + o(1)]\,, \quad x \to -\infty\,. \end{aligned} \tag{2}$$

These conditions uniquely determine the solutions $y_j^\pm$.

The solutions y_1^+, y_1^- (y_2^+, y_2^-) describe waves travelling to the right (to the left). There are the identities

$$\overline{y_1^-(x)} = y_2^-(x)\,, \quad \overline{y_1^+(x)} = y_2^+(x)\,.$$

For fixed λ any solution y can be represented in the form

$$y(x) = c_1^- y_1^-(x) + c_2^- y_2^-(x) = c_1^+ y_1^+(x) + c_2^+ y_2^+(x)\,,$$

where $c_j^\pm$ are constants. The *scattering matrix* is the (2×2)-matrix $S(\lambda) = S_{jk}(\lambda))$ defined by the relation

$$\begin{bmatrix} c_1^+ \\ c_2^- \end{bmatrix} = S \begin{bmatrix} c_1^- \\ c_2^+ \end{bmatrix}\,. \tag{3}$$

The solutions y_1^-, y_2^+ can be interpreted as waves travelling from $-\infty$ and from $+\infty$ towards the centre (converging waves), and the solutions y_1^+, y_2^- as going out from the centre to $\pm\infty$, being divergent waves which are scattered under the influence of the potential $q(x)$. The S-matrix describes the result of the scattering process, and it has the following two basic properties.

1) The S-matrix is unitary, that is

$$|s_{11}|^2 + |s_{12}|^2 = |s_{21}|^2 + |s_{22}|^2 = 1\,, \quad s_{11}\bar{s}_{21} + s_{12}\bar{s}_{22} = 0\,.$$

2) $s_{11} = s_{22}$ (this is equivalent to the relation $S(-\lambda) = S^*(\lambda)$, where S^* is the Hermitian conjugate matrix).

1.2 Coefficients of Transmission and Reflection. Suppose that the solution y has the form

$$y(x) = T_+ y_1^+(x) = y_1^-(x) + R_+ y_2^-(x). \tag{4}$$

For $x \gg 1$, y is a wave travelling to the right and, for $x \ll 1$, it is the sum of an incoming wave y_1^- and a reflected wave $R_+ y_2^-$. The numbers T_+ and R_+ come from the elements of the S-matrix:

$$T_+ = s_{11}, \quad R_+ = s_{21}. \tag{5}$$

Then $|T_+|^2$ and $|R_+|^2$ are called the *coefficients of transmission* and *reflection* respectively. Since the S-matrix is unitary it follows that $|T_+|^2 + |R_+|^2 = 1$ (conservation of energy). Similarly, if a wave travels to the left then

$$y(x) = T_- y_2^-(x) = y_2^+(x) + R_- y_1^+(x),$$

where

$$T_- = s_{22}, \quad R_- = s_{12}, \quad |T_-|^2 + |R_-|^2 = 1.$$

1.3 The Problem on the Half-Line. Let us consider equation (1) on the half-line $\mathbb{R}^+ : 0 < x < \infty$, and let us impose one of the following boundary conditions at $x = 0$:

$$y(0) = 0, \quad y'(0) = 0, \quad y'(0) + ay(0) = 0, \tag{6}$$

where a is real. The solution y satisfying one of the boundary conditions (6) has the form

$$y(x) = c_1(\lambda) y_1(x) + c_2(\lambda) y_2(x),$$

where $y_j = y_j^+$. The function

$$S(\lambda) = -c_2(\lambda)/c_1(\lambda) \tag{7}$$

is called the *scattering amplitude*. The wave $c_2 y_2$ plays the role of the incoming wave travelling to the left. Since we can take y real then $\overline{c_1(\lambda)} = c_2(\lambda)$ and therefore

$$|S(\lambda)| = 1. \tag{8}$$

1.4 Over-Barrier Reflection. Let $q(x) > 0$ for $x \in \mathbb{R}$ and suppose that it decays sufficiently fast as $|x| \to \infty$ that all the integrals

$$\int_{-\infty}^{\infty} |q^{(j)}(x)| dx, \quad j = 1, 2, \ldots,$$

converge. We find the asymptotic behaviour of the S-matrix as $\lambda \to \infty$.

Equation (1) has solutions

$$\tilde{y}^{+}_{1,2}(x,\lambda) = q^{-1/4}(x)\exp\{\pm i\sqrt{S(0,x)}\}\exp\left\{i\int_{+\infty}^{x} a_{1,2}(t,\lambda)dt\right\},$$

where

$$S(0,x) = \int_0^x \sqrt{q(t)}dt\,, \quad a_{1,2} = \sum_{j=1}^{\infty}(\pm\lambda)^{-j}\beta_j(x)\,.$$

The functions $\beta_j(x)$ are determined in § 10, (5). These asymptotic expansions are dual, that is, they are applicable as $\lambda \to \infty$ uniformly in $x \in \mathbb{R}$ and as $x \to \infty$ uniformly in $\lambda \geqslant \lambda_0 > 0$ (§ 6). The second F.S.S. (y_1^-, y_2^-) has the same form, the only difference being that the integral is taken from $-\infty$ to x and that its asymptotic series is applicable as $x \to -\infty$.

We have

$$\tilde{y}_1^+(x,\lambda) = A(\lambda)\tilde{y}_1^-(x,\lambda) + B(\lambda)\tilde{y}_2^-(x,\lambda)\,.$$

Putting $x = 0$ into this identity and into the differentiated identity we obtain

$$A(\lambda) = \exp\left\{-i\int_{-\infty}^{\infty} a_1(x,\lambda)dx\right\}, \quad B(\lambda) = O(\lambda^{-\infty})\,.$$

Let us express the solution $\tilde{y}_j^{\pm}$ in terms of $y_j^{\pm}$. Then we find the elements s_{11}, s_{12} of the scattering matrix using (4) and (5). From (2) we have

$$\int_0^x \sqrt{q(t)}dt = x\sqrt{q_\pm} + B_\pm + o(1)\,, \quad B_\pm = \int_0^{\pm\infty}(\sqrt{q(x)} - \sqrt{q_\pm})dx\,,$$

where $o(1)$ is infinitely small as $x \to \pm\infty$, respectively.

Comparing the asymptotic formulae for y_j^+, $\tilde{y}_j^+ (y_j^-, \tilde{y}_j^-)$ as $x \to \infty$ (respectively as $x \to -\infty$) we obtain

$$\tilde{y}^{+}_{1,2} = e^{\pm i\lambda B_+}y^{+}_{1,2}\,, \quad \tilde{y}^{-}_{1,2} = e^{\pm i\lambda B_-}y^{-}_{1,2}\,.$$

Recall that each solution of (1) is uniquely determined by its asymptotic behaviour as $x \to \infty$ or as $x \to -\infty$. Finally we obtain

$$s_{11} = \exp\left\{i\lambda\left(\int_0^{\infty}(\sqrt{q(x)} - \sqrt{q_+})dx + \int_{-\infty}^{0}(\sqrt{q(x)} - \sqrt{q_-})dx\right)\right.$$
$$\left. + i\int_{-\infty}^{\infty}\sum_{j=1}^{\infty}\lambda^{-j}\beta_j(x)dx\right\}, \quad s_{21} = O(\lambda^{-\infty})\,. \qquad (9)$$

The elements s_{11}, s_{12} are expressed in terms of elements which have already been computed.

2. The Scattering Amplitude in the Presence of Absorption. We consider the equation

$$y'' + (k^2 + ikr(x))y = 0 \tag{10}$$

on the half-line $x > 0$ with the boundary condition $y(0) = 0$. Here $k > 0$ is a parameter, $r(x) > 0$ for $0 < x < l$, $r(x) \equiv 0$ for $x \geqslant l$, $r(x) \in C^\infty$ for $x > 0$ and $r(x)$ has a singularity at $x = 0$:

$$\lim_{x\to 0} r(x) = \infty, \quad \int_0^t \sqrt{r(x)}dx = \infty. \tag{11}$$

For $x > l$ we have

$$y(x,\ k) = A(k)e^{-ikx} - B(k)e^{ikx}.$$

The function $s(k) = B(k)/A(k)$ is called the *scattering amplitude*. The physical interpretation of the problem is as follows. A plane wave is incident from infinity onto an absorbtion layer which fills the interval $(0, l)$. It is required to find the asymptotic behaviour of the scattering amplitude $s(k)$ as $k \to +\infty$.

It follows from (11) that there is a unique (to within a constant multiple) solution $y(x, k)$ of (10) such that $y(0, k) = 0$. The scattering amplitude is expressed in terms of y:

$$s(k) = e^{-2ikl}\frac{iky(l,k) + y'(l,k)}{iky(l,k) - y'(l,k)}. \tag{12}$$

It follows from (10) that $s(\overline{-k}) = \overline{s(k)}$.

We investigate the asymptotic expansion and the analyticity in k of the solution $y(x, k)$ where $r(x)$ satisfies the conditions of paragraph 1.5 of § 6. Let $Q(x, k) = -k^2 - ikr(x)$. Let us find for which complex k the conditions (1), (2) from § 3 are satisfied. Since $r(x) \geqslant 0$ we have $Q \neq 0$ if $k \in D_0$, where D_0 is the complex k-plane with a cut along the imaginary half-line $(-i\infty, 0]$. Let D_x be the half-plane Im $k \geqslant -r(x)/2$ with a cut along the line $[-ir(x)/2, 0]$, and let $w = \sqrt{Q(x,k)}$ be the branch of the root in D_x such that Re $w \to +\infty$ as Im $k \to +\infty$. The function w is a one-to-one map of D_x onto the half-plane Re $w > 0$ such that Re $\sqrt{Q(x,k)} \geqslant 0$ for $k \in D_x$ and, in particular, Re $\sqrt{Q(x,k)} \geqslant 0$ for Im $k \geqslant 0$. Therefore $y(x, k)$ has the asymptotic series (8) of § 7, where $j = 2$ and q must be replaced by $-1 - ik^{-1}r(x)$, as $|k| \to \infty$ with Im $k \geqslant 0$. Moreover $y(x, k)$ is holomorphic in k in D_0 for each fixed x, as follows from § 3, paragraph 1.5, so that $s(k)$ is meromorphic in D_0.

The asymptotic behaviour of the scattering amplitude is determined by the behaviour of the function $r(x)$ in a neighbourhood of $x = l$ (as in the Redge problem, § 10). There are two possibilities.

A. $r(x)$ has a zero of finite order $n \geqslant 1$. Then

$$s(k) = 2^{-n-3} i^{-n} k^{-n-2} e^{-2ikl}[1 + O(k^{-1/2})]$$

as $|k| \to \infty$ with $\operatorname{Im} k \geqslant 0$. This formula is proved directly by substituting the asymptotic series expansion for y into (12). It is clear that the smoother the join of the absorbtion layer $0 < x < l$ with the region $x > l$, the smaller the coefficient of reflection $|s(k)|$.

B. $r(x)$ has a zero of infinite order. Then $s(k) = 0(k^{-\infty})$. To obtain more precise information we have to turn to a system of integral equations for y as in the Redge problem. From (14) of § 2 we have

$$y = Q^{-1/4}(u_1 + u_2),$$

where

$$u_1 = 1 + \int_0^x \alpha_1(u_1 + u_2)dt,$$
$$u_2 = -\int_0^x \alpha_1 e^{2S(x,t)}(u_1 + u_2)dt.$$

Here α_1 can be expressed in terms of the function Q by formula (9) of § 1,

$$S(t,x) = \int_t^x \sqrt{Q(t,k)}dt, \quad \operatorname{Re}\sqrt{Q} \geqslant 0, \quad \operatorname{Im} k \geqslant 0.$$

Since $u_1 = 1 + 0(k^{-1/2})$ and $u_2 = 0(k^{-1/2})$, all the bounds being given for $\operatorname{Im} k \geqslant 0$, from (12) we obtain

$$s(k) = -e^{-2ikl} u_2(l,\ k)[1 + O(k^{-1/2})].$$

Further

$$u_2(l,k) = -\int_0^l [1 + O(k^{-1/2})]\alpha_1 e^{-2S(t,x)}dt$$

and the problem reduces to calculating the asymptotic behaviour of this integral. If $r(x) = \exp\{-f(x)\}$ where $f(x) \to +\infty$ sufficiently regularly as $x \to l - 0$ (for example if $f(x) = A(l-x)^{-\alpha}$, $A > 0$, $\alpha > 0$) then the asymptotic behaviour for $k = i\sigma$, $\sigma \to +\infty$, can be calculated by Laplace's method. Then

$$s(k) \sim \frac{\sqrt{2\pi}/f'(t_0))^2}{8\sigma^3\sqrt{f''(t_0)}} \exp\{2\sigma t_0 - f(t_0\}, \quad \sigma \to +\infty.$$

Here t_0 is a saddle point for which $f'(t_0) = 2\sigma$ and $t_0 \to l$ as $\sigma \to +\infty$.

3. Adiabatic Invariant of a Linear Harmonic Oscillator. Let us consider the equation

$$\ddot{x} + \omega^2(\varepsilon t)x = 0 \tag{13}$$

on the real line $\mathbb{R} = (-\infty, +\infty)$ where $\varepsilon > 0$ is a small parameter, $\omega(t) > 0$, $\omega(t) \in C^\infty(\mathbb{R})$ and there exist the finite limits

$$\lim_{x \to \pm\infty} \omega(t) = \omega_\pm > 0 \, .$$

Equation (13) describes a linear harmonic oscillator whose frequency of oscillation $\omega(\varepsilon t)$ changes slowly with time.

An *adiabatic invariant* of a physical system is a value which changes slowly under slow (but not necessarily small) changes of the parameters of the system. In other words, the adiabatic invariant is an approximate conservation law or an approximate first integral. For equation (13) the *Ehrenfest adiabatic invariant* is the ratio of the energy of the oscillator to its frequency:

$$J(t, \varepsilon) = \frac{\dot{x}^2 + \omega(\varepsilon t)x^2}{2\omega(\varepsilon t)} \, , \tag{14}$$

where $x = x(t, \varepsilon)$ is the solution of equation (13). The value

$$J(\varepsilon) = J(+\infty, \varepsilon) - J(-\infty, \varepsilon) \tag{15}$$

is called the *total variation of the adiabatic invariant.*

Remark. For $\varepsilon = 1$ (13) has exact first integral (the Lewis invariant)

$$I = \frac{1}{2}[\rho^{-2}x^2 + (\rho\dot{x} - \dot{\rho}x)^2]$$

where $\rho(t)$ is the solution of the non-linear equation

$$\ddot{\rho} + \omega^2(t)\rho = -\rho^{-3} \, .$$

The connection between the last equation and equation (13) turns out to be useful in a number of ways.

3.1 The Adiabatic Invariant and the Scattering Problem. The substitution $\tau = \varepsilon t$ reduces equation (13) to the form

$$\varepsilon^2 \frac{d^2\psi}{d\tau^2} + \omega^2(\tau)\psi = 0 \, , \tag{16}$$

where $\psi(\tau, \varepsilon) = x(t, \varepsilon)$. This equation differs from (1) only in notation. Let $\omega(t)$ tend to a limit sufficiently quickly as $t \to \pm\infty$, so that the following condition is satisfied:

1) The integrals

$$\int^{+\infty} |\omega(t) - \omega_+|dt\,, \quad \int_{-\infty} |\omega(t) - \omega_-|dt\,, \quad \int_{-\infty}^{\infty} (|\dot\omega(t)|^2 + |\ddot\omega(t)|)dt$$

converge. Then for each fixed $\varepsilon > 0$ equation (16) has two F.S.S. with the following asymptotic formulae:

$$\psi_{1,2}^+ \sim \omega_+^{-1/2} \exp\{\pm i\omega_+\tau/\varepsilon\}\,, \quad \tau \to +\infty\,,$$
$$\psi_{1,2}^- \sim \omega_-^{-1/2} \exp\{\pm i\omega_-\tau/\varepsilon\}\,, \quad \tau \to -\infty\,,$$

which can be twice differentiated. Any solution of (16) can be represented as

$$\psi = c_1^+\psi_1^+ + c_2^+\psi_2^+ = c_1^-\psi_1^- + c_2^-\psi_2^-\,, \tag{17}$$

where the $c_j^\pm$ depend only on ε and the formula for the total variation of the adiabatic invariant is

$$J(\varepsilon) = 2(c_1^+c_2^+ - c_1^-c_2^-)\,. \tag{18}$$

3.2 Bounds for $J(\varepsilon)$. By construciton $J(t,\varepsilon)$ is a homogeneous operator, $J(\varepsilon)$ is the homogeneous quadratic functional defined on the solutions of (13). Therefore, to calculate the asymptotic behaviours for $J(t,\varepsilon)$ and $J(\varepsilon)$ we must normalize the solutions. We normalize a solution ψ of the form (17) as follows. We fix numbers c_1^-, c_2^-, not depending on ε, that is, we give the asymptotic behaviour of ϕ as $\tau \to -\infty$. Then

$$J(\varepsilon) = 2|s_{11}|^{-2}[(|s_{12}|^2 + |s_{21}|^2)c_1^-c_2^- - s_{21}(c_1^-)^2 - \bar s_{21}(c_2^-)^2]\,. \tag{19}$$

If the solution x is real, then $c_1^- = c_2^- = c$ and

$$J(\varepsilon) = 2|s_{11}|^{-2}[(|s_{12}|^2 + |s_{21}|^2) - 2\mathrm{Re}\,(s_{21}c^2)]\,.$$

With the condition (1) $J(\varepsilon) = 0(\varepsilon)$ as $\varepsilon \to +0$. If also $\omega(t) - \omega_+$ and $\omega(t) - \omega_-$ lie respectively in the Schwarz spaces $S(\mathbb{R}^+)$ and $S(\mathbb{R}^-)$, then

$$J(\varepsilon) = O(\varepsilon^\infty), \quad \varepsilon \to +0\,.$$

In Chap. 3, § 8 it will be shown that if a function $\omega(t)$ is holomorphic in a neighbourhood of the real axis then $J(\varepsilon)$ decreases exponentially as $\varepsilon \to +0$.

Chapter 3. Second-Order Equations in the Complex Plane

In this chapter we consider equations of the form

$$w'' + p(z, \lambda)w' + q(z, \lambda)w = 0$$

with entire or meromorphic coefficients. The fundamental problem of asymptotic theory is the construction of the asymptotic behaviour of the fundamental system of solutions as $\lambda \to \infty$ in the whole complex z-plane. We also consider a series of concrete problems in spectral analysis and mathematical physics.

§ 1. Stokes Lines and the Domains Bounded by them

1. Local Structure of Stokes Lines. Let us consider the equation

$$w'' - \lambda^2 q(z)w = 0\,, \tag{1}$$

where $\lambda > 0$ is a parameter and $q(z)$ is holomorphic in a domain D of the complex z-plane. We write

$$S(z_0, z) = \int_{z_0}^{z} \sqrt{q(t)}dt \tag{2}$$

and we investigate the local structure of the level curves l_c : Re $S(z, z_0) = c$. Let $q(z_0) \neq 0$ and let U be the disc $|z - z_0| < r$ of radius $0 < r \ll 1$. The function $\sqrt{q(x)}$ splits in U into two holomorphic branches. We fix one of them. Then formula (2) determines a function that is holomorphic in U (the path of the integration lies in U), and this is an element (germ) of an analytic function. Extending it analytically we obtain the multi-valued analytic function $S(z_0, z)$ whose singular points are the zeros of $q(z)$ (the branch points). The family of level curves $\{l_c\}$, $-\infty < c < \infty$, is defined uniquely although S can be infinite-valued. Let us investigate the local structure of the level curves. If $q(z_0) \neq 0$ and $r \ll 1$, then by (2) we can express z in terms of S : $z - z_0 = \phi(S)$, where $\phi(S)$ is holomorphic at $S = 0$ and $\phi'(0) \neq 0$. Therefore locally the family $\{l_c\}$ is constructed as a family of parallel segments and small arcs of l_c are analytic curves.

The critical points of the family $\{l_c\}$ are the zeros of the function $q(z)$ and are the *turning points* of (1).

Let z_0 be a turning point. The maximal connected component of the level curve

$$\operatorname{Re} S(z_0, z) = 0 \tag{3}$$

with initial point z_0, and having no other turning points, is called a *Stokes line* of equation (1). A Stokes line is an analytic curve. In a similar way we define the Stokes line for the equation

$$w'' - Q(z, \lambda)w = 0\,,$$

where the parameter λ can be complex:

$$\operatorname{Re} \int_{z_0(\lambda)}^{z} \sqrt{Q(t, \lambda)}dt = 0\,.$$

Here again z_0 is a turning point.

Stokes lines and the domains bounded by them play a fundamental role in the study of the asymptotic behaviour of a solution in the large. The union of all Stokes lines is called the *Stokes graph* and is denoted by Φ, and its connected components are called the *Stokes complexes*.

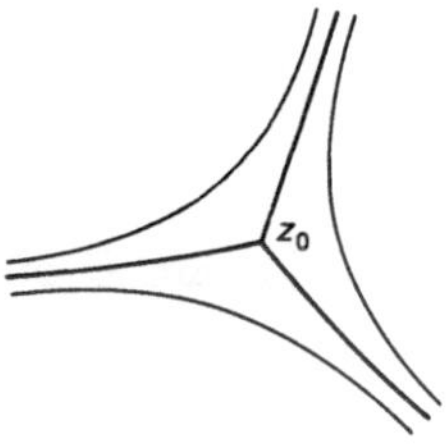

Fig. 1

Fig. 2

Remark. There are some variations in the definition of the Stokes line. For instance in [Heading 1] the Stokes lines are defined by the equation $\operatorname{Im} S(z_0, z) = 0$ and the Stokes lines defined above are called the conjugate Stokes lines or anti-Stokes lines.

Let z_0 be an nth-order turning point. Then for $z \approx z_0$

$$q(z) \sim a(z - z_0)^n\,, \quad a \neq 0\,,$$

and

$$S(z_0, z) \sim \frac{2\sqrt{a}}{n+2}(z - z_0)^{n/2+1}\,.$$

Therefore $n+2$ Stokes lines emanate from an nth-order turning point and the angle between adjacent lines at z_0 is $2\pi/(n+2)$. Three Stokes lines emanate from a simple turning point ($n=1$) and the angle between adjacent lines is $2\pi/3$ (Fig. 1).

Let z_0 be a simple pole of $q(z)$. Then $S(z_0, z) \sim a\sqrt{z-z_0}$ as $z \to z_0$, where $a \neq 0$. Equation (3) determines locally one curve with starting point z_0 (Fig. 2), which is also called a *Stokes line*. Accordingly a first order pole is called a *turning point of order* -1. Examples of Stokes lines are shown in Fig. 3.

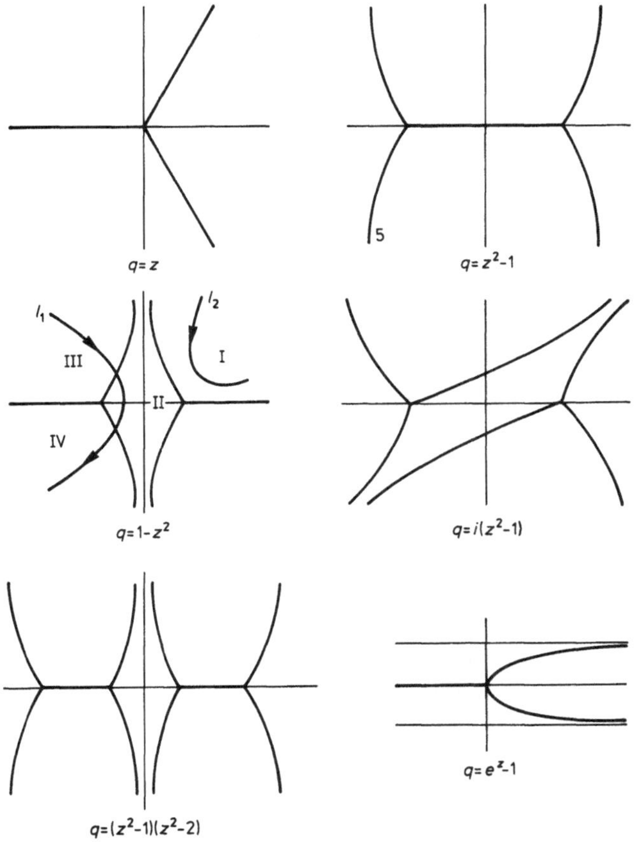

Fig. 3

2. Global Structure of Stokes Lines

2.1 $q(z)$ an Entire Function. We have the following theorem.

Theorem . *Let D be a bounded, simply-connected domain with piecewise-smooth boundary Γ and let $q(z)$ be holomorphic in $D \cup \Gamma$, except possibly at* *a finite number of poles lying in D. If* Re $S(z_0, z) \equiv 0$ *on* Γ, *where* $z_0 \in \Gamma$, *then* $q(z)$ *has not less than two poles in D.*

Let l be the level curve Re $S(z_0, z) = 0$, $z_0 \in l$, that is, the maximal connected component of this set. Since $q(z)$ is an entire function the curve l cannot contain a closed component. Let l_0 be a maximal connected component of the curve l not containing a turning point. Then l_0 is a simple, non-closed curve and its end can only be at $z = \infty$ or at a turning point. The function Im $S(u, z_0)$ is strictly monotonic along l_0, so that $S(z_0, z)$ is a one-to-one map of the curve l_0 onto the vertical line L : Re $S = 0$, $a <$ Im $S < b$, in the complex S-plane.

We introduce the condition:

1. If l_0 is not a Stokes line then $a = -\infty$, $b = +\infty$.

This condition is satisfied if $q(z)$ is a polynomial, but it is not necessarily satisfied if $q(z)$ is entire.

Examples. 1. Let $q(z) = e^{2z}$. Then one of the branches of the function $S(0, z)$ is e^z. This function maps the level curve l_k : Im $z = k\pi + \pi/2$, k an integer, onto the ray $(0,\ i(-1)^k\infty)$. In this example $q(z) \to 0$ if $z \in l_k$ and $z \to \infty$, Re $z \to -\infty$. It is natural to call the point $z = \infty$ a turning point (of infinite order) and the line l_k a Stokes line.

In what follows we assume that condition 1 is satisfied

2. Suppose that $q(z)$ is a polynomial:

$$q(z) = a_0 z^n + a_1 z^{n-1} + \ldots + a_n\,, \qquad a_0 = r_0 e^{i\phi_0} \neq 0\,.$$

Then for $|z| \gg 1$

$$S(z_0, z) = \frac{2\sqrt{a_0}}{n+2} z^{n/2+1}\left(1 + \sum_{j=1}^{\infty} b_j z^{-j}\right), \qquad b_1 = \frac{(n+2)a_1}{2na_0}\,.$$

This series converges for $|z| \gg 1$. The level curve Re $S(z_0, z) = 0$, not containing turning points, is a simple infinite curve having two asymptotes, which coincide with some of the following rays.

$$l_k : z = \rho e^{i\phi_k} - \frac{a_1}{na_0}\,, \qquad 0 < \rho < \infty\,,$$

$$\phi_k = \frac{(2k+1)\pi - \phi_0}{n+2}\,, \qquad k = 0, 1, \ldots, n+1\,.$$

An infinite Stokes line has one asymptote, one of the rays l_k. The structure of the level curves Im $S(z_0, z) =$ const is similar; their asymptotes are given by the rays

$$l'_k : z = \rho \exp\left\{i\left(\phi_k - \frac{\pi}{n+2}\right)\right\} - \frac{a_1}{na_0}\,, \qquad 0 < \rho < \infty\,.$$

These results are also true for non-integer n, $n > -2$.

2.2 $q(z)$ a Meromorphic Function. We assume that equation (1) has turning points. If $q(z)$ is a rational function then equation (1) has no turning points (in particular no first order poles) only in the following cases:

$$q = a\,, \quad q = a(z - z_0)^{-2}\,, \quad q = a(z - z_0)^{-2}(z - z_1)^{-2}\,,$$

where a and z_j are constants. In all these cases equation (1) is integrable in terms of elementary functions.

Let O be the set of all singular points of (1) on the Riemann sphere. Then O consists of the finite poles of $q(z)$ and the point $z = \infty$, if $\zeta = 0$ is a singular point of $\tilde{q}(\zeta) = \zeta^{-4}q(\zeta^{-1})$ (Chap. 1, § 1).

The classification of finite singular points is as follows.

1. Irregular singular points – poles of order $n \geqslant 3 (I_n)$.
2. Regular singular points of type I – first order poles (R_1).
3. Regular singular points of type II – second order poles (R_2).

The point $z = \infty$ belongs to 1, 2 or 3 according as $\zeta = 0$ is a pole of order $n \geqslant 3$, or $n = 1$, or $n = 2$ for $\tilde{q}(\zeta)$ respectively. If $\tilde{q}(0) \neq 0$ then $z = \infty$ is a non-critical point. If $\tilde{q}(\zeta)$ has a zero of order n at $\zeta = 0$ then $z = \infty$ is a turning point of order n. If $\zeta = 0$ is an essential singular point of $\tilde{q}(\zeta)$ then $z = \infty$ is an essential singular point of (1).

As a rule the singular points of (1) are branch points of infinite order for all its non-trivial solutions and therefore it is natural to consider this solution on the universal covering $\tilde{C}_z$ of the set $C_z \backslash 0$. There is a natural projection $p : \tilde{C}_z \to C_z \backslash 0$. Throughout what follows we shall use the notation: $\tilde{M}$ is a set lying in $\tilde{C}_z$, and M is its projection.

We assume that condition 1 of paragraph 2.1 is satisfied, but now for the curve $\tilde{l}$.

Condition 1 is satisfied for rational functions $q(z)$. If the level curve l is not a Stokes line, $z_0 \in l$, then $\tilde{l}$ is an infinite, non-closed curve (the curve l can be closed, see Example 2). The function $S(z_0, z)$ is a one-to-one map of $\tilde{l}$ onto the imaginary axis Re $S = 0$ in the complex S-plane.

Examples. 1. Let $z_0 \neq \infty$ be a pole of $q(z)$ of order $n \geqslant 3$. Thus

$$q(z) = (z - z_0)^{-n} \sum_{j=0}^{\infty} a_j (z - z_0)^j\,, \quad a_0 = r_0 e^{i\phi_0} \neq 0\,. \tag{4}$$

Then for $|z - z_0| \ll 1$

$$S(z_1, z) = \sqrt{a}(z - z_0)^{-n/2+1} \sum_{j=0}^{\infty} b_j (z - z_0)^j\,,$$

where

$$b_0 = \frac{2}{2-n}, \quad b_1 = -\frac{a_1}{na_0}.$$

This series converges for $|z - z_0| \ll 1$. The level curve l : Re $S(z_1, z) = 0$, not containing a turning point, is a simple curve with starting and finishing point z_0 for $|z_1 - z_0| \ll 1$ (Fig. 4). Its ends touch two of the rays

$$l_k : (2-n) \arg (z - z_0) + \phi_0 = (2k+1)\pi, \quad k = 0, \pm 1, \ldots$$

at z_0. The structure of the level curves Im S = const. is similar. These results are also true for n non-integer, $n > 2$.

Fig. 4

2. Let $z_0 \neq \infty$ be a second order pole, that is, $q(z)$ has the form (4) with $n = 2$. The local topological structure of the level curves Re S = const. is the same as for the function $q(z) = a_0 z^{-2}$.

(1) $a_0 > 0$: the level curves are rays centred at $z = 0$.
(2) $a_0 < 0$: the level curves are circles $|z|$ = const..
(3) Im $a_0 \neq 0$: the level curves are logarithmic spirals $\alpha \ln r - \beta\phi$ = const. which spiral around $z = 0$ where $a_0 = \alpha + i\beta$.

The structure of a level curve in the neighbourhood of a first-order pole was investigated in paragraph 1.

The level curves Re S = const. can be extremely complicated, even when $q(z)$ is a rational function. For example, the closure $[\Phi]$ of the Stokes graph Φ can have interior points.

3. Let

$$q_\alpha(z) = e^{2i\alpha}(z-a_1)^{-1}(z-a_2)^{-1}(z-a_3)^{-1}(z-a_4)^{-1},$$

where α, a_j are real, $a_1 < a_2 < a_3 < a_4$, $0 < \alpha < \pi$, and $\tan\alpha$ is irrational. Observe that $z = \infty$ is non-singular and is not a turning point. If l is an arbitrary level curve Re $S(z_0, z)$ = const., then its closure $[l]$ is the whole complex z-plane. Indeed for $\alpha = 0$, the function $S(a_1, z)$ is a one-to-one map of the upper half-plane Im $z \geqslant 0$ onto a rectangle Π in the complex S-plane with sides parallel to the coordinate axes. For $\alpha \neq 0$ the upper half-plane is mapped onto the rectangle Π_α which is obtained from Π by rotation through an angle α about the origin. We extend the function S analytically across the intervals formed by the points a_j on the real axis. Then S is a one-to-one map of $\tilde{C}_z$ onto the whole complex S-plane, covered by equal rectangles Π_α, $\Pi'_\alpha, \ldots$. The level curve l is mapped to the vertical line $L : \operatorname{Re} S = \text{const}$. If we combine all the rectangles which intersect L with Π_α then its segments are everywhere dense in Π_α since $\tan \alpha$ is irrational. Consequently $[l] = C_z$.

In this example $[\Phi] = C_z$ and the set of all interior points $[\tilde{\Phi}]$ of $[\Phi]$ consists of one connected component. The structure of $[\tilde{\Phi}]$ was investigated in [Jenkins].

A. $[\tilde{\Phi}]$ can consist of an arbitrary finite number of connected components. Further $\partial[\tilde{\Phi}]$ contains only turning points and singular points of type R_1.

B. If equation (1) has not more than three distinct singular points or has no points of type R_1 then $[\tilde{\Phi}]$ is empty so that $[\Phi] = \Phi$.

3. Domains Bounded by Stokes Lines. If $q(z)$ is an entire or meromorphic function then $S(z_0, z)$ is generally infinite-valued and maps the complex C_z-plane onto an infinite-sheeted surface. In particular if $q(z)$ is a polynomial of degree n then $S(z_0, z)$ is an elliptic integral for $n = 3$ or 4 and a hyperelliptic integral for $n \geqslant 5$. Let D be a simply-connected domain and let $q(z)$ be a holomorphic function with no zeros in D. Then the function $\sqrt{q(z)}$ splits in D into two holomorphic branches; this is also true for $S(z_0, z)$ if $z_0 \in D$ and the integral is taken along a path lying in D. In what follows we will always be concerned with one of these branches of the function D.

3.1 $q(z)$ an Entire Function. Suppose that the following condition is satisfied:

2. $[\Phi] = \Phi$.

This condition is satisfied if $q(z)$ is a polynomial, and there are no known entire functions for which it is not satisfied.

The Stokes lines divide up the complex plane into domains of the following two types:

I. D is of *half-plane type*. The function $S(z_0, z)$ is a one-to-one map of D onto a half-plane of the form $\operatorname{Re} S > a$ or $\operatorname{Re} S < a$.

II. D is of *band type*. The function $S(z_0, z)$ is a one-to-one map of D onto a band of the form $a < \operatorname{Re} \ S < b$.

Domains of both types are simply connected. The boundary of a domain of half-plane (band) type consist of one (two) connected compounds. Examples are shown in Fig. 3.

Example. Suppose that $q(z)$ is a polynomial of degree n. Then the Stokes lines divide up the z-plane into $n + 2$ domains of half-plane type and N domains of band type, where $0 \leqslant N \leqslant n - 1$. If there are no finite Stokes lines then $N = n - 1$. The level line $\operatorname{Re} S = \text{const}$ lying in a domain of half-plane type has the two adjacent rays l_k and l_{k+1} as asymptotes.

3.2 $q(z)$ a Meromorphic Function. In this case the domain $C_z \backslash [\Phi]$ consists of domains of half-plane, band, annulus and disk types.

III. D is of *annulus type*. The function $w = e^s$ is a one-to-one map of $\tilde{D}$ onto the annulus $0 < a < |w| < b < \infty$. The boundary of such a domain consists of two connected components (Fig. 5).

IV. The domain D is of *disk type*. The function $w = e^s$ is a one-to-one map of $\tilde{D}$ onto a domain of the form $0 < |w| < a$ or $a < |w| < \infty$. The boundary of such a domain consists of a connected curve and the point z_0 which is a second order pole (Fig. 6).

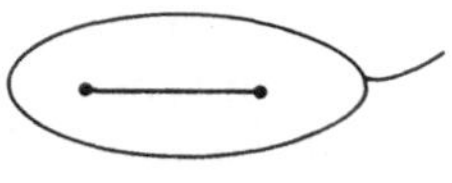

Fig. 5

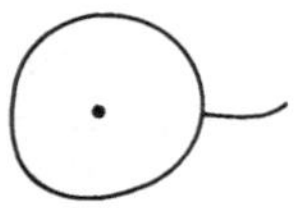

Fig. 6

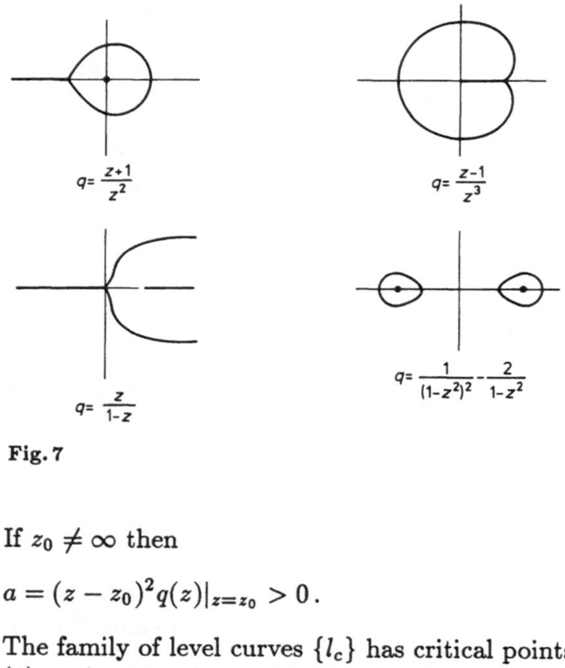

Fig. 7

If $z_0 \neq \infty$ then

$$a = (z - z_0)^2 q(z)|_{z=z_0} > 0\,.$$

The family of level curves $\{l_c\}$ has critical points of the following types:

(1) z_0 is a turning point of order k, $k = -1, 1, 2, \ldots$. If $z_0 \neq \infty$ and $k = -1$, then z_0 is a first order pole, that is, a regular singular point of equation (1).

(2) z_0 is a second order pole, finite of infinite (a regular singular point of (1)). Suppose that $z_0 \neq \infty$. Then there are two possibilities:

(a) $a > 0$: a small punctured neighbourhood U of the point is covered by one domain of disk type;

(b) $a \notin (0, \infty)$: U is covered by the closure of a finite number of domains of band type.

(3) z_0 is an irregular singular point of order $k \geqslant 3$. If $z_0 \neq \infty$ then z_0 is a pole of $q(z)$ of order k. In this case U is covered by the closure of k domains of half-plane type and a finite number of domains of band type.

Examples of Stokes lines can be seen in Fig. 7.

§ 2. WKB-Bounds in the Complex Plane

1. Canonical Paths. Let D be a simply connected domain on the Riemann sphere, and let $Q(z)$ be holomorphic with no zeros in D. We consider equation

$$w'' - Q(z)w = 0 \tag{1}$$

and we fix a branch of the function $S(z_0, z) = \int_{z_0}^{z} \sqrt{Q(t)}dt$ in D. In Chap. 2, § 2, the WKB-bounds for a solution of (1) on the real line were obtained. Equation (1) was reduced to a system of integral equations and a fundamental role in the derivation of the WKB-bounds was the fact that on any integration path (x, b), $x < b$, we have Re $S(t, x) \leqslant 0$, $t \in (x, b)$, so that $|\exp S(t, x)| \leqslant 1$. This fact can be formulated in another way: Re $S(x_0, x)$ is non-decreasing along the integration path. The generalization of this property to paths in the complex plane is the idea of a canonical path.

Let $\gamma = \gamma(z_0, z^\star)$ be a piecewise-smooth curve lying in a domain D and connecting the points $z_0 \in D$ and $z^\star \in \partial D$. The curve γ is called a *canonical path* if Re $S(z_0, z)$ is non-decreasing along γ as z moves from z_0 to $z^\star$. In particular, canonical paths are arcs of the level lines Re S = const, Im S = const, containing no turning points. We generally choose z^* to be a singular point of (1).

We denote by $S(D)$ and $S(\gamma)$ the images of the D and γ under the mapping $S = S(z_0, z)$. If γ is a canonical path then $S(\gamma)$ has the property that its intersection with any vertical line Re S = const is either empty or consists of one connected component.

Examples 1. Let D be a domain of half-plane type, $z_0 \in D$ and the branch $\sqrt{Q(z)}$ is chosen so that Re $S(z_0, z) > 0$ in D. The pre-image of any ray $S(\gamma)$ lying in $S(D)$ is a canonical path γ. If $z_1 \in D$ then any two canonical paths $\gamma_0(z_1, z^\star)$ and $\gamma_1(z_1, z^\star)$ in D can be continuously deformed onto each other (on the Riemann sphere) so that the intermediate paths $\gamma_t(z_1, z^\star)$, $0 < t < 1$ are canonical. Such paths are called *equivalent* (or *S-homotopic*).

Let $z_1, z_2 \in D$, let $\gamma(z_1, z^\star)$ be a canonical path and let Re$S(z_0, z_2) \geqslant$ Re $S(z_0, z_1)$ for definiteness. Then we can join the points z_1, z_2 by a canonical path $\gamma_1(z_1, z_2)$ so that the path $\gamma\gamma_1$ is canonical (Fig. 8). We say that all canonical paths in D are equivalent (or S-homotopic).

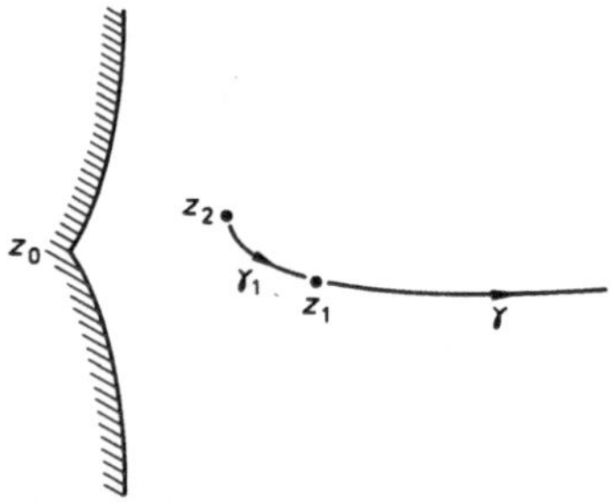

Fig. 8

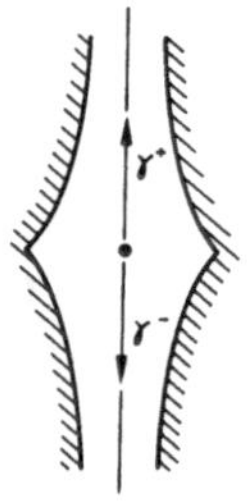

Fig. 9

Suppose for simplicity that ∂D contains a finite number of turning points. We remove ε-neighbourhoods of these points from $S(D)$ and denote the pre-image of the domain so obtained by D_ε. Then all canonical paths in D_ε are equivalent.

2. Let D be a domain of band type (§ 1). Then all infinite canonical paths in D split into two classes $\gamma^\pm$ of S-homotopic paths. Thus Im $S = \pm\infty$ along a path of class $\gamma^\pm$ (Fig. 9).

Remarks. Let us consider a domain of band type (see Fig. 9). On the Riemann sphere this domain has one infinitely distant boundary point $z = \infty$, a singular point of (1). However, from the point of view of asymptotic theory, we must assume that ∂D contains two infinitely distant points $z = \pm\infty$; Im $S \to \pm\infty$ as $z \to z^\pm$. To be precise, an infinitely distant boundary point is defined by an equivalence class of canonical paths.

2. Fundamental Theorem. Suppose that there is a canonical path connecting the points $z \in D$ and $z^* \in \partial D$ for (1). We write

$$\rho(z, D) = \inf_{\gamma(z,z^*)} \int_{\gamma(z,z^*)} |\alpha_1(t)||dt|, \tag{2}$$

where the infimum is taken over all canonical paths joining the points z and z^*, and $\alpha_1(t)$ is given by formula (9) of Chap. 2 § 1. Suppose that the following conditions are satisfied:

(1) for each point $z \in D$ there is a canonical path $\gamma(z, z^*)$;

(2) $\rho(z, D) < \infty$ for any point $z \in D$ (for a fixed class of S-homotopic paths).

Then we have

Theorem . *Equation (1) has a solution $w(z)$ such that for $z \in D$ there are the bounds*

$$\left|\frac{w(z)}{Q^{-1/4}(z)\exp\{-S(z_0, z)\}} - 1\right| \leqslant 2(e^{2\rho(z,D)} - 1),$$
$$\left|\frac{w'(z)}{Q^{-1/4}(z)\exp\{-S(z_0, z)\}} + 1\right|$$
$$\leqslant \frac{1}{4}\left|\frac{Q'(z)}{Q^{3/2}(z)}\right| + 4\left[1 + \frac{1}{4}\left|\frac{Q'(z)}{Q^{3/2}(z)}\right|\right][e^{2\rho(z D)} - 1], \tag{3}$$

where $z_0 \in D$ is an arbitrary fixed point.

This theorem is due to G. Birkhoff [Birkhoff]. Its proof is word for word the same as that of the theorem in § 2 of Chap. 1. We choose a canonical path $\gamma(z, z^\star)$ for the contour of integration on which Re $S(t, z) \leqslant 0$, so that

$$|\exp\{2S(t, z)\}| \leqslant 1.$$

As in the real case the asymptotic formulae for w and w' come from (3). Let γ be a canonical path with $z \in \gamma$. Then its arc $\gamma(z_1, z^\star)$ is a canonical path. It follows from condition 2) that

$$\lim_{z \to z^\star} \int_{\gamma(z, z^\star)} |\alpha_1(t)||dt| = 0. \tag{4}$$

Consequently, there is the asymptotic formula for $z \in \gamma$, $z \to z^\star$

$$w(z) \sim Q^{-1/4}(z)\exp\{-S(z_0, z)\}. \tag{5}$$

This relation is satisfied if $z \to z^\star$ along an arbitrary canonical path. In what follows "$z \to z^\star$" will mean "$z \to z^\star$ along some canonical path".

Suppose that we have

$$\lim_{z \to z^\star} Q'(z)Q^{-3/2}(z) = 0. \tag{6}$$

Then it follows from the bounds in (3) that

$$w'(z) \sim -Q^{-1/4}(z)\exp\{-S(z_0, z)\}, \quad z \to z^\star. \tag{7}$$

Condition 1) is satisfied for the domains given in examples 1 and 2 above. In the following examples, we discuss the convergence of the integral

$$\rho(z) = \int_z^{z^\star} |\alpha_1(t)||dt|, \quad z \neq z^\star.$$

We suppose that $z^\star \neq \infty$ and we put $z^\star = 0$ for simplicity.

Examples. 1. Let $z = 0$ be a pole of order $n \geqslant 3$ for $Q(z)$. Then $\rho(z) < \infty$ if the integral is taken over the closed interval $[0, z]$. This is true when $Q(z) \sim az^\alpha$, $a \neq 0$, $\alpha < -2$ for $z \to 0$ in a sector D with a vertex at $z = 0$, and this asymptotic formula can be twice differentiated. The integral $\rho(z) = \infty$ for $\alpha \geqslant -2$. In particular, $\rho(z) = \infty$ if $z = 0$ is a regular singular point of equation (1).

2. Let $Q(z)$ be a polynomial, $z^\star = \infty$, and the integral $\rho(z)$ is taken along a ray. Then $\rho(z) < \infty$. This is true when $Q(z) = az^\alpha$ as $z \to \infty$ in the sector D if $a \neq 0$ and $\alpha > -2$. If $\alpha \leqslant -2$ then $\rho(z) = \infty$.

If the conditions of the theorem are satisfied, together with condition (6), then equation (1) has a solution $\tilde{w}(z)$ such that as $z \to z^\star$

$$\tilde{w}(z) \sim Q^{-1/4}(z) \exp\{S(z_0, z)\}, \quad \tilde{w}'(z) \sim Q^{1/4}(z) \exp\{S z_0, z)\}.$$

The solutions $w(z)$ and $\tilde{w}(z)$ form a F.S.S.

3. Boundary Conditions for the Solutions. In singular eigenvalue problems, in scattering problems and in other problems, the boundary conditions for the desired solutions are given at infinity or at a singular point of the equation. Let us formulate their analogues in the complex plane, assuming that the conditions of the theorem are satisfied together with condition (6). Let $\gamma = \gamma(z_0, z^\star)$ be a canonical path. Then there are two possibilities:

$$\lim_{z \to z^\star,\, z \in \gamma} \operatorname{Re} S(z_0, z) = +\infty, \tag{8a}$$

and

$$\lim_{z \to z^\star,\, z \in \gamma} \operatorname{Re} S(z_0, z) = a, \quad 0 \leqslant a < \infty, \tag{8b}$$

which leads to two types of boundary conditions.

I. Condition for decreasing solutions. Let $w(z)$ be the solution constructed in the theorem of paragraph 2. Then for case (8a)

$$\lim_{z \to z^\star,\, z \in \gamma} w(z) = 0. \tag{9}$$

Since $\lim_{z \to z^\star,\, z \in \gamma} \tilde{w}(z) = \infty$ any solution satisfying condition (9) has the form (const.) $w(z)$.

II. The radiation condition. In case (8b) the solution $w(z)$ constructed in the theorem of paragraph 2 satisfies

$$\lim_{z \to z^\star,\, z \in \gamma} \frac{w'(z)}{\sqrt{Q(z)}\, w(z)} = -1. \tag{10}$$

This condition produces a unique solution.

In both cases the solutions for which the asymptotic formula (5) holds is unique, that is, the solution is uniquely determined by its asymptotic behaviour.

All the results derived above carry over to an equation of the form

$$(P(z)w')' - Q(z)w = 0$$

(see Chap. 2, § 2).

§ 3. Equations with Polynomial Coefficients. Asymptotic Behaviour of a Solution in the Large

1. Statement of the Problem. We consider equations of the form

$$w'' + p(z,\lambda)w' + q(z,\lambda)w = 0\,, \tag{1}$$

where p and q are polynomials in z with coefficients depending on a large parameter $\lambda > 0$. The fundamental problem of asymptotic theory as applied to equation (1) is this: to find the asymptotic behaviour of a F.S.S. as $\lambda \to \infty$ in the whole complex z-plane.

This problem has been investigated fully only in the case where the dependence of the coefficients on the parameter has a simple form. We describe an algorithm for the solution of the fundamental problem for the equation

$$w'' - \lambda^2 q(z)w = 0\,. \tag{2}$$

We bring in a finite collection of unbounded domains $\{D_j\}$ (they are called *canonical*), the union of which covers the whole complex z-plane, excluding neighbourhoods of turning points. We construct a special F.S.S. (called *elementary*) $(u_j,\ v_j)$ the asymptotic behaviour of which is known everywhere in D_j. Any solution $w(z,\lambda)$ of equation (2) can be represented in the form

$$w(z\,,\lambda) = \alpha_j u_j + \beta_j v_j = \alpha_k u_k + \beta_k v_k\,,$$

where α_j and β_j depend only on λ. We have

$$\begin{bmatrix}\alpha_k\\ \beta_k\end{bmatrix} = \Omega_{jk}(\lambda)\begin{bmatrix}\alpha_j\\ \beta_j\end{bmatrix}\,. \tag{3}$$

The matrix $\Omega_{jk}(\lambda)$ is called the *transition matrix* from the F.S.S. (u_j, v_j) to the F.S.S. (u_k, v_k) and does not depend on w. It is clear that

$$\Omega_{jl} = \Omega_{kl}\Omega_{jk}\,, \quad \Omega_{jk} = \Omega_{kj}^{-1}\,.$$

It can be proved that any transition matrix from one elementary F.S.S. to another is the product of a finite number of simpler transition matrices belonging to one of four types. The asymptotic behaviours of the simpler matrices can be computed – see paragraph 3.

Thus the solution of the fundamental problem breaks down into three problems.

1. The topological problem. To find the Stokes lines for equation (1).

2. The analytical problem. To find the asymptotic behaviour of elementary F.S.S.

3. The algebraic problem. To multiply the transition matrices.

Problem (1) is essentially computational. For a specific equation we can find turning points and construct the Stokes lines with a computer. The solution to problem 2 was given earlier. The asymptotic behaviour of the F.S.S. near a turning point is considered in Chap. 4.

The algorithm given above is quite general, and it is applicable both when $q(z)$ is an entire or mermomorphic function and when the dependence of the coefficients in the equation on a parameter is more complex.

2. Elementary Fundamental Systems of Solutions of Equation (2).

2.1 Canonical Domains. The turning points of (1) do not depend on λ. Let us introduce the notation

$$S(z_0,\ z) = \int_{z_0}^{z} \sqrt{q(t)}dt \tag{4}$$

and let $\lambda > 0$. Then the Stokes lines are determined by the equation

$$\text{Re } S(z_0,\ z) = 0\,,$$

where z_0 is a turning point, not depending on λ (see (3) of § 2).

A domain D in the complex z-plane is called *canonical* if $S(z_0, z)$ is a one-to-one map of D onto the whole complex plane with a finite number of vertical cuts (see Fig. 10). The domain D is simply connected and contains no turning points and ∂D consists of Stokes lines (the pre-images of the sides of the cuts). A canonical domain is the union of two domains of half-plane type ("left" and "right") and several domains of band type.

Let us remove from $S(D)$ left (right) ε-neigbourhoods of the cuts and ε-neighbourhoods of the turning points (Fig. 10); we denote the pre-image of the domain so obtained by $D_\varepsilon^+(D_\varepsilon^-)$. For each $z^\star \in D_\varepsilon^+$ there is an infinite canonical path $\gamma^+(z^\star)$ such that Re $S(z_0, z) \to +\infty$ for $z \in \gamma^+$, $z \to \infty$. For each $z^\star \in D_\varepsilon^-$ there is an infinite path $\gamma^-(z^\star)$ such that $-\gamma^-(z^\star)$ is canonical and Re $S(z_0, z) \to -\infty$ for $z \in \gamma^-$, $z \to \infty$.

2.2 The Principal Asymptotic Term of the Solution. Let D be a canonical domain with $\lambda > 0$ fixed. We recall that any solution of equation (2) is an entire function of z. Equation (2) has a unique (to within a constant multiple) solution $w_1(z, \lambda)$ such that

$$\lim w_1(z,\ \lambda) = 0\,, \quad z \to \infty\,, \qquad \text{Re } S(z_0,\ z) \to +\infty\,, \tag{5}$$

and a unique (to within a constant multiple) solution $w_2(z, \lambda)$ such that

$$\lim w_2(z, \lambda) = 0, \quad z \to \infty, \quad \text{Re } S(z_0, z) \to -\infty. \tag{5'}$$

For $\lambda \gg 1$ these solutions form a F.S.S. For w_1 we take the solution w in the theorem of § 2.

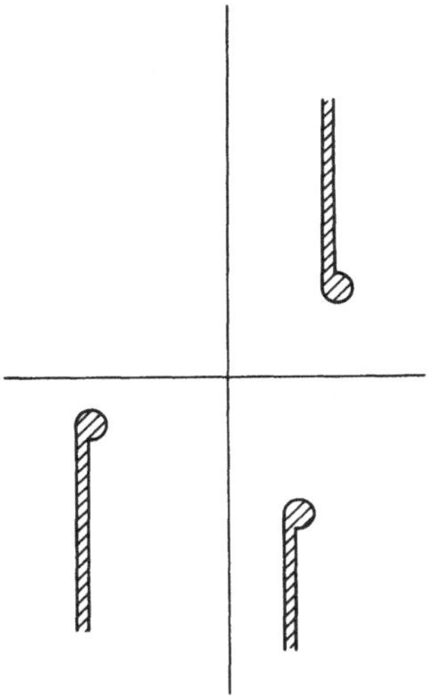

Fig. 10

Let us denote

$$\tilde{w}_{1,2}(z, \lambda; z_0) = q^{-1/4}(z) \exp\{\mp\lambda S(z_0, z)\}, \quad \rho^{\pm}(z) = \int_{\gamma^{\pm}(z)} |\alpha_1(t)||dt|, \tag{6}$$

where the branch of $\sqrt[4]{q(z)}$ in D is fixed. From (3) of § 2 we have

$$\left| \frac{w_1(z, \lambda)}{\tilde{w}_1(z, \lambda; z_0)} - 1 \right| \leqslant 2(e^{\lambda^{-1}\rho^{+}(z)} - 1)$$

for $z \in D_\varepsilon^+$ and $\lambda > 0$. Fix $\lambda_0 > 0$. Then for $\lambda \geqslant \lambda_0$ and $z \in D_\varepsilon^+$ we have

$$w_1(z, \lambda) = \tilde{w}_1(z, \lambda; z_0)[1 + \lambda^{-1}\phi_1(z, \lambda)], \tag{7}$$

where

$$|\phi_1(z, \lambda)| \leqslant C\rho^{+}(z),$$

and C does not depend on λ. Consequently

$$w_1(z, \lambda) \sim \tilde{w}_1(z, \lambda; z_0).$$

Moreover these asymptotic behaviours are *dual*. That is, it is true if

a) $\lambda \to +\infty$ uniformly for $z \in D_\varepsilon^+$;

b) $z \to \infty$ so that Re $S(z_0, z) \to \infty$ uniformly for $\lambda \geq \lambda_0 > 0$.

Similarly there is a solution w_2 such that

$$w_2(z, \lambda) = \tilde{w}(z, \lambda; z_0)[1 + \lambda^{-1}\phi_2(z, \lambda)], \tag{8}$$

where

$$|\phi_2(z, \lambda)| \leqslant \varepsilon_2(z).$$

The asymptotic behaviour $w_2 \sim \tilde{w}_2$ is also dual. It is valid if

a) $\lambda \to +\infty$ uniformly for $z \in D_\varepsilon^-$;

b) $z \to \infty$ so that Re $S(z_0, z) \to -\infty$ uniformly for $\lambda \geqslant \lambda_0 > 0$.

There are analogous dual asymptotic behaviours for all derivatives with respect to z and λ.

If $\lambda \geqslant \lambda_0 \gg 1$ then (w_1, w_2) is the F.S.S. Then each of the solutions tends to zero at "its own infinity" (see (5) and (5′)) and increases at the "opposite infinity".

We note an important particular case. Let D be a canonical domain and let all the cuts in $[S(D)]$ be directed to one side, downwards for definiteness. Then

$$\lim_{z \to \infty,\ z \in D} \phi_j(z, \lambda) = 0, \quad j = 1, 2,$$

if Im $S(z_0, z) \to \infty$ or $|\text{Re } S(z_0, z)| \to \infty$ uniformly for $\lambda \geqslant \lambda_0$, where $\lambda_0 > 0$ is arbitrary but fixed.

2.3 Maximal Domains of Applicability of WKB-Asymptotic Behaviour. Let D_0 be a domain of half-plane type, let $z_0 \in \partial D$ be a turning point and let the branch of $S(z_0, z)$ in D_0 be chosen so that Re $S > 0$ for $z \in D_0$. Equation (1) has a solution w_1 such that

$$w_1(z, \lambda) = \tilde{w}_1(z, \lambda; z_0)[1 + O(\lambda^{-1})] \tag{9}$$

as $\lambda \to \infty$ uniformly for $z \in D_{0\varepsilon}$ where the domain $D_{0\varepsilon}$ is obtained by removing from D_0 neighbourhoods of the boundary turning points. Moreover $w_1 \to 0$ for $z \in D_0$ and $z \to \infty$ for each fixed $\lambda > 0$. We pose the problem: to find the maximal domain in the complex z-plane in which the asymptotic expansion (9) remains true.

Examples. 1. Let $q(z) = z$, and let D_0 be the domain $|\arg z| < \pi/2$. From the turning point $z = 0$ there arise three Stokes lines l_1, l_2 and l_3 – the rays arg $z = \pm\pi/3$ and arg $z = \pi$. Let D be the complex z-plane cut along the ray l_3. Then the function $S(0, z) = (2/3)z^{3/2}$ is a one-to-one map of the domain D onto a domain consisting of three half-planes. Therefore for each point $z \in D$ there is a canonical path $\gamma^+(z)$ which ends in D_0 and Re $S \to +\infty$ along $\gamma^+(z)$. Suppose that D_ε is obtained from D by removing an ε-neighbourhood of the cut l_3. Then (9) holds in D_ε because of the theorem from § 2. Also, $\varepsilon > 0$ can be chosen arbitrarily small and not depending on λ.

The asymptotic formula (9) is not applicable in any larger domain. In fact we will assume that we can "rub out" a part of the cut l_3, that is, we can add to D_ε an ε-neighbourhood of an interval of the form $[x_1, x_2]$, $x_j < 0$. The function $\tilde{w}_1$ has a branch point $z = 0$ and is therefore not single-valued in the domain $\tilde{D}_\varepsilon$, but the solution w_1 is single-valued. We remark also that $\tilde{D}_\varepsilon$ does not satisfy the conditions of the theorem in § 2, since for any point z, lying on one of the sides of the cut l_3, there is no canonical path $\gamma^+(z)$.

2. Let $q(z) = z^n$ where $n \geqslant 2$ is an integer; then $z = 0$ is a multiple turning point. Let D_0 be the domain $|\arg z| < \pi/(n+2)$. Then a canonical path $\gamma^+(z)$ exists for an arbitrary point z in D: $|\arg z| < 3\pi/(n+2)$ and does not exist for points lying on ∂D. The asymptotic expansion (9) can be used in D_ε, which is obtained from D by removing an ε-neighbourhood of ∂D. The solution w_1 is expressed in terms of Bessel functions and the maximality of D_ε follows from their known asymptotic behaviour.

3. Let $q(z) = z^2 - 1$, where D is the domain $I \cup II \cup III$ (see Fig. 3). For any point $z \in D$ ($z \in \partial D$) there exists (does not exist) a canonical path and (9) can be used in D_ε. The maximality of D_ε follows from the known asymptotic behaviour of Weber functions.

Suppose that $q(z)(\not\equiv \text{const})$ is a polynomial such that

1) all its zeros are simple;
2) equation (2) has no finite Stokes lines.

Let us construct the maximal domain for which (9) can be used. From 2) ∂D_0 consists of two Stokes lines l_1, l_2 and from the turning point $z_0 \in \partial D_0$ there arises a further Stokes line l_3. Let us make a cut along l_3 and remove a neighbourhood of it from the complex plane. If $q(z)$ is linear then the domain D_ε so obtained is maximal (Example 1). In the contrary case D_0 borders on a domain D_1 of band type (Fig. 11) and ∂D_1 contains a connected component consisting of the Stokes lines l_4, l_5 and a turning point z_1. From z_1 there emerges yet another Stokes line l_6. Let us remove an ε-neighbourhood of l_6 from the complex plane. If the degree of $q(z)$ is two then the domain D_ε which is obtained is maximal; in the contrary case we extend this process. At each step a cut is made along this Stokes line, which does not lie in a previous adjoined domain (of half-plane or band type).

Let conditions 1), 2) be satisfied. Then the maximal domain D_ε for which we can apply(9) is the whole complex plane, from which neighbourhoods of some Stokes lines have been removed.

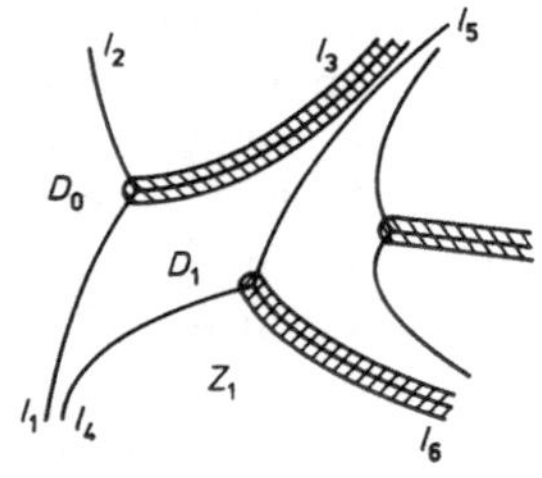

Fig. 11

Further, $\varepsilon > 0$ can be chosen sufficiently small, not depending on λ. It is convenient to express it thus: D is a domain with cuts along the Stokes curves. A cut is made along one of the three Stokes lines which emanate from a turning point (see Fig. 11). An algorithm for the construction of D is given above. Here the closure $[D]$ is the whole complex z-plane. If either of conditions 1), 2) is not satisfied then $C_z \backslash [D]$ is a domain (Examples 2 and 3).

Let us state some properties of $w_1(z, \lambda)$ for $\lambda \gg 1$ fixed.

1. Behaviour of the solution w_1 as $z \to \infty$. By construction Re $S(z_0, z) < 0$ for $z \notin D_0 \cup \partial D_0$. Let $D_1 \neq D_0$ be a domain of half-plane type and let S be a sector in D_1. Then Re $S(z_0, z) \to -\infty$ for $z \in S$ and $z \to \infty$ so that

$$w_1(z, \lambda) \to \infty\,, \quad z \to \infty\,, \quad z \in S\,.$$

In this way, as $z \to \infty$ the solution w_1 decreases exponentially in D_0 and increases exponentially in all the remaining domains of half-plane type.

2. Zeros of w_1. It follows from (9) that w_1 has no zeros in the maximal domain D for which the asymptotic formula is applicable. This solution has infinitely many zeros near each of the Stokes lines along which cuts are made.

2.4 The Solution w_1 as an Entire Function. Let

$$q(z) = a_0 z^n + a_1 z^{n-1} + \ldots + a_n\,, \quad \text{Re } a_0 > 0\,, \quad n \geqslant 1\,.$$

Then w_1 is an entire function of completely regular growth, the order of the growth is $n/2 + 1$ and of type $2\lambda |a_0|/(n+2)$.

The solution w_1 depends on the parameters $\lambda, a_0, \ldots, a_n$. For fixed $\lambda > 0$, w_1 is a holomorphic function of a_0 in the domain Re $a_0 > 0$ and is an entire function of the variables z, $a_1, \ldots, a_n$.

2.5 Asymptotic Expansions of the Solutions. Let D be a canonical domain and let w_1 and w_2 be solutions of the form (7), (8) respectively. Then, for $\lambda \to \infty$ and $z \in (D^+ \cap D^-)$, we have

$$\begin{aligned} w_1(z,\ \lambda) &\sim q^{-1/4}(z)\exp\{-\lambda S(z_0,\ z)\} \\ &\quad \times \exp\left\{-\sum_{k=1}^{\infty}(-\lambda)^{-k}\int_{\gamma^+(z)}\alpha_k(t)dt\right\}, \\ w_2(z,\ \lambda) &\sim q^{-1/4}(z)\exp\{\lambda S(z_0,\ z)\} \\ &\quad \times \exp\left\{-\sum_{k=1}^{\infty}\lambda^{-k}\int_{\gamma^-(z)}\alpha_k(t)dt\right\}. \end{aligned} \tag{10}$$

The functions $\alpha_k(z)$ were obtained in Chap. 3, § 3. The first (second) asymptotic expansion remains in force when $z \in D$ and $z \to \infty$ so that $\operatorname{Re} S(z_0, z) \to +\infty(-\infty)$. Both series can be differentiated in z and λ an arbitrary number of times.

The asymptotic expansion for w_1 can also be written in the form

$$w_1(z,\ \lambda) = \tilde{w}_1(z,\ \lambda;\ z_0)\left[1 + \sum_{k=1}^{\infty}\lambda^{-k}a_k(z)\right].$$

Here the $a_k(z)$ are determined by the relation

$$1 + \sum_{k=1}^{\infty}\lambda^{-k}a_k(z) = \exp\left\{-\sum_{k=1}^{\infty}(-\lambda)^{-k}\int_{\gamma^+(z)}\alpha_k(t)dt\right\},$$

where equality is understood in the sense of equality of the formal power series in λ^{-1}. For $\lambda \geqslant \lambda_0 \gg 1$ and $z \in D_\varepsilon^{-1}$ we have

$$\left|\frac{w_1(z,\ \lambda)}{\tilde{w}_1(z,\ \lambda;\ z_0)} - 1 - \sum_{k=1}^{N}\lambda^{-k}a_k(z)\right| \leqslant C_N(1+|z|))^{-\alpha_N},$$

where

$$\alpha_N = (N+1)(n/2+1).$$

Here $N \geqslant 1$ is arbitrary and n is the degree of the polynomial $q(z)$. The functions $a_k(z)$ are holomorphic in D and $a_k(z) = 0(z^{-k(n/2+1)})$ as $z \to \infty$, $z \in D$. Similar bounds occur for w_2 and for the derivatives of w_1 and w_2.

2.6 Elementary F.S.S. Let D be a canonical domain, l a Stokes line in D and $z_0 \in l$ a turning point (the beginning of l). An elementary F.S.S. $(u(z),\ v(z))$ is uniquely determined by the triple $(l,\ z_0,\ D)$. Let us select the branch of $S(z_0, z)$ in D such that

$$\operatorname{Im} S(z_0\ z) > 0, \quad z \in l. \tag{11}$$

The solutions u, v have the asymptotic behaviour

$$\begin{aligned} u(z) &\sim cp^{-1/4}(z)\exp\{\lambda S(z_0,\ z)\}, \\ v(z) &\sim cp^{-1/4}(z)\exp\{-\lambda S(z_0,\ z)\}. \end{aligned} \tag{12}$$

Here c is a normalizing constant given by

$$|c| = 1, \quad \lim_{z \to z_0,\ z \in l} \arg\,[c q^{-1/4}(z)] = 0. \tag{13}$$

The asymptotic formulae (12) can be used for u and v for fixed $\lambda > 0$ and $z \to \infty$ in a domain D such that Re $S(z_0, z) \to +\infty$ and $-\infty$ respectively. By construction, u and v are proportional to the solutions w_2 and w_1 in paragraph 2.2 and are uniquely determined by their asymptotic behaviour. We observe that Re $S(z_0, z) > 0$ (< 0) on the right (left) of l, near to l.

3. Transition Matrices. There are four types of transition matrices called *elementary*.

1) $(l, z_1, D) \to (l, z_2, D)$. Only the direction of the Stokes line is changed; this transition exists only for a finite Stokes line.

2) $(l_1, z_1, D) \to (l_2, z_2, D)$. Here the rays $S(l_1)$ and $S(l_2)$ are directed to one side.

3) $(l, z_0, D_1) \to (l, z_0, D_2)$. Only the canonical domain is changed.

4) $(l_1, z_0, D_1) \to (l_2, z_0, D_2)$ where l_1, l_2 are adjacent Stokes lines arising from the same turning point.

Any transition matrix $(l_1, z_1, D_1) \to (l_2, z_2, D_2)$ from one elementary F.S.S. to another is the product of a finite number of elementary transition matrices.

3.1 The Transition Matrix for $(l, z_1, D) \to (l, z_2, D)$. This matrix has the form

$$\Omega = e^{i\phi_0} \begin{bmatrix} 0 & e^{-i\lambda\alpha} \\ e^{i\lambda\alpha} & 0 \end{bmatrix}, \tag{14}$$

where

$$\alpha = |S(z_1\,, z_2)|\,, \quad e^{i\phi_0} = c_2/c_1\,.$$

3.2 The Transition Matrix for $(l_1, z_1, D) \to (l_2, z_2, D)$. Suppose that the rays $S(l_1)$ and $S(l_2)$ are directed to one side and l_2 lies to the left of l_1. Then

$$\Omega = e^{i\phi_0} \begin{bmatrix} e^{-\lambda a} & 0 \\ 0 & e^{\lambda a} \end{bmatrix}, \tag{15}$$

where

$$a = S(z_1\ z_2)\,, \mathrm{Re}\ a > 0\,, \quad e^{i\phi_0} = c_2/c_1\,.$$

Formulae (14) and (15) are exact but not asymptotic.

3.3 The Transition Matrix for $(l, z_0, D_1) \to (l, z_0, D_2)$. Here only the canonical domain containing l is changed. For instance (see Fig. 3, $q = 1 - z^2$), for D_1 we can take the union of the domains I, II and III (together with the corresponding Stokes lines); for D_2 we can take the union of I, II and IV.

Let D be the connected component of $D_1 \cap D_2$ containing l. Then $S(D)$ is the band $-a_- < \operatorname{Re} S < a_+$ with vertical cuts at $a_\pm > 0$. We have

$$\Omega = \begin{bmatrix} 1+\eta_{11} & \omega_{12} \\ \omega_{21} & 1+\eta_{22} \end{bmatrix},$$

where

$$\omega_{12} = O(\exp\{-2\lambda(a_+ - \varepsilon)\}, \quad \omega_{21} = O(\exp\{-2\lambda(a_- - \varepsilon)\}), \tag{16}$$

and $\varepsilon > 0$ is arbitrary. If $a_- = -\infty$ then $\eta_{11} = \omega_{21} = 0$; if $a_+ = +\infty$ then $\omega_{12} = \eta_{22} = 0$.

Remark. It follows from (16) that if we need to compute the transition matrices only to within $0(\lambda^{-1})$, then it is sufficient to indicate the pair (l, z_0) and we are not concerned with the choice of canonical domain.

We next give the asymptotic expansions for the elements $\omega_{11}(\lambda)$ and $\omega_{12}(\lambda)$. Let l_1 be an infinite contour which starts in D_1, where $\operatorname{Re} S \to -\infty$ and ends in D_2 where $\operatorname{Re} S \to -\infty$; l_2 is determined in the same way but with S replaced by $-S$. Then

$$\omega_{jj}(\lambda) = \exp\left\{\sum_{k=1}^{\infty} \lambda^{-k} \int_{l_j} \alpha_k(z)dz\right\}, \quad j = 1, 2. \tag{17}$$

The contours l_1, l_2 are depicted in Fig. 11.

3.4 The Transition Matrix for $(l_1, z_0, D_1) \to (l_2, z_0, D_2)$. Let z_0 be a turning point of order n, let Stokes lines l_1 and l_2 go out from z_0, and let l_2 lie to the left of l_1. Then

$$\Omega = \exp\left\{-\frac{i\pi n}{2(n+2)}\right\} \begin{bmatrix} 0 & 1 \\ 1 & 2i \sin\frac{\pi n}{2(n+2)} \end{bmatrix} + O(\lambda^{-1}). \tag{18}$$

A more exact bound for the remainder term $w_{22}(\lambda)$ is not known if $n > 1$. If z_0 is a simple turning point then

$$\Omega = e^{-i\pi/6} \begin{bmatrix} 0 & 1 \\ 1 & i \end{bmatrix} + O(\lambda^{-1}). \tag{19}$$

To obtain the asymptotic expansions for the transition matrices we must define the choice of canonical domains more precisely. Let z_0 be a simple turning point, let l_1, l_2, l_3 be Stokes lines starting at z_0, and let l_{j+1} lie to the left of l_j (indexed thus: $4 = 1, \ldots$). Choose the canonical domain D_j so that the part of D_j on the left of l_j coincides with the part of D_{j+1} on the right of l_{j+1} and denote by $\Omega_{j,j+1}$ the transition matrix $(l_j, z_0, D_j) \to (l_{j+1}, z_0, D_{j+1})$. We will call these domains and the corresponding elementary F.S.S. compatible. Then

$$\Omega_{j,j+1} = e^{-i\pi/6} \begin{bmatrix} 0 & \alpha_{j,j+1}^{-1} \\ 1 & i\alpha_{j+1,j+2} \end{bmatrix} , \tag{20}$$

with

$$\alpha_{12}\alpha_{23}\alpha_{31} = 1 ,$$

which follows from the identity $\Omega_{31}\Omega_{23}\Omega_{12} = I$. We have

$$\alpha_{j,j+1}(\lambda) = \exp\left\{ \sum_{k=1}^{\infty} (-\lambda)^{-k} \int_{\gamma_j\ j+t} \alpha_k(t)dt \right\} . \tag{21}$$

The infinite contour $\gamma_{j,j+1}$ lies in $D_j \cup D_{j+1}$ starting in D_{j+1}, where Re $S \to +\infty$, and finishing in D_j where Re $S \to -\infty$. The branch of $\sqrt{q(z)}$ is chosen in the same way as for the F.S.S. (u_j, v_j). The contour γ_{12} is indicated in Fig. 12.

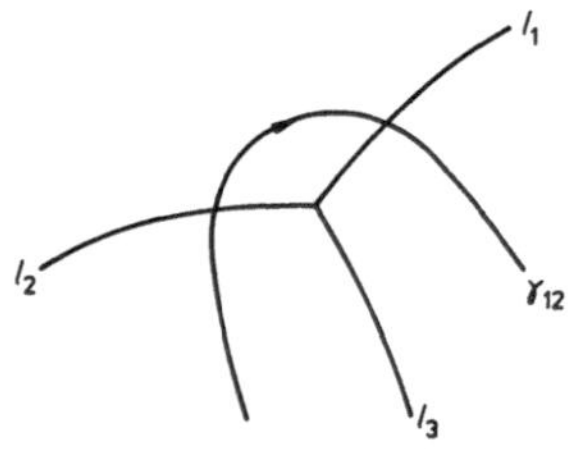

Fig. 12

Let z_0 be a turning point of order n and let $\Omega_{j,j+1}$ be the transition matrix $(l_j, z_0, D_{j+1}) \to (l_{j+1}, z_0, D_j)$. Then

$$\Omega_{j,j+1} = \exp\left\{ -\frac{i\pi n}{2(n+2)} \right\} \begin{bmatrix} 0 & \alpha_{j,j+1}^{-1} \\ 1 & \beta_{j,j+1} \end{bmatrix}$$

and formula (21) holds for $\alpha_{j,j+1}$.

3.5 Polynomials with Real Coefficients. Here the transition points and Stokes lines for equation (2) have additional properties. Let us recall the notation that $M^\star$ is the set symmetric to M about the real axis.

1. The turning points and Stokes lines are symmetric about the real axis.

2. If x_1, x_2 are real turning points and $q(x) < 0$ for $x \in l = (x_1, x_2)$, then l is a Stokes line and $\phi_0 = 0$ in (14).

3. Let x_0 be a simple real turning point and let l_0, l_1, l_2 be the Stokes lines emanating from x_0. Then one of the Stokes lines (let it be l_0) is an interval of the real line, $l_2 = l_1^\star$ and the curve $l_1 \backslash x_0$ lies in the upper half-plane Im $z > 0$. The curves $l_1 \backslash x_0$ and $l_2 \backslash x_0$ do not intersect the real axis.

4. Let x_1 and x_2 be simple real turning points, $x_2 < x_1$, and let $q(x) < 0$ for $x_2 < x < x_1$. We denote by l_1 and l_2 the Stokes lines which emanate from the points x_1, x_2 and lie in the upper half-plane. Then $\phi_0 = -\pi/6$ in (15).

5. Suppose that the Stokes line l intersects the real axis at x_0, where $q(x_0) \neq 0$. Then l is a finite Stokes line, $l = l^*$ (Fig. 13), $q(z) = (z^2 - a^2)(z^2 + b^2)(z^2 + c^2)$, where $a, b, c > 0$.

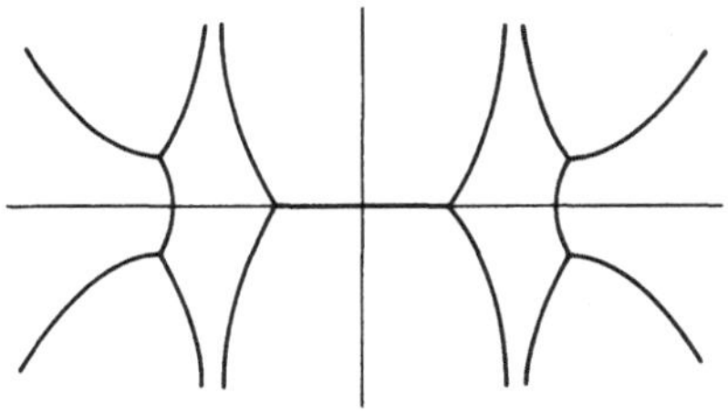

Fig. 13

6. Let x_0 be a simple real turning point, and let $q(x) > 0$ for $x > x_0$. We choose the branch $\sqrt{q(x)} > 0$ for $x > x_0$. Then $S(x_0, x) = \int_{x_0}^{x} \sqrt{q(t)}dt > 0$ for $x > x_0$, and $S(x_0, +\infty) = +\infty$. Therefore the half-line $x > x_0$ intersects a domain D_0 of half-plane type. Let l and l^* be Stokes lines emanating from x_0 and lying respectively in the half-planes Im $z > 0$, Im $z < 0$. There is a domain D such that the function S is single-sheeted in D,

$$D \supset (x_0, +\infty), \quad D \supset D_0, \quad D = D^*, \quad \partial D \supset l \cup l^*$$

and $S(D)$ is the half-plane Re $S > 0$ with a finite number of vertical cuts.

7. Let $q(x) < 0$ on the real axis $\mathbb{R}$. Then $\mathbb{R}$ is contained in a domain of band type, $D = D^*$. In fact suppose that $\sqrt{q(x)} = i|\sqrt{q(x)}|$. Then $S(0, x)$ is a one-to-one map of $\mathbb{R}$ onto the imaginary axis.

8. Let $q(x_1) = q(x_2) = 0$, $x_1 < x_2$, and let $q(x) > 0$ for $x_1 < x < x_2$. Denote by l_1, l_2 the Stokes lines going out from x_1, x_2 respectively and which lie in the half-plane Im $z > 0$. Then there exists a domain D such that $\partial D \supset l_1 \cup l_2$, $D = D^*$, S is single-sheeted in D and $S(D)$ is a band of the form $a <$ Re $< b$ with a finite number of vertical cuts.

9. Let $q(x_0) = 0$ and $q(x) < 0$ for $x > x_0$. Then the ray $l = (x_0, \infty)$ is a Stokes line. There is a canonical domain D such that $D \supset l$, $D = D^*$ and all the boundary cuts of $S(D)$ are directed to the side opposite to $S(l)$ (Fig. 14).

Since $q(x)$ is real we can improve some of the formulae for the transtition matrices.

10. Let x_0 be a simple real turning point and let $q(x) > 0$ for $x_0 < x < b \leqslant \infty$. We index the Stokes lines $l_0, l_1, l_{1'}$, emanating from x_0 so that $l_0 = (a, x_0)$, Im $z > 0$ on l_1 and $l_{1'} = l_{1'}^*$. Let us select canonical domains $D_0, D_1, D_{1'}$ so that

$$D_0 = D_0^\star, \quad D_{1'} = D_1^\star, \quad D_0 \supset l_0, \quad D_1 \supset (x_0,\ b),$$

and introduce elementary F.S.S. corresponding to (l_j, x_0, D_j), $j = 0, 1, 1'$. Then in the same notation as in (20) we have

$$\alpha_{10}\overline{\alpha}_{01'} = 1, \quad |\alpha_{1'1}| = 1. \tag{22}$$

11. Let $b = \infty$ in the conditions of paragraph 10. Then

$$\alpha_{11'} = \exp\left\{\sum_{k=1}^{\infty} \lambda^{-k} \int_{l_{11'}} \alpha_k(z)dz\right\},$$

where the contour $l_{11'}$ is indicated in Fig. 15. The branch of $\sqrt{q(z)}$ is chosen so that $\sqrt{q(z)} = i|\sqrt{q(z)}|$ on the intersection of the contour with the real line.

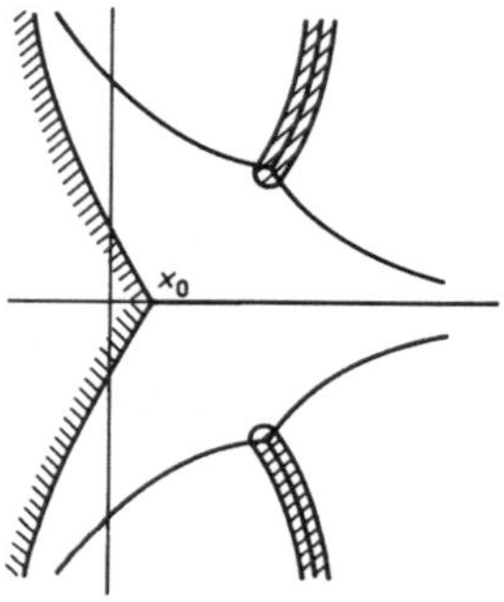

Fig. 14

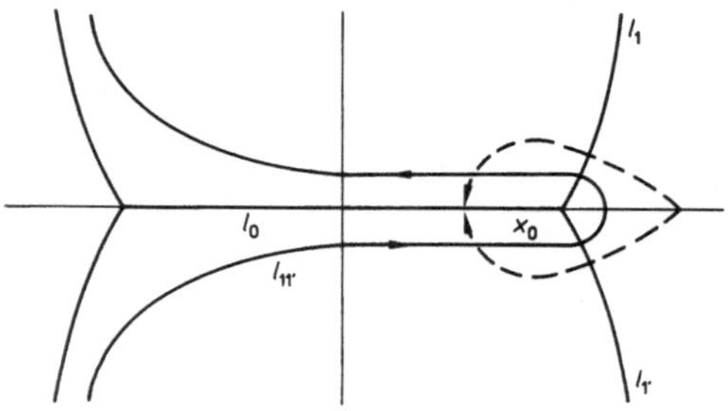

Fig. 15

12. Suppose that the conditions of paragraph 11 are satisfied, and let w_1 be a real solution of equation (2) such that $w_1(\infty, \lambda) = 0$. Then

$$w_1(x,\ \lambda) \sim cq^{-1/4}(x)\exp\{-\lambda S(x_0,\ x)\}, \quad \lambda \to \infty,$$

where c is a real constant, $\sqrt{q(x)} > 0$, and $\sqrt[4]{q(x)} > 0$ for $x > x_0$. The asymptotic behaviour of the solution w_1 can be formally obtained on the Stokes line (a, x_0) in the following way.

Extending the asymptotic behaviour analytically from the half-line $x > x_0$ to l_0 bypassing the turning point x_0 from above (from below), we obtain the values $w_1^+(w_1^-)$. Then the paths from which the extensions originate are shown in Fig. 15. Then the asymptotic behaviour of w_1 is

$$w_1(x,\ \lambda) = \frac{1}{2}[w_1^+(x,\ \lambda) + w_1^-(x,\ \lambda)].$$

This is an easily remembered rule.

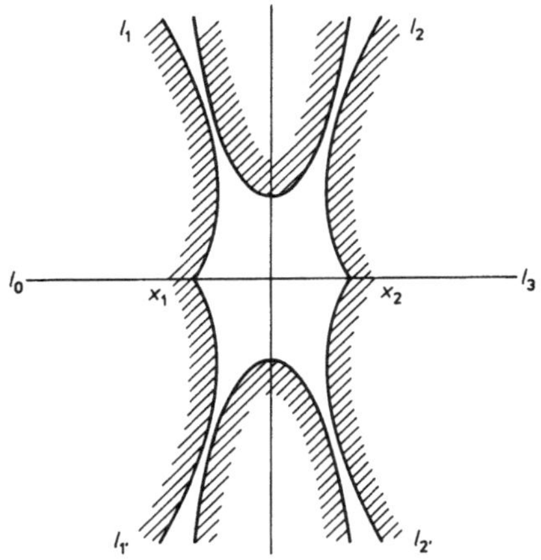

Fig. 16

13. Let $q(x)$ be a real-valued function, $x_1 < x_2$, $q(x_1) = q(x_2) = 0$ and $q(x) > 0$ for $x_1 < x < x_2$. Let x_1 and x_2 be simple turning points. Then there arise from x_1 Stokes lines $l_0, l_1, l_1^\star$, and from x_2 Stokes lines $l_2, l_2^\star, l_3$ (Fig. 16), Here Im $z > 0$ for $z \in l_1$, $z \in l_2$, and the Stokes lines l_0, l_3 are intervals of the real axis. We introduce elementary F.S.S. (u_j, v_j) corresponding to the triplet (l_j, x_j, D_j), $0 \leqslant j \leqslant 3$, where $x_j = x_1$ or $x_j = x_2$ depending on which of the turning points is the end of the Stokes lines l_j, and $(u_{j'}, v_{j'}), j = 1, 2$, corresponding to the triplet $(l_j^\star, x_j, D'_j)$. We choose canonical domains so that $D_0 = D_0^\star$, $D_3 = D_3^\star$, and

$$v_0(z, \lambda) \equiv \overline{u_0(\overline{z}, \lambda)}, \quad v_3(\overline{z}, \lambda) \equiv \overline{u_3(\overline{z}, \lambda)}.$$

We choose the remaining canonical domains by compatibility (paragraph 3.4) so that, in particular, the identity $\alpha_{32}\alpha_{22'}\alpha_{2'3} = 1$ is satisfied. There is a domain D, whose boundary contains the Stokes lines $l_1, l_1^\star, l_3, l_3^\star$ and which is mapped in a one-to-one manner by S onto the band $a < \text{Re } S < b$ with a finite number of vertical cuts, and moreover $D = D^\star$. We put

$$D_2 = D_0^+ \cup D \cup D_3^+ \cup l_1 \cup l_2, \quad D_1 = D_2,$$
$$D'_2 = D_0^+ \cup D \cup D_3^- \cup l_1 \cup l_2^\star, \quad D_1 = D_2^\star,$$

where $D_j^\pm = D_j \cap \{\text{Im } z \gtrless 0\}$. Then [Fedoryuk 10]

$$|\alpha_{11'}| = (1+\delta)^{-1/2}, \quad |\alpha_{22'}| = (1+\delta)^{1/2},$$
$$\alpha_{1'1} = \alpha_{32}\overline{\alpha_{2'3}}, \quad \text{Im } (\alpha_{10}/\overline{\alpha}_{32}) = 0, \tag{23}$$

$$\delta = e^{-2\lambda\xi}[1 + O(\lambda^{-1})], \quad \xi = \int_{x_1}^{x_2} \sqrt{q(x)}dx > 0.$$

These formulae are of interest even though $\alpha_{11'}(\lambda)$, $\alpha_{22'}(\lambda)$ are found only to within $0(\lambda^{-N})$ for arbitrary N, because the values of the moduli can be found with exponential precision. In actual fact

$$\begin{bmatrix} u_0 \\ v_0 \end{bmatrix} = \Omega(\lambda) \begin{bmatrix} u_3 \\ v_3 \end{bmatrix},$$

$$\omega_{22}(\lambda) = \overline{\omega_{11}(\lambda)}, \quad \omega_{21}(\lambda) = \overline{\omega_{12}(\lambda)}.$$

Further, from the identity

$$\Omega(\lambda) = [(\Omega_{10}\Omega_{21}\Omega_{32})^T]^{-1},$$

we find

$$\omega_{11} = ie^{\lambda\xi}[-(\alpha_{1'1}\alpha_{2'3})^{-1} + \alpha_{10}e^{-2\lambda\xi}], \quad \omega_{12} = e^{\lambda\xi}\alpha_{32}\alpha_{\overline{1'1}}^{-1},$$
$$\omega_{21} = e^{\lambda\xi}\alpha_{\overline{2'3}}^{-1}, \quad \omega_{22} = ie^{\lambda\xi}\alpha_{32},$$

which leads to the relations

$$\alpha_{1'1} = \alpha_{32}\overline{\alpha_{2'3}}, \quad 1+\delta = |\alpha_{22'}|^2,$$
$$\delta = e^{-2\lambda\xi}\alpha_{10}(\overline{\alpha_{32}})^{-1}.$$

Here we have used the identity $\alpha_{32}\alpha_{22'}\,\alpha_{2'3} = 1$. Consequently the variable δ is real, so that

$$\mathrm{Im}\,(\alpha_{10}/\overline{\alpha}_{32}) = 0, \quad |\alpha'_{22}| = \sqrt{1+\delta}e^{i\phi_2(\lambda)},$$

and we can obtain the rest of the relations in (23) similarly.

4. Arbitrary Dependence of the Coefficients on a Parameter. We consider the equation

$$w'' - Q(z,\ \lambda)w = 0, \tag{24}$$

with

$$Q(z,\ \lambda) = \sum_{j=0}^{n} a_{n-j}(\lambda)z^j, \quad a_0(\lambda) \neq 0,$$

where the $a_k(\lambda)$ are continuous functions of λ for $\lambda \geqslant \lambda_0 > 0$. Here the turning points and the Stokes lines of the equation depend on λ. Suppose that the conditions which follow are satisfied.

1) For $\lambda \geqslant \lambda_0$ all the zeros $z_1(\lambda), \ldots, z_n(\lambda)$ of the polynomial Q are simple.
2) There is a function $N(\lambda) > 0$ with $N(\infty) = \infty$ such that the circles

$$O_k(\lambda): \ |z - z_k(\lambda)| \leqslant |Q'_z(z(\lambda),\ \lambda)|^{-1/3}N(\lambda)$$

do not intersect for $\lambda \geqslant \lambda_0$.

We introduce the notation

$$S(z,\ \lambda;\ z_0) = \int_{z_0}^{z} \sqrt{Q(t,\ \lambda)}dt\,,$$

and

$$\tilde{w}_{1,2}(z,\ \lambda;\ z_0) = Q^{-1/4}(z,\ \lambda)\exp\{\mp S(z,\ \lambda;\ z_0)\}\,.$$

Let $D_0(\lambda)$ be a canonical domain,

$$D(\lambda) = D_0(\lambda)\Big\backslash \bigcup_{k=1}^{n} O_k(\lambda)\,.$$

Then for $\lambda \geqslant \lambda_0 \gg 1$ equation (24) has solutions w_1 and w_2 such that

$$\begin{aligned} \left|\frac{w_j(z,\ \lambda)}{\tilde{w}_j(z,\ \lambda;\ z_0)} - 1\right| &\leqslant A_j(z)N^{-3/2}(\lambda)\,, \\ \left|\frac{w'(z,\ \lambda)}{Q^{1/2}(z,\ \lambda)\tilde{w}_j(z,\ \lambda;\ z_0)} \pm 1\right| &\leqslant A_j(z)N^{-3/2}(\lambda)\,. \end{aligned} \tag{25}$$

Here $j = 1, 2$, the plus sign is taken for $j = 1$, and $A_j(z) \to 0$ as $z \to \infty$, $z \in D(\lambda)$, if Re $S \to +\infty$ for $j = 1$, and Re $S \to -\infty$ for $j = 2$. In particular

$$w_j(z,\ \lambda) \sim \tilde{w}_j(z,\ \lambda;\ z_0)\,, \quad \lambda \to \infty\,,$$

uniformly in $z \in D(\lambda)$.

The formula in (19) for the transition matrices remains true, the only difference being that the remainder term of order $0(\lambda^{-1})$ must be replaced by $0(N^{-3/2}(\lambda))$.

Example . Let $q(z)$ be a polynomial of degree $n \geqslant 2$, and let $Q(z, \lambda) = q(z) - \lambda$. Then we can put $N(\lambda) = \lambda^{1/3-2\varepsilon/3}$, where $\varepsilon > 0$ is arbitrary. Thus the remainder terms in (25) have order $0(\lambda^{-1/2+\varepsilon})$.

5. Additional Parameters. Let us consider the equation

$$w'' - \lambda^2 q(z,\ \omega)w = 0\,, \tag{26}$$

where

$$q(z,\ \omega) = a_0(\omega)z^n + a_1(\omega)z^{n-1} + \ldots + a_n(\omega)\,.$$

Here $\lambda > 0$ is a large parameter, ω is a complex parameter, and the $a_j(\omega)$ are polynomials with $a_0(\omega) \not\equiv 0$. We investigate the uniformity of the asymptotic formuae in the parameter ω. Let us introduce the notation

$$S(a,\ b;\ \omega) = \int_a^b \sqrt{q(z,\ \omega)}dz\,. \tag{27}$$

5.1 The Stokes Graph. The *Stokes graph* $\Phi(\omega_0)$ is the union of all the Stokes lines of equation (26) for $\omega = \omega_0$. Two graphs $\Phi(\omega_1)$ and $\Phi(\omega_2)$ are said to be ***equivalent*** if there is a homeomorphism $\phi: \Phi(\omega_1) \to \Phi(\omega_2)$ under which turning points are mapped to turning points. The value of the parameter ω_0 is called ***regular*** if there is a neighbourhood $U \ni \omega_0$ such that all the Stokes graphs for $\omega \in U$ are equivalent; the value of ω_0 is called ***singular*** otherwise. We describe the structure of the set I of all singular points of ω. We have

$$q(z, \omega) = a_0(\omega)(z - b_1(\omega))^{n_1} \dots (z - b_k(\omega))^{n_k},$$

where $n_j \geqslant 1$ is an integer, $n_1 + \dots + n_k = n$ and $b_j(\omega)$ is an algebraic function. Let I_1 be the set of all zeros of the coefficient $a_0(\omega)$, I_2 the set of all values of ω such that $b_j(\omega) = b_l(\omega)$, $j \neq l$ and I_3 the set of all values of ω such that for some $j, l, j \neq l$

$$\operatorname{Re} S(b_j(\omega), b_l(\omega), \omega) = 0. \tag{28}$$

These definitions do not depend on the choice of branches for $b_j(\omega)$. Thus $I = I_1 \cup I_2 \cup I_3$ and the set I is closed. The sets I_1 and I_2 consist of a finite number of points, and the set I_3 consists of a finite number of analytic curves which are called singular.

Recall that any connected component of a Stokes graph is called a *Stokes complex*. A Stokes complex is called simple (compound) if it contains precisely one (at last two) turning points. If $\omega_0 \in I_3$ then there is a finite Stokes line joining the points $b_j(\omega)$ and $b_l(\omega)$ so that there is a compound Stokes complex. Such complexes play a significant role in problems on eigenvalues (§ 5).

Examples. 1. $q = -z^2 + \omega$ (a harmonic oscillator). Here I_1 is empty, I_2 is the point $\omega = 0$, and I_3 is the line $\operatorname{Im} \omega = 0$.

2. $q = -z(z - \omega)^2$ (equation (26) describes the linear density waves in a spiral galaxy). Here I_1 is empty, I_2 is the point $\omega = 0$ and I_3 is the union of the four rays of $\operatorname{Im} \omega^{5/2} = 0$.

3. $q = -a(\omega)(z - b_1(\omega))^{n_1}(z - b_2(\omega))^{n_2}$ where a, b_1, and b_2 are polynomials. Here I_1 is the set of roots of the equation $a(\omega) = 0$, I_2 is the set of roots of the equation $b_1(\omega) = b_2(\omega)$ and I_3 is the set given by the equation

$$\operatorname{Re}\left[e^{i\pi n_2/2}(a(\omega))^{1/2}(b_2(\omega) - b_1(\omega))^{(n_1+n_2+2)/2}\right] = 0.$$

4. $q = e^{i\phi_0} z^n - \omega$. The set I_1 is empty, I_2 is the point $\omega = 0$ and I_3 consists of the rays $\arg \omega = \psi$ such that

$$2\psi - \phi_0 + 2(k + j)\pi/n = \pi l.$$

Here j, k, l are integers, $k \neq j$, $0 \leqslant j$, $k \leqslant n - 1$.

5. $q = \omega^2 p(z)$, where $p(z)$ is a polynomial, and $z_1, \dots, z_l$ are all its distinct zeros. The set I_1 is the point $\omega = 0$, I_2 is empty and I_3 consist of a finite number of rays $\arg \omega = \psi$ such that

$$\operatorname{Re}\left(e^{i\psi}\int_{z_j}^{z_k}\sqrt{p(t)}dt\right)=0$$

for some $j,\ k,\ j\neq k$.

The set of regular points of $C\backslash I$ consists of a finite number of connected components (domains). Let ω_1 and ω_2 be regular values lying in one connected component Ω, let γ be a simple curve which contains the points ω_1 and ω_2 and which lies in Ω. If $D(\omega_1)$ is a canonical domain for $\omega=\omega_1$ then for ω moving along γ we obtain a family of equivalent canonical domains $\{D(\omega)\}$ depending continuously on the parameter.

5.2 Analyticity of the WKB-Asymptotic Behaviour with Respect to a Parameter. Let ω_0 be a regular value, with $D(\omega_0)$ a canonical domain, and let $\Omega\supset\omega_0$ be the maximal domain consisting of regular points. We denote

$$\tilde{w}_{1,2}(z,\ \lambda,\ \omega;\ z_0)=q^{-1/4}(z,\ \omega)\exp\{\mp\lambda S(z_0,\ z,\ \omega)\}\,.$$

There exists a F.S.S. of the form (7), (8) for the domain $D(\omega_0)$ for $\lambda\gg 1$:

$$w_j(z,\ \lambda,\ \omega)\sim\tilde{w}_j(z,\ \lambda,\ \omega;\ z_0)\,,\quad j=1,\ 2\,, \tag{29}$$

for $\omega=\omega_0$. Suppose that the curve γ lies in Ω and connects the points ω_0 and ω_1. We extend the solution ω_j and the WKB-approximation $\tilde{\omega}_j$ analytically along γ with respect to the parameter ω. Then formula (29) remains in force as $\lambda\to\infty$, $z\in D(\omega)$ for arbitrary $\omega\in\gamma$. In particular the functions on both sides of (29) are analytic in Ω. If $K\subset\Omega$ is compact then the asymptotic behaviour of the solutions is uniform in $\omega\in K$. This is true also for higher approximations (10). All the asymptotic series mentioned can be differentiated in z,λ and ω an arbitrary number of times. Correspondingly the formulae for the transition matrices remain in force.

Let ω_0 be a singular point, $\gamma\ni\omega_0$ is a curve, all the points of which, excepting ω_0, are regular, and let $\{D(\omega)\}$ be a continuous family of canonical domains ($\omega\in\gamma,\ \omega\neq\omega_0$). Then the limiting domain $D(\omega_0)$ is canonical. However the Stokes graphs $\Phi(\omega),\ \ \omega\neq\omega_0$, and $\Phi(\omega_0)$ are not equivalent and formula (29) is not applicable for $\omega=\omega_0$. There is an important particular case where formula (29) remains true. Let L be a singular curve not containing points of the type I_1, I_2 and let Ω_1,Ω_2 be the connected components of the set of regular points adjoining L. Then (29) is preserved for the analytic continuation along an arbitrary curve γ lying in $\Omega_1\cup L\cup\Omega_3$.

All the above results remain in force if the coefficients $a_0(\omega),\ldots,a_n(\omega)$ are holomorphic in the closure $[G]$ of a bounded domain of the complex ω-plane. Moreover they can be generalized in an obvious way to the case of several complex parameters $\omega=(\omega_1,\ldots,\omega_m)$.

§ 4. Equations with Entire or Meromorphic Coefficients

1. Equations with Entire Coefficients. We consider the equation

$$w'' - \lambda^2 q(z) w = 0\,, \tag{1}$$

where $q(z)$ is an entire function and $\lambda > 0$ is a large parameter. Any solution of equation (1) is an entire function of z. The results derived in § 3 for polynomials $q(z)$ carry over in the main to this case also but there are exceptions as well. We will point out the main differences between polynomial and entire transcendental functions from the point of view of WKB-approximations. Let l be the semi-infinite level curve Re $S(z_0, z) = 0$ (or Im $S(z_0, z) = 0$) with initial point z_0. If $q(z)$ is a polynomial then

$$I = \int_l |\alpha_1(z)||dz| < \infty\,.$$

If $q(z)$ is an entire function then this integral may diverge. For instance if $q(x)$ is a real periodic or almost periodic function and $|q(z)| \geqslant \delta > 0$ on the real axis then $l = [0, \infty]$ is the level curve Im $(0, x) = 0$, and $I = \infty$ when q is not a constant. Another example when $z = \infty$ a turning point of infinite order occurs when $q(z) = e^{2z}$. Here $l = (-\infty, 0]$, and $I = \infty$. In both examples the WKB-approximations are inapplicable.

A canonical domain D is defined in the same way as in § 3, paragraph 1, the only difference being that $[S(D)]$ may contain an infinite number of vertical cuts. In precisely the same way the domains $D_\varepsilon^\pm$ and D_ε are defined, but the sizes of the ε-neighbourhoods are different for each cut (if there are infinitely many of them and they are condensed/concentrated). Let D be a canonical domain and suppose that the following condition is satisfied (§ 2, (16)):

$$\sup_{z \in D_\varepsilon} \rho^\pm(z) < \infty\,. \tag{2}$$

Then equation (1) has solutions $w_1(z, \lambda)$ and $w_2(z, \lambda)$ for which formulae (7), (8) from § 3 are satisfied for $z \in D_\varepsilon^+ (z \in D_\varepsilon^-)$ and $\lambda \geqslant \lambda_0$. The ideas of elementary F.S.S., of transition matrices and all the formulae for them are preserved (to within $0(\lambda^{-1})$). If instead of condition (2) there is the condition

$$\sup_{z \in D_\varepsilon} \left(\int_{\gamma_z^+} |\alpha_k(t)||dt|\,, \quad \int_{\gamma_z^-} |\alpha_k(t)||dt| \right) < \infty\,, \quad k = 1,\, 2,\, \dots\,, \tag{3}$$

then the asymptotic expansion (10) from § 3 is true for w_1 and w_2. All the other results of § 3 (except paragraphs 4–6) are also preserved if condition (3) is satisfied in the corresponding canonical domains.

2. $q(z)$ a Meromorphic Function. In this case the solutions of equation (1), as a rule, are not single-valued functions; and the singular points of the equation are branch points of the solutions.

We use the same classification of singular points for equation (1) as in § 1, paragraph 2.2. Suppose that $q(z)$ is rational.

2.1 Equations Without Regular Singular Points. Here $[\Phi] = \Phi$ in the Stokes graph and the Stokes lines partition the complex plane into a finite number of domains of half-plane and band types (§ 1). These domains and the canonical domain constructed from them are simply-connected, so that in each of them we can choose a single-valued solution to equation (1). All the results of § 3 are valid in this case. Using the asymptotic expansions for the transition matrices we can calculate the asymptotic behaviour as $\lambda \to \infty$ of the generators of the monodromy group for equation (1).

Example . The equation

$$w'' - \lambda^2 \frac{z^2 - 1}{z^3} w = 0$$

has two irregular singular points $z = 0$ and $z = \infty$. Fix a point $a \neq 0$ and $\neq \infty$, and take a F.S.S. $(w_1(z),\ w_2(z))$ which is holomorphic at this point. We then extend this F.S.S. analytically along the simple closed curve γ which has $z = 0$ as an interior point. We then obtain the F.S.S. $(\tilde{w}_1(z),\ \tilde{w}_2(z))$ that is holomorphic at the point a and

$$\begin{bmatrix} \tilde{w}_1(z) \\ \tilde{w}_2(z) \end{bmatrix} = U(\lambda) \begin{bmatrix} w_1(z) \\ w_2(z) \end{bmatrix},$$

where $U(\lambda)$ is a 2×2 matrix. In this example the monodromy group G for the equation has the one generator U and consists of the matrices U^n, $n = 0,\ \pm 1,\ \pm 2, \ldots$. Recall that the monodromy groups corresponding to different F.S.S. given at the point a are similar (Chap. 1, § 1).

We introduce three elementary F.S.S. $(u_j,\ v_j)$ corresponding to the triples $(l_j,\ z_j,\ D_j)$ where $z_1 = -1$ and $z_2 = z_3 = 1$ (the Stokes lines can be seen in Fig. 17).

We pass from the F.S.S. (u_1, v_1) to (u_2, v_2), then to $(u_3,\ v_3)$, and finally back to (u_1, v_1) (Fig. 17). The transition matrix

$$U = \Omega_{31} \Omega_{23} \Omega_{12}$$

so obtained is the desired generator of the monodromy group. We have (§ 3, (15), (20))

$$\Omega_{23} = e^{i\pi/6} \begin{bmatrix} -i + O(\lambda^{-1}) & 1 \\ 1 + O(\lambda^{-1}) & 0 \end{bmatrix}, \quad \Omega_{12} = \frac{c_1}{c_2} \begin{bmatrix} e^{\lambda \alpha_1} & 0 \\ 0 & e^{-\lambda \alpha_1} \end{bmatrix},$$

$$\Omega_{31} = \frac{c_3}{c_1}\begin{bmatrix} e^{\lambda\alpha_2} & 0 \\ 0 & e^{-\lambda\alpha_2} \end{bmatrix},$$

where the c_j are normalizing constants (§ 3), and

$$\alpha_j = \int_{\gamma_j} \sqrt{q(z)}dz\,, \mathrm{Re}\ \alpha_j > 0\,, \quad j = 1,\ 2\,.$$

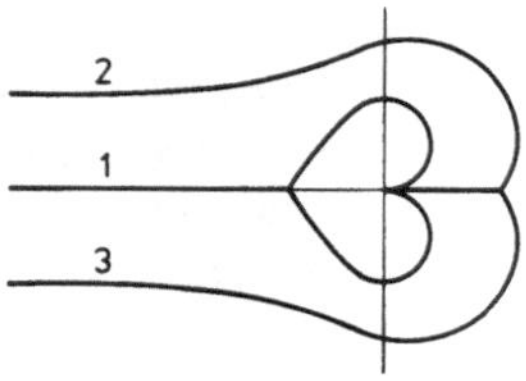

Fig. 17

The contour γ_1 connects the points -1, 1 and lies in the upper half-plane, while the contour γ_2 connects the points $1, -1$ and lies in the lower half-plane. Therefore

$$\alpha_1 = \alpha_2\,, \quad \alpha_1 + \alpha_2 = \int_\gamma \sqrt{q(z)}dz = \alpha\,,$$

where γ is a simple closed curve beginning and ending at $z = 1$. Also, γ contains $z = 0$ as an interior point, and is positively oriented. We note that $\sqrt{q(z)}$ takes different values at the beginning and end of γ. We have

$$U(\lambda) = \begin{bmatrix} [1 + O(\lambda^{-1})]e^{\lambda\alpha} & -i \\ -i + O(\lambda^{-1}) & 0 \end{bmatrix}.$$

The eigenvalues of U are

$$\mu_{1,2}(\lambda) = \pm[1 + O(\lambda^{-1})]\exp\left\{\pm\lambda\int_\gamma\sqrt{\frac{z^2-1}{z^3}}dz\right\}.$$

The integral along the contour γ equals $\frac{1}{2}(1 - i)B$ (3/2, 1/4).

2.2 Equations with Regular Singular Points of Type R_2, 1. Here the function $q(z)$ can have second order poles. If $a \neq 0$ is such a pole then

$$q(z) = -p(z)/(z-a)^2\,, \quad p(a) > 0\,.$$

The Stokes lines partition the complex plane into a finite number of domains of half-plane and band types. Further, a neighbourhood of a point of type R_2, 1 is covered by the closure of a finite number of domains of band type. Therefore all the results derived in § 3 about transition matrices continue to hold.

The next result is new here, and it concerns the selection of a solution using a boundary condition at a point a of type R_2. Let D_0 be a domain of band type. In addition let two of the Stokes lines l, $\tilde{l}$ bounding D_0 end at an R_2 point and let $D \supset D_0$ be a canonical domain. We introduce the new parameter

$$\mu = \sqrt{\lambda^2 + \frac{1}{4p(a)}}, \tag{4}$$

as in Chap. 2, § 6, (11), where $\mu > 0$. Equation (1) has two solutions w_1 and w_2 such that

$$w_{1,2}(z, \lambda) = [q(z)]^{-1/4} \exp\{\pm\mu S(z_0, z)\}[1 + \mu^{-1}\phi_{1,2}(z, \mu)]. \tag{5}$$

These asymptotic formulae are dual. Fix $\lambda_0 > 0$. Then for $\lambda \geqslant \lambda_0$ and $z \in D_\varepsilon$

$$|\phi_{1,2}(z, \mu)| \leqslant C.$$

If $\lambda > 0$ is fixed and $z \to a$ in D_0, then

$$|\phi_{1,2}(z, \mu)| \leqslant C|z - a|,$$

and we have

$$w_{1,2}(z, \lambda) \sim C_{1,2}(\lambda)(z - a)^{1/2 \pm i\mu\sqrt{p(a)}},$$

so that these solutions are strongly oscillatory. If w is a solution of equation (1) such that

$$w(z, \lambda) \sim (z - a)^{1/2 + i\mu\sqrt{p(a)}}, \quad z \to a, \quad z \in D_0, \tag{6}$$

then $w \equiv C(\lambda)w_1$. Similarly, w_2 can be defined using a boundary condition at the point a. Condition (6) can be replaced by the requirement that

$$\frac{w'(z, \lambda)}{w(z, \lambda)} \sim \left(\frac{1}{2} + i\mu\sqrt{p(a)}\right)(z - a)^{-1}, \quad z \to a, \quad z \in D_0.$$

These results allow us to move away from a singular point of type R_2, that is, to expand the asymptotic behaviour of a solution to those domains in which the standard WKB approximations can be used and then to connect the asymptotic behaviours.

§ 5. Asymptotic Behaviour of the Eigenvalues of the Operator $-d^2/dx^2 + \lambda^2 q(x)$. Self-Adjoint Problems

1. Statement of the Problem. We consider the equation

$$ly \equiv y'' - \lambda^2 q(x) y = 0, \tag{1}$$

where $q(x)$ is continuous and real on the real axis and λ is a spectral parameter.

1.1 The Problem on the Whole Axis. Suppose that $\lim_{x\to\pm\infty} q(x) = q_\pm (> 0)$ exist as finite or infinite limits. Then the spectrum of the operator l considered in $L_2(\mathbb{R})$ is purely discrete and consists of a countable set of positive eigenvalues $\{\lambda_n\}$ (we will not consider the series $\{-\lambda_n\}$). We index them in increasing order:

$$0 < \lambda_0 < \lambda_1 < \ldots < \lambda_n < \ldots, \qquad \lim_{n\to\infty} \lambda_n = +\infty.$$

The eigenfunctions $y_n(x) = y(x, \lambda_n)$ decrease exponentially as $|x| \to \infty$, and the condition $y \in L_2(\mathbb{R})$ can be replaced by the boundary conditions

$$y(-\infty, \lambda) = 0, \quad y(+\infty, \lambda) = 0. \tag{2}$$

It is required to investigate the asymptotic behaviour of λ_n as $n \to \infty$.

1.2 The Problem on the Half-Line $\mathbb{R}^+ = [0, +\infty)$. We impose the boundary condition

$$ay(0, \lambda) + by'(0, \lambda) = 0, \tag{3}$$

where a and b are real constants and $(a, b) \neq (0, 0)$. Let $\lim_{x\to+\infty} q(x) = q_+$, $0 < q_+ \leqslant \infty$. Then the spectrum of the problems (1), (3) is purely discrete and has the same properties as in problem (1), (2). The condition $y \in L_2(\mathbb{R}^+)$ can be replaced by the boundary condition

$$y(+\infty, \lambda) = 0. \tag{4}$$

2. The Problem on the Whole Axis with Two Turning Points

2.1 Asymptotic Behaviour of the Eigenvalues. Suppose that $q(x)$ has precisely two real zeros $x_1 < x_2$, both simple. Then $q(x) < 0$ for $x_1 < x < x_2$ and $q(x) > 0$ for $x < x_1$, $x > x_2$. Suppose that $q(x)$ is a polynomial of degree $n \geqslant 2$. The asymptotic behaviour of the eigenvalues can be calculated in this case using the theorems from § 2 without the ideas of canonical domain or transition matrices.

For each fixed $\lambda > 0$ equation (1) has unique (to within a constant multiple) solutions $y_1(x, \lambda)$ and $y_2(x, \lambda)$ such that $y_1(-\infty, \lambda) = 0$ and

$y_2(+\infty, \lambda) = 0$. If λ is an eigenvalue then $y_2(x, \lambda) = Cy_1(x, \lambda)$ because of (2). We take two different points z^- and z^+. Then we obtain the equation for the eigenvalues

$$\frac{y_1(z^+,\ \lambda)\, y_2(z^-,\ \lambda)}{y_2(z^+,\ \lambda)\, y_1(z^-,\ \lambda)} = 1\,. \tag{5}$$

This equation is exact, that is, it can be used for all $\lambda > 0$.

I. The solution y_2. Since $q(x) > 0$ for $x > x_2$ equation (1) has a solution with asymptotic expansion

$$y_2(x,\ \lambda) = q^{-1/4}(x)\exp\left\{-\lambda S(x_2,\ x) + \sum_{k=1}^{\infty}(-\lambda)^{-k}\int_{+\infty}^{x}\alpha_k(t)dt\right\} \tag{6}$$

as $\lambda \to +\infty$ uniformly in $x \geqslant x_0 + \delta$ where $\delta > 0$ is arbitrary. Here $\sqrt{q(x)} > 0$ and $\sqrt[4]{q(x)} > 0$ for $x > x_2$. We will make explicit the domain of applicability of this asymptotic expansion for z complex and close to x. The interval $[x_1, x_2]$ is a Stokes line. From the turning point x_j there arise two Stokes lines l_j, $l'_j = l^*_j$, $j = 1, 2$ (Fig. 18), and Im $z > 0$ for $z \in l_j$, $z \neq x_j$. Recall that M^* is the image of the set M under reflection in the real line. If Π is a small band of the form $|\text{Im}\ z| < \varepsilon$ then the asymptotic expansion in (6) is applicable in the domain Π^+ which is obtained from Π by removing neighbourhoods of the Stokes lines l_0, l_1, l'_1 in a domain lying to the left of l_1 and l'_1 (see Fig. 18). For an arbitrary point $z \in \Pi^+$ there is an infinite canonical path $\gamma \supset [x_2 + \delta, +\infty]$ (see Fig. 18). For z^+ we take a point close to l_0 and above l_0, for instance $z^+ = (x_1 + x_2)/2 + i\varepsilon$, $0 < \varepsilon \ll 1$. Then we put $z^- = \overline{z^+}$.

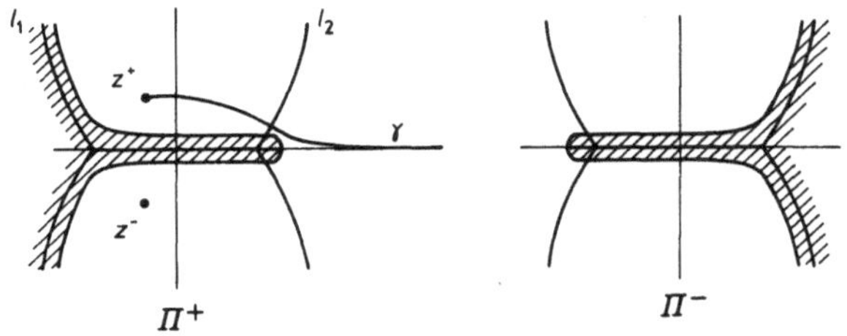

Fig. 18

II. The solution y_1. Equation (1) has a solution with asymptotic expansion

$$y_1(x,\ \lambda) = q^{-1/4}(x)\exp\left\{\lambda S(x_1,\ x) + \sum_{k=1}^{\infty}\lambda^{-k}\int_{-\infty}^{x}\alpha_k(t)dt\right\} \tag{7}$$

as $\lambda \to +\infty$ uniformly in $x \leqslant x_1 - \delta$, where $\delta > 0$ is arbitrary. Here $\sqrt{q(x)} > 0$ and $\sqrt[4]{q(x)} > 0$ for $x < x_1$. The domain of applicability of the asymptotic series in (7) close to the x-axis is Π^- (see Fig. 18). In particular it is applicable at the points $z^{\pm}$.

Substituting (6) and (7) into (5), we obtain an equation for the eigenvalues. Here we must carefully consider the choice of branch for all the multi-valued functions. Let U be a small simply-connected complex neighbourhood of the segment l_0. Then each of the functions $\sqrt{q(z)}$ and $a_k(z)$ breaks down in $V = U \backslash l_0$ into two single-valued holomorphic branches. Denote by $(\sqrt{q(x)})_j$ the branch appearing in the asymptotic series for y_j, and for convenience replace $q^{-1/4}(z)$ by

$$\exp\left\{-\frac{1}{4}\int_{z_1}^{z} \frac{q'(t)}{q(t)}dt\right\},$$

where $z_1 = x_1 - \delta$, $z_2 = x_2 + \delta$ and $\delta > 0$.

The branches $(\sqrt{q(z)})_1$ and $(\sqrt{q(z)})_2$ are chosen so that $(\sqrt{q(x)})_1 > 0$ for $x < x_1$ and $(\sqrt{q(x)})_2 > 0$ for $x > x_2$. Therefore $(\sqrt{q(x)})_2 < 0$ for $x < x_1$ and $(\sqrt{q(x)})_2 = i|\sqrt{q(x)}|$ on the upper part of the cut. Also

$$(\sqrt{q(z)})_1 = -(\sqrt{q(z)})_2, \quad (\alpha_k(z))_1 = (-1)^k(\alpha_k(z))_2, \quad z \in V.$$

Let C be a simple closed curve going around the interval l_0 and oriented counter-clockwise (Fig. 19). Substituting the asmptotic expansion (6), (7) into equation (5), we obtain an equation of the form

$$\exp\left\{\sum_{k=-1}^{\infty} d_k\lambda^{-k}\right\} = 1.$$

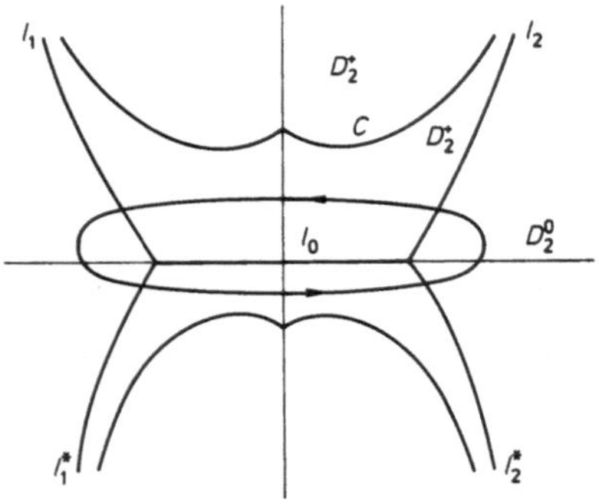

Fig. 19

Then, with integrals taken along paths in V, we have

$$\begin{aligned} d_{-1} &= S_1(x_1, z^+) + S_2(x_2, z^+) \quad - S_2(x_2, z^+) - S_1(x_1, z^-) \\ &= [-S_2(x_1, z^+) + S_2(x_2, z^+)] \quad + [S_2(x_1, z^-) - S_2(x_2, z^-)]. \end{aligned}$$

The first expression in the square bracket equals the integral $S_2(x_2, x_1)$ taken along the upper side of l_0, so that it equals $i\xi_0$, where

$$\xi_0 = \int_{x_1}^{x_2} |\sqrt{q(x)}| dx . \tag{8}$$

The second term also equals $i\xi_0$ and therefore $d_{-1} = 2i\xi_0$ or

$$d_{-1} = \oint_C \sqrt{q(z)} dz .$$

Here and below $\sqrt{q(z)} = (\sqrt{q(z)})_2$, that is

$$\sqrt{q(z)} > 0 , \quad z = x > x_2 . \tag{9}$$

Then

$$d_0 = -\frac{1}{4} \oint_C \frac{q'(z)}{q(z)} dz = -\pi i ,$$

$$d_k = \left[\int_{-\infty}^{z^+} (\alpha_k(z))_1 dz - (-1)^k \int_{+\infty}^{z^+} (\alpha_k(z))_2 dz\right]$$
$$+ \left[\int_{+\infty}^{z^-} (-1)^k (\alpha_k a(z))_1 dz - \int_{-\infty}^{z^-} (\alpha_k(z))_1 dz\right] = (-1)^k \oint_C \alpha_k(z) dz ,$$

where the branch of $\sqrt{q(z)}$ is chosen according to (9). Finally we obtain an equation for the eigenvalues

$$\exp\left\{2i\lambda\xi_0 - i\pi + \sum_{k=1}^{\infty} (-1)^k \oint_C \alpha_k(z) dz\right\} = 1 ,$$

the left hand side of which is an asymptotic series. All the numbers d_k are purely imaginary. Consequently there is the asymptotic expansion

$$\lambda_n \xi_0 = n\pi + \frac{\pi}{2} + \sum_{k=1}^{\infty} \beta_k \lambda_n^{-k} , \quad n \to \infty , \tag{10}$$

where

$$\beta_k = \frac{i}{2} (-1)^k \oint_C \alpha_k(z) dz .$$

Applying the Bürman-Lagrange formula we obtain an asymptotic series for λ_n in powers of n^{-1} as $n \to \infty$:

$$\lambda_n = \frac{n\pi}{\xi_0} + \frac{\pi}{2\xi_0} + \sum_{k=1}^{\infty} \gamma_k (\pi n)^{-k} , \tag{11}$$

where

$$\gamma_k = \frac{1}{k!}\left(\frac{d}{d\mu}\right)^{k-1} [f(\mu)]^k|_{\mu=0}$$

and

$$f(\mu) = \xi_0 - \frac{\pi\mu}{2} - \mu\sum_{k=1}^{\infty} \beta_k\mu^k .$$

In spite of the fact that the series $f(\mu)$ generally diverges, the formal substitution $\mu = 0$ gives a finite expression for γ_k. It follows from (11) that

$$\lambda_n = \left(\int_{x_1}^{x_2} |\sqrt{q(x)}|dx\right)^{-1} \left(n\pi + \frac{\pi}{2}\right) + \frac{i}{64\pi n}\oint_C q'^2(z)q^{-5/2}(z)dz + O\left(\frac{1}{n^2}\right) .$$

It follows from the above considerations that the asymptotic expansions (10), (11) are true under the following assumptions on $q(x)$:

1) $q(x) \in C^\infty(\mathbb{R})$, $\lim_{x\to\pm\infty} q(x) = q_\pm > 0$;
2) the integrals $\int_{-\infty} |\alpha_k(x)|dx$, $\int^{+\infty} |\alpha_k(x)|dx$ converge for all $k = 1, 2, \ldots$;
3) the function $q(z)$ is holomorphic in a complex neighbourhood of the interval $[x_1, x_2]$.

If the integrals in condition 2) converge for $k = 1, 2, \ldots, N$ then there is an asymptotic expansion of the form (10) with remainder term $O(\lambda_n^{-N})$.

Let us make some remarks.

1. The requirement that $q(z)$ should be analytic (condition 3)) is superfluous: the asymptotic expansions (10), (11) hold under conditions 1) and 2) only. In this case

$$\beta_k = i(-1)^{k+1}\int_{x_1}^{x_2} \alpha_k(t)dt ,$$

where $\sqrt{q(x)} = i|\sqrt{q(x)}|$ and the integrals β_k are regularized in a suitable way.

2. The functions $\alpha_{2k}(z)$ are holomorphic in a domain U except for the poles at x_1 and x_2, and β_{2k} are rational functions of the values of the derivatives of $q(z)$ at x_1 and x_2. If $q(z)$ is a polynomial then the β_{2k} are algebraic functions of its coefficients.

3. If $q(z)$ is a polynomial then the function $S = \int \sqrt{q(z)}dz$ is an Abel integral (elliptic for $n = 3, 4$ and hyperelliptic for $n > 5$) which is associated with equation (1) in the obvious way. The number

$$d_{-1} = \oint_C \sqrt{q(z)}dz$$

is the purely imaginary period of the Abel integral S. It follows from (11) that the principal asymptotic term is given by this period: $\lambda_n \sim n\pi i/d_{-1}$. If $q(x)$ has any finite number of turning points then the principal asymptotic term is also given by the purely imaginary period of S.

2.2 The Harmonic Oscillator. We consider the equation

$$y'' - \lambda^2(x^2 - a^2)y = 0, \quad a > 0.$$

As is known, the eigenvalues are determined by the relationships

$$\lambda_n \int_{-a}^{a} \sqrt{a^2 - x^2}\,dx = n\pi + \frac{\pi}{2}, \quad n = 0, 1, 2, \dots. \tag{12}$$

The method of paragraph 2.1 allows us to calculate all the λ_n exactly. In this case $x_1 = -a$ and $x_2 = a$, and there are no Stokes lines apart from l_j and l'_j, $j = 0, 1, 2$. Let D_2 be the domain bounded by l_1, l_0, l'_2, and let $S_2 = \int_a^z \sqrt{q(t)}dt$, where $\sqrt{q(x)} > 0$ for $x > a$. Then $S_2(D_2)$ is the complex S-plane with a cut along the ray $(-i\infty, 0)$. We have

$$y_1(z, \lambda) = q^{-1/4}(z)e^{-\lambda S_2}[1 + \varepsilon_1(z, \lambda)],$$

where $\varepsilon_1(z, \lambda) \to 0$ for $z \in D_2$ and $z \to \infty$, and $|\mathrm{Re}\, S_2| \to \infty$ uniformly in $\lambda \geqslant \lambda_0 > 0$ (§ 3, paragraph 2). Let D_1 be the domain bounded by l_0, l_2 and l'_1, and let $S_1 = \int_{-a}^z \sqrt{q(t)}dt$, where $\sqrt{q(x)} > 0$ for $x < -a$. Then

$$S_1(D_1) = S_2(D_2), \quad y_2(z, \lambda) = q^{-1/4}e^{\lambda S_1}[1 + \varepsilon_2(z, \lambda)],$$

where $\varepsilon_2(z, \lambda) \to 0$ for $z \in D_1$, and $z \to \infty$, and $|\mathrm{Re}\, S_1| \to \infty$ uniformly in $\lambda \geqslant \lambda_0 > 0$. If λ is an eigenvalue then $y_1(z, \lambda) \equiv Cy_2(z, \lambda)$. Let z be in the domain D, bounded by l_0, l_1 and l_2. Then

$$1 + \varepsilon_1(z, \lambda) = Ce^{i\lambda\xi_0}(1 + \varepsilon_2(z, \lambda)).$$

Letting z go to infinity in D (for instance $z = iy$, $y \to \infty$) we obtain $1 = Ce^{i\lambda\xi_0}$. Taking the domains D_j^* instead of D_j and D^* instead of D, we obtain $1 = Ce^{-1\lambda\xi_0}$. Eliminating C we obtain the equation for the eigenvalues $e^{2i\lambda\xi_0} = 1$, from which (12) follows.

2.3 Bohr-Sommerfeld Quantization Rule. We consider the eigenvalue problem for the Schrödinger equation

$$-\frac{h^2}{2m}\psi'' + (U(x) - E)\psi = 0 \tag{13}$$

with a real potential $U(x)$. Here E is the real spectral parameter and $h > 0$ is a small parameter.

Suppose that $U(x)$ has the form of a "potential well", that is, $U(x)$ has precisely one minimum point x_0, and $U'(x) > 0$ or < 0 according as $x > x_0$ or $x < x_0$. Suppose that $U(x_0) = 0$. Then there exist finite or infinite limits

$U_\pm = U(\pm\infty) > 0$. Put $J = [E_1^0, E_2^0]$, where $0 < E_1^0 < E_2^0 < \min(U_+, U_-)$. For each $E \subseteq J$ equation (12) has precisely two, and moreover simple, turning points $x_1(E) < x_2(E)$.

For $E \in J$ suppose that $q = 2m(U(x) - E)$ satisfies conditions of the type (2), (3) of paragraph 2.1: $U(x)$ is holomorphic in a neighbourhood of $I = [x_1(E_2^0),\ x_2(E_2^0)]$ and the integrals $\int_{-\infty}^{a} |a_k(x)| dx$, $\int_b^{+\infty} |a_k(x)| dx$ converge for $a < x_1(E_2)$, $b > x_2(E_2)$. Here the $\alpha_k(x, E)$ are computed in terms of the function $q = 2m(U(x) - E)$. The same arguments as in paragraph 2.1 lead to the asymptotic expansion

$$\int_{x_1(E)}^{x_2(E)} \sqrt{2m(E - U(x))}dx = h\left(n\pi + \frac{\pi}{2}\right) + \sum_{k=1}^{\infty} \beta_k(E) h^{k+1}, \tag{14}$$

where

$$\beta_k(E) = \frac{i}{2}(-1)^k \oint_C \alpha_k(z, E) dz\,.$$

This relationship is an equation for E, from which we can find the asymptotic behaviour of the eigenvalues $E_n(h)$ under the following conditions:

$$h \to 0, \quad E_1^0 \leqslant E \leqslant E_2^0, \quad f(E_1^0) < \left(n + \frac{1}{2}\right) h < f(E_2^0).$$

The principal asymptotic term is determined from the equation

$$f(E) \equiv \int_{U(x)<E} \sqrt{2m(E - U(x))}dx = \pi\left(n + \frac{1}{2}\right) h\,. \tag{15}$$

Formula (15) is called the *Bohr-Sommerfeld quantization rule*.

The function $f(E)$ has the following properties:

1. If $U(x) \in C^\infty(\mathbb{R})$ then $f(E) \in C^\infty(E_1^0, E_2^0)$ for $0 < E_1^0 < E_2^0$, and is strictly monotonic increasing. If x_0 is a non-degenerate minimum point ($U''(x_0) > 0$) then $f(E) \in C^\infty[0, E_2^0]$.

The latter statement is not true if $U''(x_0) = 0$. For instance, when $U(x) = x^4$, we have $f(E) = \text{const.}\ E^{3/4}$.

2. If $U(x)$ is holomorphic in a complex neighbourhood of the interval $I = [x_1(E_2^0), x_2(E_2^0)]$ then $f(E)$ is holomorphic in a complex neighbourhood of the interval $J = [E_1^0, E_2^0]$, $E_1^0 > 0$. If in addition $U''(x_0) > 0$ then this is also true for $E_1^0 = 0$.

This property of $f(E)$ follows from the formula

$$f(E) = \oint_C \sqrt{2m(E - U(z))}dz\,.$$

Let us make some remarks.

1. Let $0 \leqslant E_1^0 \leqslant E \leqslant E_2^0$, $h \to 0$, $n \to \infty$ so that $nh \to$ const. Then the principal asymptotic term of $E_n(h)$ is

$$E_n^0(h) = f^{-1}(h(n + \frac{1}{2})).$$

Under these conditions there is the asymptotic expansion

$$E_n(h) = E_n^0(h) + c_1 h^2 + c_2 h^4 + \dots$$

The coefficients c_n can be calculated by perturbation methods.

2. If $U''(x_0) > 0$ then (14) is valid when $0 \leqslant E \leqslant E_2^0$. This follows from the fact that for $E \approx 0$ equation (13) has no turning points close to the real line and distinct from $x_1(E)$, $x_2(E)$. Therefore we can obtain the asymptotic behaviour of the lower energy levels $E_n(h)$ from (14). Suppose that $h \to 0$ and $0 \leqslant n \leqslant n_0$, where n_0 is fixed. Then

$$E_n(h) = \pi \left(n + \frac{1}{2}\right) U''(x_0)h + O(h^2).$$

3. For fixed E with $0 < E_1^0 \leqslant E \leqslant E_2^0$, the functions $\alpha_{2k}(z, E)$ are holomorphic in a complex neighbourhood of the interval $[x_1(E), x_2(E)]$, excluding the poles at $x_1(E)$, $x_2(E)$, and the β_{2k} are rational functions of the derivatives of $U(x)$ at these points. If $U(x)$ is a polynomial then the β_{2k} are algebraic functions of E and the coefficients of $U(x)$.

2.4 Asymptotic Behaviour of the Eigenfunctions. Suppose that $q(z)$ satisfies conditions (1) – (3) of paragraph 2.1, let U_ε be a complex ε-neighbourhood of the interval $[x_1, x_2]$ and let $q(z)$ be holomorphic in U_ε. We denote by U the union of the domain $U_\varepsilon \backslash U_{\varepsilon_0}$, $0 < \varepsilon_0 < \varepsilon$, and the half-lines $x \leqslant x_1 - \varepsilon$, $x \geqslant x_2 + \varepsilon$. The asymptotic expansion of the eigenfunction $y_n(x)$ as $x \to \infty$ on the half-line $x \geqslant x_2 + \varepsilon$ is given by formula (6), where $\lambda = \lambda_n$, and it is uniform in x. All the functions in the asymptotic expansion are multi-valued functions of z for $z \in U$. Nevertheless, as was shown in paragraph 2.1, the expansion (6) for $y_n(z) = y(z, \lambda_n)$ is valid everywhere in U; that is, for $\lambda = \lambda_n$ the right hand side of (6) is a single-valued function for $z \in U$. We obtain the asymptotic expansion for $y_n(x)$ as $n \to \infty$ on the half-line $x \leqslant x_1 - \varepsilon$ by analytically extending (6) from the half-line $x \geqslant x_2 + \varepsilon$. This was proved for the principal asymptotic term by Birkhoff [Birkhoff].

We choose the branches of $\sqrt{q(x)}$ and $\sqrt[4]{q(x)}$ that are positive for $x < x_1$. Then, as $n \to \infty$,

$$y_n(x) = c_n q^{-1/4}(x) \exp\left\{\lambda_n S(x_1, x) + \sum_{k=1}^{\infty} \lambda_n^{-k} \int_{-\infty}^{x} \alpha_k(t)dt\right\}, \tag{16}$$

where

$$c_n = -i \exp\left\{-i\lambda_n \xi_0 + \sum_{k=1}^{\infty} \lambda_n^{-k} \int_{C_+} \alpha_k(t)dt\right\}$$

and $x \leqslant x_1 - \varepsilon_0$. Here C_+ is a contour consisting of the half-lines $(-\infty,\ x_1 - \varepsilon_0)$, $(x_2 + \varepsilon_0, \infty)$ and part of a contour C joining the points $x_1 - \varepsilon_0$, $x_2 + \varepsilon_0$ and lying in the upper half-plane. This asymptotic series can be differentiated in x an arbitrary number of times. We point out that the constant c_n is real since $y_n(x)$ is real, and that $c_n = (-1)^n[1 + 0(n^{-1})]$.

We now find the asymptotic series for $y_n(x)$ on the interval $I = [x_1 + \varepsilon_0,\ x_2 - \varepsilon_0]$. If we extend analytically the right hand side of (6) from the half-line $[x_2, \infty)$ to a point x lying on the upper or lower side of the cut I, we obtain the values $y_2^+(x, \lambda_n)$ or $y_2^-(x, \lambda_n)$ respectively, and their half-sum (§ 3, paragraph 2.1) is the required asymptotic series. Since $y_2^-(x, \lambda_n) = \overline{y_2^+(x, \lambda_n)}$, we have

$$y_n(x) = |q(x)|^{-1/4} \operatorname{Re} \left(\exp\left\{ i\lambda \int_{x_2}^{x} |\sqrt{q(t)}|dt + i\frac{\pi}{4} + \sum_{k=1}^{\infty} (-\lambda_n)^{-k} \int_{+\infty}^{x} \alpha_k(t)dt \right\} \right). \tag{17}$$

The branch of $\sqrt{q(x)}$ is chosen so that $\sqrt{q(x)} > 0$ for $x > x_2$, and the contour of integration bypasses x_2 from above. The principal asymptotic term is

$$y_n(x) = |q(x)|^{-1/4} \left[\cos\left(\left(n\pi + \frac{\pi}{2} \right) \xi_0^{-1} \int_{x_2}^{x} |\sqrt{q(t)}|dt + \frac{\pi}{4} \right) \right] + O(n^{-1}),$$

so that $y_n(x)$ is strongly oscillatory on $[x_1, x_2]$. The asymptotic formula for the eigenfunctions of the form (6), (16), (17) are also valid for equation (13). The principal terms of the asymptotic expansions have the form

$$\begin{aligned} y_n(x) &= (U(x) - E_n)^{-1/4} \exp\left\{ -h^{-1} \int_{x_2(E)}^{x} \sqrt{U(t) - E_n}dt \right\}, \\ &\quad x > x_2(E), \\ y_n(x) &= (E_n - U(x))^{-1/4} \cos\left(h^{-1} \int_{x_2(E)}^{x} \sqrt{-U(t) + E_n}dt + \frac{\pi}{4} \right), \\ &\quad x_1(E) < x < x_2(E), \\ y_n(x) &= (U(x) - E_n)^{-1/4} \exp\left\{ h^{-1}(-1)^n \int_{x_1(E)}^{x} \sqrt{U(t) - E_n}dt \right\}, \\ &\quad x < x_1(E). \end{aligned}$$

All the roots in these formulae are positive.

2.5 Application of Transition Matrices. Let $q(z)$ be a polynomial satisfying the conditions of paragraph 2.1. We will obtain formula (10) for the eigenvalues by using the asymptotic expansions of transition matrices (§ 3). This method may appear more complex than that of paragraph 2.1 for the case under consideration. However it is more universal and we can apply it, in particular, to the case where equation (1) has an arbitrary finite number of real turning points.

From § 3, paragraphs 3, 5, 6, it follows that there is a domain D_2^0 such that $S(x_2, z)$ is single-sheeted in D_2^0,

$$D_2^0 \supset (x_2, +\infty), \quad \partial D_2^0 \supset l_1 \cup l_{1'}, \quad (D_2^0)^* = D_2^0,$$

and $S(D_2^0)$ is the half-plane Re $S > 0$ with a finite number of vertical cuts. In Fig. 19 D_2^0 is the union of D_1, D_2 and the Stokes line 1. If $q(z)$ is an entire function satisfying condition 1) (§ 1, paragraph 2) then $S(D_2^0)$ can contain infinitely many vertical cuts.

We choose a canonical domain $D_2 \supset (x_2, +\infty)$ such that $D_2 \supset D_2^0$ and $D_2 = D_2^*$. Then $\partial D_2 \supset l_1 \cup l_0 \cup l_{1'}$ (see Fig. 19). Suppose that $D_2^{\pm} = D_2 \backslash [D_2^0] \cap \{\pm \text{Im } z \gtrless 0\}$; then D_2^+, D_2^- do not intersect, they lie respectively in the upper and lower half-planes, and $(D_2^+)^* = D_2^-$.

In an analogous way we construct $D_1^0 \supset (-\infty, 1)$. We put

$$D_1 = D_1^0 \cup l_1 \cup l_2^* \cup D_2^+ \cup D_2^- ;$$

Then $D_1^* = D_1$ and $D_1 \backslash D_2 = D_2^+ \cup D_2^-$. We take the canonical domain

$$D_0 = D_2^* \cup l_0 \cup D_2^- = D_4$$

and the elementary F.S.S. (u_j, v_j) corresponding to the triple (l_j, x_j, D_j), where $x_4 = x_1$. Suppose that $y_2(x, \lambda)$ is a solution which has asymptotic expansion (6). Then $y_2(z, \lambda) \equiv e^{-i\pi/12} v_2(z, \lambda)$. We will express y_2 in terms of the F.S.S. (u_1, v_1), that is, we will extend its asymptotic behaviour from l_2 to l_1. We have $y_1 = e^{-i\pi/12}(au_1 + bu_1)$, where

$$\begin{bmatrix} a \\ b \end{bmatrix} = \Omega_{21} \begin{bmatrix} 0 \\ 1 \end{bmatrix} = \Omega_{41} \Omega_{04} \Omega_{20} \begin{bmatrix} 0 \\ 1 \end{bmatrix}$$

and Ω_{jk} is the transition matrix from the jth-system to the kth-system. The solutions $u_1(x, \lambda)$ and $v_1(x, \lambda)$ respectively increase and decrease exponentially as $x \to -\infty$ and if λ is an eigenvalue then $b(\lambda) = 0$. This is the equation for the eigenvalues.

We will calculate $b(\lambda)$ using the asymptotic expansion for transition matrices. We take two more elementary F.S.S. $(u_{j'}, v_{j'})$, $j = 1, 2$, corresponding to $(l_{j'}, x_j, D_{j'})$ where $D_{j'} = D_j$. Then the F.S.S. which correspond to the Stokes lines emanating from x_1 are compatible (§ 3, paragraph 3.4) and similarly for x_2.

We have (§ 3, paragraphs 3.1, 3.4)

$$\Omega_{20} = e^{-i\pi/6} \begin{bmatrix} 0 & -\frac{1}{\alpha_{20}} \\ 1 & i\alpha_{02'} \end{bmatrix},$$

$$\Omega_{41} = e^{-i\pi/6} \begin{bmatrix} 0 & -\frac{1}{\alpha_{41}} \\ 1 & i\alpha_{11'} \end{bmatrix},$$

$$\Omega_{04} = \begin{bmatrix} 0 & e^{-i\lambda\xi_0} \\ e^{i\lambda\xi_0} & 0 \end{bmatrix},$$

so that

$$\Omega_{21} = e^{-i\pi/3} \begin{bmatrix} 0 & -\beta^{-1}\alpha_{02'}\alpha_{41}^{-1} \\ \beta^{-1} & i(\beta^{-1}\alpha_{02'} + \beta\alpha_{11'}\alpha_{20}^{-1}) \end{bmatrix} . \tag{18}$$

and

$$\beta = e^{i\lambda\xi_0} .$$

Taking into account the identity $\alpha_{02'}\alpha_{2'2}\alpha_{20} = 1$, we obtain the equation for the eigenvalues

$$e^{2i\lambda\xi_0}\alpha_{2'2}\alpha_{11'} = 1 .$$

Since (§ 3)

$$\alpha_{jk} = \int_{l_{jk}} \alpha_k(z)dz ,$$

where the contours $l_{11'}$ and $l_{2'2}$ are shown in Fig. 15 (and have both ends at infinity respectively in the domains D_1, D_2), there is the asymptotic expansion

$$(\alpha_{11'}\alpha_{2'2})^{-1} = \exp\left\{\sum_{k=1}^{\infty} \oint_C \alpha_k(z)dz\right\} ,$$

where the branch of $\sqrt{q(x)}$ is chosen in accordance with (9). We again obtain equation (10) for the eigenvalues.

3. The Problem on the Whole Axis with Many Turning Points.

3.1 Example. Let

$$q(x) = \prod_{j=1}^{2N}(x - x_j) , \quad x_1 < x_2 < \ldots < x_{2N} .$$

Equation (1) has $2N$ simple real turning points x_j and N finite Stokes lines $\tilde{l}_j = [x_{2j-1}, x_{2j}]$ lying on the real axis. From the turning point x_j there arise two infinite Stokes lines l_j and l_j^*, where l_j lies in the half-plane $\operatorname{Im} z > 0$. The Stokes lines decompose the complex z-plane into $2N + 2$ domains of half-plane type and $N - 1$ domains of band type.

Domains of *half-plane type*: a) D^-, $\partial D^- = l_1 \cup l_1^*$; b) D_j, $\partial D_j \supset l_{2j-1} \cup \tilde{l}_j \cup l_{2j}$, $1 \leqslant j \leqslant N$; c) D^+, $\partial D^+ = l_{2N} \cup l_{2N}^*$; d) D_j^*, $1 \leqslant j \leqslant N$.

Domains of *band type*: G_j, $\partial G_j = l_{2j} \cup l_{2j}^* \cup l_{2j+1} \cup l_{2j+1}^*$. (The case $N = 2$ is shown in Fig. 3. In particular D^+ contains the half-line $x > x_{2N}$, and D^- contains the half-line $x < x_1$.)

Equation (1) has solution $y_2(x, \lambda)$ (see (6)), the properties of which were stated in paragraph 2.1. We will extend the asymptotic behaviour of this solution to the half-line $x < x_1$. We introduce the canonical domains

$$\mathcal{D}_1 = D^- \cup l_1 \cup l_1^* \cup D_1 \cup D_1^*,$$
$$\mathcal{D}_{2N} = D^+ \cup l_{2N} \cup l_{2N}^* \cup D_N \cup D_N^*,$$

so that $\mathcal{D}_1 \supset (-\infty, x_1)$ and $\mathcal{D}_{2N} \supset (x_{2N}, +\infty)$, and these domains are symmetric about the real axis. We put

$$\mathcal{D}_{2N-1} = D_N \cup \tilde{l}_N \cup D_N^* = \mathcal{D}_{2N-1}^*$$

and construct the domains $\mathcal{D}_{2N-3}, \ldots, \mathcal{D}_3$ in precisely the same way. Further, suppose that

$$\mathcal{D}_{2N-3} = D_N \cup G_{N-1} \cup D_{N-1} \cup l_{2N-1} \cup l_{2N-2}.$$

Let us bring in the elementary F.S.S. (u_j, v_j) corresponding to the triple (l_j, x_j, D_j), (u_j^*, v_j^*) corresponding to (l_j^*, x_j, D_j^*), and $(\tilde{u}_j, \tilde{v}_j)$ corresponding to $(x_{2j}, \tilde{l}_j, \mathcal{D}_{2j-1})$. We have (paragraph 2.5)

$$y_2(z, \lambda) = e^{-i\pi/12} v_{2N}(z, \lambda) = e^{-i\pi/12}(au_1 + bv_1).$$

The eigenvalues are determined from the equation $b(\lambda) = 0$. We have

$$\Omega_{2N,1} = \Omega_{2,1} \ldots \Omega_{2N-1,2N-2}\Omega_{2N,2N-1}.$$

The matrices $\Omega_{2j,2j-1}$ have the form (18) where

$$\xi \to \xi_j = \int_{x_{2j-1}}^{x_{2j}} |\sqrt{q(x)}|dx, \quad \Omega_{2j-1,2j-2} = e^{-i\pi/2+\lambda\eta_j} \begin{bmatrix} e^{-2\lambda\eta_j} & 0 \\ 0 & 1 \end{bmatrix},$$

$$\eta_j = \int_{x_{2j-1}}^{x_{2j}} \sqrt{q(x)}dx > 0$$

(see § 3, (15)). Therefore

$$b(\lambda) = c(\lambda)[(\Omega_{2N,2N-1})_{22}(\Omega_{2N-2,2N-3})_{22} \cdots (\Omega_{21})_{22} + O(e^{-2\lambda\eta_0})]$$

as $\lambda \to \infty$, where

$$\eta_0 = \min_{1 \leqslant j \leqslant N-1} \eta_j,$$

and $c(\lambda) \neq 0$. Making use of (18) and restricting ourselves to the principal terms, we arrive at the equation for the eigenvalues

$$\prod_{j=1}^{N} [\cos(2\lambda\xi_j) + O(\lambda^{-1})] = O(e^{-2\lambda\eta_0}).$$

Therefore there are N series $\{\lambda_n^{(k)}\}$ of eigenvalues

$$\lambda_n^{(k)} = (n + 1/2)\pi\xi_k^{-1} + O(n^{-1}), \quad n \to \infty, \tag{19}$$

where $k = 1, 2, \ldots, N$. For each of the series there is an asymptotic expansion of the form (5), where $C = C_j$ is a closed contour going around the interval $[x_{2j-1}, x_{2j}]$. Amongst these series there can be those that have the same asymptotic expansion.

3.2 The General Case. Suppose the polynomial $q(x)$ has real zeros $x_1 < \ldots < x_{2N}$ and they are all simple. Then formula (19) is true. The method of proof is the same as in paragraph 3.1; the difference is only in the choice of canonical domains. Namely, for D^- we take the domain described in § 3, paragraph 3.5, No. 6 and we choose D^+ similarly. For G_j we take a domain not of band type, but a domain whose boundary contains the Stokes lines l_{2j}, $l_{2j}^\star$, l_{2j+1}, $l_{2j+1}^\star$ and which is mapped onto the band $a < \operatorname{Re} S < b$ with a finite number of vertical cuts (§ 3, paragraph 3.5, No. 8). The domains D_j of half-plane type, such that $\partial D_j \supset \tilde{l}_j$, are constructed in the same way as in paragraph 2.5.

The asymptotic expansions (19) are true when $q(x) \in C^\infty(\mathbb{R})$ is holomorphic in a complex neighbourhood of the interval $[x_1, x_2]$ and the integrals

$$\int_{-\infty} |\alpha_k(t)| dt, \quad \int^{+\infty} |\alpha_k(t)| dt, \quad k = 0, 1, 2, \ldots,$$

converge (see paragraph 2.1). The remarks stated in paragraph 2.1 remain in force.

4. $q(x)$ is an Even Function. In this case any eigenfunction is either even or odd. Suppose that $q(x)$ has precisely four real turning points $x_1 < x_2 < 0 < x_3 < x_4$, all simple ($x_1 = -x_4$, $x_2 = -x_3$), and suppose that $q(x)$ is a polynomial for simplicity. Then the spectrum consists of two series of eigenvalues $\{\lambda_n^+\}, \{\lambda_n^-\}$. Their asymptotic expansion in powers of n^{-1} are identical and have the form (11), where the contour C bypasses the interval $[x_3, x_4]$, so that

$$\lambda_n^+ - \lambda_n^- = O(n^{-\infty}), \quad n \to \infty.$$

It turns out that the difference $\lambda_n^+ - \lambda_n^-$ is exponentially small as $n \to \infty$, and this leads to the exponentially small splitting of the spectrum. We denote

$$\xi_0 = \int_{x_1}^{x_2} |\sqrt{q(x)}| dt, \quad \eta_0 = \int_{x_2}^{x_3} \sqrt{q(x)} dx > 0.$$

The principal asymptotic term of the spectrum is

$$\lambda_n^\pm = (n + 1/2)\pi\xi_0^{-1} + O(n^{-1}).$$

There is also the Landau-Lifschitz formula

$$|\lambda_n^+ - \lambda_n^-| = \xi_0^{-1} \exp\{-\eta_0 \lambda_n^\pm\}[1 + O(n^{-1})]. \tag{20}$$

The proof of this formula was given in [Fedoryuk 5] and is based on the following fact. If $q(z)$ is an even function then the turning points and the Stokes lines are distributed symmetrically relative to the origin of coordinates. Let $(u(z), v(z))$ be the elementary F.S.S. corresponding to the triple (l, z_0, D). Then $(u(-z), v(-z))$ is the elementary F.S.S. corresponding to $(-l, -z_0, -D)$. Let us consider an example: $q(x) = (x - x_1)(x - x_2)(x - x_3)(x - x_4)$. The proof of (20) for the general case differs only in insignificant technical details. The Stokes lines are shown in Fig. 3. Let $y_2(x, \lambda)$ be a solution of the form (6), let canonical domains $D_4 \supset l_4$ and $D_0 \supset l_0$ be chosen in the same way as in paragraph 2.5 and the canonical domain $D_3 \supset l_3$ is taken as $D_3 = D_0^+ \cup l_3 \cup l_{2'} \cup (-D_0^+) \cup G$, where G is the domain bounded by l_2, $l_{2'}$, l_3, $l_{3'}$. Then $D_3 = -D_3$.

We will extend the asymptotic expansion of y_2 from l_4 not to l_1 but to $l_{1'}$. We take the elementary F.S.S. $(u_j(z), v_j(z))$ corresponding to the triple (l_j, x_j, D_j) for $j = 4, 3, 2', 1'$ $(x_{j'} = x_j)$. Then

$$\Omega_{41'} = \Omega_{2'1'} \Omega_{32'} \Omega_{43}.$$

We have $(u_{j'}(z), v_{j'}(z)) = (u_k(-z), v_k(-z))$, if $j = 1$, $k = 4$, and $j = 2$, $k = 3$. Therefore

$$\Omega_{2'1'} = \Omega_{43}^{-1}.$$

We have

$$\Omega_{43} = e^{-i\pi/3} \begin{bmatrix} 0 & a \\ b & c \end{bmatrix}, \quad \Omega_{32'} = \begin{bmatrix} 0 & e^{-\lambda\eta_0} \\ e^{\lambda\eta_0} & 0 \end{bmatrix},$$

where a, b, c have forms similar to (18). Since $y_2(x, \lambda) = e^{-i\pi/12} v_1(x, \lambda)$ and $u_{1'}(x, \lambda)$ is exponentially increasing as $x \to -\infty$, the eigenvalues are determined from the equation $(\Omega_{41'})_{12} = 0$, which reduces to the form $c^2 = a^2 \exp(-2\lambda\eta_0)$. Consequently,

$$c(\lambda) = \pm a(\lambda) e^{-\lambda\eta_0}. \tag{21}$$

We denote the series of roots corresponding to the plus (minus) sign by $\lambda_n^+(\lambda_n^-)$. Since

$$a(\lambda_n^\pm) = i(-1)^{n+1}[1 + O(n^{-1})], \quad c'(\lambda_n^\pm) = 2i\lambda_n^\pm(-1)^{n+1},$$

then (20) follows from (21).

The case where $q(x)$ is even and has eight real zeros was studied in [Fedoryuk 5].

5. The Problem on the Half-Line. We consider the problem of the eigenvalues on the half-line $x > 0$ for equation (1) with boundary conditions

$$y(0,\ \lambda) = 0\,, \quad y(\infty,\ \lambda) = 0\,.$$

We will restrict ourselves to the simplest case where $q(x)$ is a polynomial having, for $x \geqslant 0$, precisely one simple turning point $x_0 > 0$. As in paragraph 2.1, the eigenfunction $y(x,\lambda) = (\text{const.})\ y_2(x,\lambda)$ and the eigenvalues are determined from the equation $y_2(0,\lambda) = 0$. Let x_1 be fixed, with $0 < x_1 < x_0$. Then for $0 \leqslant x \leqslant x_1$ the asymptotic behaviour of y_2 has the form (17) where x_2 must be replaced by x_0. There is therefore the asymptotic expansion, similar to (10),

$$\lambda_n \xi_0 = n\pi + \frac{\pi}{4} + \sum_{k=1}^{\infty} \beta_k \lambda_n^{-k}\,, \quad n \to \infty\,,$$

where

$$\xi_0 = \int_0^{x_0} |\sqrt{q(x)}|dx\,, \quad \beta_k = \frac{i}{2}(-1)^k \oint_C \alpha_k(z)dz\,.$$

Here C is a simple closed contour starting and finishing at $z = 0$, going around the interval $0 \leqslant x \leqslant x_0$ and positively oriented. The branch of $\sqrt{q(z)}$ for $z \in C$ is chosen in accordance with (9).

In precisely in the same way, the asymptotic behaviour of the discrete spectrum is calculated with a boundary condition of the form (3) and for an arbitrary fininte number of turning points on the half-line $x > 0$.

§ 6. Asymptotic Behaviour of the Discrete Spectrum of the Operator $-y'' + \lambda^2 q(x)y$. Non-Self-Adjoint Problems

1. Statement of the Problem. We consider the equation

$$y'' - \lambda^2 q(x)y = 0\,, \tag{1}$$

where $q(x)$ is a complex-valued function and we consider the problem of the eigenvalues on the whole line or on the half-line $[0,+\infty)$. The statement of these problems is the same as in § 5. We consider the case where the spectrum is purely discrete, and study the asymptotic behaviour of the eigenvalues $\{\lambda_n\}$ as $|\lambda_n| \to \infty$. In addition, we assume that $q(z)$ is analytic is a neighbourhood of the real line or half-line.

This problem is non-self-adjoint, since $q(x)$ takes complex values. We point out that no results are known concerning the asymptotic behaviour of the spectrum of (1) where $q(x)$ is a complex-valued function.

We consider, for example, the eigenvalue problem

$$y(0) = 0, \quad y(1) = 0$$

on the interval $[0, 1]$ for equation (1). Let $q(x) \in C^\infty(I)$ and let $\arg q(x) \not\equiv$ const. Denote by $y(x, \lambda)$ the solution of (1) with the Cauchy condition $y(0, \lambda) = 0$, $y'(0, \lambda) = 1$; then the eigenvalues λ_n are the roots of the equation $y(1, \lambda) = 0$. The function $y(1, \lambda)$ is entire, of first-order growth and can be shown to have infinitely many zeros $\{\lambda_n\}$. In the case where $q(x)$ is real-valued, the numbers λ_n^2 are real, and we can find their asymptotic behaviour with the assumption that $q(x)$ has a finite number of zeros on the interval $[0, 1]$. However, if $\arg q(x)$ is not constant, the asymptotic behaviour of the eigenvalues is unknown. It is even unknown whether they are grouped near a finite number of rays in the complex λ-plane for $|\lambda_n| \gg 1$. It is entirely probable that this is not so and that there are simply no asymptotic formulae. To some extent this hypothesis is confirmed in this paragraph. It turns out that the distribution of eigenvalues with large indices in the complex λ-plane and their asymptotic formulae are determined by the position of the Stokes lines relative to the real axis; for polynomial $q(x)$ the corresponding conditions are necessary and sufficient.

2. The Problem on the Whole Line. Let $q(z)$ be a polynomial of degree $m \geqslant 2$:

$$q(z) = a_0 z^m + a_1 z^{m-1} + \ldots + a_m . \tag{2}$$

In § 5 it was pointed out that the asymptotic behaviour of the spectrum for the self-adjoint problem is connected with the existence of a finite Stokes line. If the polynomial $q(z)$ does not have the form $(az + b)^m$, then there exists a finite number of values $\arg \lambda = \psi_k$ for which there is a finite Stokes line. However, for such a Stokes line to generate an infinite series of eigenvalues concentrated near the ray $\arg \lambda = \psi_k$, some conditions on the topology of the Stokes line must be satisfied.

2.1 The Stokes Complex Joining $+\infty$ with $-\infty$. For the definition of the Stokes complex see § 1. We put $\arg \lambda = \psi$, $\arg a_0 = \phi_0$ and suppose that $\exp\{i(2\psi + \phi_0)\} \notin (-\infty, 0)$ if m is even, and $\exp\{i(2\psi + \phi_0)\} \notin (-\infty, +\infty)$ if m is odd. Then there are domains $D^-(\lambda)$ and $D^+(\lambda)$ of half-plane type such that $D^-(\lambda)$ contains a half-line of the form $(-\infty, a]$ and $D^+(\lambda)$ contains a half-line of the form $[b, +\infty)$.

By definition the Stokes complex $K(\lambda)$ captures $+\infty(-\infty)$ if one of the domains into which $K(\lambda)$ splits the complex z-plane contains $D^+(\lambda)$, $(D^-(\lambda))$. The Stokes complex $K(\lambda)$ joins $+\infty$ with $-\infty$ if it captures both $+\infty$ and $-\infty$ (Fig. 20). We introduce the conditions:

(1) all the zeros of the polynomial $q(z)$ are simple;

(2) for each fixed $\arg \lambda$, equation (1) has no more than one finite Stokes line.

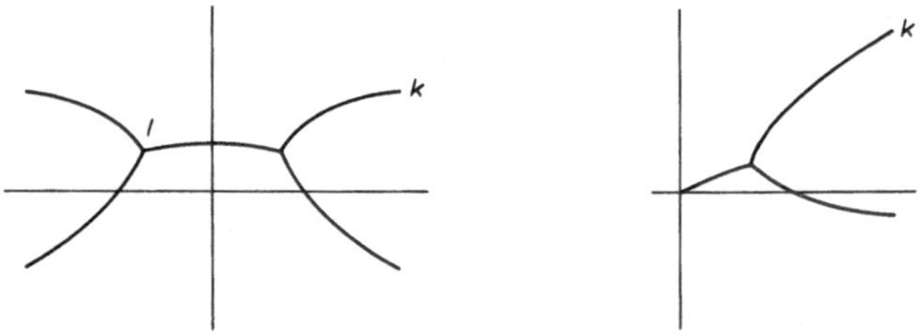

Fig. 20

It is not difficult to see that this is a typical situation. The eigenfunction satisfies the boundary conditions

$$y(-\infty,\ \lambda) = 0\,, \quad y(+\infty,\ \lambda) = 0\,. \tag{3}$$

We fix $\arg\lambda = \psi_0$ and suppose that $\exp\{i(2\psi_0 + \phi_0)\}$ is a non-negative number. Then for each fixed $\lambda = \rho\exp(i\psi_0)$, $\rho > 0$, equation (1) has a solution such that

$$y_2(x,\ \lambda) \sim q^{-1/4}(x)\exp\{-\lambda S(x_0,\ x)\}\,, \quad x \to +\infty\,. \tag{4}$$

Here $x_0 \gg 1$ and $\mathrm{Re}[\lambda S(x_0, x)] > 0$ for $x > x_0$, so that $y_2(+\infty, \lambda) = 0$. A second linearly independent solution exponentially increases as $x \to +\infty$, so that the eigenvalues are determined from the equation $y_2(-\infty, \lambda) = 0$. Let S_δ be the sector $|\arg\lambda - \psi_0| \leqslant \delta \ll 1$. Then (4) is applicable as $x \to +\infty$ and for $\lambda \in S_\delta$ fixed, $\lambda \neq 0$. This asymptotic behaviour is dual: it is applicable as $\lambda \to \infty$, $\lambda \in S_\delta$ uniformly in $x \geqslant x_0$.

Let ψ_0 be such that there is no Stokes complex joining $+\infty$ with $-\infty$. Then there can only be a finite number of spectral points in S_δ for $\delta \ll 1$. We will show this in the case where there is no finite Stokes line. Then $y_2(z, \lambda)$ decreases exponentially for $\lambda \in S_\delta$, $|\lambda| \geqslant \lambda_0 \gg 1$, as $|z| \to \infty$ in $D^+(\lambda)$, and increases exponentially in all other domains of half-plane type including also the domain $D^-(\lambda)$ (§ 3). The general case is considered in [Fedoryuk 5].

Let $z_\pm$ be turning points lying on $\partial D^\pm$. We denote

$$\xi_0 = \int_{z_-}^{z_+} \sqrt{q(z)}dz\,,$$

where the contour of integration does not pass through any other turning points. Then there is a unique value $\arg\lambda = \psi_0$ such that $\mathrm{Re}\,(\xi_0\exp(i\psi_0)) = 0$; in addition, there is a finite Stokes line l joining z_- with z_+. Suppose that $\exp\{i(2\psi_0 + \phi_0)\}$ is non-negative for m even and is not real for n odd. Then the Stokes complex containing l joins $+\infty$ with $-\infty$.

The problem (1), (3) has an infinite series of eigenvalues $\{\lambda_n\}$ with asymptotic expansion

$$\lambda_n \oint_C \sqrt{q(z)}dz + \sum_{k=1}^{\infty}(-\lambda_n)^k \oint_C \alpha_k(z)dz = \pi i(2n+1)\,, \quad n \to \infty\,. \tag{5}$$

Here C is a simple closed contour, going around the line l. It is clear that there is a series $\{-\lambda_n\}$. Moreover, the problem (1), (3) can have a finite number of points of the discrete spectrum. The proof can be derived using the same methods as in § 2, paragraphs 2.1, 2.5.

The principal asymptotic term has the form

$$\lambda_n = \pi i(2n+1)\left[\oint_C \sqrt{q(z)}dz\right]^{-1} + O(n^{-1})\,.$$

We note that $\oint_C \sqrt{q(z)}dz$ is one of the periods of the Abel integral $\int \sqrt{q(z)}dz$.

As in § 5 we can calculate the asymptotic behaviour of the eigenfunctions and obtain the asymptotic expansion for λ_n in powers of n^{-1}. The eigenfunction $y(x, \lambda_n)$ has n zeros situated close to the line l.

These results can be generalized to the case where there are $k(>1)$ Stokes complexes, joining $+\infty$ and $-\infty$ as for the case considered in § 5, paragraph 5.

2.2 The Problem on the Half-Line. We consider the eigenvalue problem for equation (1) on the half-line $[0, \infty)$ with boundary condition at zero:

$$ay(0,\ \lambda) + by'(0,\ \lambda) = 0\,, \tag{6}$$

where $(a, b) \neq (0, 0)$. Let $q(z)$ be a polynomial of the form (2), where $n \geqslant 1$. If $\exp\{i(2\psi_0 + \phi_0)\} \notin (-\infty, 0]$ then there is a domain $D^+(\lambda)$ of half-plane type containing a half-line of the form (x_0, ∞), $x_0 > 0$. From the definition the Stokes complex, $D^+(\lambda)$ connects 0 and $+\infty$ if

1) there exists a domain $D(\lambda) \supset D^+(\lambda)$ such that $\partial D(\lambda) \subset K(\lambda)$;
2) there exists a Stokes line $l(\lambda) \in \partial D(\lambda)$ passing through $z = 0$.

The simplest possibility is represented in Fig. 20.

Suppose that $q(z)$ satisfies conditions 1), 2) or paragraph 2.1. As in paragraph 2.1 we can prove that if for $\arg \lambda = \psi_0$ there is no Stokes complex joining 0 and ∞, then there can only be a finite number of eigenvalues in the sector $S_\delta : |\arg \lambda - \psi_0| \leqslant \delta \ll 1$. Suppose that for $\arg \lambda = \psi_0$ such a Stokes complex exists, and we will restrict ourselves to the boundary condition $y(0, \lambda) = 0$. Then there is an infinite series of eigenvalues $\{\lambda_n\}$ with asymptotic behaviour given by formula (22) of § 5, where the contour C does not go around the interval $[0, x_0)$, but instead around the Stokes line l.

The above results carry over to this case where $q(z)$ is holomorphic in a neighbourhood of the real line or of the half-line $[0, +\infty)$.

3. The Sturm-Liouville Equation. We consider the equation

$$-y'' + q(x)y = \lambda y, \tag{7}$$

where $q(z)$ is a polynomial of the form (2) with complex coefficients. We investigate the eigenvalue problem on the whole line and on the half-line. Let $a_0 = \rho_0 \exp(i\theta)$, $\rho_0 > 0$, and we assume that: 1) $|\theta| < \pi$; 2) $\theta \neq 0$ if m is odd for the problem on the half-line. Then the spectrum for the problem on the whole line and on the half line is purely discrete.

With the change of variable

$$x = \varepsilon^{-1}\zeta, \quad \varepsilon = |\lambda|^{-1/m}$$

equation (7) is reduced to a form similar to (1):

$$y'' - \mu^2 Q(\zeta, \varepsilon)y = 0 \tag{8}$$

$$\mu = |\lambda|^{1/2+1/m}, \quad Q(\zeta, \varepsilon) = a_0\zeta^m - e^{i\phi} + \sum_{k=1}^{m-1} a_k \varepsilon^k \zeta^{m-k}.$$

Here $\phi = \arg\lambda$, $-\pi < \phi \leqslant \pi$. It is clear that for small $|\varepsilon|$ the structure of the turning points of (8) is roughly the same as for $\varepsilon = 0$, that is, as for the equation

$$w'' - \mu^2(a_0\zeta^m - e^{i\phi})w = 0. \tag{9}$$

3.1 Topology of the Stokes Lines for (9). The turning points are

$$\zeta_k(\phi) = \rho_0^{-1/m} e^{i\psi_k}, \quad \psi_k = \frac{\phi - \theta + 2k\pi}{m}, \quad 0 \leqslant k \leqslant m-1,$$

and they are all simple and lie on the circle $|\zeta| = \rho_0^{-1/m}$. Denote

$$S(a,b) = \int_a^b \sqrt{a_0 t^m - e^{i\phi}}\, dt.$$

Let C^0 be the complex ζ-plane with cuts along the rays $\gamma_k : \zeta = \rho\exp(i\psi_k)$, $\rho_0^{-1/m} \leqslant \rho < \infty$, which emanate from the turning points and let $\sum_k$ be the sector $\psi_{k-1} < \arg\zeta < \psi_k$. We will index the turning points, rays, etc. mod m. The branch of the function $S = S(0,z)$ in C^0 is normalized by the condition

$$\sqrt{a_0 t^m - e^{i\phi}}|_{t=0} = ie^{i\phi/2}.$$

Then $S(0,z)$ is expressed in terms of the Christoffel-Schwarz integral, from which it follows that S is a one-to-one map of $\sum_k$ onto a domain $\tilde{\sum}_k$ in the complex S-plane. The boundary of this domain is piecewise linear, consisting of intervals $\tilde{\Gamma}_j = [0, P_j]$ and rays $\tilde{\Gamma}_{j'} = [P_j, \infty)$, $j = k-1, k$. The angle between Γ_j and $\tilde{\Gamma}_{j'}$ at P_j is $3\pi/2$ (the angle is taken inside $\tilde{\sum}_k$). Here P_j is the image of the turning point $\zeta_j(\phi)$:

$$P_j = c_0 \rho_0^{-1/m} e^{i\phi/2} \zeta_k(\phi), \quad \int_0^1 \sqrt{1-t^m}\, dt = \frac{\sqrt{\pi}}{2m} \frac{\Gamma(1/m)}{\Gamma(3/2+1/m)}.$$

We fix ϕ and show that, if Γ_k is not a Stokes line, then Γ_k is contained in some domain D of half-plane type and ∂D contains Stokes lines emanating from ζ_k.

Let Γ_k^+ and Γ_k^- be the sides of the cut Γ_k, lying in $\sum_{k+1}$ and $\sum_k$ respectively.

Suppose that the line l : Re S = Re P_k lies in $\tilde{\sum}_k$. Then $\tilde{\sum}_k$ contains a half-plane $\tilde{D}$ of the form Re $S >$ Re P_k or Re $S <$ Re P_k. Its pre-image is the required domain D, and the pre-image of l consists of the two Stokes lines forming ∂D.

Suppose $\tilde{\sum}_k$ does not contain l. Then it contains one of the vertical rays with initial point P_k, and Re S = Re P_k on the ray. We denote the ray by l', and the pre-image of l' is a Stokes line L'.

Let D be the domain bounded by the Stokes line and the ray L'. There is either one Stokes line L^+ starting at ζ_k lying entirely inside $\sum_k$, or two of them. We denote by D^+ the domain bounded by Γ_k^+ and L^+; if there are two Stokes lines then we choose L^+ as the one for which D^+ does not contain a Stokes line. Then the domain $D = D^+ \cup D^- \cup \Gamma_k$ contains the ray Γ_k and has no turning points; D is the desired domain.

3.2 The Stokes Complex. From 1) it follows that there exists a domain $D^+(\phi)$ of half-plane type which contains a half-line of the form $[a, +\infty)$. The bounding rays of the sector S_+ given by $\psi_- < \arg \psi < \psi_+$, where $\psi_\pm = (\pm\pi - \theta)/(m+2)$, are the asymptotes of the Stokes lines bounding $D^+(\phi)$. We now show that the turning point $\zeta_0(\phi)$ for arbitrary ϕ lies in $\partial D^+(\phi)$.

Suppose that $\psi_- < \psi_0 < \psi$; then the ray $\Gamma_0(\phi)$ lies in S_+. On the strength of 3.1 this ray is contained in a domain of half-plane type, which in this case must be $D^+(\phi)$. The assertion is therefore proved for $|\alpha| < \pi$, $\alpha = [(m+2)\phi - 2\theta]/m$. Suppose that $\alpha \in [\pi, \pi + 2\pi/m]$; then $S_+ \subset \sum_0$ and there are no turning points in S_+. Let us consider the image $\tilde{\sum}_0$ of $\sum$. We have

$$\arg P_0 = \pi/2 + \alpha \geqslant \pi, \quad \arg P_{-1} \arg P_0 - 2\pi/m,$$

so that Re $P_0 \leqslant$ Re $P_{-1} < 0$ and therefore $\tilde{\sum}_0$ contains the half-plane $\prod$: Re $S <$ Re P_0. Since $\arg S(0,x) \to \pi + \theta/2$ as $x \to +\infty$, we have Re $S(0,x) \to -\infty$ and the image of $[0,+\infty)$ for $a \gg 1$ is contained in $\prod$. Consequently the pre-image of $\prod$ is $D^+(\phi)$.

Because of conditions 1), 2) there is a domain $D^-(\phi)$ of half-plane type containing a half-line of the form $(-\infty, -a]$. If m is even then $\zeta_{m/2}(\phi) \in \partial D^-(\phi)$; if m is odd then $\zeta_{(m+1)/2}(\phi) \in \partial D^-(\phi)$ for $0 < \theta < \pi$ and

$\zeta_{(m-1)/2}(\phi) \in \partial D^-(\phi)$ for $-\pi < \theta < 0$. This follows from the above arguments.

Let m be even; then the Stokes complex joining $-\infty$ with $+\infty$ must contain the turning points $\zeta_0(\phi)$ and $\zeta_{m/2}(\phi)$. These points must lie on a finite Stokes line and, in particular, the integral $S(\zeta_0(\phi), \zeta_{m/2}(\phi))$ must be purely imaginary. It follows from this that for m even, $+\infty$ and $-\infty$ are connected if and only if

$$\phi = 2\theta/(m+2). \tag{10}$$

For m odd, $+\infty$ and $-\infty$ are connected if and only if

$$\phi = (2\theta - \pi)/(m+2), \quad 0 < \theta < \pi; \quad \phi = (2\theta + \pi)/(m+2), \quad -\pi < \theta < 0. \tag{11}$$

The points 0 and $+\infty$ are connected if and only if ϕ has the form (10) both for m even and m odd.

3.3 Asymptotic Behaviour of the Eigenvalues. The turning points of (7) have the form

$$\zeta_k(\phi, \varepsilon) = \zeta_k(\phi)\left[1 + \sum_{j=1}^{\infty} c_{jk}(\phi)\varepsilon^j\right],$$

where the series converges for small $|\varepsilon|$. The Stokes complex joining $+\infty$ and $-\infty$ (or 0 and $+\infty$) exists for $\phi = \phi_0(\varepsilon)$, where $\phi_0(\varepsilon)$ is a smooth function for small $\varepsilon \geqslant 0$, $\phi_0(0) = \phi_0$ and ϕ_0 has the form (10) or (11). The asymptotic behaviour of the eigenvalues of the problem on the whole line is given by formula (5), where λ must be replaced by μ and q by $Q(\zeta, \varepsilon)$. The contour C contains the Stokes line joining the points $\zeta_0(\phi_0)$ and $\zeta_k(\phi_0)$, where $k = m/2$ for m even, $k = (m+1)/2$ for m odd and $0 < \theta < \pi$, while $k = (m-1)/2$ for m odd and $-\pi < \theta < 0$. We have

$$\left|\oint_C \sqrt{a_0 t^m - e^{i\phi_0}}\, dt\right| = |\mathrm{Im}\,(P_0 - P_k)|,$$

and the asymptotic behaviour of the eigenvalues has the form

$$\lambda_n = A_m c_m e^{i\phi_0} n^{2m/(m+2)}\left[1 + \sum_{k=1}^{\infty} \alpha_k n^{-2k/(m+2)}\right].$$

Here ϕ_0 has the form (10) or (11),

$$A = |a_0|^{1/(m+2)}\left[\frac{\sqrt{\pi}2m\Gamma(3/2 + 1/m)}{\Gamma(1/m)c_m}\right]^{(m+2)/(2m)}$$

and $c_m = 1$ for m even, while $c_m = \cos\pi/(2m)$ for m odd.

§ 7. The Eigenvalue Problem with Regular Singular Points

1. Statement of the Problem. We consider the equation

$$(1-z^2)w'' + p(z)w' + [\lambda q(z) + r(z)]w = 0\,, \tag{1}$$

where the functions $p(z)$, $q(z)$, $r(z)$ are holomorphic in a simply-connected domain D which contains $I = [-1, 1]$. Equation (1) has two singular points $z = +1$ and $z = -1$, both regular, with characteristic exponents $(0, \rho_+)$ and $(0, \rho_-)$, where $\rho_\pm = 1 \pm p(\pm 1)/2$. The point $z = 1$ is non-singular if and only if $p(1) = q(1) = r(1) = 0$. We will exclude this case and the analogous one for $z = -1$.

Equations of the form (1) include the Legendre equation, the hypergeometric equation, the Mathieu equation, the equation for sketched angular spheroidal functions and angular Coulomb spheroidal functions with index $m = 0$ and so on. For the equations just mentioned the numbers $\rho_\pm$ are real.

Let U be a small neighbourhood of $z = 1$. Then in U there exists a F.S.S. $(w_1^+(z, \lambda),\ w_2^+(z, \lambda))$ of the following form (Chap. 1, § 2):

1) ρ_+ is not an integer. The solution w_1^+ is holomorphic in U, and

$$w_2^+(z,\ \lambda) = (1-z)^{\rho_+}\tilde{w}_2^+(z,\ \lambda)\,, \tag{2}$$

where $\tilde{w}_2^+$ is holomorphic in U. We normalize $w_1^+(1, \lambda) = \tilde{w}_2^+(1, \lambda) = 1$; then $w_1^+(z, \lambda)$, $\tilde{w}_2^+(z, \lambda)$ are holomorphic in the aggregate of the variables (z, λ) for $z \in U, \lambda \in \mathbb{C}$.

2) $\rho_+ < 0$ is an integer. The solution w_1^+ is the same as in 1), and

$$w_2^+(z,\ \lambda) = \theta_+(\lambda)w_1^+(z, \lambda)\ln(1-z) + (1-z)^{\rho_+}\tilde{w}_2^+(z,\ \lambda)\,, \tag{3}$$

where $\tilde{w}_2^+$ is holomorphic U and non-zero for $z = 1$. The coefficient $\theta_+(\lambda)$ is a polynomial in λ.

3) $\rho_+ \geqslant 0$ is an integer. The solution w_1^+ has the form (2) and is holomorphic in U, and

$$\tilde{w}_2^+(z,\ \lambda) = \tilde{w}_3^+(z,\ \lambda) + \theta_+(\lambda)w_1^+(z,\ \lambda)\ln(1-z)\,, \tag{4}$$

where $\tilde{w}_3^+$ is holomorphic in U and $\theta_+(\lambda)$ is a polynomial in λ.

In each of these cases equation (1) has a solution that is holomorphic at $z = 1$. There is an analogous F.S.S. $(w_1^-(z, \lambda),\ w_2^-(z, \lambda))$ in a neighbourhood of $z = -1$.

We call a number λ an *eigenvalue* of equation (1) (on the interval $(-1, 1)$) if there exists a solution $w(z, \lambda) \not\equiv 0$ that is holomorphic at $z = 1$ and $z = -1$. This solution is called an *eigenfunction*, and it follows from the analytic theory of differential equations that an eigenfunction is holomorphic in a domain D. We denote the set of all eigenvalues by $\sum$.

The statement of the problem on eigenvalues given here differs from the classical one where boundary conditions are given at $z = 1$ and $z = -1$. For instance, for the Legendre equation

$$(1 - z^2)w'' - 2zw' + \lambda w = 0$$

one imposes the boundary condition

$$|w(-1)| < \infty, \quad |w(1)| < \infty.$$

For this equation $\rho_- = \rho_+ = 0$ (case 3)), so that one of the solutions is holomorphic at $z = 1$, and the second has a logarithmic singularity (similarly for $z = -1$). In the general case the boundary conditions for the eigenvalue problem can be put as follows:

$$|w^{(n_-)}(-1)| < \infty, \quad |w^{(n_+)}(1)| < \infty,$$

where the numbers $n_\pm$ can be expressed in terms of $\rho_\pm$.

The spectrum $\sum$ is the set of zeros of some entire function of growth order not greater than 1/2. Therefore

$$\sum |\lambda_n|^{-1/2+\varepsilon} < \infty,$$

where the sum is taken over all $\lambda_n \in \sum \backslash \{0\}$ and $\varepsilon > 0$ is arbitrary. More precise information on the behaviour of λ_n as $n \to \infty$ clearly cannot be obtained in the general case. In this paragraph the asymptotic behaviour of the λ_n is found when $q(x) > 0$ for $x \in I$. This condition is satisfied for all the classes of equations mentioned above.

2. False Spectrum. If $\lambda \in \sum$, the corresponding eigenfunction of (1) is single-valued in D, and therefore also in D with a cut along the interval I, that is, in $D \backslash I$. Let us now pose the following problem. Suppose that (1) has a solution which is single-valued in $D \backslash I$ for some λ. Is λ an eigenvalue of (1)? It turns out that the following cases are possible:

I. ρ_- and ρ_+ are not integers. Here either $\lambda \in \sum$ or the solutions w_2^-, w_2^+ are linearly dependent, that is

$$w_2^+(z, \lambda) \equiv C(\lambda) w_2^-(z, \lambda).$$

II. ρ_- is an integer and ρ_+ is not an integer. Here either $\lambda \in \sum$ or the solutions w_2^-, w_1^+ are linearly dependent, that is

$$w_2^+(z, \lambda) \equiv C(\lambda) w_1^-(z, \lambda).$$

III. ρ_- and ρ_+ are integers. In this case $\lambda \in \sum$.

We denote by $\tilde{\sum}$ the set of all $\lambda \notin \sum$ for which (1) has a solution which is single-valued in $D \backslash I$. We call $\tilde{\sum}$ the *false spectrum*. From I–III it follows that for $\lambda \in \sum \cup \tilde{\sum}$ it is necessary and sufficient that one of the four pairs of canonical solutions $(w_1^+,\ w_1^-)$, $(w_1^+,\ w_2^-)$, (w_2^+, w_1^-), (w_2^+, w_2^-) is linearly dependent.

If $\lambda \in \sum \cup \tilde{\sum}$ then the monodromy group G of (1) in D can be completely described. In particular, this group is solvable and in case III is nilpotent.

3. Asymptotic Behaviour of the Solutions of Equation (1) in $D \backslash I$. We denote

$$\lambda = \mu^2, \quad S(a,\ z) = \int_a^z \sqrt{\frac{q(t)}{1-t^2}} dt,$$

$$f(a,\ z) = \exp\left\{-\frac{1}{2}\int_a^z \frac{p(t)}{1-t^2} dt\right\},$$

$$w_{1,2}^0(z, \mu; a) = \left(\frac{1-z^2}{q(z)}\right)^{-1/4} f(a,\ z)\exp\{\pm i\mu S(a,\ z)\}.$$

As $\mu \to \infty$ equation (1) has an F.A.S. of the form

$$w_{1,2}(z,\mu) = w_{1,2}^0(z,\mu;a)\exp\left\{\sum_{k=1}^{\infty}(\pm\mu)^{-k}\int_a^z y_k(t)dt\right\}, \tag{5}$$

where the + (−) sign is taken for w_1 (w_2). The functions $y_k(z)$ are determined from the recurrence relations

$$y_{k+1}(z) = \frac{i}{2\sqrt{\tilde{q}(z)}}\left[y_k'(z) + \tilde{p}(z)y_k(z) + \sum_{j=0}^{k} y_j(z)y_{k-j}(z)\right],$$

$$y_0(z) = -\frac{\tilde{q}'(z)}{4\tilde{q}(z)} - \frac{\tilde{p}(z)}{2},$$

$$y_1(z) = \frac{i}{2\sqrt{\tilde{q}(z)}}\left[y_0'(z) + \tilde{r}(z) + \left(\frac{\tilde{q}'(z)}{4\tilde{q}(z)}\right)^2 - \frac{\tilde{p}^2(z)}{4}\right], \tag{6}$$

where

$$\tilde{q}(z) = q(z)(1-z^2)^{-1}, \quad \tilde{p}(z) = p(z)(1-z^2)^{-1}, \quad \tilde{r}(z) = r(z)(1-z^2)^{-1}.$$

Throughout what follows we assume that

$$q(x) > 0, \quad -1 \leqslant x \leqslant 1. \tag{7}$$

Since we are only interested in the asymptotic behaviour of the solutions in a neighbourhood of I we can assume that $q(z) \neq 0$ in D. Let Π be the half-band Re $\mu > 0$, $|\text{Im}\ \mu| < A$ where $A > 0$ is fixed. We investigate the asymptotic behaviour of the solutions for $\mu \in \Pi$, $\mu \to \infty$.

3.1 The Stokes line. Let $\lambda > 0$; then the interval $I = [-1, 1]$ is a Stokes line. The level curves Re $S(a, z)$ = const, close to I, are simple closed analytic curves containing I as an interior set. The function $\sqrt{\tilde{q}(z)}$ breaks down in

$D\backslash I$ into two holomorphic branches. We take the branch of the root that is negative on the upper side of I:

$$\sqrt{\tilde{q}(x+i0)} < 0\,, \quad -1 < x < 1\,. \tag{8}$$

Then

$$\sqrt{\tilde{q}(x-i0)} > 0,\ -1 < x < 1\,,$$
$$\sqrt{\tilde{q}(x)} = -i|\sqrt{\tilde{q}(x)}|\,, \quad x > 1\,.$$

For this choice of the branch we have

$$\mathrm{Re}[iS(1,\ z)] < 0\,, \quad z \in D\backslash I\,, \tag{9}$$

where the integral is taken along a path in $D\backslash I$. The level curve Im $S(z,1) = a$ for small $a > 0$ is a simple closed curve that goes around I. We replace D by the domain Im $S(1,z) < a$, and we denote this new domain by D again.

The domain $D\backslash I$ is a domain of ring type, and $S = S(1,z)$ is infinite-valued in this domain. We denote by $\tilde{D}$ the Riemann surface of S, considered in $D\backslash I$. Thus $\tilde{D}$ is the universal covering of $D\backslash I$. Then $S(1,z)$ is single-sheeted in $\tilde{D}$ and maps $\tilde{D}$ onto the band $G\ :\ 0 < \mathrm{Im}\ S < a$. The part of ∂G on which Im $S = 0$ consists of an infinite number of equal intervals $l_0, l_1, l_{-1}, l_2, l_{-2}, \ldots$, where l_0 is the image of the upper side of I, and l_0 has the form $[0, ib]$, $b > 0$.

3.2 The Solution w_1. The solutions of (1) are generally infinite-valued functions in $D\backslash I$ and their Riemann surfaces are $\tilde{D}$. It is therefore necessary to define the notion of a solution accurately. We fix a point $a \in \partial D$, $a > 1$ and assign the germ $w(z,\mu)$ of some solution at this point. Extending this germ analytically to all paths beginning at the point a leads to a multivalued function, each germ of which is a solution of (1). We also denote this function by $w(z,\mu)$. We remove from the band G ε-neighbourhoods of the images of all turning points and we denote the pre-image of the domain so obtained by $\tilde{D}^1_\varepsilon$. For $\mu \in \Pi$, Re $\mu \geqslant a_0 \gg 1$ and for arbitrary $N \geqslant 1$, equation (1) has a solution of the form

$$w_1^N(z,\mu) = w_1^0(z,\mu;a)\exp\left\{\sum_{j=0}^{N}\mu^{-j}\int_a^z y_j(t)dt\right\}[1 + O(\mu^{-N-1})]\,, \tag{10}$$

where the bound for the remainder term is uniform in those z lying in an arbitrary compact set $K \subset \tilde{D}^1_\varepsilon$. The solution w_1^N is holomorphic in $\mu \in \Pi$, with Re $\mu \geqslant a_0 > 0$, for each fixed $z \in \tilde{D}^1_\varepsilon$. The asymptotic expansion (10) can be differentiated in z and μ an arbitrary number of times. In what follows we write w_1 instead of w_1^N.

We consider the branch of $\sqrt[4]{q(z)}$ in a neighbourhood of the point a such that

$$\sqrt[4]{\tilde{q}(a)} = e^{i\pi/4}|\sqrt[4]{\tilde{q}(a)}|\,.$$

Further, we will replace $w_1^0(z,\mu;a)$ by the function

$$w_1^0(z,\ \mu) = [\tilde{q}(z)]^{-1/4} f(a,\ z)\exp\{i\mu S(1,\ z)\}\,,$$

which leads to multiplying w_1 by a constant.

The existence of (10) follows from the fact that an arbitrary point $z \in \tilde{D}_\varepsilon^1$ can be joined to the point a by a canonical path γ. For γ we can take the pre-image of an interval which lies in G_ε and joins the images of the stated points. The initial germ of w_1 for $z = a$ is $w_1(a,\mu) = 1 + 0(\mu^{-N-1})$.

Suppose that $-1 < x < 1$. Let the curve $\alpha^+(x)$ join a and x, and let Im $z > 0$ on this curve. The value of the solution obtained by extending analytically along $\alpha^+(x)$ will be denoted by $w_1(x + i0,\mu)$. If $\alpha^-(x)$ is the curve symmetric to $\alpha^+(x)$ about the real axis, then the value w_1 obtained by extending analytically along $\alpha^-(x)$ will be denoted by $w_1(x - i0,\mu)$. It follows from (10) that for $-1+\delta \leqslant x \leqslant 1-\delta$ $(0 < \delta < 1)$

$$\begin{aligned} w_1(x+i0,\mu) &= -i|\tilde{q}(x)|^{-1/4}[1+O(\mu^{-1})] \\ &\quad \times \exp\left\{-i\mu\int_1^x |\sqrt{\tilde{q}(t)}|dt - \frac{1}{2}\int_{\alpha+}\tilde{p}(t)dt\right\}, \\ w_1(x+i0,\mu) &= |\tilde{q}(x)|^{-1/4}[1+O(\mu^{-1})] \\ &\quad \times \exp\left\{i\mu\int_1^x |\sqrt{\tilde{q}(t)}|dt - \frac{1}{2}\int_{\alpha-}\tilde{p}(t)dt\right\}. \end{aligned} \tag{11}$$

We observe that w_1 decreases exponentially for $\mu \in \Pi$, $\mu \to \infty$ at an arbitrary point $z \in \tilde{D}$, as follows from (9).

3.3 The Solution w_2. We fix a point b lying on the upper side of I and suppose that $B = S(1,b)$, where the integral is taken along $-\alpha^+(b)$. The set of all points $S \in G'_\varepsilon$ which can be joined to the point B by a curve, along which Re (iS) is non-increasing, will be denoted by G_ε^2 and its pre-image by $\tilde{D}_\varepsilon^2$. The projection of $\tilde{D}_\varepsilon^2$ onto the plane $\mathbb{Z}$ is constructed in the following way: from the closure of $D\backslash I$ we must remove neighbourhoods of $z=-1$, $z=1$ and of the lower side of I.

For $\mu \in \Pi$, Re $\mu \geqslant a_0 \gg 1$ and for arbitrary $N \geqslant 1$ equation (1) has a solution w_2^N of the form

$$w_2^N(z,\mu) = w_2^0(z,\mu;a)\exp\left\{\sum_{j=1}^N(-\mu)^{-j}\int_a^z y_j(t)dt\right\}[1+O(\mu^{-N-1})]\,, \tag{12}$$

the bound for the remainder term being uniform for z lying in an arbitrary compact set $K \subset \tilde{D}_\varepsilon^2$. The other properties of w_2^N are the same as those for w_1^N. We write w_2 for w_2^N. As above we replace $w_0^2(z,\mu;a)$ by the function

$$w_2^0(z,\ \mu) = [\tilde{q}(z)]^{-1/4} f(a,\ z)\exp\{-i\mu S(1,\ z)\}\,.$$

We observe that w_2 increases exponentially for $\mu \in \Pi$, $\mu \to \infty$ at each point $z \in \tilde{D}^2 \backslash I$ because of (9). The solutions w_1 and w_2 form a F.S.S. for (1). With the same choice of value for $\tilde{q}^{-1/4}$ (a) as in paragraph 3.2, we have

$$w_2(x+i0,\mu) = -i|\tilde{q}(x)|^{-1/4}[1+O(\mu^{-1})]\exp\left\{i\mu\int_1^x |\sqrt{\tilde{q}(t)}|dt \right. \\ \left. -\frac{1}{2}\int_{a_+}\tilde{p}(t)dt\right\}. \tag{13}$$

However the result given above concerning the asymptotic expansion of w_2 does not allow us to find the value $w_2(x-i0,\mu)$.

3.4 Transition Matrices. We next find the asymptotic behaviour of w_2 at the lower side of the cut, that is, the value $w_2(x-i0,\mu)$. We fix $x \in (-1,1)$ and we let $\alpha^\pm = \alpha^\pm(x)$ be the paths described in paragraph 3.2. Also, let $\gamma^+ = \alpha^+(\alpha^-)^{-1}$, so that γ^+ is a simple closed curve going around $z=1$ in the positive direction. We put $w = (w_1, w_2)^T$; then

$$w(x-i0,\ \mu) = T_1(\mu)w(x+i0,\ \mu).$$

We denote

$$a_\pm = \exp\{2\pi i\rho_\pm\}, \qquad A = \exp\left\{\sum_{k=0}^{\infty}\mu^{-k}\alpha_k\right\}, \\ \alpha_{2k} = -\oint_{\gamma^+} y_{2k}(t)dt, \qquad \alpha_{2k+1} = \left(\int_{\alpha^-}+\int_{\alpha^+}\right) y_{2k+1}(t)dt. \tag{14}$$

For $\mu \in \Pi$ and $\mu \to \infty$ we have

$$t^1_{11}(\mu) = O(\mu^{-\infty}), \quad t^1_{12}(\mu) = A, \\ t^1_{21}(\mu) = -A^{-1}a_+^{-1} + O(\mu^{-\infty}), \quad t^1_{22}(\mu) = 1 + a_+^{-1} + O(\mu^{-\infty}). \tag{15}$$

To prove this, we start with

$$w_1(x-i0,\ \mu) = t^1_{11}w_1(x+i0,\ \mu) + t^1_{12}w_2(x+i0,\ \mu), \\ w_1'(x-i0,\ \mu) = t^1_{11}w_1'(x+i0,\ \mu) + t^1_{12}w_2'(x+i0,\ \mu).$$

Since $\sqrt{\tilde{q}(x-i0)} = -\sqrt{\tilde{q}(x+i0)}$, we have $S(1,x-i0) = -S(1,x+i0)$ and therefore

$$\frac{w_1'(x-i0,\mu)}{w_1(x-i0,\mu)} = \frac{w_2'(x+i0,\mu)}{w_2(x+i0,\mu)} + O(\mu^{-\infty}),$$

from which it follows that $t^1_{11}(\mu) = O(\mu^{-\infty})$.

Replacing $w_1(x-i0,\mu)$ and $w_2(x+i0,\mu)$ by their asymptotic expansions (10), (12) and cancelling by $\exp(i\mu S)$, we arrive at the equation

$$t_{12}^1(\mu) = \exp\left\{\sum_{k=0}^{\infty}\left(\mu^{-k}\int_{\alpha_-} y_k(t)dt - (-\mu)^{-k}\int_{\alpha_-} y_k(t)dt\right)\right\}$$

to within an order term $O(\mu^{-\infty})$. The functions $y_{2k}(z)$ are single-valued in a neighbourhood of $z = 1$, and have a pole there, so that

$$\int_{\alpha^-} y_{2k}(t)dt - \int_{\alpha^+} y_{2k}(t)dt = \oint_{\gamma^+} y_{2k}(t)dt .$$

Further, $y_{2k+1}(z) = \sqrt{\tilde{q}(z)}\tilde{y}_{2k+1}(z)$, where $\tilde{y}_{2k+1}$ is single-valued in a neighbourhood of $z = 1$. Therefore α_{2k+1} does not depend on the choice of $x \in (-1, 1)$. Formula (15) t_{12}^1 is proved in the same way. We have

$$w^+(x - i0,\ \mu) = \tilde{T}_1(\mu)w^+(x + i0,\ \mu)\,,$$

where w^+ is a canonical F.S.S. (paragraph 1). The eigenvalues of $\tilde{T}_1(\mu)$ are 1 and a_+^{-1}, and the eigenvalues of $T_1(\mu)$ are the same since these matrices are similar. The asymptotic expansions for t_{21}^1 and t_{22}^1 follow from Vieta's theorem. Formulae (14), (15) allow us to find the asymptotic expansions of the solutions.

Remark. The formulae (14), (15) for $T_1(\mu)$ are also valid in the general case. Suppose that $p(z)$, $q(z)$, and $r(z)$ are holomorphic at $z = 1$ with $q(1) \neq 0$, and let U be a small neighbourhood of this point. For $\lambda > 0$ there arises one Stokes line l from $z = 1$, which is given by the equation Re $(i\int_1^z \sqrt{\tilde{q}(t)}dt) = 0$. For definiteness we assume that Re $(z - 1) < 0$ for $z \in l \cap U$ and $z \neq 1$, and we choose the branch of $\sqrt{q(z)}$ in $U\backslash l$ so that Re $[iS(1, z)] < 0$ in $U\backslash l$. Then in $U\backslash l$ there exists a F.S.S. $(w_1(z, \lambda),\ w_2(z, \lambda))$ for which (10), (12) are true. Here the line l plays the role of I and, by $z + i0$ and $z - i0$, are understood points lying on the upper and lower sides of l. The asymptotic formulae for $T_1(\mu)$ remain in force.

We fix a point $x+i0$, $-1 < x < 1$, lying on the upper side of I and continue analytically the F.S.S. $w(x+i0, \mu)$ along a simple curve γ^-, beginning at $x+i0$ and ending at $x - i0$, which goes around $z = -1$. Then we obtain the F.S.S. $w^1(x - i0, \mu) \neq w(x - i0, \mu)$. We have

$$w(x + i0, \mu) = T_2(\mu)w^1(x - i0, \mu)\,.$$

In the same way it can be proved that the following asymptotic formulae are valid.

$$t_{11}^2(\mu) = 1 + a_-^{-1}\,, \qquad t_{22}^2(\mu) = 0\,,$$
$$t_{12}^2(\mu) = -ABa_-^{-1}\,, \quad t_{21}^2(\mu) = -(AB)^{-1} \tag{16}$$

for $\mu \in \Pi$ and $\mu \to \infty$, to within terms of order $0(\mu^{-\infty})$. The expressions for a_- and A were given in (14), and

$$B = \exp\left\{ i\mu \oint_\gamma \sqrt{\tilde{q}(t)}dt + \sum_{k=0}^{\infty} \mu^{-k} \oint_\gamma y_k(t)dt \right\}, \tag{17}$$

where γ is a simple closed contour going around I in the positive direction. The principal term of the asymptotic series has the form

$$B = \exp\left\{ -2i\mu \int_{-1}^{1} \left| \sqrt{\frac{q(t)}{1-t^2}} \right| dt + \pi i(\rho_+ + \rho_- - 1) \right\} [1 + O(\mu^{-1})],$$

where the value of the root is positive.

4. Asymptotic Behaviour of the Spectrum. Let $\lambda \in \sum$ and let $w(z, \lambda)$ be an eigenfunction. Then in a small neighbourhood of the point $z = a$, where $a \in D$ and $a > 1$, we have

$$w(z,\ \lambda) = c(\lambda) w^T(z,\ \lambda),$$

where $c(\lambda) = (c_1(\lambda),\ c_2(\lambda))$ and $w(z, \lambda)$ is the F.S.S. $(w_1(z, \lambda),\ w_2(z, \lambda))$. For the number λ to be a point of the spectrum it is necessary and sufficient that w is single-valued in neighbourhoods of the singular points $z = 1$ and $z = -1$, so that we must have

$$c(\mu)T_1(\mu) = c(\mu), \quad c(\mu)T_2(\mu) = c(\mu).$$

From the first of these we find c (to within a multiple):

$$c = (a_+^{-1},\ -A), \tag{18}$$

and from the second we find $B = a_- a_+$. Consequently equation (1) has an infinite sequence of eigenvalues $\lambda_n = \mu_n^2$ with asymptotic expansion

$$\mu_n \oint_\gamma \sqrt{\tilde{q}(z)}dz - i \sum_{k=0}^{\infty} \mu_n^{-k} \oint_\gamma y_k(z)dz = 2\pi n + 2\pi(\rho_+ + \rho_-), \quad n \to \infty.$$

The contour γ is the same as in (16), and the branch $\sqrt{\tilde{q}(z)}$ in $D \backslash I$ is chosen as in (8). We have

$$\oint_\gamma \sqrt{\tilde{q}(z)}dz = -2 \int_{-1}^{1} |\sqrt{\tilde{q}(x)}| dx,$$

$$\oint_\gamma y_0(z)dz = \pi i \left(\frac{p(1) - p(-1)}{2} - 1 \right).$$

Hence, the principal asymptotic term has the form

$$\mu_n = \pi[n - \frac{1}{2} + \frac{1}{4}(p(-1) - p(1))] \left[\int_{-1}^{1} \sqrt{\frac{q(x)}{1-x^2}} dx \right]^{-1} + O(\frac{1}{n^2}) \tag{19}$$

as $n \to \infty$. This formula was obtained in [Dorodnitsyn] under the condition that $p(-1) > 0$, $p(1) < 0$, but without assuming that functions p, q and r are analytic. If p, q and r are real, then the n-th eigenfunction has exactly n zeros on the interval $(-1, 1)$ and formula (19) gives the asymptotic behaviour of the n-th eigenvalue. As in § 5, paragraph 2, we can obtain the asymptotic expansion of the form

$$\mu_n = \sum_{k=-1} a_k n^{-k}$$

as $n \to \infty$. It follows from formula (19) that if the function $p(x)$ is complex-valued, then the eigenvalues λ_n for $n \gg 1$ lie inside some parabola in the complex λ-plane, which contains the half line $(0, \infty)$. From (18) we obtain

$$w_n(x) = a_+^{-1} w_1(x, \lambda_n) - A w_2(x, \lambda_n),$$

which allows us to find the asymptotic expansion of the eigenfunction $w_n(x)$ on any interval of the form $[-1+\delta,\ 1-\delta)]$ with $0 < \delta < 1$. In order to find the asymptotic behaviour of the eigenfunction near to $z = \pm 1$ we can make use of results derived in Chap. 4, § 4.

5. Asymptotic Behaviour of the Pseudospectrum. For the number λ to belong to the set $\sum \cup \tilde{\sum}$ it is necessary and sufficient that

$$c(\mu) T_1(\mu) T_2(\mu) = c(\mu),$$

that is, the matrix $T_1(\mu)T_2(\mu)$ has an eigenvalue equal to unity. It follows from (15), (16) that either $B = a_- a_+$ and then $\lambda \in \sum$, or $B = 1$ and then $\lambda \in \tilde{\sum}$, if $a_- a_+ \neq 1$. When $a_- \neq 1$, $a_+ \neq 1$, $a_- a_+ = 1$ we cannot distinguish the spectrum and the pseudospectrum. From the equation $B = 1$ and (16), we find the asymptotic expansion for the points $\tilde{\lambda}_n = \tilde{\mu}_n^2$ of the pseudospectrum. The principal asymptotic term is

$$\tilde{\mu}_n = \left[\pi n + \frac{\pi}{2}\left(\frac{p(1) - p(-1)}{2} - 1\right)\right]\left[\int_{-1}^{1} \sqrt{\frac{q(x)}{1-x^2}} dx\right]^{-1} + O\left(\frac{1}{n^2}\right).$$

§8. Quasiclassical Approximation in Scattering Problems

1. Statement of the Problem. We consider the equation

$$y'' - \lambda^2 q(x) y = 0, \tag{1}$$

where $\lambda > 0$ is a parameter and $q(x)$ is real-valued. To this form we can reduce the *Schrödinger equation*

$$-\frac{h^2}{2m}\psi'' + [U(x) - E]\psi = 0,$$

describing the one-dimensional motion of a quantum-mechanical particle of mass m with energy E in a potential field with potential energy $U(x)$. We assume the existence of the limits

$$\lim_{x\to\pm\infty} q(x) = q_\pm \neq 0. \tag{2}$$

In quantum mechanics the following three problems are considered:
1. Reflection at a potential barrier of infinite width. In this case

$$q_- < 0, \quad q_+ > 0,$$

with possibly $q_+ = +\infty$ (we will not consider the case where $q_- > 0$, $q_+ < 0$). Suppose that the integral

$$\int_{-\infty} ||\sqrt{q(x)}| - |\sqrt{q_-}||dx \tag{3}$$

converges. Then for any fixed $\lambda > 0$ equation (1) has a F.S.S. $(y_1^-,\ y_2^-)$ such that

$$y_{1,2}^- \sim |q_-|^{-1/4}e^{\pm ik_- x}, \quad x \to -\infty, \quad k_- = \lambda|\sqrt{q_-}|. \tag{4}$$

The solution $y_1^-(y_2^-)$ describes a wave travelling to the right (to the left). If we take the wave travelling to the right then it is reflected at the barrier, that is, the solution has the form

$$y(x,\ \lambda) = y_1^-(x,\ \lambda) + a(\lambda)y_2^-(x,\ \lambda). \tag{5}$$

As $x \to +\infty$ the solution necessarily vanishes:

$$y(+\infty,\ \lambda) = 0. \tag{6}$$

The value $R = |a(\lambda)|^2$ is called the *coefficient of reflection* from the barrier and it is not difficult to show that $|R(\lambda)| = 1$. Thus we are only interested in the phase function $a(\lambda)$.

Suppose next that the limits $q_+ < 0$ and $q_- < 0$ are finite, that condition (3) holds and that the integral

$$\int^{+\infty} ||\sqrt{q(x)}| - |\sqrt{q_+}||dx \tag{7}$$

converges. Then (1) has a F.S.S. $(y_1^-,\ y_2^-)$ and a F.S.S. $(y_1^+,\ y_2^+)$ such that, as $x \to +\infty$

$$y_1^+(x,\ \lambda) \sim |q_+|^{-1/4}e^{ik_+ x},\ y_2^+(x,\ \lambda) \sim |q_+|^{-1/4}e^{-ik_+ x},\ k_+ = \lambda|\sqrt{q_+}|. \tag{8}$$

The statement of the scattering problem in this case is given in § 11 of Chap. 2. There are two variants of this problem.

2. The problem of passage through the potential barrier. In this case $q(x)$ has zeros.

3. The problem of over-barrier reflection. Here $q(x) < 0$ on the whole axis.

In this paragraph we calculate the asymptotic behaviour of the S-matrix as $\lambda \to +\infty$ under the assumption that $q(z)$ is holomorpic in the neighbourhood of the real axis.

In Chap. 2, § 11 it was shown that (1) has a F.A.S. of the form

$$y_{1,2}(x, \lambda) = q^{-1/4}(x) \exp\left\{ \pm\lambda S(a, x) + \sum_{k=1}^{\infty} (\pm\lambda)^{-k} \int^{x} \alpha_k(t) dt \right\}, \tag{9}$$

where the notation used is

$$S(a, x) = \int_a^x \sqrt{q(t)} dt, \quad \alpha_0(x) = -\frac{q'(x)}{4q(x)}, \tag{10}$$

the functions $\alpha_k(x)$ being determined by formulae (3) and (4) of § 3, Chap. 2. Throughout what follows it will be assumed that the following integrals converge

$$\int^{+\infty} |\alpha_k(t)| dt, \quad \int_{-\infty} |\alpha_k(t)| dt, \quad k = 1, 2, \ldots \tag{11}$$

In order to obtain the principal asymptotic term it is sufficient that these integrals converge for $k = 1$.

To derive the asymptotic formulae we first assume that $q(x)$ is an entire function and that the integrals of $|\alpha_k(z)|$ converge along those canonical paths γ which are necessary for the solution of the problem. All these formulae are true under less rigid conditions on $q(x)$. In particular, the analiticity of this function is necessary only in paragraph 4.

2. Reflection at a Barrier. The function $q(x)$ is positive for $x \geqslant a \gg 1$, and we take the branches of $\sqrt{q(x)}$ and $\sqrt[4]{q(x)}$ which are positive for $x \geqslant a$. Equation (1) has for each fixed $\lambda > 0$ a solution $y_2(x, \lambda)$ for which (9) is true as $x \to +\infty$. This solution satisfies the condition $y_2(+\infty, \lambda) = 0$ and is determined by its asymptotic behaviour to within a constant multiple. The asymptotic behaviour (9) is dual; it is applicable for $x \geqslant a$ and $\lambda \to +\infty$ uniformly in x. For $x \leqslant -a$, where $a \gg 1$, $q(x)$ is negative, and then we take the branches of the roots as:

$$\sqrt{q(x)} = i|\sqrt{q(x)}|, \quad \sqrt[4]{q(x)} = e^{i\pi/4} |\sqrt[4]{q(x)}| .$$

Equation (1) has for each fixed $\lambda > 0$ a F.S.S. $(\tilde{y}_1(x, \lambda), \tilde{y}_2(x, \lambda))$ for which (9) is true as $x \to -\infty$. This asymptotic behaviour is dual. In the definitions of these solutions the lower limit of integration is not indicated for the integral $S(a, x)$, but for the solutions $\tilde{y}_1, \tilde{y}_2$ it is the same. For arbitrary $\lambda > 0$ we have

$$\begin{aligned} y_2(x,\ \lambda) &= a_1(\lambda)\tilde{y}_1(x,\ \lambda) + a_2(\lambda)\tilde{y}_2(x,\ \lambda)\,, \\ y_2'(x,\ \lambda) &= a_1(\lambda)\tilde{y}_1'(x,\ \lambda) + a_2(\lambda)\tilde{y}_2'(x,\ \lambda)\,. \end{aligned} \tag{12}$$

The solution y_2 is real, and

$$\tilde{y}_2(x,\ \lambda) = \overline{i\tilde{y}_1(x,\ \lambda)}\,,$$

from which it follows that

$$a_1(\lambda) = \overline{ia_2(\lambda)}\,. \tag{13}$$

The solutions y_j are proportional to y_j^-, and so the problem of calculating the asymptotic behaviour of the coefficient $a(\lambda)$ from (5) is reduced to calculating the asymptotic behaviour of one of the coefficients $a_1(\lambda)$, $a_2(\lambda)$.

2.1 One Turning Point. Suppose that $q(x)$ has precisely one, simple, turning point x_0. Then $q(x) > 0$ for $x > x_0$ and $q(x) < 0$ for $x < x_0$. We put $a = x_0$ in (9) for all three of the solutions. Then for arbitrary $\lambda > 0$ we have

$$\begin{aligned} \tilde{y}_1(x,\ \lambda) &= e^{-i\pi/4} e^{i\lambda B_-(x_0)} y_1^-(x,\ \lambda)\,, \\ \tilde{y}_2(x,\ \lambda) &= e^{-i\pi/4} e^{-i\lambda B_-(x_0)} y_2^-(x,\ \lambda)\,, \end{aligned} \tag{14}$$

where

$$B_-(x_0) = -x_0|\sqrt{q_-}| + \int_{x_0}^{-\infty} (|\sqrt{q(x)}| - |\sqrt{q_-}|)dx\,,$$

which follows by comparing the asymptotic behaviour of y_j^- and $\tilde{y}_j^-$ for λ fixed and $x \to -\infty$.

In this case we can find the asymptotic behaviour of the coefficients $a_1(\lambda)$ and $a_2(\lambda)$ without resorting to the transition matrices. From the turning point x_0 there arise three Stokes lines $l = (-\infty, x_0]$, l_0, and l_0^*, where $\operatorname{Im} z > 0$ on l_0. Let D be a thin band containing the real axis. Then (9) is true in $D\backslash l$ for y_2, in $D\backslash l_0^*$ for $\tilde{y}_1$ and in $D\backslash l_0$ for $\tilde{y}_2$. From (12) we find

$$a_1(\lambda) = \frac{y_2}{\tilde{y}_1}\,\frac{(\tilde{y}_2'/\tilde{y}_2) - (y_2'/y_2)}{(\tilde{y}_2'/\tilde{y}_2) - (\tilde{y}_1'/\tilde{y}_1)}\,, \tag{15}$$

where the values of the solutions are taken at an arbitrary point z_0. Suppose that z_0 lies near the Stokes line l_0^* and $\operatorname{Im} z_0 < 0$. Because of the choice of branches of $\sqrt{q(x)}$ and $\sqrt[4]{q(z)}$ in $D\backslash l_0^*$ for y_2 we have

$$\sqrt{q(x - i0)} = -i|\sqrt{q(x)}|\,, \quad \sqrt[4]{q(x - i0)} = e^{-i\pi/4}|\sqrt[4]{q(x)}|$$

for $x < x_0$. Therefore the integrals under the exponent signs for the asymptotic expansions of the solutions $\tilde{y}_1, y_2$ are taken of functions which coincide for $z = z_0$. Hence

$$\frac{y_2'(z_0,\ \lambda)}{y_2(z_0,\ \lambda)} = \frac{\tilde{y}_1'(z_0,\ \lambda)}{\tilde{y}_1(z_0,\ \lambda)} + O(\lambda^{-\infty}).$$

Consequently, to within a term of order $O(\lambda^{-\infty})$, we have

$$a_1(\lambda) = \frac{y_2(z_0,\ \lambda)}{\tilde{y}_1(z_0,\ \lambda)} = i\exp\left\{\sum_{k=1}^{\infty}(-\lambda)^{-k}\int_{l_-}\alpha_k(z)dz\right\},$$

where the contour l_- goes from $+\infty$ to $-\infty$ and goes around x_0 from below. From (13) we find

$$a_2(\lambda) = \overline{ia_1(\lambda)} = -\exp\left\{\sum_{k=1}^{\infty}(-\lambda)^{-k}\int_{l_+}\alpha_k(z)dz\right\},$$

where the contour l_+ goes from $+\infty$ to $-\infty$ and goes around x_0 from above. The branch of $\sqrt{q(z)}$ is chosen so that $\sqrt{q(x)} > 0$ for $x > x_0$. From (15) and the formulae for $a_1(\lambda)$ and $a_2(\lambda)$, we find that the coefficient $a(\lambda)$ in (5) is

$$a(\lambda) = -i\exp\{-2i\lambda B_-(x_0)\}\exp\left\{\sum_{k=1}^{\infty}(-\lambda)^{-k}\int_{l}\alpha_k(z)dz\right\}, \tag{16}$$

where the contour l goes around the half-line $(-\infty, x_0]$ in the positive direction.

The formulae obtained also allow us to find the asymptotic expansion of y_0 on the half-line $(-\infty, x_0-\varepsilon]$, where $\varepsilon > 0$ is arbitrarily small, but not depending on λ. The principal asymptotic term has the form

$$y_0(x,\ \lambda) = 2|q(x)|^{-1/4}\left[\cos\left(\lambda\int_{x_0}^{x}|\sqrt{q(t)}|dt - \frac{\pi}{4}\right) + O(\lambda^{-1})\right].$$

This asymptotic behaviour is uniform in x.

2.2 Several Turning Points. Suppose that $q(x)$ has a finite number of turning points, all of them simple. Then there are an odd number of such points, and we denote them by $x_0 < x_1 < \ldots < x_{2m}$. The intervals $l_{0j} = [x_{2j-1}, x_{2j}]$ are Stokes lines. There is the further Stokes line $l = (-\infty, x_0]$ and from each turning point x_j there emanate another two Stokes lines l_j and l_j^* with $\operatorname{Im} z > 0$ for $z \in l_j$. The solutions $\tilde{y}_j$ are chosen as in paragraph 2.1, and we put $a = x_{2m}$ in formula (9) for y_2, so that

$$y_2(x,\ \lambda) \sim q^{-1/4}(x)\exp\{-\lambda S(x_{2m},\ x)\},\quad x \to +\infty.$$

We take the elementary F.S.S. (u_j, v_j) corresponding to $(l_j,\ x_j,\ D_j)$, $(u_j^*,\ v_j^*)$ corresponding to $(l_j^*,\ x_j,\ D_j^*)$, $(u_{0j},\ v_{0j})$, $(u_{ij},\ v_{ij})$ corresponding to $(\bar{l}_{0j},\ x_{2j},\ D_{0j})$, $(l_{0j},\ x_{2j-1},\ D_{0j})$ and $(u_{-1},\ v_{-1})$ corresponding to (l, x_0, D). We will not describe the choice of the canonical domains D_j, D_j^* and others; they are chosen in the same way as in § 5 of paragraph 3.2. We have (§ 5, paragraph 2.5)

$$y_2(z,\ \lambda) = e^{-i\pi/12} v_{2m}(z,\ \lambda),$$

and from the choice of $\tilde{y}_j$ it follows that

$$\tilde{y}_1(z,\ \lambda) = e^{-i\pi/4} v_{-1}(z,\ \lambda), \quad \tilde{y}_2(z,\ \lambda) = e^{-i\pi/4} u_{-1}(z,\ \lambda).$$

Let $\Omega = (\omega_{jk}(\lambda))$ be the transition matrix from $(u_{2m},\ v_{2m})$ to (u_{-1}, v_{-1}); then

$$y_2(x,\ \lambda) = e^{i\pi/6}[\omega_{22}(\lambda)\tilde{y}_1(x,\ \lambda) + \omega_{12}(\lambda)\tilde{y}_2(x,\ \lambda)],$$

so that

$$a_1(\lambda) = e^{i\pi/6}\omega_{22}(\lambda), \quad a_2(\lambda) = e^{i\pi/6}\omega_{12}(\lambda). \tag{17}$$

Let us put $\Omega_{2m,1}(\lambda) = (\tilde{\omega}_{jk}(\lambda))$ and denote

$$\eta_j = \int_{x_{2j}}^{x_{2j+1}} \sqrt{q(x)}dx > 0. \tag{18}$$

Since

$$\Omega = \Omega_{0,-1}\Omega_{1,0}\Omega_{2m,1}$$

and the matrices $\Omega_{0,-1},\ \Omega_{1,0}$ are determined by (20), (15) of § 3, we have

$$a_2(\lambda) = e^{i\pi/6} e^{i\lambda\eta_0} \alpha_0^{-1}(\lambda)\tilde{\omega}_{22}(\lambda), \tag{19}$$

The element $\tilde{\omega}_{22}(\lambda)$ was calculated in § 5, paragraph 3.1, and we obtain

$$a_2(\lambda) = 2^m \exp\left\{\sum_{j=0}^{m} \eta_j + i\frac{\pi}{6}(m-1)\right\}\left[\prod_{j=1}^{m} \cos \lambda\xi_j + O(\lambda^{-1})\right]. \tag{20}$$

The presence of an exponentially increasing multiplier, containing $\sum_{j=0}^{m} \eta_j$ is connected only with the fact that the contour of integration for the integral $S(a, x)$ starts at $a = x_0$ for $\tilde{y}_j$ and at $a = x_{2m}$ for y_2.

Suppose that y has the form (5). Then

$$y(x,\ \lambda) = \exp\left\{i\frac{\pi}{4} - i\lambda B_-\right\} a_1^{-1}(\lambda) y_2(x,\ \lambda). \tag{21}$$

If equation (1) has only the one turning point x_0 then $|a_1(\lambda)| = 1 + 0(\lambda^{-1})$, as was shown in paragraph 2.1, and therefore inside the potential barrier, that is, for x fixed and $x \geqslant x_0 + \varepsilon$, we have

$$|y(x,\ \lambda)| = |y_2(x,\ \lambda)|[1 + O(\lambda^{-1})]. \tag{22}$$

If we have several turning points then, as follows from (20), there are series of resonance values $\{\lambda_{jn}\}$, $j = 1, \ldots, m$, $n = 1, 2, \ldots$, of the form

$$\lambda_{jn} = (n\pi + \pi/2)\xi_j^{-1} + O(n^{-1}), \quad n \to \infty,$$

for which $|a_1(\lambda_{jn})| = 0(n^{-1})$ (recall that the exponentially increasing multiplier in (20) must be omitted). We observe that there is precisely the same asymptotic behaviour for the eigenvalues of equation (1) with potential $\tilde{q}(x)$, equal to $q(x)$ for $x_{2j-1} \leqslant x \leqslant x_{2j}$ and positive outside this interval. We will not go into the known quantum mechanical explanation of this effect [Landau].

The same asymptotic expansions are valid for the series $\{\lambda_{jn}\}$ as for the series of eigenvalues (§ 5, (11)). From (19) and the formula for $\tilde{\omega}_{22}(\lambda)$ (§ 5, paragraph 3.1) it follows that

$$|a_1(\lambda_{jn})| = O(n^{-\infty}), \quad n \to \infty .$$

A more exact result can be obtained for $m = 1$ (paragraph 6).

3. The Problem of Transmission Through a Barrier. Let (y_1^-, y_2^-) and $y_1^+, y_2^+)$ be the F.S.S. introduced in paragraph 1. We consider the problem of transmission through a barrier for the plane wave y_1^- travelling from left to right. The solution in this case has the form

$$\begin{aligned} &y = y_1^- + R_+(\lambda)y_2^- , \quad x < 0, \quad |x| \geqslant 1 \\ &y = T_+(\lambda)y_1^+ , \qquad\qquad x \geqslant 1 , \end{aligned} \tag{23}$$

that is, from the left at the barrier there are the incoming wave y_1^- and the reflected wave $R_+(\lambda)y_2^-$, and from the right the outgoing wave $T_+(\lambda)y_1^+$.

The variables $|R_+(\lambda)|^2$ and $|T_+(\lambda)|^2$ are called respectively the *coefficients of reflection* from the barrier and *of transmission* through the barrier. For all $\lambda > 0$ we have

$$|T_+(\lambda)|^2 + |R_+(\lambda)|^2 = 1 , \tag{24}$$

which is a property of the fact that the S-matrix is unitary (Chap. 2, § 11). In an analogous way we can pose the problem for a wave going from right to left:

$$\begin{aligned} &y = y_2^+ + R_-(\lambda)y_1^+ , \quad x \geqslant 1 , \\ &y = T_-(\lambda)y_2^- , \qquad\qquad x < 0, \quad |x| \geqslant 1 . \end{aligned}$$

Suppose next that all the real turning points are simple. Then their number is even: $x_0 < x_1 < \ldots < x_{2m-1}$. We will denote by $\tilde{y}_{1,2}^-$ the solutions $\tilde{y}_{1,2}$ introduced in paragraph 2 and introduce the solutions $\tilde{y}_{1,2}^+$ for which (9) holds for $\lambda > 0$ fixed and $x \to +\infty$. We put $a = x_{2m-1}$ in (9), that is,

$$\tilde{y}_{1,2}^+(x, \lambda) \sim q^{-1/4}(x) \exp\{\pm\lambda S(x_{2m-1}, x)\} , \quad x \to +\infty .$$

For $x > x_{2m-1}$, we take the following branches of the roots:

$$\sqrt{q(x)} = i|\sqrt{q(x)}| , \quad \sqrt[4]{q(x)} = e^{i\pi/4}|\sqrt[4]{q(x)}| ;$$

then for arbitrary $\lambda > 0$

$$\tilde{y}_{1,2}(x, \lambda) = e^{-i\pi/4} e^{\pm i\lambda B_+} y_{1,2}^+(x, \lambda), \tag{25}$$

$$B_+ = -x_{2m-1}|\sqrt{q_+}| + \int_{x_{2m-1}}^{+\infty} (|\sqrt{q(x)}| - |\sqrt{q_+}|)dx .$$

For each fixed $\lambda > 0$ we have

$$\tilde{y}_1^+(x, \lambda) = a_1(\lambda)\tilde{y}_1^-(x, \lambda) + a_2(\lambda)\tilde{y}_2^-(x, \lambda). \tag{26}$$

From the identities (14), (25) and from (23) we find that

$$R_+(\lambda) = \frac{a_2(\lambda)}{a_1(\lambda)} e^{-2i\lambda B_-}, \quad T_+(\lambda) = \frac{1}{a_1(\lambda)} e^{i\lambda(B_+ - B_-)}, \tag{27}$$

Thus the problem reduces to calculating the asymptotic behaviour of $a_1(\lambda)$ and $a_2(\lambda)$.

3.1 Two Turning Points. In this case we can calculate the asymptotic behaviour of the coefficients of reflection and transmission without resort to transition matrices, as in paragraph 2.1. From the turning point x_0 there arise the Stokes lines $l_{-1} = (-\infty, x_0], l_0, l_0^*$ and, from the point x_1, there arise the Stokes lines $l_4 = [x_1, +\infty), l_1, l_1^*$ where Im $z > 0$ for $z \in l_0$ and $z \in l_1$. Let D be a narrow band containing the real axis. We extend the asymptotic behaviour of $\tilde{y}_1^+(x, \lambda)$ from the half-line $x > x_1$ to the half-line $x < x_0$. The asymptotic expansion (9) for $\tilde{y}_1^+$ is applicable in $D \backslash (l_1^* \cup l_{-1})$, for $\tilde{y}_1$ in D $(l_0^* \cup l_4)$ and for $\tilde{y}_2^-$ in D $(l_0 \cup l_4)$.

We determine the coefficient $a_1(\lambda)$ from a relationship of the form (15), in which $\tilde{y}_j$ must be replaced by $\tilde{y}_j^-$ and y_2 by $\tilde{y}_2^+$. As in paragraph 2.1 we obtain

$$a_1(\lambda) = \frac{\tilde{y}_1^+(z_0, \lambda)}{\tilde{y}_1^-(z_0, \lambda)} + O(\lambda^{-\infty}),$$

where Im $z_0 < 0$ and z_0 lies near the Stokes line l_{-1} (Fig. 21). Let us make use of (9). We have

$$a_1(\lambda) = \exp\{\lambda(S_+(x_1, z_0) - S_-(x_0, z_0))\}[q_+(z_0)]^{-1/4}$$
$$\times [q_-(z_0)]^{-1/4} \exp\left\{\sum_{k=1}^{\infty} \lambda^{-k} \left| \int_{l_+} \alpha_k(t)dt - \int_{l_-} \alpha_k(t)dt \right| \right\},$$

where the + (−) sign relates to the branch corresponding to $\tilde{y}_1^+(\tilde{y}_1^-)$. Because of the choice of branch,

$$S_+(x_1, z_0) - S_-(x_0, z_0) = \int_{x_0}^{x_1} \sqrt{q(t)}dt > 0 .$$

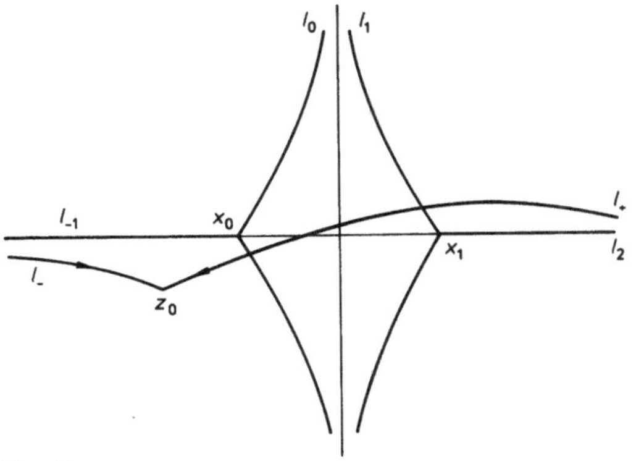

Fig. 21

Further, the contour l_+ goes from $+\infty$ to z_0 and goes around x_1 from above, and x_0 from below, since the asymptotic expansion for $\tilde{y}_1^+$ is applicable in $D\backslash(l_1^* \cup l_{-1})$. The contour l_- goes from $-\infty$ to z_0 below the real axis. Therefore the branches $q_\pm^{1/4}(z_0)$ coincide, and finally we obtain

$$a_1(\lambda) = \exp\{\lambda S(x_0, x_1)\} \times \exp\left\{\sum_{k=1}^{\infty} \lambda^{-k} \int_{l_1} \alpha_k(z)dz\right\}, \tag{28}$$

where l_1 goes from $+\infty$ to $-\infty$, passing x_1 from above and x_0 from below.

Similarly, choosing $\overline{z}_0$ in place of z_0, we obtain

$$a_2(\lambda) = -i \exp\{\lambda S(x_0,\ x_1)\} \exp\left\{\sum_{k=1}^{\infty} \lambda^{-k} \int_{l_2} \alpha_k(z)dz\right\} \tag{29}$$

where the contour l_2 joins $+\infty$ and $-\infty$ and passes both turning points x_0, x_1 from above. The branch of $\sqrt{q(z)}$ is chosen on l_1 and l_2 in the same way as for y_1^+.

The asymptotic behaviour of the coefficient $R_+(\lambda)$ follows from (27), (28), and the asymptotic behaviour of $T_+(\lambda)$ has the form

$$T_+(\lambda) = -ie^{-2i\lambda B} - \exp\left\{\sum_{k=1}^{\infty} \lambda^{-k} \int_l \alpha_k(z)dz\right\},$$

where l goes around the half-line $(-\infty, x_0]$ from below. In particular, as $\lambda \to +\infty$,

$$|R_+(\lambda)|^2 \sim \exp\{-2\lambda S(x_0,\ x_1)\},$$
$$|T_+(\lambda)|^2 = 1 + O(\exp\{-2\lambda S(x_0,\ x_1)\}).$$

Thus the coefficient of reflection from the barrier is exponentially small.

3.2 Several Turning Points. In order to find the asymptotic behaviour of $a_1(\lambda)$ and $a_2(\lambda)$ we extend the asymptotic behaviour of $\tilde{y}_1^+(x,\lambda)$ from the half-line $x > x_{2m+1}$ to the half-line $x < x_0$. In this case we must add to the Stokes lines considered in paragraph 2.2 the Stokes lines l_{2m+1}, l^*_{2m+1} (Im $z > 0$ on l_{2m+1}) and the Stokes line $l_{2m+2} = [x_{2m+1}, +\infty)$. To the elementary F.S.S. introduced in paragraph 2.2 we add the F.S.S. (u_{2m+1}, v_{2m+1}) and (u_{2m+2}, v_{2m+2}). We have

$$\tilde{y}_1^+(x,\lambda) = e^{-i\pi/4}u_{2m+2}(x, \lambda)$$

and the solutions $\tilde{y}_{1,2}$ were expressed in paragraph 2.2 in terms of the F.S.S. (u_{-1}, v_{-1}). We have

$$\Omega_{2m+2,2m} = \Omega_{2m+2,2m+1}\Omega_{2m+1,2m} ,$$

where the asymptotic behaviour of the latter two matrices have the form (15), (20) of § 3. Therefore

$$\Omega_{2m+2,2m}\begin{bmatrix}1\\0\end{bmatrix} = e^{\lambda\eta_m+i\pi/6}\begin{bmatrix}O(\lambda^{-1})\\1+O(\lambda^{-1})\end{bmatrix}$$

(for the notation for η_m see (18)). Further

$$\Omega_{2m+2,-1} = \Omega_{2m,-1}\Omega_{2m+2,2m} ,$$

and the asymptotic behaviour of $\Omega_{2m,-1}$ was calculated in paragraph 2.2. From this we find

$$\begin{aligned} a_1(\lambda) &= 2^m \exp\left\{\lambda\sum_{j=0}^{m+1}\eta_j+\phi\right\}\left[\prod_{j=1}^{m}\cos\lambda\xi_j+O(\lambda^{-1})\right] ,\\ a_2(\lambda) &= 2^m \exp\left\{\lambda\sum_{j=0}^{m}\eta_j+\phi\right\}\left[\prod_{j=1}^{m}\cos\lambda\xi_j+O(\lambda^{-1})\right] . \end{aligned} \tag{30}$$

In this case the value of the transmission coefficient of does not have the order term $\exp\{-2\lambda\sum_{j=0}^m\eta_j\}$ which is determined by the width of the barrier, but is considerably smaller, because of the presence of the multiplier in the square brackets in the formula for $a_1(\lambda)$. In particular, it can turn out that $|T_+(\lambda_n)|^2 = 1+0(\lambda_n^{-1})$ for some sequence $\{\lambda_n\}$ with $\lim\lambda_n = +\infty$. This case will be considered in paragraph 5.

4. Over-Barrier Reflection. In this case $q(x) < 0$ for all x. In § 11 of Chap. 2 we found the asymptotic behaviour of $T_+(\lambda)$ and that $R_+(\lambda) = 0(\lambda^{-\infty})$, $\lambda \to +\infty$. We will prove that the coefficient of reflection from the barrier is exponentially small. Now $S(0,x)$ is a one-to-one map of the real axis $\mathbb{R}$ onto the imaginary axis and therefore there exists a domain D of band type containing $\mathbb{R}$. In addition $D = D^*$ and ∂D consists of two connected components ∂D^+, ∂D^-, symmetric about $\mathbb{R}$. We suppose that Im $z > 0$ on ∂D^+.

4.1 One turning point on ∂D^+. Let us denote this point by z_0, so that $\partial D^- \ni \bar{z}_0$, and let z_0 be a simple turning point. There are three Stokes lines emanating from z_0: l_1, l_2, and l_3, where l_1 has the half line $(-\infty, 0)$ as its asymptotic direction and l_3 has the half line $(0, +\infty)$ as its asymptotic direction (Fig. 22). Let us use the transition matrices for simplicity.

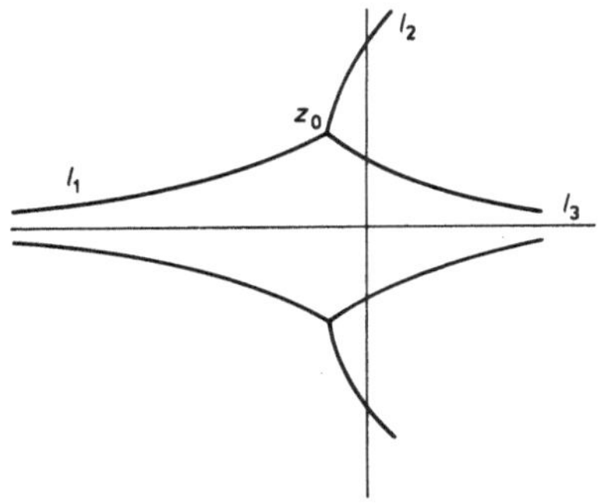

Fig. 22

Let us introduce the elementary F.S.S. (u_j, v_j) corresponding to (l_j, z_0). We can not specify the choice of canonical domains since the transition matrices are needed only to within $O(\lambda^{-1})$. We have

$$u_1(z, \lambda) \sim c_1 q^{-1/4}(z) \exp\{\lambda S(z_0, z)\}.$$

This asymptotic behaviour is dual: it is applicable for $z \in D$ fixed, $\lambda \to +\infty$ and for $\lambda > 0$ fixed, $z \in D$, Re $z \to +\infty$. Therefore u_1 is proportional to y_+. Since Im $(z_0, z) > 0$ for $z \in l_1$ by definition of an elementary F.S.S. (§ 3), we have for real x

$$S(z_0, x) = S(z_0, 0) + S(0, x),$$
$$B = \text{Re } S(z_0, 0) > 0,$$
$$S(0, x) = i \int_0^x |\sqrt{q(t)}| dt.$$

Since D is simply-connected we can select a holomorphic branch of $q^{1/4}(z)$ in D. We choose the branch so that $q^{1/4}(x) = e^{i\pi/4}|q(x)|^{1/4}$ for real x. From the definition of an elementary F.S.S. (§ 3) we have

$$c_1 = e^{\phi_1}, \quad \phi_1 = \lim_{z \to z_0, \ z \in l_2} \arg q^{1/4}(z).$$

Therefore for arbitrary $\lambda > 0$

$$u_1(z,\lambda) = e^{i\phi_1 + i\pi/4} e^{\lambda B + i\lambda A_+} y_1^+(z,\lambda), \tag{31}$$

where

$$A_+ = \int_0^{\infty} (|\sqrt{q(x)}| - |\sqrt{q_+}|)dx .$$

Let us express u_3 and v_3 in terms of y_1 and y_2^-. We have

$$S(z_0,\ x) = S(z_0,\ 0) + S(0,\ x),$$

where this time

$$\operatorname{Re} S(z_0,\ 0) < 0, \quad \operatorname{Im} S(0,\ x) < 0 .$$

Therefore

$$\begin{aligned} u_3(z,\ \lambda) &= e^{-i\pi/4 + i\phi_3} e^{-\lambda B - i\lambda A_-} y_2^-(z,\ \lambda), \\ v_3(z,\ \lambda) &= e^{-i\pi/4 + i\phi_3} e^{\lambda B + i\lambda A_-} y_1^-(z,\ \lambda), \end{aligned} \tag{32}$$

where

$$A_- = \int_0^{-\infty} (|\sqrt{q(x)}| - |\sqrt{q_-}|)dx ,$$

$\phi_3 = \lim_{z \to z_0, z \in l_3} \arg q^{1/4}(z)$. We express u_1 in terms of u_3 and v_3. Since

$$\Omega_{13} = \begin{bmatrix} 1 \\ 0 \end{bmatrix} = \Omega_{12}\Omega_{23} \begin{bmatrix} 1 \\ 0 \end{bmatrix} = e^{-i\pi/3} \begin{bmatrix} 1 \\ i \end{bmatrix} + O(\lambda^{-1}),$$

we have

$$\begin{aligned} u_1(z,\ \lambda) = [e^{-i\pi/3} &+ O(\lambda^{-1})] u_3(z,\ \lambda) \\ &+ i[e^{-i\pi/3} + O(\lambda^{-1})] v_3(z,\ \lambda) . \end{aligned}$$

From this formula and from (31)–(32) we obtain

$$R_+(\lambda) = -i \exp\{-2\lambda(B + iA_-)\}[1 + O(\lambda^{-1})] . \tag{33}$$

Thus $R_+(\lambda)$ decreases exponentially since $\operatorname{Re} B > 0$. We can give a more elegant form for B, namely,

$$2B = \oint_C \sqrt{q(z)}dz ,$$

where C is a simple closed curve containing z_0 and $\overline{z}_0$ as interior points.

4.2 Two Turning Points on ∂D^+. Suppose that there are precisely two, and moreover simple, turning points z_1 and z_2 on ∂D^+ and suppose for definiteness that a Stokes line l_1 emanates from z_1 having the line $(0,\ +\infty)$ as its asymptotic direction. Then from z_2 there emerges the Stokes line l_3 having the half-line $(-\infty, 0)$ as its asymptotic direction. The arc l_2 of the curve ∂D^+ joining z_1 and z_2 is also a Stokes line. Further, from z_1 there emerges another Stokes line l_4, and from z_2 the Stokes line l_5, which lie exterior to D.

We now express the F.S.S. $(u_1,\ v_1)$ in terms of the F.S.S. $(u_3,\ v_3)$. To do this we change from l_1 to l_2, where the transition matrix has the form (19) of § 3. Then we replace the initial point z_1 of l_2 by z_2 (the transition matrix has the form (14) of § 3) and then change from l_2 to l_3 ((19) of § 3). Finally we obtain

$$\Omega_{31} = e^{i(\pi/3+\phi_0)} \begin{bmatrix} -i(\alpha+\alpha^{-1}) & \alpha \\ \alpha^{-1} & 0 \end{bmatrix}, \quad \alpha = e^{i\lambda\xi_0}, \\ \xi_0 = \left| \int_{z_1}^{z_2} \sqrt{q(z)}dz \right| . \tag{34}$$

The explicit form of ϕ_0 is not significant. From this formula and from (31)–(32) we obtain

$$R_+(\lambda) = -2i[\cos\lambda\xi_0 + O(\lambda^{-1})]\exp\{-i\lambda\xi_0 - 2\lambda B_2 - 2i\lambda A_-\}, \tag{35}$$

where

$$B_2 = \int_{z_2}^{0} \sqrt{q(z)}dz, \quad \text{Re } B_2 > 0 .$$

Observe that $\xi_0 = 1/2 \text{ Im } \oint_C \sqrt{q(z)}dz$ where C is a closed contour going around the turning points z_1 and z_2.

We can obtain an asymptotic series in place of $0(\lambda^{-1})$ in this formula. In this case the phenomenon of resonance arises. The coefficient of reflection $R_+(\lambda)$, as is clear from (35), is exponentially small; but for a value of λ_n of the form

$$\lambda_n = \left(n\pi + \frac{\pi}{2}\right)\xi_0^{-1} .$$

it becomes even smaller. It can be shown that the expression $\cos\lambda\xi_0 + 0(\lambda^{-1})$ has order $0(\tilde{\lambda}_n^{-\infty})$ for values of $\tilde{\lambda}_n$ of the form $\tilde{\lambda}_n = \lambda_n + \sum_{k=1}^{\infty} a_k n^{-k}$.

4.3 A Simple Pole on ∂D^+. Let D^+ have a simple pole z_1. Since from z_1 there arises precisely one Stokes line l_0, this line must end at the turning point $z_0 \in \partial D^+$. We will assume that z_0 is a simple turning point and that ∂D^+ contains no other turning points or poles (Fig. 23). From z_1 there also arise the Stokes lines l_1 and l_3 described in paragraph 4.1.

We express the F.S.S. (u_1, v_1) in terms of the F.S.S. (u_3, v_3). To do this it is necessary to make the following transitions (where the number in the brackets is the formula from § 3 which gives the form of the transition matrix): from l_1 to l_0 ((19); but we must take the inverse matrix, since the transition is perfomed clockwise), from $(l_0,\ z_1)$ to (l_0, z_0) (14), from the right side of l_0

to the left side ((18), $n = 1$, the inverse matrix), from (l_0, z_0) to (l_0, z_1) (14) and from l_0 to l_3 ((19), the inverse matrix). Therefore

$$\Omega_{31} = e^{i\pi/3} \begin{bmatrix} 2 - \alpha^2 - \alpha^{-2} & -i\alpha^2 \\ -i\alpha^{-2} & 0 \end{bmatrix}, \quad \alpha = e^{i\lambda\xi_0}, \quad \xi_0 = \left| \int_{z_0}^{z_1} \sqrt{q(z)}dz \right| . \tag{36}$$

Expressing u_1 in terms of u_3, v_3 and taking into account (31)–(33), we obtain

$$R_+(\lambda) = 2i[\cos 2\lambda\xi_0 + O(\lambda^{-1})] \exp\{-2\lambda(B + A_- + \xi_0)\}, \tag{37}$$

where $B = S(z_0, 0)$ and Re $B > 0$. Here we also have resonant values of λ of the form

$$\lambda_n = \frac{1}{2}\left(\pi n + \frac{\pi}{2}\right)\xi_0^{-1} .$$

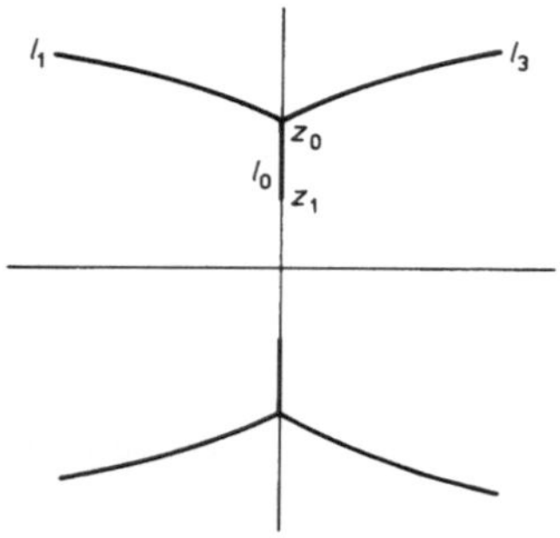

Fig. 23

5. Adiabatic Invariant. We consider the equation

$$\ddot{x} + \omega^2(\varepsilon t)x = 0, \tag{38}$$

where $\omega(t) > 0$ for real t, $\omega \in C^\infty(\mathbb{R})$ and there exist finite limits $\lim_{t\to\pm\infty} \omega(t) = \omega_\pm > 0$. Equation (38) has the form (1) where $\varepsilon = \lambda^{-1}$ and $\omega^2 = -q$. The adiabatic invariant $J(t, \varepsilon)$ was introduced in § 11, Chap. 2 and formula (18) was obtained, which expresses the total variation of $J(\varepsilon) = J(+\infty, \varepsilon) - J(-\infty, \varepsilon)$ in terms of the S-matrix. Therefore $J(\varepsilon)$ decreases exponentially for $\varepsilon \to +0$ with the conditions of paragraphs 4.1–4.3. Also the asymptotic formulae for $J(\varepsilon)$ come from the asymptotic formula for $T_+(\lambda)$ and from formulae (33), (35), (37) for $R_+(\lambda)$.

6. The Problem of the Largest Total Transmission Through a Barrier. Suppose that $q(x)$ satisfies the conditions of paragraph 3, so that $D_+(\lambda) = |T_+(\lambda)|^2$ is the coefficient of transmission through a barrier for a wave travelling to the right. If $q(x)$ has two simple turning points then

$D_+(\lambda)$ is monotonic decreasing for $\lambda \gg 1$. If there are more than two turning points then $D_+(\lambda)$ is non-monotonic for $\lambda \gg 1$ and has local maxima λ_n. The asymptotic behaviour of $D_+(\lambda_n)$ can be computed when there are four simple turning points $x_1 < x_2 < x_3 < x_4$ [Fedoryuk 10]. In this case there are the Stokes lines l_j, l_j^*, $1 \leqslant j \leqslant 4$, emanating from the turning points x_j, $\operatorname{Im} z > 0$ for $z \in l_j$, and three real Stokes lines $l_0 = (-\infty, x_1)$, $l_0' = (x_2, x_3)$, $l_5' = (x_4, +\infty)$. As $q(x)$ is real, the canonical domains $D_0 \supset l_0$ and $D_5 \supset l_5$ can be chosen symmetrically about the real axis and such that all the cuts in $S(D_j)$ are directed to the side opposite to the ray $S(l_j)$. We join these domains by chains of canonical domains as in § 5 of paragraph 3. Let us introduce elementary F.S.S. (u_j, v_j) and $(u_{j'},\ v_{j'})$, and denote

$$\xi = \int_{x_2}^{x_3} |\sqrt{q(x)}|dx\,, \quad \eta_1 = \int_{x_1}^{x_2} \sqrt{q(x)}dx\,, \quad \eta_2 = \int_{x_3}^{x_4} \sqrt{q(x)}dx\,, \tag{39}$$

$$\delta_j = e^{-2\lambda\eta_j}\,, \quad \gamma = e^{i\lambda\xi}\,.$$

We consider the transition matrices

$$\Omega_{32} = e^{-i\pi/3}(a_{jk})\,, \quad \Omega_{30} = -ie^{\lambda\eta_1}(b_{jk})\,, \quad \Omega_{50} = -e^{-\lambda(\eta_1+\eta_2)}(c_{jk})\,. \tag{40}$$

We have

$$\begin{aligned}
&c_{11} = b_{12}\,, \quad c_{12} = \delta_2\alpha_{51}^{-1}b_{11} + i\alpha_{41'}b_{12}\,,\\
&c_{21} = b_{22} = \bar{c}_{12}\,, \quad c_{22} = \delta_2\alpha_{54}^{-1}b_{21} + i\alpha_{44'}b_{22} = \bar{c}_{11}\,,\\
&a_{11} = 0\,, \quad a_{21} = \gamma^{-1}\,, \quad a_{12} = \gamma(\alpha_{30'}\alpha_{0'2})^{-1}\,,\\
&a_{22} = i\alpha_{30'}^{-1}(\gamma\alpha_{22'} + \gamma^{-1}\alpha_{3'3}^{-1})\,,\\
&b_{11} = \alpha_{10}^{-1}a_{21}\,, \quad b_{12} = \alpha_{10}^{-1}a_{22}\,, \quad b_{21} = i\alpha_{01'}a_{21}\,,\\
&b_{22} = \delta_1 a_{12} + i\alpha_{01'}a_{21}\,.
\end{aligned} \tag{41}$$

From the identity

$$u_5(x) = \delta_1\delta_2(c_{11}u_0(x) + c_{21}v_0(x))$$

we obtain

$$D_+ = \sqrt{q_+/q_-}\,\delta_1\delta_2|c_{21}|^2\,, \quad R_+ = |c_{11}|^2|c_{21}|^{-2}\,. \tag{42}$$

Let us examine whether it is possible that the barrier is completely transparent, that is, $D_+(\lambda) = 1$ for some λ. Then $R_+(\lambda) = 0$ so that $c_{11} = 0$ and

$$\gamma\alpha_{22'} + \gamma^{-1}\alpha_{3'3}^{-1} = 0\,. \tag{43}$$

From § 3, (23) it follows that

$$\begin{aligned}
&\alpha_{22'} = (1+\tilde{\delta}_1)^{1/2}e^{i\phi_1}\,, \quad \alpha_{3'3} = (1+\tilde{\delta}_2)^{-1/2}e^{i\phi_2}\\
&\tilde{\delta}_j = \delta_j[1 + O(\lambda^{-1})]\,, \quad \phi_j = 1 + O(\lambda^{-1})
\end{aligned}$$

and the $\phi_j(\lambda)$ are real. The equation $c_{11} = 0$ takes the form

$$\exp\{i(2\lambda\xi + \phi_1 + \phi_2)\} = -\sqrt{\frac{1+\tilde{\delta}_2}{1+\tilde{\delta}_1}}. \tag{44}$$

From this it follows that if $\eta_1 \neq \eta_2$ then $D_+(\lambda) < 1$ for $\lambda \gg 1$, that is, total transmission through the barrier is impossible.

The case where $q(x)$ is an even function deserves special attention. Equation (1) has even and odd solutions for $\lambda > 0$, so that the eigenvalues of $C(\lambda)$ are ± 1 and therefore $\overline{c_{11}(\lambda)} = -c_{11}(\lambda)$. Therefore, by (43), the number $A = \alpha_{10}^{-1}\alpha_{30}^{-1}(\gamma\alpha_{22'} + \gamma^{-1}\alpha_{3'3}^{-1})$ is real, and $A = 2\cos\lambda\xi + O(\lambda^{-1})$. Hence the equation $A = 0$, and so also the equation $D_+(\lambda) = 1$, has infinitely many solutions. Clearly this is possible only if $q(x+T)$ is even for some T.

We compute the value $D_+(\lambda)$ at local maxima. From (42) we have

$$D_+(\lambda) = \left[1 + \sqrt{\frac{q_-}{q_+}}|c_{11}|^2 e^{2\lambda(\eta_1+\eta_2)}\right]^{-1}. \tag{45}$$

We will represent (44) as

$$e^{i\psi(\lambda)} = \chi(\lambda) - 1.$$

Since $\alpha_{jk} = 1 + 0(\lambda^{-1})$ for all j, k, we have

$$|c_{11}|^2 = \left[4(1-\chi)\cos^2\frac{\psi}{2} + \chi^2\right][1 + O(\lambda^{-1})].$$

At extremal points of $D_+(\lambda)$ we have

$$(\eta_1 + \eta_2)|c_{11}|^2 - \chi'\cos\psi - (1-\chi)(\psi'\sin\psi + \chi') = 0$$

and since $\psi'(\lambda) = 2\xi + 0(\lambda^{-2})$ we have $\cos\psi \sim -1$ at the maximum points λ_n. Hence for λ_n we obtain the asymptotic expansion

$$2\lambda_n\xi + \sum_{k=1}^{\infty}\lambda_n^{-k}\oint_C \alpha_k(z) = 2n\pi + \pi, \quad n \to \infty, \tag{46}$$

where C is a simple closed countour, going around the segment $[x_2, x_3]$ and positively orientated. Also, $\sqrt{q(x)} > 0$ for $x \in C$, $x > x_3$, and the functions $\alpha_k(z)$ are determined by formula (3) of § 3, Chap. 2. The principal asymptotic term is

$$\lambda_n = -\xi^{-1}[\pi n + \pi/2] + O(n^{-1}).$$

If $\eta_1 < \eta_2$ then $\chi \sim \varepsilon_1/2$ at $\lambda = \lambda_n$, so that

$$\psi = 2\pi n - \pi \sim \frac{\eta_2 - \eta_1}{4}e^{-4\lambda_n\eta_2}.$$

Consequently as $n \to \infty$

$$D_+(\lambda_n) = 4\sqrt{\frac{q_+}{q_-}} \exp\{-2\lambda_n|\eta_1 - \eta_2|\}[1 + O(n^{-1})], \quad \eta_1 \neq \eta_2, \tag{47}$$

and hence $D_+(\lambda_n)$ is exponentially small. However the value $D_+(\lambda_n)$ is exponentially large compared to the case where there are only two turning points and the barrier has the same width, i.e.

$$\eta = \int_{x_1}^{x_2} |\sqrt{q(x)}| dx = \eta_1 + \eta_2,$$

since in this case $D_+(\lambda_n) \sim \exp\{-\lambda_n(\eta_1 + \eta_2)\}$.

Let $\eta_1 = \eta_2 = \eta$. Then the extremal points are given by the equation

$$\psi'\beta + \chi(2\xi\chi + \chi') + O(\chi^3) = 0,$$

and since $\chi = e^{-2\lambda\eta}[1 + 0(\lambda^{-1})]$, we have $\beta = 0(\chi^2\lambda^{-1})$. From this we find that

$$D_+(\lambda_n) = 1 + O(n^{-2}), \quad \eta_1 = \eta_2, \tag{48}$$

as $n \to \infty$, where $0(n^{-2}) \leqslant 0$. Here there is almost total transmission for $\lambda = \lambda_n$.

If $\eta_1 \neq \eta_2$ then $D_+(\lambda_n) < 1$. However there are always complex values $\tilde{\lambda}_n$ with exponentially small imaginary part such that $D_+(\tilde{\lambda}_n) = 1$. That is,

$$\begin{aligned} \tilde{\lambda}_n - \lambda_n &= \frac{i\delta_2}{4\xi}[1 + O(n^{-1})], \quad \eta_2 < \eta_1, \\ \tilde{\lambda}_n - \lambda_n &= -\frac{i\delta_1}{4\xi}[1 + O(n^{-1})], \quad \eta_1 < \eta_2. \end{aligned} \tag{49}$$

To prove this, we note that the solutions $y_j^\pm(x, \lambda)$ are holomorphic in λ in a band of the form $0 < \operatorname{Re} \lambda < \infty$, $|\operatorname{Im} \lambda| < c$, and the same is true for the elements $s_{jk}(\lambda)$ of the scattering matrix. Therefore equation (43) has a complex solution $\tilde{\lambda}_n$ close to λ_n and

$$\begin{aligned} \psi(\tilde{\lambda}_n) &= 2\pi n + \pi + \frac{i}{2}(\delta_2 - \delta_1) + o(|\delta_1| + |\delta_2|), \\ \psi(\tilde{\lambda}_n) - \psi(\lambda_n) &= (\tilde{\lambda}_n - \lambda_n)\psi'(\lambda_n)[1 + O(n^{-1})], \end{aligned}$$

and since $\psi'(\lambda_n) = 2\xi + 0(n^{-1})$, (49) follows from (43).

7. Quasistationary Levels. We consider the problem of transmission through a barrier (paragraph 1, problem 2) and we seek those λ for which there is a solution of the form

$$y = y_1^+ , \quad y = Ay_2^- .$$

The solution y for $|x| \gg 1$ is a divergent wave and the corresponding values of λ are called *quasistationary levels*. This solution does not exist if λ is real since the S-matrix is non-singular. However there are infinitely many complex quasistationary levels $\tilde{\lambda}_n$ with exponentially small imaginary part. If $q(x)$ satisfies the conditions of paragraph 5 then

$$\begin{aligned} \tilde{\lambda}_n - \lambda_n &= \frac{i}{4\xi} e^{-2\lambda_n \eta_1}[1 + O(n^{-1})] , \quad \eta_2 < \eta_1 , \\ \tilde{\lambda}_n - \lambda_n &= \frac{i}{4\xi} e^{-2\lambda_n \eta_2}[1 + O(n^{-1})] , \quad \eta_1 < \eta_2 , \\ \tilde{\lambda}_n - \lambda_n &= \frac{i}{2\xi} e^{-2\lambda_n \eta}[1 + O(n^{-1})] , \quad \eta_1 = \eta_2 = \eta . \end{aligned} \tag{50}$$

Here the λ_n are determined by (46) and all the O-terms are purely imaginary. In fact the $\tilde{\lambda}_n$ are determined by the equation $c_{21} = 0$, which has the form

$$\gamma \alpha_{22'} + \gamma^{-1} \alpha_{3'3}^{-1} = \varepsilon_2 ,$$

so that

$$e^{i\phi} + \chi = \delta_1 \gamma^2 (1 + \delta_1)^{-1/2} (\alpha_{01'} \alpha_{0'2})^{-1} . \tag{51}$$

The equation $e^{i\psi} + 1 = 0$ has infinitely many solutions λ_n of the form (46). Let us put $\tilde{\lambda}_n = \lambda_n + \lambda'_n$. If $\eta_1 < \eta_2$ then (51) has the form

$$e^{i\psi} + 1 = -\frac{\delta_1}{2}[1 + O(\lambda^{-1})] ;$$

and if $\eta_1 = \eta_2$ then it has the form

$$e^{i\psi} + 1 = -\delta[1 + O(\lambda^{-1})] ,$$

from which (50) follows.

8. Scattering for Energies Close to the Maximum Potential Energy. The standard model is the Schrödinger equation

$$-\frac{h^2}{2m}\psi'' + (E - V(x))\psi = 0 ,$$

where $h > 0$ is a small parameter and $V(\pm\infty) = 0$. Suppose that the potential $V(x)$ has the single maximum x_0 with $V''(x_0) < 0$. Then for values of E close to $E_0 = V(x_0)$ there are two close turning points, real for $E < E_0$ and complex for $E > E_0$. Here the asymptotic expansions of the solutions for

x close to x_0 are expressed in terms of Weber functions (Chap. 4, § 7) and this is the only way we can find the asymptotic behaviour of $T_\pm(h, E)$ and $R_\pm(h, E)$ as $h \to 0$ uniformly in E, close to E_0. However we can find the asymptotic behaviour of the ratio $T_\pm/R_\pm$ without recourse to the standard equation. We consider the equation

$$y'' - \lambda^2 q(x, \alpha)y = 0\,, \tag{52}$$

where α is a parameter and $\alpha \in J = [-\alpha_0, \alpha_0]$. Suppose that $q(x, \alpha) \in C^\infty(\mathbb{R} \times J)$,

$$q(x, 0) \leqslant 0\,, \quad q(0, 0) = 0\,, \quad q''_{xx}(0, 0) < 0\,, \quad q'_\alpha(0, 0) > 0\,,$$

so that $q(x, \alpha) = ax^2 + b\alpha + c\alpha x \dots$ for small $|x|$, $|\alpha|$. The other conditions on $q(x, \alpha)$ are the same as in the problem of transmission through a barrier (paragraph 1); we require only the uniformity of these conditions in $\alpha \in J$. The number $\alpha_0 > 0$ is assumed to be sufficiently small.

Let $\alpha > 0$ be fixed; then (52) has two complex turning points $z_0(\alpha)$, $\overline{z_0(\alpha)}$ and the Stokes lines have the same form as in Fig. 21. Let $\tilde{y}$ be a solution such that

$$\tilde{y}(z, \lambda, \alpha) \sim c_0 q^{-1/4}(z, \alpha) \exp\{\lambda S(z_0(\alpha), z)\}$$

for $z \in l_3(\alpha)$ and $z \to \infty$, and suppose that this asymptotic behaviour is dual in z and in λ. The branch S is chosen so that

$$\operatorname{Im} S(z_0(\alpha), z) > 0\,, \quad z \in l_3(\alpha)\,; \quad \lim_{z \to z_0(\alpha)} c_0 \arg q^{-1/4}(z, \alpha) = 0\,.$$

Then for all $\lambda > 0$ and $\alpha < 0$

$$\tilde{y}(x, \lambda, \alpha) = \exp\{i\pi/12 + \lambda c + i\lambda B_+(\alpha)\} y_1^+(x, \lambda, \alpha)\,,$$

where

$$c = S(z_0(\alpha), 0)\,, \quad \operatorname{Re} c > 0\,,$$

$$B_+(\alpha) = \int_0^\infty [\sqrt{|q(x, \alpha)|} - \sqrt{|q_+(\alpha)|}]dx\,.$$

If $\alpha < 0$ is fixed then the asymptotic expansion for $\tilde{y}$ is applicable anywhere in a neighbourhood of the real axis except in some neighbourhood of the Stokes lines $l_1(\alpha)$ and $l_2(\alpha)$. But as $\alpha \to 0$ the turning points $z_0(\alpha)$ and $\overline{z_0(\alpha)}$ merge, so that the half-line $(-\infty, 0]$ must be excluded from the domain of applicability of the asymptotic behaviour. Nevertheless, the asymptotic behaviour is applicable at a point $\tilde{z}$ such that $\operatorname{Re} \tilde{z} < 0$, $\operatorname{Im} \tilde{z} > 0$ (it is not applicable at $\overline{\tilde{z}}$).

Let us bring in the solutions

$$\tilde{y}_j(z, \lambda, \alpha) \sim c_j q^{-1/4}(z, \alpha) \exp\{\lambda S_j(z_j(\alpha), z)\}\,, \quad j = 1, 2\,.$$

The asymptotic expansions for $\tilde{y}_{1,2}$ are valid as $z \to \infty$ respectively for $z \in l_1(\alpha)$ and $z \in l^\star(\alpha)$,

$$\operatorname{Im} S_1(z_0(\alpha),\, z) < 0\,, \quad z \in l_1(\alpha)\,; \quad \operatorname{Im} S_2(\overline{z_0(\alpha)},\, z) > 0\,, \quad z \in l_1^\star(\alpha)\,,$$

and along the corresponding Stokes line $\lim_{z\to z_j(\alpha)} \arg[c_j q^{-1/4}(z,\alpha)] = 0$. We have

$$\tilde{y}_1(x,\lambda,\alpha) = \exp\left\{-\frac{i\pi}{12} + \lambda c + i\lambda B_-(\alpha)\right\} y_1^-(x,\lambda,\alpha)\,,$$

$$\tilde{y}_2(x,\lambda,\alpha) = \exp\left\{\frac{i\pi}{12} + \lambda \bar{c} - i\lambda B_-(\alpha)\right\} y_2^-(x,\lambda,\alpha)\,,$$

where

$$B_-(\alpha) = \int_0^\infty [\sqrt{|q(x,\alpha)|} - \sqrt{|q_-(\alpha)|}]dx\,.$$

Further, for $\lambda > 0$ and $\alpha > 0$

$$\tilde{y}(z,\ \lambda,\ \alpha) = A(\lambda,\ \alpha)\tilde{y}_1(z,\ \lambda,\ \alpha) + B(\lambda,\ \alpha)\tilde{y}_2(z,\ \lambda,\ \alpha)\,.$$

At $\tilde{z}$ the solutions $\tilde{y}$ and $\tilde{y}_2$ are exponentially large for $\lambda \gg 1$, and the solution $\tilde{y}_1$ is exponentially small. By standard methods we find that

$$B(\lambda,\alpha) = \frac{\tilde{y}(\tilde{z},\ \lambda,\ \alpha)}{\tilde{y}_2(z,\ \lambda,\ \alpha)} = [-i + O(\lambda^{-1})]e^{2\lambda c}\,. \tag{53}$$

However, we have not succeeded in finding the coefficient A. We can now find the ratio

$$R_+/T_+ = B\exp\{-i\lambda(B_+(\alpha) + B_-(\alpha))\}\,. \tag{54}$$

Since $|R_+|^2 + |T_+|^2 = 1$, we have

$$|R_+| = \exp\left\{-\lambda \int_{\overline{z_0(\alpha)}}^{z_0(\alpha)} \sqrt{q(t)}dt\right\}[1 + O(\lambda^{-1})]\,, \tag{55}$$

where the right-hand side is exponentially small for $\alpha < 0$ fixed and $\lambda \to +\infty$. The formulae (54), (55) remain true also for $a \leqslant 0$. In particular, for $a = 0$, we have

$$|R_+(0,\alpha)| = \frac{1}{\sqrt{2}} + O(\lambda^{-1})\,, \quad |T_+(0,\alpha)| = \frac{1}{\sqrt{2}} + O(\lambda^{-1})\,.$$

§ 9. Sturm-Liouville Equations with Periodic Potential

1. Basic Properties of the Solutions. We consider the equation

$$w'' + \lambda^2 q(x) w = 0\,, \tag{1}$$

where $q(x)$ is a continuous periodic function with period $T > 0$ and $\lambda > 0$ is a parameter. For fixed λ we have the following basic theorem.

Theorem (Floquet-Lyapunov). *Equation (1) has a fundamental system of equations either of the form*

$$w_1(x,\ \lambda) = e^{\mu_1 x} p_1(x,\ \lambda)\,, \quad w_2(x,\ \lambda) = e^{\mu_2 x} p_2(x,\ \lambda) \tag{2a}$$

or of the form

$$w_1(x,\ \lambda) = e^{\mu_1 x} p_1(x,\ \lambda)\,,\ w_2(x,\ \lambda) = e^{\mu_2 x}[x p_1(x,\ \lambda) + p_3(x,\ \lambda)]\,, \tag{2b}$$

where the $p_j(x, \lambda)$ are periodic functions with period T.

The numbers $\mu_1(\lambda)$ and $\mu_2(\lambda)$ are called the *characteristic exponents*, and the numbers $\rho_j(\lambda) = \exp\{T\mu_j(\lambda)\}$ are called *multipliers*. The case (2b) can occur only when $\mu_1(\lambda) = \mu_2(\lambda)$.

Let $Y(x, \lambda)$ be a fundamental matrix for (1):

$$Y(x, \lambda) = \begin{bmatrix} y_1(x,\ \lambda) & y_2(x,\ \lambda) \\ y_1'(x,\ \lambda) & y_2'(x,\ \lambda) \end{bmatrix}\,,$$

where $y_1(x, \lambda)$ and $y_2(x, \lambda)$ form a F.S.S. for (1). Then the multipliers are the roots of the equation

$$\det[Y(x_0 + T,\ \lambda) - \rho Y(x_0,\ \lambda)] = 0\,,$$

where x_0 is arbitrary.

In the sequel, it is assumed that $q(x)$ is real-valued, so that if $w(x, \lambda)$ is a solution of (1) then $\overline{w(x, \lambda)}$ is also a solution. Suppose that these solutions form a F.S.S.. Then the multipliers are determined from the equation

$$\rho^2 - 2a\rho + 1 = 0\,,$$

where

$$a = \operatorname{Re}\left\{\left|\frac{w(0,\ \lambda)}{w'(0,\ \lambda)}\frac{\overline{w(T,\ \lambda)}}{\overline{w'(0,\ \lambda)}}\right|\right\}\left\{\left|\frac{w(0,\ \lambda)}{w'(0,\ \lambda)}\frac{\overline{w(0,\ \lambda)}}{\overline{w'(0,\ \lambda)}}\right|^{-1}\right\}. \tag{3}$$

Therefore $\rho_1\rho_2 = 1$, and if ρ_1, ρ_2 are non-real then $\overline{\rho}_2 = \rho_1$ and $|\rho_1| = |\rho_2| = 1$.

The number λ lies in a zone of stability if all the solutions of (1) are bounded on the real axis, and it lies in a zone of instability (lacuna) otherwise.

Since

$$\rho_{1,2} = a \pm \sqrt{a^2 - 1}\,,$$

the zones of instability are determined by the inequality $|a| \geqslant 1$. In this case both mulitpliers are real and form the pair ρ, ρ^{-1}. If λ lies in a zone of stability, that is, $|\rho| < 1$, then the multipliers are complex conjugates and form the pair $\exp(i\phi)$, $\exp(-i\phi)$, $0 < \phi < \pi$.

2. Bounds for the Width of the Lacunae. Let Δ_n be the width of the nth lacuna. Then it is known that $\Delta_n = o(n^{-1})$ as $n \to \infty$. Further, if $q(x) > 0$ and $q(x) \in C^m(\mathbb{R})$, then then $\Delta_n = O(n^{-m-1})$ as $n \to \infty$. We will show that if $q(x) > 0$ and $q(x) \in C^\infty(\mathbb{R})$ then $\Delta_n = 0(n^{-\infty})$ as $n \to \infty$. Equation (1) has a solution $w(x, \lambda)$, for which there is the asymptotic expansion as $\lambda \to \infty$, uniform on each finite interval (Chap. 1, § 10):

$$w(x,\lambda) = \exp\left\{\int_0^x y(t,\lambda)dt\right\}, \quad y(x,\lambda) = \sum_{k=-1}^{\infty} \lambda^{-k}\alpha_k(x)\,. \tag{4}$$

All the $\alpha_k(x)$ are periodic with period T, the $\alpha_{2k}(x)$ are real, the α_{2k+1} are purely imaginary, and

$$\alpha_{-1}(x) = i\sqrt{q(x)}\,, \quad \alpha_0(x) = -\frac{1}{4}\frac{d}{dx}\ln q(x)\,.$$

We now put $y(x,\lambda) = y_1(x,\lambda) + iy_2(x,\lambda)$ and show that

$$\tilde{y}_1(T,\lambda) = \int_0^T y_1(x,\lambda)dx = O(\lambda^{-\infty})\,, \quad \lambda \to \infty\,. \tag{5}$$

Omitting terms of order $O(\lambda^{-\infty})$, we have

$$y' + y^2 + \lambda^2 q = 0\,, \quad y_1' + y_1^2 - y_2^2 + \lambda^2 q = 0\,, \quad y_2' + 2y_1y_2 = 0\,,$$

so that

$$\int_0^T y_1(x,\lambda)dx = -\frac{1}{2}\int_0^T \frac{y_2'(x,\lambda)}{y_1'(x,\lambda)}dx = 0\,.$$

From (3), (4) we find that

$$a(\lambda) = e^{\tilde{y}_1(T,\lambda)}\cos\tilde{y}_2(T,\lambda) + O(\lambda^{-\infty}) = \cos\tilde{y}_2(T,\lambda) + O(\lambda^{-\infty})\,. \tag{6}$$

If λ_{1n} and λ_{2n} are the end-points of the lacuna Δ_n, then at both points $a(\lambda) = 1$ or $a(\lambda) = -1$. Let $\tilde{\lambda}_n$ be the mid-point of the lacuna. Then there is the asymptotic expansion

$$\tilde{y}_2(T,\tilde{\lambda}_n) = n\pi + O(\lambda_n^{-\infty})\,,$$

from which we get the asymptotic expansion for $\tilde{\lambda}_n$ in odd powers of n^{-1}:

$$\tilde{\lambda}_n = n\pi \left[\int_0^T \sqrt{q(x)}dx \right]^{-1} + \sum_{k=0}^{\infty} C_k n^{-2k-1} . \tag{7}$$

From (6) it follows that $\Delta_n = O(n^{-\infty})$. Furthermore, we have found the asymptotic behaviour of the multipliers:

$$\rho_{1,2}(\lambda) = \cos \tilde{y}_2(T, \lambda) \pm \sqrt{\cos^2 \tilde{y}_2(T, \lambda) - 1 + O(\lambda^{-\infty})} + O(\lambda^{-\infty}),$$

the principal term of which is

$$\rho_{1,2}(\lambda) = \cos\left[\lambda \int_0^T \sqrt{q(x)}dx \right] \pm \sqrt{\cos^2 \left[\int_0^T \sqrt{q(x)}dx \right] + O(\lambda^{-1})} \\ + O(\lambda^{-1}) .$$

3. Lacunae for Analytic Potentials. If $q(x)$ is holomorphic in some neighbourhood of the real line then the width of the nth-lacuna, generally speaking, decreases exponentially as $n \to \infty$. We will show this, given some assumptions on $q(z)$.

3.1 Stokes Lines. Suppose that $q(z)$ is an entire function, positive on the real axis and $q(z) \not\equiv$ const.. Then, with

$$S(0, z) = \int_0^z \sqrt{q(t)}dt$$

and $\sqrt{q(x)} > 0$ for real x, $iS(0, z)$ is a one-to-one mapping of the real axis onto the imaginary axis in the complex S-plane. Therefore the real axis is contained in a domain of band type and $D = D^*$, this domain being symmetric about the real axis. The boundary of D consists of two connected components Γ^+, Γ^-, and we suppose that Im $z > 0$ on Γ^+ for definiteness. The curves $\Gamma^\pm$ are invariant under a shift by T parallel to the real axis. On the curve $\Gamma_T^+ = \Gamma^+ \cap \{0 \leqslant \text{Re } z \leqslant T\}$ there is at least one turning point for equation (1). The asymptotic behaviour of the width of the nth-lacuna Δ_n is determined by the number of turning points lying on Γ^+ and by their multiplicity. We will consider the main case: on Γ^+ there is precisely one, and moreover simple, turning point z_0. From this point there arise three Stokes lines; suppose that $l_1, l_2 \in \Gamma^+$ and l_1 lies to the left of $l_2, l_3 \notin \Gamma^+$.

3.2 Fundamental System of Solutions. We choose the branch of $\sqrt{q(z)}$ on l_1 so that

$$\text{Im } [iS(z_0,\ z)] > 0 , \quad z \in l_1 .$$

As in §§ 3, 5 we can prove the existence of a domain D_1 such that $\partial D_1 \supset l_2 \cup l_3$. The function $iS(z_0, z)$ is a one-to-one mapping of D_1 onto the half-plane

$\operatorname{Re}(iS) > 0$, with a finite or infinite number of vertical cuts. We denote by D_2 the domain obtained from D_1 by a shift of T. Suppose that $q(z)$ satisfies the conditions of § 4, paragraph 1, in D_1. These conditions are satisfied, in particular, by trigonometric polynomials that are positive on the real axis. Then (§ 4) equation (1) has a solution of the form

$$w(z,\ \lambda) = q^{-1/4}(z)\exp\{-i\lambda S(z_0,\ z)\}[1+\lambda^{-1}\varepsilon_1(z,\ \lambda)] \tag{8}$$

for each fixed $\lambda > 0$, where $\varepsilon_1(z,\lambda) \to 0$ if $z \to \infty$ in D_1, so that $\operatorname{Re}[iS(z_0,z)] \to +\infty$. This solution is determined by the condition $\lim_{z\to\infty} w = 0$, $z \in D_1$, to within a constant multiple. As shown in § 3, the asymptotic expansion (8), as $\lambda \to +\infty$, is valid everywhere in the complex z-plane, excluding neighbourhoods of some of the Stokes lines, that is, a domain such as $|\varepsilon_1(z,\lambda)| \leqslant c$ for $\lambda \geqslant \lambda_0 \gg 1$. In particular, (8) is valid as $\lambda \to +\infty$ in $D_1 \cup D \cup D_1^\star \cup D_2 \cup D_2^\star$, with neighbourhoods of l_2 and $l_2^\star$ removed. In a similar way we can determine the solution $w_2(z,\lambda)$, associated with D_2:

$$w_2(z,\ \lambda) = q^{-1/4}(z)\exp\{-i\lambda S(z_0+T,\ z)\}[1+\lambda^{-1}\varepsilon_2(z,\ \lambda)]\,. \tag{9}$$

The branch of $q^{-1/4}(z)$ is chosen so that $q^{1/4}(x) > 0$ for real x.

We show that for each fixed $\lambda > 0$ there is the identity

$$w_1(z,\ \lambda) = w_2(z+T,\ \lambda)\,. \tag{10}$$

Let $z \to \infty$ in D_1 so that $\operatorname{Re}[iS(z_0,z)] \to +\infty$; then $w_1(z,\lambda) \to 0$ and since $z+T \in D_2$ we have $w_2(z,\lambda) \to 0$. Consequently, $w_1(z+T,\lambda) = Aw_2(z,\lambda)$. Dividing both sides by $q^{-1/4}(z)$ and taking into account the choice of branches of S, we obtain

$$1+\lambda^{-1}\varepsilon_1(z+T,\lambda) = A[1+\lambda^{-1}\varepsilon_2(z+T,\lambda)]\,.$$

Letting z go to infinity in D as indicated above and taking into account that $\varepsilon_1, \varepsilon_2 \to 0$, we arrive at the identity (10).

Let us take the F.S.S. $\{w_j(z,\lambda),\ \overline{w_j(\overline{z},\lambda)}\}$, $j = 1,2$, and let $W_j(z,\lambda)$ be the corresponding fundamental matrices. We have

$$W_1(z,\ \lambda) = W_2(z,\ \lambda)\Omega(\lambda)\,. \tag{11}$$

Because of the choice of F.S.S., the elements $\omega_{jk}(\lambda)$ of $\Omega(\lambda)$ satisfy

$$\omega_{22}(\lambda) = \overline{\omega_{11}(\lambda)}\,, \quad \omega_{21}(\lambda) = \overline{\omega_{12}(\lambda)}\,.$$

Since $W_1(z,\lambda) = W_2(z+T,\lambda)$, we have

$$W_2(z+T,\lambda) = W_2(z,\ \lambda)\Omega(\lambda)\,.$$

Now the multipliers are the eigenvalues of the monodromy matrix $\Omega(\lambda)$ and they are determined from the equation

$$\rho^2 - 2\rho \operatorname{Re}\omega_{11} + |\omega_{11}|^2 - |\omega_{12}|^2 = 0\,.$$

Since $\rho_1\rho_2 = 1$ we have

$$|\omega_{11}(\lambda)|^2 = |\omega_{12}(\lambda)|^2 + 1\,. \tag{12}$$

We next find the asymptotic behaviour of ω_{11} and ω_{12} as $\lambda \to +\infty$. We have

$$w_1(z,\ \lambda) = \omega_{11}w_2(z,\ \lambda) + \omega_{21}(\lambda)w_3(z,\ \lambda),\quad w_3(z,\ \lambda) = \overline{w_2(\bar{z},\lambda)}\,.$$

Let us agree to denote by $f_j(z)$ the branch of multivalued function $f(z)$ corresponding to w_j.

Let $z \to \infty$ in $D_2^\star$ such that Re $[iS(z_0,z)] \to -\infty$; then $w_1 \to \infty$, $w_2 \to \infty$, $w_3 \to 0$. It follows that

$$\omega_{11}(\lambda) = \lim_{z\to\infty} \frac{w_1(z,\ \lambda)}{w_2(z,\ \lambda)}\,.$$

Replacing the solutions by their asymptotic series, we obtain

$$\omega_{11}(\lambda) = \left[\frac{q_2(z)}{q_1(z)}\right]^{1/4} \exp\{i\lambda[S_2(z_0+T,z) - S_1(z_0,z)]\}[1+O(\lambda^{-1})]\,,$$

where we can take an arbitrary point of $D_2^\star$ for z, since the ratio of the roots and the exponent are independent of z. Because of the choice of branch of $q_j^{1/4}(z)$ the ratio of these roots is unity. The branches S_1 and S_2 coincide in $D_2^\star \cup D \cup \partial D$ and therefore

$$S_2(z_0+T,\ z) - S_1(z_0,\ z) = \int_{z_0+T}^{z_0} \sqrt{q(t)}dt > 0\,,$$

where the integral is taken over l_2. Since $q(x)$ is periodic and real, this integral equals

$$\alpha = \int_0^T \sqrt{q(x)}dx\,. \tag{13}$$

Finally we obtain

$$\omega_{11}(\lambda) = e^{i\lambda\alpha}[1+O(\lambda^{-1})]\,.$$

It is not difficult to obtain the asymptotic expansion for $\omega_{11}(\lambda)$ in powers of λ^{-1}:

$$\omega_{11}(\lambda) = \exp\{iy_2(T,\lambda)\} + O(\lambda^{-\infty})\,,$$

where y_2 is as in (6).

Next let $z \to \infty$ in D_2 so that Re $[iS(z_0,z)] \to -\infty$; then $w_1 \to \infty$, $w_2 \to \infty$, $w_3 \to \infty$. It follows that

$$\omega_{21}(\lambda) = \lim_{z\to\infty} \frac{w_1(z,\ \lambda)}{w_3(z,\ \lambda)}\,.$$

It then follows from the definition of w_3 that, for $z \in D_2$ fixed and $\lambda \to +\infty$,

$$w_3(z,\ \lambda) = q_2^{-1/4}(z)\exp\{i\lambda S_2(\bar{z}_0 + T, z)\}[1 + O(\lambda^{-1})] .$$

Let us fix $z_1 \in D_2$ which lies above l_1, and is close to it. We obtain the value of $q^{1/4}(z)$ by extending analytically the branch of the root from $x = 0$ along the path γ_j. Since (8) is inapplicable on l_2, γ_1 passes on the left of z_0. The path γ_2 passes z_0 on the right since the asymptotic expansion for the solution w_3 is applicable on l_2. Consequently, $[q_2(z_1)/q(z_1)]^{1/4} = i$. Further,

$$S_1(z_0, z) + S_2(\bar{z}_0 + T, z) = \int_{\bar{z}_0+T}^{z_0+T} \sqrt{q(t)}dt + \alpha = \beta + \alpha , \tag{14}$$

where

$$\beta = \int_{\bar{z}_0}^{z_0} \sqrt{q(t)}dt > 0 .$$

Finally we obtain

$$\omega_{21}(\lambda) = i\exp\{-\lambda\beta - i\lambda\alpha\}[1 + O(\lambda^{-1})] . \tag{15}$$

From (12)–(15) we find that

$$\omega_{11}(\lambda) = \exp\{i(\lambda\alpha + \phi(\lambda))\}\sqrt{1 + \exp\{-2\lambda\beta\}(1 + \psi(\lambda))} , \tag{16}$$

where $\phi(\lambda)$ and $\psi(\lambda)$ are asymptotic expansions in powers of λ^{-1} beginning with λ^{-1}, and the functions $\phi(\lambda)$ and $\psi(\lambda)$ are real.

4. Asymptotic Behaviour of Δ_n. Let the nth-lacuna be $(\lambda_{1n}, \lambda_{2n})$, $\lambda_{1n} < \lambda_{2n}$; then the asymptotic expansions for λ_{jn} in powers of n^{-1} coincide and have the form (7). We restrict ourselves to the case where n is even. Then

$$\text{Re}\ \omega_{11}(\lambda_{1n}) = \text{Re}\ \omega_{11}(\lambda_{2n}) = 1 ,$$

and therefore

$$\cos[\alpha\lambda_{jn} + \phi(\lambda_{jn})] = 1 - \frac{1}{2}\exp\{-2\beta\lambda_{jn}\}[1 + O(n^{-1})] , \quad j = 1, 2 . \tag{17}$$

Let $\tilde{\lambda}_n$ be the point of the lacuna at which $\cos[\alpha\lambda + \phi(\lambda)] = 1$. Then

$$\lambda_{1n} = \tilde{\lambda}_n - \alpha_n , \quad \lambda_{2n} = \tilde{\lambda}_n + \beta_n , \quad \alpha_n > 0 , \quad \beta_n > 0.$$

Expanding the left hand side of (17) in its Taylor series at $\tilde{\lambda}_n$ and adding these expressions, we obtain

$$\Delta_n = \frac{2}{\alpha}\left\{-\frac{n\pi\beta}{\alpha}\right\}[1 + O(n^{-1})] , \tag{18}$$

where α, β are as in (13), (14).

5. Supplements. Suppose that there are two simple turning points z_1 and z_2 on the curve $\Gamma_T^+ = \Gamma^+ \cap \{0 \leqslant \operatorname{Re} z \leqslant T\}$. Then

$$\Delta_n = \frac{2|\cos(n\pi\gamma/\alpha) + O(n^{-1})|}{\alpha\sqrt{\beta+1}} \exp\left\{-\frac{n\pi\beta}{\alpha}\right\}.$$

Here α has the same form as above,

$$\beta = \int_{\bar{z}_1}^{z_1} \sqrt{q(z)}dz\,, \quad \gamma = \int_{z_1}^{z_2} \sqrt{q(z)}dz > 0\,.$$

Let $q(z)$ be meromorphic, and suppose that there is one simple turning point z_0 and one simple pole z_1 on Γ_T^+, where Γ_T^+ has the same form as above. Suppose also that the Stokes line joining z_0 and z_1 lies in $[D]$. Then

$$\Delta_n = \frac{8|\sin(n\pi\gamma/\alpha) + O(n^{-1})}{\alpha^2\sqrt{1+\beta}} \exp\left\{-\frac{n\pi\beta}{\alpha}\right\},$$

where α, β are as in (13), (15), and

$$\gamma = \int_{z_0}^{z_1} \sqrt{q(t)}dt > 0\,.$$

These formulae are proved by means of transition matrices which are used to calculate monodromy matrix $\Omega(\lambda)$. A resonance feature arises in both of these cases.

Chapter 4. Second-Order Equations with Turning Points

In this chapter we consider equations of the form

$$y'' + \lambda^2 q(x,\ \lambda^{-1})y = 0\,, \quad \lambda \to \infty\,,$$

which have turning points. We derive asymptotic formulae for the solutions that are applicable in a real or complex neighbourhood of a turning point.

§ 1. Simple Turning Points. The Real Case

1. Formal Asymptotic Solutions. We consider the equation

$$y'' - \lambda^2 q(x)y = 0\,, \tag{1}$$

where $\lambda > 0$ is a large parameter, and $x \in I = [a, b]$. Suppose that the following conditions are satisfied:

1) the function $q(x)$ is real with $q(x) \in C^\infty(I)$;
2) equation (1) has one, and moreover simple, turning point x_0, $a < x_0 < b$.

Then $q(x_0) = 0$ and $q'(x_0) \neq 0$; we assume that $q'(x_0) > 0$ for definiteness. For small $|x - x_0|$ equation (1) can be approximated by the equation $y'' - \lambda^2 q'(x_0)(x - x_0)y = 0$, the solutions of which are the Airy functions $w(\lambda^{2/3}(q'(x_0))^{2/3}(x - x_0))$.

The F.A.S. of equation (1) is sought in the following form, due to F. Olver [Olver 1]:

$$\begin{gathered} y = Aw(\lambda^{2/3}\xi(x)) + \lambda^{-1/3}Bw'(\lambda^{2/3}\xi(x))\,, \\ A = \sum_{n=0}^{\infty} \lambda^{-n}A_n(x)\,, \quad B = \sum_{n=0}^{\infty} \lambda^{-n}B_n(x)\,. \end{gathered} \tag{2}$$

Here $\xi(x)$, $A_n(x)$, $B_n(x)$ are unknown functions, and $w(t)$ is a solution of the Airy equation $w'' - tw = 0$. Substituting (2) into equation (1) and equating the coefficients of the functions w and w' to zero, we obtain

$$\begin{aligned} \lambda^2 A({\xi'}^2\xi - q) + 2\lambda B'\xi'\xi + \lambda B(\xi'\xi)' + A'' &= 0\,, \\ \lambda^2 B({\xi'}^2\xi - q) + \lambda(2A'\xi' + A\xi'') + B'' &= 0\,. \end{aligned}$$

Substituting into this system the asymptotic expansions for $A(x)$ and $B(x)$, and equating to zero the coefficients of powers of λ^{-1}, we obtain the equation for $\xi(x)$

$$\xi'^2(x)\xi(x) = q(x)\,. \tag{3}$$

We put

$$\xi(x) = \left(\frac{3}{2}\int_{x_0}^{x} \sqrt{q(t)}dt\right)^{2/3} = \left(\frac{3}{2}S(x_0,x)\right)^{2/3}\,. \tag{4}$$

The function $\xi(x)$ is real, $\xi(x) \in C^\infty(I)$, sgn $\xi(x) = $ sgn $(x - x_0)$ and

$$\xi(x) \sim [q'(x_0)]^{2/3}(x - x_0)\,, \qquad x \to x_0\,.$$

We obtain the recurrence system of equations for the functions A_n and B_n

$$\begin{gathered} 2\sqrt{\xi'\xi}(B_n\sqrt{\xi'\xi})' + A''_{n-1} = 0\,, \\ 2\sqrt{\xi'}(A_n\sqrt{\xi'})' + B''_{n-1} = 0\,, \\ n = 0,\ 1,\ \ldots\,,\quad A_{-1} = B_{-1} = 0\,. \end{gathered}$$

In particular $A_0(x) = c[\xi'(x)]^{-1/2}$. The functions $A_n(x)$ and $B_n(x)$ are required to be smooth for $x \in I$, and hence

$$A_{2n+1}(x) = 0\,, \quad B_{2n}(x) = 0\,, \quad n = 0,\ 1,\ \ldots\,.$$

From the recurrence relations we obtain

$$\begin{aligned} A_n(x) &= \frac{1}{\sqrt{\xi'(x)}}\left[c_n - \sqrt{\frac{1}{2}}\int_{x_0}^{x}\frac{B''_{n-1}(t)}{\sqrt{\xi'(t)}}dt\right]\,, \\ B_n(x) &= \frac{1}{2\sqrt{\xi(x)\xi'(x)}}\int_{x_0}^{x}\frac{A''_{n-1}(t)}{\sqrt{\xi(t)\xi'(t)}}dt\,, \end{aligned} \tag{5}$$

where the c_n are constants. In these formulae $\sqrt{\xi'(x)} > 0$, and for definiteness $\sqrt{\xi(x)} > 0$ for $x > x_0$ and $\sqrt{\xi(x)} = i|\sqrt{\xi(x)}|$ for $x < x_0$ (in fact the choice of the branch of $\sqrt{\xi(x)}$ is not important). We note that the lower limit of the integral in (5) is the turning point x_0. For any other choice of this limit the functions A_n and B_n will have a singularity at x_0. Finally we obtain a F.A.S of equation (1) in the form

$$\begin{aligned} y &= \left[\frac{1}{\sqrt{\xi'(x)}} + \sum_{n=1}^{\infty}\frac{A_{2n}(x)}{\lambda^{2n}}\right] w(\lambda^{2/3}\xi(x)) \\ &\quad + \lambda^{-4/3}\sum_{n=0}^{\infty}\frac{B_{2n+1}(x)}{\lambda^{2n}} w'(\lambda^{2/3}\xi(x))\,. \end{aligned} \tag{6}$$

Another form of F.A.S. has been suggested by T. Cherry [Cherry]:

$$y = Aw(\lambda^{2/3}\xi), \quad A = \sum_{n=0}^{\infty} A_n(x)\lambda^{-n}, \quad \xi = \sum_{n=0}^{\infty} \xi_n(x)\lambda^{-n}. \tag{7}$$

Substituting this expression into equation (1) and equating the coefficients of w and w' to zero, we obtain the relations

$$\lambda^2 A(\xi\xi'^2 - q) + A'' = 0, \quad 2A'\xi' + A\xi'' = 0.$$

Replacing the functions A and ξ by their asymptotic expansions and equating the coefficients of powers of λ^{-1} to zero, we obtain the recurrence system of equations

$$\xi_0\xi_0'^2 - q = 0, \quad 2A_0'\xi_0' + A_0\xi_0'' = 0,$$

$$A_0\xi_0'(\xi_1\xi_0' + 2\xi_0\xi_1') = 0. \; \ldots,$$

from which we can find successively ξ_0, A_0, ξ_1, A_1, ... The functions $\xi_0(x)$ and $A_0(x)$ coincide with the above functions $\xi(x)$ and $A(x)$, but the formulae for successive terms of the expansion (7) turn out to be more complicated than the formulae (5). Other ways for constructing the F.A.S. are given in paragraph 3.3.

2. Airy Functions. An *Airy function* is a solution of the *Airy equation*

$$y'' - xy = 0. \tag{8}$$

2.1 Integral Representations and Series. The solutions of the Airy equation are expressed in terms of the Bessel functions of order 1/3:

$$y(x) = \sqrt{x} J_{1/3}\left(i\frac{2}{3}x^{3/2}\right).$$

But because the Airy equation plays an important role in applications and in the asymptotic theory of linear differential equations we consider the Airy functions independently.

In the literature there are two different notations for Airy functions: $\mathrm{Ai}(x)$, $\mathrm{Bi}(x)$ and $v(x)$, $w_1(x)$, $w_2(x)$ (the Fock notation). We use the latter notation. The connection between these functions is as follows:

$$\mathrm{Ai}(x) = \frac{v(x)}{\sqrt{\pi}}, \quad \mathrm{Bi}(x) = \frac{1}{2\sqrt{\pi}}(w_1(x) + w_2(x)).$$

The functions $\mathrm{Ai}(x)$, $\mathrm{Bi}(x)$ are real for real x.

The functions $v(x)$, $w_1(x)$, $w_2(x)$ have integral representations

$$v(x) = \frac{1}{2\sqrt{\pi}} \int_{-\infty}^{\infty} e^{i(xt+t^3/3)} dt \,,$$

$$w_1(x) = \frac{1}{\sqrt{\pi}} \left(\int_{\infty e^{-2\pi i/3}}^{0} + \int_0^{\infty} \right) e^{xt-t^3/3} dt \,,$$

$$w_2(x) = \frac{1}{\sqrt{\pi}} \left(\int_{\infty e^{2\pi i/3}}^{0} + \int_0^{\infty} \right) e^{xt-t^3/3} dt \,.$$

All Airy functions are entire functions of x. We have the identities

$$v(z) = \frac{w_1(z) - w_2(z)}{2i} \,, \quad \overline{w_2(z)} = w_1(\overline{z}) \,.$$

In particular, for real x the function $v(x)$ is real and $\overline{w_2(x)} = w_1(x)$.

If $y(x)$ is a solution of the Airy equation then the function $y(e^{2\pi i/3}x)$ is also a solution. This leads to the identities

$$w_1(xe^{2\pi i/3}) = e^{i\pi/3} w_2(x) \,, \quad w_2(xe^{-2\pi i/3}) = 2e^{i\pi/6} v(x) \,.$$

The function v has series expansion

$$v(z) = \frac{1}{3^{2/3}\sqrt{\pi}} \sum_{n=0}^{\infty} \frac{\Gamma((n+1)/3)}{n!} \sin\left[\frac{2}{3}(n+1)\pi\right] (3^{1/3}z)^n \,,$$

which converges for all z. We also note that

$$w_1(0) = \frac{2\sqrt{\pi}e^{-i\pi/6}}{3^{2/3}\Gamma(2/3)} \,, \quad w_1'(0) = \frac{2\sqrt{\pi}e^{-i\pi/6}}{3^{4/3}\Gamma(4/3)} \,,$$

$$v(0) = \operatorname{Im}\, w_1(0) \,, \qquad v'(0) = \operatorname{Im}\, w_1'(0) \,.$$

2.2 Asymptotic Expansions. For real $x \to +\infty$ we have the asymptotic expansions

$$\begin{aligned} w_1(x) &= \frac{1}{\sqrt{\pi}} x^{-1/4} e^{2x^{3/2}/3} \sum_{n=0}^{\infty} a_n x^{-3n/2} \,, \\ v(x) &= \frac{1}{2\sqrt{\pi}} x^{-1/4} e^{-2x^{3/2}/3} \sum_{n=0}^{\infty} (-1)^n a_n x^{-3n/2} \,, \end{aligned} \tag{9}$$

where

$$a_n = \frac{\Gamma(3n+1/2)}{(2n)!} 9^{-n} \,.$$

These expansions can be differentiated in x an arbitrary number of times. For real $x \to -\infty$ there are the asymptotic expansions

$$w_1(x) = \frac{e^{i\pi/4}}{\sqrt{\pi}}(-x)^{-1/4} \exp\left\{i\left(\frac{2}{3}(-x)^{3/2} + \frac{\pi}{4}\right)\right\} \sum_{n=0}^{\infty} i^{-n} a_n (-x)^{-3n/2},$$

$$v(x) = \frac{1}{\sqrt{\pi}}(-x)^{-1/4} \operatorname{Im}\left\{\exp\left\{i\left(\frac{2}{3}(-x)^{3/2} + \frac{\pi}{4}\right)\right\} \times \sum_{n=0}^{\infty} i^{-n} a_n (-x)^{-3n/2}\right\}, \tag{10}$$

where $\sqrt{-x} > 0$, $\sqrt[4]{-x} > 0$, and the a_n are as in (9).

In particular, the function $v(x)$ decreases exponentially as $x \to +\infty$ and oscillates as $x \to -\infty$:

$$v(x) \sim \frac{1}{2\sqrt{\pi}} x^{-1/4} e^{-2x^{3/2}/3}, \quad x \to +\infty,$$

$$v(x) = \frac{1}{\sqrt{\pi}} |x|^{-1/4} \left[\sin\left(\frac{2}{3}|x|^{3/2} + \frac{\pi}{4}\right) + O(|x|^{-3/2})\right], \quad x \to -\infty.$$

The functions $w_1(x)$ and $w_2(x)$ oscillate rapidly for $x < 0$ and increase exponentially as $x \to +\infty$.

Next we derive the asymptotic formulae for the Airy functions for complex z. Let l_1, l_2 and l_3 be the rays $\arg z = \pi$, $\arg z = \pi/3$ and $\arg z = -\pi/3$ (see Fig. 3). These rays are the Stokes lines. We remove from the complex plane a sector S_2 containing the Stokes line l_2; that is $|\arg z - \pi/3| \geqslant \varepsilon > 0$ outside S_2. The asymptotic behaviour of the function $w_1(z)$ outside S_2 is given by formula (9) where the branches of $\sqrt[4]{z}$ and $\sqrt{z}$ are positive for positive z. Inside the sector S_2

$$w_1(z) = -w_1(ze^{2\pi i/3}) - w_1(ze^{-2\pi i/3}).$$

2.3 Zeros of Airy Functions. All the zeros of $v(z)$ and its derivatives are real, simple and lie on the half-line $(-\infty, 0)$. All the zeros of $w_1(z)$ and $w_1'(z)$ lie on the ray $\arg z = \pi/3$, while all the zeros of $w_2(z)$ and $w_2'(z)$ lie on the ray $\arg z = -\pi/3$; that is, the zeros of all Airy functions and their derivatives lie on the Stokes lines l_1, l_2 and l_3.

The zeros t_s and t_s' of the functions $v(x)$ and $v'(x)$ have the asymptotic behaviour

$$t_s = \left[\frac{3}{2}\pi\left(s - \frac{1}{4}\right)\right]^{2/3} + O(s^{-4/3}),$$

$$t_s' = \left[\frac{3}{2}\pi\left(s - \frac{3}{4}\right)\right]^{2/3} + O(s^{-4/3}).$$

3. Asymptotic Behaviour of the Solutions

3.1 Equation (1). Suppose that conditions 1) and 2) are satisfied. Then for arbitrary $N \geqslant 1$ equation (1) has a solution of the form

$$y_0(x,\lambda) = \left[\frac{1}{\sqrt{\xi'(x)}} + \sum_{n=1}^{N} \frac{A_{2n}(x)}{\lambda^{2n}} + O\left(\frac{1}{\lambda^{2N+2}}\right)\right] v(\lambda^{2/3}\xi(x))$$
$$+ \lambda^{-4/3}\left[\sum_{n=0}^{\infty} \frac{B_{2n+1}(x)}{\lambda^{2n+1}} + O\left(\frac{1}{\lambda^{2N+2}}\right)\right] v'(\lambda^{2/3}\xi(x)), \qquad (11)$$

Here v is the Airy-Fock function, and the coefficients A_n, B_n of the asymptotic series are determined from (5) where $c_n = 0$ for $n \geqslant 1$. The bounds for the remainder terms are uniform in $x \in I$. The asymptotic series (11) can be differentiated in x and λ an arbitrary number of times while preserving the uniformity in x of the bounds for the remainder terms.

This remark applies to all the asymptotic expansions discussed below. The principal asymptotic term has the form

$$y_0(x,\lambda) = \left[\frac{1}{\sqrt{\xi'(x)}} + O\left(\frac{1}{\lambda^2}\right)\right] v(\lambda^{2/3}\xi(x)) - \frac{1}{2\lambda^{4/3}}\left[\frac{1}{\sqrt{\xi'(x)\xi(x)}}\right.$$
$$\left.\times \int_{x_0}^{x} \left(\frac{1}{\sqrt{\xi'(t)}}\right)'' \frac{dt}{\sqrt{\xi'(t)\xi(t)}} + O\left(\frac{1}{\lambda^2}\right)\right] v'(\lambda^{2/3}\xi(x)). \qquad (12)$$

In particular, at the turning point,

$$y_0(x_0,\lambda) \sim \frac{\sqrt{\pi}}{3^{2/3}\Gamma(2/3)(q'(x_0))^{2/3}}. \qquad (13)$$

The solution y_0 is rapidly oscillating for $x \in x_0$, and the principal asymptotic term is

$$y_0(x,\lambda) = \frac{1}{\sqrt{\xi'(x)}} v(\lambda^{2/3}\xi(x)) + O\left(\frac{1}{\lambda^{4/3}}\right), \quad x \leqslant x_0.$$

The function $v(x)$ has infinitely many zeros $t_k, \ldots < t_k < t_{k-1} < \ldots < t_1 < 0$. Therefore, close to x_k, one must include the term containing v' in the principal asymptotic term; see (12). For $t \geqslant 0$ we have $v(t) \neq 0$ and the principal asymptotic term has the form

$$y_0(x,\lambda) = v(\lambda^{2/3}\xi(x)) \frac{1}{\sqrt{\xi'(x)}} \left[1 + O\left(\frac{1}{\lambda^{2/3}}\right)\right].$$

Equation (1) also has solutions of the form

$$y_j(x,\lambda) = \left[\frac{1}{\sqrt{\xi'(x)}} + \sum_{n=1}^{N} \frac{A_{2n}(x)}{\lambda^{2n}} + O\left(\frac{1}{\lambda^{2N+2}}\right)\right] w_j(\lambda^{2/3}\xi(x))$$
$$+ \lambda^{-4/3}\left[\sum_{n=0}^{N} \frac{B_{2n}(x)}{\lambda^{2n}} + O\left(\frac{1}{\lambda^{2N+2}}\right)\right] w_j'(\lambda^{2/3}\xi(x)),$$
$$j = 1, 2, \qquad (14)$$

where w_1, w_2 are Airy functions. The principal asymptotic term is

$$y_j(x, \lambda) = w_j(\lambda^{2/3}\xi(x))[1 + O(\lambda^{-2/3})].$$

The solutions y_j are rapidly oscillating for $x \leqslant x_0$ and increase exponentially for $x \geqslant x_0 + \delta$, $\delta > 0$, and $\lambda \to +\infty$. We can choose the solutions to be complex conjugates: $\overline{y_2(x, \lambda)} = y_1(x, \lambda)$.

Airy functions can be replaced by their asymptotic expansions under the condition $\lambda^{2/3}|\xi(x)| \gg 1$; that is, for $|x - x_0| \gg \lambda^{-2/3}$. Therefore the thickness of the boundary layer in which formulae (11) and (14) cannot be simplified has order $O(\lambda^{-2/3})$.

3.2 Connection Formulae. Let $\delta > 0$ be fixed and independent of λ. Then for $x \notin (x_0 - \delta, x_0 + \delta)$ and $\lambda \to +\infty$ there are asymptotic formulae for y_j of the same form as those in Chap. 2, § 3:

$$y_1(x, \lambda) = \frac{\lambda^{-1/6}}{\sqrt{\pi}} q^{-1/4}(x) e^{\lambda S}[1 + \lambda^{-1}\phi_1^+(x, \lambda)], \quad x \geqslant x_0 + \delta, \tag{15}$$

$$y_1(x, \lambda) = \frac{\lambda^{-1/6}}{\sqrt{\pi}} e^{i\pi/4} |q(x)|^{-1/4} e^{i\lambda|S|}[1 + \lambda^{-1}\phi_1^-(x, \lambda)], \quad x \leqslant x_0 - \delta,$$

$$y_0(x, \lambda) = \frac{\lambda^{-1/6}}{2\sqrt{\pi}} q^{-1/4}(x) e^{-\lambda S}[1 + \lambda^{-1}\phi_0(x, \lambda)], \quad x \geqslant x_0 + \delta,$$

$$y_0(x, \lambda) = \frac{\lambda^{-1/6}}{2\sqrt{\pi}} |q(x)|^{-1/4} \exp\left\{i\lambda|S| - \frac{i\pi}{4}\right\}[1 + \lambda^{-1}\psi_+(x, \lambda)]$$

$$- \exp\left\{-i\lambda|S| + \frac{i\pi}{4}\right\}[1 + \lambda^{-1}\psi_-(x, \lambda)], \quad x \leqslant x_0 - \delta. \tag{16}$$

In these formulae $S = \int_{x_0}^{x} \sqrt{q(t)}dt$, $S > 0$ for $x > x_0$, and $\psi_j^\pm$, $\psi_\pm$ are asymptotic series in powers of λ^{-1}. Formulae (15), (16) are called *connection formulae* : if the asymptotic expansions of the solution is known on one side of a turning point we can use them to find the asymptotic expansion on the other side.

Let $I = (-\infty, b]$, let conditions 1), 2) be satisfied, and let $q(x)$ be a polynominal. Then there are asymptotic expansions of the form (11), (14) as $\lambda \to +\infty$ which are uniform in $x \in I$. Further, these asymptotic expansions are dual, that is, they hold for $\lambda > 0$ fixed, and $x \to -\infty$. This is also true when all the integrals $\int_{-\infty} |\alpha_k(x)|dx$, $k = 1, 2, \ldots$, converge.

3.3 Additional Parameters. We consider the equation

$$y'' - \lambda^2 q(x, \alpha) y = 0 \tag{17}$$

on the interval I. Here α is a real parameter and $\alpha \in J = [-\alpha_0, \alpha_0]$, $\alpha_0 > 0$. We introduce the conditions:

(1) $q(x, \alpha)$ is real and $q \in C^\infty(I \times J)$;

(2) $q(x_0, 0) = 0$, $q'_x(x_0, 0) \neq 0$, $q'_\alpha(x_0, 0) \neq 0$ and $q(x, 0) \neq 0$ for $x \in I$, $x \neq 0$.

Suppose that $q'_x(x_0, 0) > 0$ and $q'_\alpha(x_0, 0) < 0$ for definiteness. Then for small α_0 (17) has a unique, and moreover simple, turning point $x = x_0(\alpha)$ on I where $x_0(0) = 0$ and $x_0(\alpha) \in C^\infty(J)$. We seek an F.A.S. of (17) in the form (2), where A_n, B_n, ξ are functions of (x, α). Then we obtain

$$\xi(x, \alpha) = \left(\frac{3}{2}\int_{x_0(\alpha)}^{x} \sqrt{q(t, \alpha)}dt\right)^{2/3}.$$

The coefficients A_n and B_n are determined from formulae (5), with $C_n = 0$, and $x_0 = x_0(\alpha)$. If α_0 is sufficiently small then A_n, B_n, $\xi \in C^\infty(I \times J)$. Equation (17) has a solution of the form

$$y_0(x, \lambda, \alpha) = \left[\frac{1}{\sqrt{\xi'_x(x, \alpha)}} + \sum_{n=1}^{N} \frac{A_n(x, \alpha)}{\lambda_n} + O\left(\frac{1}{\lambda^{n+1}}\right)\right] v(\lambda^{2/3}\xi(x,\ \alpha))$$
$$+ \lambda^{-1/3}\left[\sum_{n=0}^{N} \frac{B_n(x, \alpha)}{\lambda^n} + O\left(\frac{1}{\lambda^{N+1}}\right)\right] v'(\lambda^{2/3}\xi(x,\ \alpha)) \quad (18)$$

and the solutions y_1 and y_2 have the same form but with w_1 and w_2 in place of v. The bounds for the remainder terms are uniform in $(x, \alpha) \in I \times J$. The difference between formulae (6) and (17) is that the coefficients A_{2n+1} and B_{2n} vanish in the first case.

In many problems there arises an equation of the form

$$y'' - \lambda^2 q(x,\ \lambda^{-1})y = 0. \quad (19)$$

Let J be the interval $0 \leqslant \varepsilon \leqslant \varepsilon_0$ and suppose that the following conditions are satisfied:

1) the function $q(x, \varepsilon)$ is real and $q \in C^\infty(I \times J)$;

2) $q(x, \varepsilon)$ has an asymptotic expansion

$$q(x, \varepsilon) = \sum_{j=0}^{\infty} q_j(x)\varepsilon^j, \quad \varepsilon \to +0,$$

which is uniform in $x \in I$. Here $q_j(x) \in C^\infty(I)$ and the asymptotic series can be differentiated in x an arbitrary number of times;

3) the function $q_0(x)$ satisfies conditions 1) and 2) from paragraph 1.

Then for $\varepsilon_0 \ll 1$, $\varepsilon \in J$, equation (19) has unique turning point $x = x_0(\varepsilon)$, where $x_0(\varepsilon) \to x_0$ as $\varepsilon \to +0$ and $x_0(\varepsilon)$ has asymptotic expansion

$$x_0(\varepsilon) = x_0 + \sum_{j=1}^{\infty} \varepsilon^j x_j, \quad \varepsilon \to +0, \quad x_1 = -\frac{q_1(0)}{q'_0(0)}.$$

Equation (19) has solutions y_0, y_1, y_2 which have asymptotic expansions of the form (18) for $\alpha = \lambda^{-1}$. The function $\xi(x,\varepsilon)$ must satisfy equation (3); we write it in the form

$$\xi(x,\varepsilon) = \left(\frac{3}{2}\int_{x_0(\varepsilon)}^{x+x_0(\varepsilon)-x_0} \sqrt{q(t,\varepsilon)}dt\right)^{2/3}.$$

Then as $\lambda \to +\infty$ uniformly in $x \in I$

$$\xi(x,\ \varepsilon) = \xi_0(x)[1 + \lambda^{-1}\xi_1(x) + O(\lambda^{-2})],$$

$$\xi_0(x) = \left(\frac{3}{2}\int_{x_0}^{x} \sqrt{q(t,0)}dt\right)^{2/3},$$

$$\xi_1(x) = \frac{1}{3}\xi_0^{-1}(x)[q_0'(0)]^{-1}\int_{x_0}^{x} \frac{q_0'(0)q_1(t) - q_x'(t,\ 0)q_1(0)}{t - x_0}dt.$$

We can construct a F.A.S. of equation (19) in another way. We seek a F.A.S. in the form (2), and the function $\xi(x)$ is determined by formula (4) where $q = q_0(x)$. Then we obtain the recurrence relation for the coefficients A_n and B_n

$$2\sqrt{\xi'\xi}(B_n\sqrt{\xi'\xi})' + A_{n-1}'' - \sum_{k=1}^{n+1} A_{n+1-k}q_k = 0,$$

$$2\sqrt{\xi'}(A_n\sqrt{\xi'})' + B_{n-1}'' - \sum_{k=1}^{n+1} B_{n+1-k}q_k = 0,$$

$$n = 0,\ 1,\ \ldots,\quad A_{-1} = B_{-1} = 0.$$

The functions $A_n(x)$, $B_n(x)$ are required to be smooth on the interval I. For $n = 0$ we obtain the system

$$2\sqrt{\xi'\xi}(B_0\sqrt{\xi'\xi})' - q_1A_0 = 0,$$

$$2\sqrt{\xi'}(A_0\sqrt{\xi'}) - q_1B_0 = 0,$$

and from these we find that

$$A_0(x) = \frac{c_0}{\sqrt{\xi'(x)}}\cosh\left[\int_{x_0}^{x}\alpha(t)dt\right],$$

$$B_0(x) = \frac{c_1}{\sqrt{\xi'(x)\xi(x)}}\sinh\left[\int_{x_0}^{x}\alpha(t)dt\right],$$

where

$$\alpha(x) = \frac{q_1(x)}{2\sqrt{q_0(x)}}.$$

The higher approximations are obtained similarly.

3.4 The Asymptotic Formulae of A.A. Dorodnitsyn [Dorodnitsyn]. We consider the equation

$$y'' + [\lambda^2 q(x) + p(x)]y = 0 \tag{20}$$

on the interval $I = [a, b]$, $a < 0 < b$, where $q(x)$ is real-valued and $p(x)$ is complex-valued. Let

$$q(x) = xr(x)\,, \quad r(x) > 0\,, \quad x \in I\,,$$

and $r(x) \in C^1(I)$, $p(x) \in C(I)$. For $x \in I$ equation (20) has the unique, and moreover simple, turning point $x = 0$.

We consider the F.S.S. (U_1, U_2) of the Airy equation

$$U'' + tU = 0$$

such that

$$U_1(0) = 1\,, \quad U_1'(0) = 0\,, \quad U_2(0) = 0\,, \quad U_2'(0) = 1\,.$$

These functions can be expressed in terms of $\mathrm{Ai}(t)$ and $\mathrm{Bi}(t)$:

$$U_1(t) = \frac{3^{2/3}\Gamma(2/3)}{2}\left[\mathrm{Ai}(t) + \frac{\mathrm{Bi}(t)}{\sqrt{3}}\right],$$

$$U_2(t) = \frac{3^{1/3}\Gamma(1/3)}{2}\left[-\mathrm{Ai}(t) + \frac{\mathrm{Bi}(t)}{\sqrt{3}}\right].$$

Let $\xi(x)$ have the form (4) where $x_0 = x$,

$$\phi(x, \lambda) = \lambda^{2/3}\xi(x)\,, \quad f(x) = p(x) + \sqrt{\xi'(x)}\frac{d^2}{dx^2}\left(\frac{1}{\sqrt{\xi'(x)}}\right).$$

Equation (20) has F.S.S. (y_1, y_2) for $\lambda \geqslant \lambda_0 \gg 1$ such that

$$y_j(x, \lambda) = \frac{1}{\sqrt{\xi'(x)}}U_j(\lambda^{2/3}\xi(x))\left[1 + \int_0^x \frac{f(t)}{\sqrt{|q(t)|}}dt + O\left(\frac{1}{\lambda^{4/3}}\right)\right], \quad x \leqslant 0\,. \tag{21}$$

For $x \geqslant 0$ we have

$$\begin{aligned} y_1(x, \lambda) = {} & \frac{1}{\sqrt{\xi'(x)}}U_1(\lambda^{2/3}\xi(x)) + \frac{1}{\lambda\sqrt{\xi'(x)}}\left[\frac{1}{2\sqrt{3}}U_1(\lambda^{2/3}\xi(x)) \right. \\ & \left. - \frac{\sqrt[3]{3}\Gamma^2(2/3)}{2\pi}U_2(\lambda^{2/3}\xi(x))\right]\int_0^x \frac{f(t)}{\sqrt{r(t)}}dt \\ & + \frac{f(x)}{\lambda^{4/3}[\xi'(x)]^{5/2}}[U_1(\lambda^{2/3}\xi(x))V_{12}(\lambda^{2/3}\xi(x)) \\ & - U_2(\lambda^{2/3}\xi(x))V_{11}(\lambda^{2/3}\xi(x))] + o(\lambda^{-1/3}U(\lambda^{2/3}\xi(x)))\,, \end{aligned} \tag{22}$$

where the notation used is

$$U(t) = \sqrt{U_1^2(t) + U_2^2(t)}\,,$$
$$V_{11}(t) = Y_{11}(t) - \frac{3^{1/3}\Gamma(2/3)}{\pi}\sqrt{t}\,,$$
$$V_{12}(t) = Y_{12}(t) - \sqrt{\frac{t}{3}}\,, \quad V_{22}(t) = Y_{22}(t) + 1 - \frac{\Gamma^2(1/3)}{\pi 3^{1/3}}\sqrt{t}\,,$$

$$Y_{11}(t) = U_1'^2 + tU_1^2(t)\,,$$
$$Y_{12}(t) = U_1'(t)U_2'(t) + tU_1(t)U_2(t)\,,$$
$$Y_{22}(t) = U_2'^2(t) - 1 + tU_2^2(t)\,. \tag{23}$$

For $x \geqslant 0$, the solution y_2 has the form

$$\begin{aligned} y_2(x,\lambda) = {} & \frac{U_2(\lambda^{2/3}\xi(x))}{\sqrt{\xi'(x)}} + \frac{1}{\lambda\sqrt{\xi'(x)}} \\ & \times \left[\frac{\Gamma^2(1/3)}{2\pi 3^{1/3}} U_1(\lambda^{2/3}\xi(x)) - \frac{1}{2\cdot 3^{1/2}} V_2(\lambda^{2/3}\xi(x))\right] \int_0^x \frac{f(t)}{\sqrt{q(t)}}\,dt \\ & + \frac{1}{\lambda^{4/3}\sqrt{\xi'(x)}} \left[\frac{f(x)}{(\xi'(x))^2} U_1(\lambda^{2/3}(\xi(x))V_{22}(\lambda^{2/3}\xi(x)) \right. \\ & \left. - U_2(\lambda^{2/3}\xi(x))V_{12}(\lambda^{2/3}\xi(x)) - \frac{f(0)}{(\xi'(0))^2} U_1(\lambda^{2/3}\xi(x))\right] \\ & + o(\lambda^{-4/3}U(\lambda^{2/3}\xi(x)))\,. \end{aligned} \tag{24}$$

These asymptotic formulae, and also the analogous asymptotic expansions for the derivatives y_1' and y_2', were obtained in [Dorodnitsyn].

4. More General Dependence on a Parameter. Let us consider the equation

$$y'' - [\lambda^2 f(x,\ \lambda) + g(x,\ \lambda)]y = 0 \tag{25}$$

on the interval $I:\ a_1 < a < a_2$, where a_1 and a_2 can be finite or infinite. Suppose that for each fixed $\lambda \geqslant \lambda_0 > 0$ the following conditions are satisfied:

1) $f(x,\lambda)(x - x_0)^{-1}$, $x_0 \in I$, is real-valued, non-vanishing and lies in $C^2(I)$;

2) $g(x,\lambda)$ is complex-valued and $g \in C^2(I)$.

Equation (25) has the unique, and moreover simple, turning point $x = x_0$.

4.1 Principal Bounds. Let $f(x,\lambda) > 0$ for $x > x_0$, to be definite. The functions

$$V(t) = 2^{5/6}3^{1/6}\pi^{1/2}\ \mathrm{Ai}((3/2)^{2/3}t)\,,$$
$$\overline{V}(t) = 2^{5/6}3^{-1/3}\pi^{1/2}\ \mathrm{Bi}((3/2)^{2/3}t)$$

form an F.S.S. of the standard equation

$$y'' - (3/2)^2 ty = 0\,,$$

and, further, $V(0) = \overline{V(0)}$. We introduce the auxiliary functions ξ, f, Ω:

$$\xi(x,\lambda) = -\left(\int_{x_0}^{x} |f(t,\lambda)|^{1/2}dt\right)^{2/3}, \qquad a_1 < x \leqslant x_0\,,$$

$$\xi(x,\lambda) = \left(\int_{x_0}^{x} |f(t,\lambda)|^{1/2}dt\right)^{2/3}, \qquad x_0 < x \leqslant a_1\,,$$

$$\tilde{f}(x,\ \lambda) = \left(\frac{2}{3}\right)^2 |f(x,\ \lambda)||\xi|^{-1}, \qquad \Omega(t) = 1 + \sqrt{|t|}\,,$$

and the error control function

$$h(x,\lambda) + \left[\tilde{f}^{-1/4}(x,\lambda)\frac{d^2}{dx^2}\tilde{f}^{-1/4}(x,\lambda) - g(x,\lambda)\tilde{f}^{-1/2}(x,\lambda)\right]\frac{1}{\Omega(\lambda^{2/3}\xi)}\,.$$

The function $\xi(x,\lambda)$ is strictly monotonic increasing for $x \in I$. For the standard equation we have $x_0 = 0$, $g \equiv 0$, $\xi = x$, $\tilde{f} \equiv 1$, and $h \equiv 0$.

Now consider the equation

$$\overline{V}(t) = V(t)\tan\pi/6\,.$$

There are no non-negative roots, and we let $-t_0$ be the negative root with smallest modulus, so that $t_0 \sim 0.279$. We introduce the weight function

$$E(t) = \sqrt{\overline{V}(t)/V(t)}\,, \quad t \geqslant -t_0\,; \quad E(t) = \sqrt{\tan\pi/6}\,, t \leqslant -t_0\,,$$

which is continuous and non-increasing on the real line. Put

$$M(t) = \left(\tan\frac{\pi}{6}V^2(t) + \cot\frac{\pi}{6}\overline{V}^2(t)\right)^{1/2},$$

$$N(t) = \left(\cot\frac{\pi}{6}V'^2(t) + \cot\frac{\pi}{6}\overline{V}'^2(t)\right)^{1/2},$$

for $t \leqslant t_0$ and

$$M(t) = (2V(t)\overline{V(t)})^{1/2}\,,$$

$$N(t) = \left[\frac{V'^2(t)\overline{V}^2(t) + \overline{V}'^2(t)V^2(t)}{V(t)\overline{V(t)}}\right]^{1/2}$$

for $t \geqslant -t_0$. We note that

$$M(t) \sim 2\cdot 3^{-1/4}|t|^{-1/4}\,, \quad N(t) \sim (3/2)M(t)\,, \quad |t| \to \infty\,.$$

Next we write

$$\rho(x_1, x_2) = \left| \int_{x_1}^{x_2} |h(x, \lambda)| dx \right| .$$

Then equation (25) has solutions w_1 and w_2 such that

$$\begin{aligned} w_1(x,\lambda) &= \tilde{f}^{-1/4}(x,\lambda)[\overline{V}(\lambda^{2/3}\xi) + \varepsilon_1] , \\ w_2(x,\lambda) &= \tilde{f}^{-1/4}(x, \ \lambda)V(\lambda^{2/3}\xi) + \varepsilon_2] . \end{aligned} \tag{26}$$

There are bounds for the remainder terms

$$\frac{|\varepsilon_j(x,\lambda)|}{M(\lambda^{2/3}\xi)} , \quad \frac{|\frac{\partial}{\partial x}\varepsilon_j(x,\lambda)|}{\lambda^{2/3}\tilde{f}^{1/2}(x,\lambda)N(\lambda^{2/3}\xi)} \leqslant \frac{\sigma_j}{\rho} E^{\pm 1}(\lambda^{2/3}\xi) \left[\exp\left\{ \frac{\delta}{\lambda^{2/3}} \rho(a_j, x) \right\} - 1 \right] . \tag{27}$$

Here $j = 1, 2$, and the plus sign is taken for $j = 1$. Also

$$\begin{aligned} \delta &= \sup_{t \in R} \left(\frac{1}{3} \sin \frac{\pi}{3} \Omega(t) M^2(t) \right) , \\ \sigma_1 &= \sup_{t \in R} \left(\frac{1}{3} \sin \frac{\pi}{3} \Omega(t) |\overline{V}(t)| E^{-1}(t) M(t) \right) , \\ \sigma_2 &= \sup_{t \in R} \left(\frac{1}{3} \sin \frac{\pi}{3} \Omega(t) |V(t)| E(t) M(t) \right) . \end{aligned}$$

These bounds can be simplified using the inequalities

$$\sigma_1 \leqslant \rho, \ \sigma_2 \leqslant \rho, \ |E(t)M(t)| \leqslant |\overline{V}(t)| , \ |E^{-1}(t)M(t)| \leqslant |V(t)| ,$$

to give

$$\begin{aligned} |\varepsilon_1(x,\lambda) &\leqslant |\overline{V}(\lambda^{2/3}\xi)| \left[\exp\left\{ \frac{\rho}{\lambda^{2/3}} \rho(a_1, x) \right\} - 1 \right] , \\ |\varepsilon_2(x,\lambda) &\leqslant |V(\lambda^{2/3}\xi)| \left[\exp\left\{ \frac{\rho}{\lambda^{2/3}} \rho(a_2, x) \right\} - 1 \right] . \end{aligned} \tag{28}$$

From this it is clear that if $\lambda^{-2/3}\rho(a_1, a_2) = o(1)$ and $\lambda \to +\infty$, then formulae (26) and (27) take the form

$$\begin{aligned} & w_j(x,\lambda) = \tilde{f}^{-1/4}(x, \ \lambda)[V_j(\lambda^{2/3}\xi) + o(|V_j(\lambda^{2/3}\xi)|)] , \\ & V_1 = \overline{V} , \quad V_2 = V . \end{aligned}$$

The bounds (26) and (27) are analogues of the WKB-bounds (Chap. 2, § 2). These bounds and all the subsequent results of Section 4 are due to F. Olver [Olver 2].

4.2 Connection Formulae. Suppose that the following additional conditions are satisfied:

3) $\int_{x_0}^{x} \sqrt{|f(t,\lambda)|}dt$ diverges for $x \to a_1 + 0$ and for $x \to a_2 - 0$;
4) the integral

$$\int_{a_1}^{a_2} \left| |f|^{-1/4} \frac{d^2}{dx^2} |f|^{-1/4} - g|f|^{-1/2} \right| dx \quad \text{converges.}$$

Then for each fixed $\lambda \geqslant \lambda_0 > 0$ equation (25) has solutions $\tilde{w}_1$ and $\tilde{w}_2$ such that

$$\tilde{w}_1(x,\lambda) = |f(x,\lambda)|^{-1/4} \left[\cos\left(\lambda \int_x^{x_0} |f(t,\lambda)|^{1/2} dt \right) + \frac{\pi}{4} + o(1) \right],$$

$$x \to a_1 + 0,$$

and

$$\tilde{w}_2(x,\lambda) = f^{-1/4}(x,\lambda) \exp\left\{ -\lambda \int_{x_0}^{x} \sqrt{f(t,\lambda)} dt \right\} [1 + o(1)],$$

$$x \to a_2 - 0. \tag{29}$$

These solutions are determined by their asymptotic behaviours in a unique way. The connection formulae describe the asymptotic behaviour of $\tilde{w}_1$ as $x \to a_2 - 0$ and of $\tilde{w}_2$ as $x \to a_1 + 0$:

$$\tilde{w}_1(x,\lambda) = (1+k) f^{-1/4}(x,\lambda) \exp\left\{ \lambda \int_{x_0}^{x} \sqrt{f(t,\lambda)} dt \right\} [1 + o(1)],$$

$$x \to a_2 - 0, \tag{30}$$

and

$$\tilde{w}_2(x,\lambda) = 2(1+\gamma)|f(x,\lambda)|^{-1/4} \left[\cos\left(\lambda \int_x^{x_0} |f(t,\lambda)|^{1/2} dt - \frac{\pi}{4} - \beta \right) \right.$$

$$\left. + o(1) \right], \quad x \to a_1 + 0.$$

Here k, γ and β are constants such that

$$\begin{aligned} |k| &\leqslant \sqrt{2}\frac{\sigma_1}{\rho} \left[\exp\left\{ \frac{\rho}{\lambda^{2/3}} \rho(a_1,a_2) \right\} - 1 \right], \\ |\gamma|, \frac{2|\rho|}{\pi} &\leqslant \frac{\sigma_2}{\rho} \left[\exp\left\{ \frac{\rho}{\lambda^{2/3}} \rho(a_1,a_2) \right\} - 1 \right]. \end{aligned} \tag{31}$$

4.3 Asymptotic Bounds for the Remainder Terms. Let the function f be independent of λ, so that (25) has the form

$$y'' - [\lambda^2 f(x) + g(x,\ \lambda)]y = 0. \tag{32}$$

Suppose that the above conditions are satisfied and that for $\lambda \geqslant \lambda_0 > 0$ there are the bounds

$$|g(x,\lambda)| \leqslant c\lambda^{\alpha}, \quad x \in I, \quad \int_{a_1}^{a_2} |g(x,\lambda)||f(x)|^{-1/2}dx \leqslant c\lambda^{\alpha},$$

where $\alpha < 1$. Then

$$\rho(a_1,\ a_2) \leqslant c\lambda^{\alpha_0 - 1/3}, \quad \alpha_0 = \max(\alpha,\ 0),$$

and from (28) it follows that

$$w_j(x,\lambda) = \tilde{f}^{-1/4}(x)[V_j(\lambda^{2/3}\xi) + O(\lambda^{\alpha_0 - 1}V_j(\lambda^{2/3}\xi))].$$

The constants k, γ and β which appear in the connection formulae (30) have order $O(\lambda^{\alpha_0-1})$ as $\lambda \to +\infty$.

§ 2. A Simple Turning Point. The Complex Case

1. Asymptotic Behaviour of the Solutions in the Neighbourhood of a Turning Point. We consider the equation

$$w'' - \lambda^2 q(z)w = 0, \tag{1}$$

where $q(z)$ is holomorphic in a neighbourhood U of the point $z = 0$, and

$$q(0) = 0, \quad q'(0) \neq 0. \tag{2}$$

The point $z = 0$ is a simple turning point of (1). The asymptotic formulae for the solution on an interval of the real line are given in § 1, and it turns out that they are also applicable in a complex neighbourhood of this point. The results are stated below. In addition we consider an alternative approach to the turning point problem to that considered in § 1, namely that of [Wasow 7].

1.1 The Standard System. Let us consider the system of two equations

$$\varepsilon w' = A(z,\ \varepsilon)w, \tag{3}$$

where the matrix function $A(z,\varepsilon)$ is holomorphic in (z,ε) in the domain $D\ :\ |z| \leqslant r,\ |\varepsilon| \leqslant \varepsilon_0$, so that

$$A(z,\varepsilon) = \sum_{n=0}^{\infty} A_n(z)\varepsilon^n.$$

The numbers r and ε_0 are assumed to be sufficiently small. Let

$$A_0(z) = \begin{bmatrix} 0 & 1 \\ z & 0 \end{bmatrix}. \tag{4}$$

It turns out that, by using the transformation

$$w = T(z,\ \varepsilon)u\,, \tag{5}$$

system (3) can be reduced to the form

$$\varepsilon u' = A_0(z)u\,. \tag{6}$$

The solutions of the system so obtained are

$$u_1 = w(\varepsilon^{-2/3}z)\,, \quad u_2 = \varepsilon^{1/3}w'(\varepsilon^{-2/3}z)\,,$$

where $w(t)$ is a solution of the Airy equation $w'' - tw = 0$. The matrix $T(z,\varepsilon) \in C^\infty(D)$ for each fixed $\varepsilon \in [0,\varepsilon_0]$. Also, it is holomorphic in z for $|z| \leqslant r$ and can be expanded in an asymptotic series

$$T(z,\varepsilon) = \sum_{n=0}^{\infty} T_n(z)\varepsilon^n\,, \quad \varepsilon \to +0\,, \tag{7}$$

uniformly in z. Moreover $\det T_0(0) = 1$.

We now show how to calculate the matrices $T_n(z)$; in this way we will obtain the asymptotic expansions for the solutions of (3) in the whole complex neighbourhood of the turning point $z = 0$. Substituting (5) and (7) into (3), we obtain from (6) the recurrence system of equations

$$\begin{aligned} &A_0T_0 - T_0A_0 = 0\,, \\ &A_0T_n - T_nA_0 = \frac{d}{dz}T_{n-1} - \sum_{k=1}^{n} A_kT_{n-k}\,, \quad n > 1\,. \end{aligned} \tag{8}$$

It follows from the form of the matrix $A_0(z)$ that any solution of the first equation in (8) has the form

$$T_0(z) = t_1(z)I + t_2(z)A_0(z)\,, \tag{9}$$

where $t_1(z)$ and $t_2(z)$ are arbitrary functions.

Further, the matrix equation

$$A_0(z)X(z) - X(z)A_0(z) = F(z)\,,$$

where $A_0(z)$ has the form (4), is solvable if and only if

$$\mathrm{Sp}\ F(z) = 0\,, \quad \mathrm{Sp}(F(z)A_0(z)) = 0^*$$

This fact is proved in an elementary way. So the second equation in (8) for $n = 1$ is solvable if and only if

$$\mathrm{Sp}\left(\frac{dT_0}{dz} - A_1T_0\right) = 0\,, \quad \mathrm{Sp}\left(A_0\frac{dT_0}{dz} - A_0A_1T_0\right) = 0\,,$$

* Here $\mathrm{Sp}(M)$ is the trace of the matrix M.

which leads to a system of equations for $t_1(z)$ and $t_2(z)$:

$$\begin{aligned} t_1' &= at_1 + bt_2\,, & zt_2' &= bt_1 + \left(za - \frac{1}{2}\right) t_2\,, \\ a(z) &= \frac{1}{2}\mathrm{Sp}\ A_1(z)\,, & b(z) &= \frac{1}{2}\mathrm{Sp}\ (A_0(z)A_1(z))\,. \end{aligned} \tag{10}$$

This system can be rewritten as $zt' = B(z)t$, where

$$B(0) = \begin{bmatrix} 0 & 0 \\ b(0) & -1/2 \end{bmatrix}.$$

The eigenvalues of $B(0)$ are $\lambda_1 = 0$ and $\lambda_2 = -1/2$, and so system (10) has a solution $t(z)$ that is holomorphic at $z = 0$ (Chap. 1,§ 2). This solution can be normalized by the condition $t_1(0) = 1$. The matrix $T_0(z)$ so obtained is holomorphic at $z = 0$ and $\det T_0(0) = 1$ (see (9)). In a similar way it can be proved that the matrices $T_1(z)$, $T_2(z)\ldots$ can be chosen so that they are holomorphic at $z = 0$ and satisfy system (8).

Let us integrate system (10). We make the substitution

$$t_1(z) = A(z)\tilde{t}_1(z)\,, \quad t_2(z) = z^{-1/2}A(z)\tilde{t}_2(z)\,,$$

$$A(z) = \exp\left\{\int_0^x a(t)dt\right\}.$$

Then we obtain the system

$$\tilde{t}_1' = f\tilde{t}_2\,, \quad \tilde{t}_2' = f\tilde{t}_1\,, \quad f(z) = z^{-1/2}b(z)\,.$$

The functions $\tilde{t}_1(z)$ and $\tilde{t}_2(z)$ satisfy the equation

$$y'' = \frac{f'(z)}{f(z)}y' + f^2(z)y\,,$$

any solution of which has the form

$$y(z) = c_1 \exp\left\{\int_0^z f(t)dt\right\} + c_2 \exp\left\{-\int_0^z f(t)dt\right\}.$$

If $c_1 = c_2$ then $y(z)$ is holomorphic at $z = 0$; if $c_1 = -c_2$ then $z^{-1/2}y(z)$ is holomorphic at $z = 0$. Finally we obtain

$$t_1(z) = A(z)\cosh B(z)\,, \quad t_2 = A(z)z^{-1/2}\sinh B(z)\,, \tag{11}$$

where

$$\begin{aligned} A(z) &= \exp\left\{\frac{1}{2}\int_0^z \mathrm{Sp}\, A_1(t)dt\right\}, \\ B(z) &= \frac{1}{2}\int_0^z t^{-1/2}\,\mathrm{Sp}(A_0(t)A_1(t))dt\,. \end{aligned} \tag{12}$$

For $|z| \leqslant r$ and $\varepsilon \to +0$ the standard system (3) has solutions of the form

$$\begin{aligned} w_1(z,\varepsilon) &= [t_1(z) + O(\varepsilon)]w(\varepsilon^{-2/3}z) + \varepsilon^{1/3}[t_2(z) + O(\varepsilon)]w'(\varepsilon^{-2/3}z) \\ w_2(z,\varepsilon) &= [zt_2(z) + O(\varepsilon)]w(\varepsilon^{-2/3}z) + \varepsilon^{1/3}[t_1(z) + O(\varepsilon)]w'(\varepsilon^{-2/3}z) . \end{aligned} \tag{13}$$

Here $w(t)$ is an arbitrary solution of the Airy equation and the bounds for the remainder terms are uniform in z for $|z| \leqslant r$. The functions $t_1(z)$ and $t_2(z)$ are determined by formulae (11) and (12). There exist asymptotic expansions in powers of ε for these solutions.

1.2 The Notion of a Simple Turning Point for System (3). Let us consider system (3) where $A(z,\varepsilon)$ satisfies all the conditions of paragraph 1.1 except that $A_0(z)$ has the form (4). Let $z = 0$ be a turning point of (3); that is, the roots of the characteristic equation

$$p^2 - p \operatorname{Sp} A_0(z) + \det A_0(z) = 0$$

coincide for $z = 0$. Let $D(z)$ be the discriminant of this equation:

$$D(z) = (a_{11}(z) - a_{22}(z))^2 + 4a_{12}(z)a_{21}(z) .$$

Then $D(0) = 0$, and the turning point $z = 0$ is called *simple* if $D'(0) \neq 0$. Thus

$$D(0) = 0 , \quad D'(0) \neq 0 . \tag{14}$$

We make the substitution

$$w(z) = \exp\left\{-\frac{1}{2}\int_0^z \operatorname{Sp} A_0(t)dt\right\} \tilde{w}(z) .$$

Then system (3) takes the form

$$\tilde{w}'(z) = \tilde{A}(z,\varepsilon)\tilde{w} , \quad \tilde{A}(z,\varepsilon) = A(z,\varepsilon) - \frac{1}{2}\operatorname{Sp} A_0(z)I , \tag{15}$$

in which $\operatorname{Sp} \tilde{A}_0(z) = 0$. It is not difficult to show that condition (14) is preserved; that is, $\tilde{D}(0) = 0$, $\tilde{D}'(0) \neq 0$. We have

$$\tilde{A}_0(z) = \begin{bmatrix} a(z) & b(z) \\ c(z) & -a(z) \end{bmatrix} ,$$

$$a^2(0) + b(0)c(0) = 0 .$$

The eigenvalues of $\tilde{A}_0(0)$ are both zero. Also $\tilde{A}_0(0) \neq 0$ since, in the contrary case, $a(0) = b(0) = c(0) = 0$ and $\tilde{D}(z)$ has a zero of multiplicity at least 2 at $z = 0$. This contradicts the condition $\tilde{D}'(0) \neq 0$. Therefore $\tilde{A}_0(0)$ reduces to the Jordan normal form

$$J = \begin{bmatrix} 0 & 1 \\ 0 & 0 \end{bmatrix} .$$

We now show that there exists a matrix $T(z)$ that is holomorphic at $z = 0$ with $\det T(0) \neq 0$ and

$$T^{-1}(z)\tilde{A}_0(z)T(z) = \begin{bmatrix} 0 & 1 \\ -\tilde{D}(z) & 0 \end{bmatrix} \equiv \tilde{B}_0(z). \tag{16}$$

It follows from this relation that

$$T(z) = \begin{bmatrix} ay + bt & y \\ cy - at & t \end{bmatrix}, \tag{17}$$

where $y(z)$ and $t(z)$ are defined as follows, according as $a(0)$ is zero or non-zero.

1) Let $a(0) = 0$; then either $b(0) \neq 0$ and $c(0) = 0$, or $b(0) = 0$ and $c(0) \neq 0$. We restrict ourselves to the first possibility. We can then put

$$t(z) = 1, \quad y(z) = 0.$$

2) Let $a(0) \neq 0$; then $b(0) \neq 0$ and $c(0) \neq 0$. We put

$$y(z) = 1, \quad t(z) = 0.$$

In both cases $\det T(0) \neq 0$.

The substitution $\tilde{w}(z) = T(z)v(z)$, where $T(z)$ has the form (17), reduces (15) to the form $v' = \tilde{B}(z, \varepsilon)v$, where $\tilde{B}_0(z)$ has the form (16). We make the transformation

$$v = P(z)u, \quad \xi = \xi(z),$$

where

$$P(z) = \begin{bmatrix} 1 & 0 \\ 0 & \xi(z) \end{bmatrix}, \quad \xi(z) = \left(\frac{3}{2}\int_0^x \sqrt{-D(t)}dt\right)^{2/3}. \tag{18}$$

By condition (14), $D(z) \sim az$ as $z \to 0$, where $a \neq 0$, and hence

$$\xi(z) \sim (-D'(0))^{1/3} z, \quad z \to 0.$$

We fix the value of the cube root. Then the branch obtained for $\xi(z)$ is holomorphic and single-sheeted at $z = 0$, so that $\xi(z)$ is a one-to-one map of a small neighbourhood U of $z = 0$ onto a small neighbourhood V of $\xi = 0$. The system therefore takes the form

$$\frac{du}{d\xi} = B(\xi, \varepsilon)u, \quad B_0(\xi) = \begin{bmatrix} 0 & 1 \\ \xi & 0 \end{bmatrix}, \tag{19}$$

where the matrix $B(\xi, P)$ has the same analytic properties for the variables ξ, ε as $A(z, \varepsilon)$ has for the variables z, ε. The solutions of system (19) have the form (13) where z must be replaced by ξ. We give the connection between w and u:

$$\begin{bmatrix} w_1 \\ w_2 \end{bmatrix} = \exp\left\{-\frac{1}{2}\int_0^x \mathrm{Sp}\ A_0(t)dt\right\} \begin{bmatrix} ay+bt & \xi \\ cy-at & t\xi \end{bmatrix} \begin{bmatrix} u_1 \\ u_2 \end{bmatrix}, \tag{20}$$

$a = (a_{11} - a_{22})/2, \quad b = a_{12}, \quad c = a_{21}$

and $a_{jk}(z)$ are the elements of $A_0(z)$. The functions $y(z)$, $t(z)$ and $\xi(z)$ are as defined above.

1.3 Equation (1). The substitution $\xi = \xi(z)$, $w = \tilde{w}\sqrt{\xi'}$, where $\xi(z)$ has the form (18), reduces equation (1) to

$$\tilde{w}''_{\xi\xi} - [\lambda^2\xi + f(\xi)]\tilde{w} = 0, \quad f(\xi) = \frac{1}{2}\frac{\xi'''}{\xi'^2} - \frac{3}{4}\frac{\xi''^2}{\xi'^4}. \tag{21}$$

This equation is equivalent to a system of the form (3), (4):

$$\varepsilon\begin{bmatrix} \tilde{w} \\ \varepsilon\tilde{w}' \end{bmatrix}' = \left[\begin{bmatrix} 0 & 1 \\ \xi & 0 \end{bmatrix} + \varepsilon^2 f(\xi)\begin{bmatrix} 0 & 0 \\ 1 & 0 \end{bmatrix}\right]\begin{bmatrix} \tilde{w} \\ \tilde{w}' \end{bmatrix}, \tag{22}$$

where $\varepsilon = \lambda^{-1}$. The asymptotic formulae obtained in §1 (for example (12) and (14)) are applicable in a small neighbourhood of the turning point $z = 0$.

2. The Global Asymptotic Behaviour of the Solutions. Let us consider equation (1) where $q(z)$ is a polynomial of degree $n \geqslant 1$ for simplicity, and $z = 0$ is a simple turning point. Three Stokes lines l_1, l_2 and l_3 emanate from the point $z = 0$, where l_2 lies on the left of l_1 and l_3 is on the left of l_2.

Let D be a canonical domain containing the Stokes line l_1, and $\partial D \supset l_2 \cup l_3$. Let U be a small neighbourhood of $z = 0$. The function $\xi = \xi(z)$ maps D in a one-to-one way onto a sector S in the complex plane with a vertex at $\xi = 0$, having an angle of $4\pi/3$ and with a finite number of cuts – the images of the bounding Stokes lines. We normalize the function $\xi(z)$ by the condition that S is the sector $\pi/3 < \arg\xi < 5\pi/3$. The lines l_1, l_2 and l_3 are mapped respectively to the rays $\arg\xi = \pi$, $\arg\xi = 5\pi/3$ and $\arg\xi = \pi/3$. Let D_ε be the domain D from which neighbourhoods of all the bounding Stokes lines are removed (Chap. 3, §3, paragraph 2.1). Then $\tilde{D} = D_\varepsilon \cap U$ is a small neighbourhood of $z = 0$.

Equation (1) has F.S.S. in $\tilde{D}$

$$w_j(z,\lambda) = w_j(\lambda^{2/3}\xi(z))A_{jN}(z,\lambda) + \lambda^{-1/3}w_j'(\lambda^{2/3}\xi(z))B_{jN}(z,\lambda), \tag{23}$$

where $w_1(t)$ and $w_2(t)$ are Airy-Fock functions (§1) and the choice of branch of $\xi(z)$ is indicated above. Further,

$$A_{jN}(z,\lambda) = 1 + \sum_{k=1}^{N}\lambda^{-k}A_k(z) + \lambda^{-N-1}R_{jN}(z, \lambda),$$

$$B_{jN}(z,\lambda) = \sum_{k=0}^{N}\lambda^{-k}B_k(z) + \lambda^{-N-1}\tilde{R}_{jN}(z, \lambda).$$

The functions $A_k(z)$ and $B_k(z)$ are the same as those in (2) of § 1. For $z \in \tilde{D}$ there are the bounds

$$|A_k(z)| \leqslant c_k(1+|z|)^{-\alpha_k}, \quad |B_k(z)| \leqslant c_k(1+|z|)^{-\alpha_k-(n+2)/6},$$

where

$$\alpha_k = \frac{1}{2}\left(\frac{n-1}{3}+k\right).$$

The bounds for the remainder terms for $z \in \tilde{D}$ and $\lambda \geqslant \lambda_0 > 1$, are

$$|R_{jN}| \leqslant c_{N+1}(1+|z|)^{-\alpha_{N+1}},$$
$$|\tilde{R}_{jN}| \leqslant c_{N+1}(1+|z|)^{-\alpha_{N+1}-(n+2)/6}.$$

The number $N \geqslant 0$ can be taken arbitrarily and these asymptotic expansions can be differentiated in z and λ any number of times.

The qualitative behaviour of the solutions is as follows. If $z \in D$, $z \to \infty$ and $\pi < \arg \xi(z) < 5\pi/3$, then $w_1(z,\lambda)$ decreases exponentially for each fixed $\lambda > 0$. For $z \to \infty$ and $\pi/3 < \arg \xi(z) < \pi$, $w_1(z,\lambda)$ grows exponentially. These assertions are also true for $w_2(z,\lambda)$ if we interchange the sectors.

There is also an asymptotic expansion of the form (23) when $q(z)$ is an entire function and the conditions of § 4, Chap. 3 are satisfied.

§ 3. Some Standard Equations

1. Solutions of the First Kind. We consider the equation

$$w'' - \frac{m^2}{4} z^{m-2} w = 0, \quad m > 0. \tag{1}$$

Branches of all functions of the form z^α are chosen in the plane with a cut on the half-line $(-\infty, 0]$ so that $z^\alpha > 0$ for $z \in (0, +\infty)$. The number m need not be an integer.

For $m = 3$ this is the Airy equation and it arises in the study of the asymptotic behaviour of solutions of second order equations in the neighbourhood of a simple turning point (§ 1). For $m \geqslant 4$ an integer, the point $z = 0$ is a turning point of order $m-2$ for (1), and for $m = 1$ it is a turning point of order -1 and is simultaneously a regular singular point of type R_1.

Equation (1) has solution

$$U(z) = \sqrt{2z/\pi}\, K_{1/m}(z^{m/2}), \tag{2}$$

where K is the Macdonald function [Leung 4]. We state the asymptotic formula for the solution $U(z)$. We have

$$U(0) = \frac{2^{(2-m)/(2m)}}{\pi^{1/2}} \Gamma\left(\frac{1}{m}\right). \tag{3}$$

The principal asymptotic term for $U'(z)$ as $z \to 0$ is

$$\begin{aligned} &-\frac{2^{(2-5m)/(2m)}}{\pi^{1/2}} \Gamma\left(\frac{1}{m} - 1\right) z^{m-1}, \quad 0 < m < 1; \\ &\frac{\ln z}{(8\pi)^{1/2}}, \quad m = 1; \quad \frac{2^{-(2+m)/(2m)}}{\pi^{1/2}}, \quad m > 1. \end{aligned} \tag{4}$$

If $z \to \infty$, $|\arg z| \leqslant (3-\delta)\pi/m$, then

$$U(z) \sim z^{1/2-m/4} \exp\{-z^{m/2}\}, \quad U'(z) \sim -\frac{m}{2} z^{m/4-1/2} \exp\{-z^{m/2}\}, \tag{5}$$

where δ is an arbitrary constant such that $0 < \delta < 1$.

In the sector $|\arg z| \leqslant 2\pi/m$ there are more exact bounds:

$$\begin{aligned} &U(z) = z^{1/2-m/4} \exp\{-z^{m/2}\}[1 + \theta_0(z)], \\ &\frac{d}{dz}(z^{m/4-1/2} U(z)) = -\frac{m}{2} z^{m/4-1/2} \exp\{-z^{m/2}\}[1 + \theta_1(z)]. \end{aligned} \tag{6}$$

For the remainder terms the bounds are

$$|\theta_j(z)| \leqslant \theta(z^{m/2}), \quad j = 0,\ 1, \tag{7}$$

where

$$\theta(z) = \exp\left\{\left|\frac{\pi(4-m^2)}{8m^2 z}\right|\right\} - 1.$$

In particular for $j = 0, 1$ we have

$$\theta_j(z) = O(z^{-m/2}), \quad z \to \infty, \quad |\arg z| \leqslant 2\pi/m.$$

The solution $U(x)$ decreases exponentially for real x as $x \to \infty$, and for $m = 3$ it is expressed in terms of the Airy-Fock function $v(x)$ (§ 1). Any solution of (1) not proportional to $U(x)$ increases exponentially as $x \to \infty$.

2. Solutions of the Second Kind. The functions

$$U_j(z) = U(ze^{-2j\pi i/m}), \quad j = 0, \pm 1, \dots, \tag{8}$$

are solutions of (1) and tend to zero as $z \to \infty$, $z \in S_j$, where S_j is the sector

$$(2j-1)\pi/m \leqslant \arg z \leqslant (2j+1)\pi/m. \tag{9}$$

The rays $\arg z = (2j-1)\pi/m$ are the Stokes lines for equation (1). There are the asymptotic formulae

$$U_j(z) = \exp\left\{ij\left(\frac{\pi}{2} - \frac{\pi}{m}\right)\right\} z^{1/2-m/4} \exp\{-(ze^{-2\pi ji/m})^{m/2}\}[1 + \theta_j^0(z)],$$

$$\frac{d}{dz}[z^{m/4-1/2}U_j(z)] = -\frac{m}{2}\exp\left\{-ij\left(\frac{\pi}{2}+\frac{\pi}{m}\right)\right\} z^{m/2-1/4} \times \exp\{-(ze^{-2\pi ji/m})^{m/2}\}[1+\theta_j^1(z)] \tag{10}$$

in the sector

$$(2j-2)\pi/m \leqslant \arg z \leqslant (2j+2)\pi/m\,. \tag{11}$$

The bounds in (7) occur for the remainder terms $\theta_j^k(z)$.

The Wronskian of a pair of such solutions is

$$W(U_j(z), U_k(z)) = im\ \exp\{-(j+k)\pi i/m\}\lambda_{jk}\,, \tag{12}$$

where

$$\lambda_{jk} = \frac{\sin((k-j)\pi/m)}{\sin(\pi/m)}\,.$$

This formula is true for all m and we note that, if $1/m$ is an integer, then

$$\lambda_{jk} = (-1)^{k-j-1}(k-j)\,.$$

The solutions $U_j(z)$ and $U_k(z)$ form a F.S.S. if

1) $1/m$ is an integer and $k \neq j$;
2) $1/m$ is not an integer and $k \neq j \pmod m$.

If $m \geqslant 2$ is an integer then there are precisely m sectors S_j of the form (9) and precisely m different solutions $U_j(z)$.

3. Connection Formulae. Any triplet of solutions $U_j(z)$, $U_k(z)$, $U_l(z)$ is connected by the relation

$$\sin\frac{(k-l)\pi}{m}e^{j\pi i/m}U_j(z) + \sin\frac{(l-j)\pi}{m}e^{k\pi i/m}U_k(z) + \sin\frac{(j-k)\pi}{m}e^{l\pi i/m}U_l(z) = 0\,. \tag{13}$$

We note the important special case:

$$U_j(z) = \mp e^{(k-j\pm 1)\pi i/m}\lambda_{jk}U_{k\pm 1}(z) \pm e^{(k-j)\pi i/m}\lambda_{j,k\pm 1}U_k(z)\,. \tag{14}$$

The connection formulae, together with the formulae (10), allow us to find the asymptotic behaviour of any solution $U_j(z)$ as $z \to \infty$ in any sector S_j.

§ 4. Multiple and Fractional Turning Points

1. Formal Asymptotic Solutions. We consider equation

$$y'' - [\lambda^2 q(x) + r(x)]y = 0 \tag{1}$$

on the interval $I = (0, a]$, $a > 0$. Let us introduce the conditions:

1) $q(x)$ is real-valued, $q(x) \in C^\infty(I)$, and $r(x) \in C^\infty$ for $x \in [0, a]$;
2) $q(x) = x^{m-2} q_1(x)$, $m > 0$, where $q_1(x)$ does not vanish for $x \in [0, a]$.

The number m need not be an integer. If $m \geqslant 3$ is an integer, then $x = 0$ is a turning point of order $m - 3$. We extend this definition to all values of $m > 0$.

For small $x > 0$ equation (1) can be approximated by the equation

$$y'' - \lambda^2 x^{m-2} q_1(0) y = 0 .$$

We therefore seek a formal asymptotic solution to equation (1) in the same form as in § 1, (2):

$$\begin{gathered} y = Aw(\lambda^{2/m}\xi) + \lambda^{-1+2/m} Bw'(\lambda^{2/m}\xi) , \\ A = \sum_{n=0}^{\infty} \lambda^{-n} A_n(x) , \quad B = \sum_{n=0}^{\infty} \lambda^{-n} B_n(x) , \end{gathered} \tag{2}$$

where $\xi(x)$, $A_n(x)$ and $B_n(x)$ are unknown functions. For $w(t)$ we take the solution of the standard equation (1) in § 3:

$$w'' = \frac{m^2}{4} t^{m-2} w . \tag{3}$$

Substituting (2) into (1) and equating to zero the coefficients for w and w', we obtain

$$\lambda^2 A \left(q - \frac{m^2}{4} \xi'^2 \xi^{m-2} \right) + \lambda \frac{m^2}{4} (2B'\xi'\xi^{m-2} + B(\xi'\xi^{m-2})') \\ + (A'' - rA) = 0 ,$$

$$\lambda^2 B \left(\frac{m^2}{4} \xi'^2 \xi^{m-2} - q \right) + \lambda(2A'\xi' + A\xi'') + B'' = 0 .$$

We choose $\xi(x)$ so that the coefficients of $\lambda^2 A$ vanish, to obtain an equation for $\xi(x)$. We take this function in the form

$$\xi(x) = \left[\frac{m}{2} \int_0^x \sqrt{q(t)} dt \right]^{2/m} = (S(0, x))^{2/m} , \tag{4}$$

where $S(0, x)$ is the function introduced in Chap. 2. § 1. We have

$$\xi(x) \sim \left(\frac{2}{m} q_1(0)\right)^{1/m} x\,, \quad x \to +0\,. \tag{5}$$

The choice of branch of $\xi(x)$ is determined by the choice of the value of $(q_1(0))^{1/m}$. Then $\xi(x)$ is inifinitely differentiable for $0 \leqslant x \leqslant a$, and $|\xi(x)|$ is strictly monotonic increasing.

For $A_n(x)$ and $B_n(x)$ we obtain the recurrence system of equations

$$\begin{aligned} &2\sqrt{f}(B_n\sqrt{f})' + A''_{n-1} - rA_{n-1} = 0\,, \\ &2\sqrt{\xi'}(A_n\sqrt{\xi'})' + B''_{n-1} - rB_{n-1} = 0\,, \end{aligned} \tag{6}$$

where

$$f = \frac{m^2}{4}\xi'(x)\xi^{m-2}(x) = \frac{q(x)}{\xi'(x)}\,.$$

This system is of the same type as the system in § 1, so that

$$\begin{aligned} A_0(x) &= \frac{c_0}{\sqrt{\xi'(x)}}\,, \\ A_n(x) &= \frac{1}{2\sqrt{\xi'(x)}}\left[c_n + \int_0^x \frac{B''_{n-1}(t) - r(t)B_{n-1}(t)}{\sqrt{\xi'(t)}}dt\right], \\ B_n(x) &= -\frac{1}{2}\sqrt{\frac{\xi'(x)}{q(x)}}\int_0^x \frac{A''_{n-1}(t) - r(t)A_{n-1}(t)}{\sqrt{q(t)}}\sqrt{\xi(t)}dt\,. \end{aligned} \tag{7}$$

Suppose now that the function $q_1(z)$ is holomorphic in a small neighbourhood U of $z = 0$, and let $\mathbb{R}^-$ be the half-line $(-\infty, 0]$. The branch of the function z^{m-2} for non-integer m in $U\backslash\mathbb{R}^-$ is chosen to be positive for $z \in U$ and $z > 0$. Then the function $\xi(z)$ constructed above is holomorphic in $U\backslash\mathbb{R}^-$. If m is an integer then $\xi(z)$ is holomorphic in U.

2. Oscillatory Solutions. We consider the equation

$$y'' + [\lambda^2 x^{m-2}p(x) + q(x)]y = 0 \tag{8}$$

under the following conditions.

(1) $m > 0$, $p(x) > 0$ for $x \in I = [0, a]$, $a > 0$;
(2) $p(x) \in C^2(I)$, $q(x) \in C^1(I)$.

We state the asymptotic formulae due to A.A. Dorodnitsyn [Dorodnitsyn]. In place of $\xi(x)$ we take the function

$$\omega(x) = \left(\frac{m}{2}\int_0^x \sqrt{t^{m-2}p(t)}dt\right)^{2/m}\,. \tag{9}$$

2.1 Solution of the Standard Equation. The equation

$$U'' + t^{m-2}U = 0$$

has F.S.S.

$$\begin{aligned} U_1(t) &= m^{-1/m}\Gamma\left(\frac{m-1}{m}\right)\sqrt{t}J_{-1/m}\left(\frac{2}{m}t^{m/2}\right), \\ U_2(t) &= m^{-1/m}\Gamma\left(\frac{m+1}{m}\right)\sqrt{t}J_{1/m}\left(\frac{2}{m}t^{m/2}\right). \end{aligned} \tag{10}$$

For $m = 1/n$, $n = 1, 2, \ldots$, we take

$$U_1(t) = -\frac{\pi n^n}{(n-1)!}\sqrt{t}N_n(2nt^{1/(2n)}). \tag{11}$$

These solutions have series expansions

$$\begin{aligned} U_2(t) &= t\left[1 - \frac{t^m}{1!m(m+1)} + \frac{t^{2m}}{2!m^2(m+1)(2m+1)}\right. \\ &\quad \left. - \frac{t^{3m}}{3!m^3(m+1)(2m+1)(3m+1)} + \ldots\right], \\ U_1(t) &= 1 - \frac{t^m}{1!m(m-1)} + \frac{t^{2m}}{2!m^2(m-1)(2m-1)} \\ &\quad - \frac{t^{3m}}{3!m^3(m-1)(2m-1)(3m-1)} + \ldots, \quad m \neq \frac{1}{n}, \\ U_1(t) &= 1 + \frac{n^2}{1!(n-1)}t^{1/n} + \frac{n^4t^{2/n}}{2!(n-1)(n-2)} \\ &\quad + \ldots - \frac{n^{2(n-1)}t^{1-1/n}}{(n-1)!(n-1)!} + \frac{n^{2n}t}{n!(n-1)!}\left(1 + \frac{1}{2} + \ldots + \frac{1}{n}\right) \\ &\quad - \ldots - \frac{2n^{2n}}{(n-1)!}\left(\frac{1}{2n}\ln t + \ln n + C\right)t \\ &\quad \times \left[\frac{t}{n!} - \frac{n^2}{1!(n-1)!}t^{1/n} + \ldots\right], \quad m = \frac{1}{n}. \end{aligned} \tag{12}$$

where C is the Euler constant. As $t \to \infty$ we have the asymptotic formulae

$$\begin{aligned} U_1(t) &= m^{(m-2)/(2m)}\Gamma\left(\frac{m-1}{m}\right)\frac{1}{\sqrt{\pi}}t^{-(m-2)/4}\left\{\cos\theta_-(t)[1 + O(t^{-m})]\right. \\ &\quad \left. + \frac{m^2-4}{16m}t^{-m/2}\sin\theta_-(t)[1 + O(t^{-m})]\right\}, \quad m \neq \frac{1}{n}, \\ U_2(t) &= m^{-(m-2)/(2m)}\Gamma\left(\frac{1}{m}\right)\frac{1}{\sqrt{\pi}}t^{-(m-2)/4}\left\{\sin\theta_+(t)[1 + O(t^{-m})]\right. \\ &\quad \left. - \frac{m^2-4}{16m}t^{-m/2}\cos\theta_+(t)[1 + O(t^{-m})]\right\}, \end{aligned} \tag{13}$$

where

$$\theta_{\pm}(t) = \frac{2}{m}t^{m/2} \pm \frac{\pi(m-2)}{4m}.$$

For $m = 1/n$ we have

$$U_1(t) = -\frac{\sqrt{\pi}n^{n-1/2}}{(n-1)!}t^{1/2-1/(4n)}\left[\sin\left(2nt^{1/(2n)} - \frac{1}{2}\pi n - \frac{\pi}{4}\right)\right.$$
$$\times [1 + O(t^{-1/n})] + \frac{4n^2-1}{16n}t^{-1/(2n)}\cos\left(2nt^{1/(2n)} - \frac{1}{2}\pi n - \frac{\pi}{4}\right)$$
$$\left.\times [1 + O(t^{-1/n})]\right]. \tag{14}$$

Let us introduce the notation

$$Y_j(x) = \frac{1}{\sqrt{\omega'(x)}}U_j(\lambda^{2/m}\omega(x)),$$
$$U = \sqrt{U_1^2 + U_2^2}, \quad Y' = \sqrt{Y_1'^2 + Y_2'^2}, \tag{16}$$
$$f(x) = q(x) + \sqrt{\omega'(x)}\frac{d^2}{dx^2}\frac{1}{\sqrt{\omega'(x)}}.$$

2.2 Method of Study. Setting $y(x) = Y_j(x) + u_j(x)$ we obtain the integral equation for $u_j(x)$

$$u_j(x) = \lambda^{-2/m}\int_0^x K(x,\tau)Y_j(\tau)d\tau + \lambda^{-2/m}\int_0^x K(x,\tau)u_j(\tau)d\tau,$$
$$j = 1, 2,$$

with kernel of the form

$$K(x, \tau) = f(\tau)[Y_1(x)Y_2(\tau) - Y_1(\tau)Y_2(x)].$$

It follows from the asymptotic behaviour of $U_1(t)$ and $U_2(t)$ that these functions are bounded on the half-line $0 \leqslant t < \infty$ for $m \geqslant 2$ and that the integral equation can be solved by the method of successive approximations. If $0 < m < 2$ then $U_1(t)$ and $U_2(t)$ are unbounded on $0 \leqslant t < \infty$, but the functions $\lambda^{(m-2)/2m}U_{1,2}(\lambda^{2/m}w(x))$ are bounded for $\lambda > 0$, $x \geqslant 0$. In this case we can also apply the method of successive approximations to the integral equation.

The first approximations involve integrals of the form

$$Y_{jk}(x) = \int_0^x U_j(t)U_k(t)dt. \tag{17}$$

The asymptotic formulae for the solutions depend essentially on whether these integrals converge for $x > 0$ and whether the values of $Y_{jk}(\infty)$ are finite or infinite. We must distinguish five cases:

(1) $m > 4$; (2) $m = 4$, (3) $2 \leqslant m < 4$; (4) $1 < m \leqslant 2$; (5) $0 < m \leqslant 1$.

2.3 $m > 4$. Equation (8) has F.S.S. of the form

$$y_j(x) = Y_j(x) + \lambda^{-4/m}\left\{\frac{f(x)}{[\omega'(x)]^2}[Y_1(t)V_{j2}(t) - Y_2(t)V_{j1}(t)] + \frac{f(0)}{[\omega'(0)]^2}[C_{j2}Y_1(t) - C_{j1}Y_2(t)]\right\} + o(\lambda^{-4/m}U(t)), \tag{18}$$

where

$$t = \lambda^{2/m}\omega(x), \quad V_{jk}(t) = Y_{jk}(t) - Y_{jk}(\infty), \quad C_{jk}(t) = Y_{jk}(\infty),$$

$$\begin{aligned} C_{11} &= \pi \operatorname{cosec} \frac{\pi}{m} m^{-(m-2)/m} \frac{\Gamma\left(\frac{m-1}{m}\right)\Gamma\left(\frac{m-4}{m}\right)}{\Gamma^2\left(\frac{m-2}{m}\right)\Gamma\left(\frac{m-3}{m}\right)}, \\ C_{12} = C_{21} &= \pi \operatorname{cosec} \frac{\pi}{m} m^{-2(m-2)/m} \frac{\Gamma\left(\frac{2}{m}\right)\Gamma\left(\frac{m-4}{m}\right)}{\Gamma\left(\frac{m-3}{m}\right)\Gamma\left(\frac{m-2}{m}\right)\Gamma\left(\frac{m-1}{m}\right)}, \\ C_{22} &= m^{(3m-4)/m} \frac{\Gamma^2\left(\frac{1}{m}\right)\Gamma\left(\frac{3}{m}\right)\Gamma\left(\frac{m-4}{m}\right)}{\Gamma^2\left(\frac{m-2}{m}\right)\Gamma\left(\frac{m-1}{m}\right)}. \end{aligned} \tag{19}$$

Analogous formulae are true for y_j': in (18) we must replace Y_j by Y_j' and U by Y' in the remainder term.

2.4 $2 \leqslant n < 4$. Here the asymptotic behaviour of the F.S.S. has the form

$$y_j(x) = Y_j(x) + \frac{1}{\lambda}\int_0^x \frac{f(t)dt}{\sqrt{p(t)}}[C_{j1}Y_1(x) + C_{j2}Y_2(x)] + O[\lambda^{-4/m}U(\lambda^{2/m}\omega(x))], \quad j = 1, 2,$$

where

$$\begin{aligned} C_{11} &= -C_{22} = \frac{1}{2}\cot\frac{\pi}{m}, \\ C_{12} &= -\frac{1}{2\pi}m^{-(m-2)/m}\Gamma^2\left(\frac{m-1}{m}\right), \\ C_{21} &= \frac{1}{2\pi}m^{-(m-2)/m}\Gamma^2\left(\frac{1}{m}\right). \end{aligned} \tag{20}$$

Asymptotic formulae for y_j' are obtained from these formulae on replacing Y_j by Y_j' and U by Y' in the remainder term.

2.5 $1 < m \leqslant 2$. Here the asymptotic formulae (19) are preserved, the only difference is that the remainder terms have the form $O(\lambda^{-2}U(\lambda^{2/m}\omega(x)))$ and $O(\lambda^{-2}Y'(x))$ for the solution and its derivative respectively.

2.6 $0 < m \leqslant 1$. In this case

$$y_1(x) = Y_1(x) + O(\lambda^{-1}U(\lambda^{2/m}\omega(x))). \tag{21}$$

We have the same formulae for the solution $y_2(x)$ as in paragraph 2.5, if $1/m$ is not an integer. If $m = 1/n$, where $n \geqslant 1$ is an integer, then

$$y_2(x) = Y_2(x) + \frac{1}{\lambda}\frac{(n!)^2}{2\pi n^{2n-1}}Y_1(x)\int_0^x \frac{f(t)dt}{\sqrt{p(t)}} + O(\lambda^{-2}U(\lambda^{2n}\omega(x))). \quad (22)$$

2.7 $m = 4$. In this case the integrals $Y_{jk}(t)$ diverge both for $t = 0$ and as $t \to \infty$, which leads to a complication of the asymptotic formulae [Dorodnitsyn]. In paragraph 3 we will give another variant of these formulae.

If $p(x)$ and $q(x)$ are infinitely differentiable for $x \in I$, then using the method of paragraph 2.2 we can obtain an asymptotic series for solutions of the form (2).

3. Integer Values of $m \geqslant 1$ [Olver 2, Olver 3]

3.1 The Standard Equation and its Solution. There are three possiblities.

I. $m > 0$ is even and the standard equation has the form

$$w'' = \frac{m^2}{4}t^{m-2}w. \quad (23)$$

II. $m > 0$ is even and the standard equation has the form

$$w'' = -\frac{m^2}{4}t^{m-2}w. \quad (24)$$

III. $m > 0$ is odd and the standard equation has the form (23).

We consider these cases in detail.

I. The standard equation (23) has F.S.S. $\{U_m(t), U_m(-t)\}$ where

$$U_m(t) = \sqrt{\frac{2t}{\pi}}K_{1/m}(t^{m/2}), \quad t > 0, \quad (25)$$

and K is the Macdonald function. The Wronskian of these solutions is

$$W(U_m(t), U_m(-t)) = m \operatorname{cosec} \pi/m.$$

We mention the two special cases where $m = 2$ and $m = 4$. Here

$$U_2(t) = e^{-t}, \quad U_4(t) = \sqrt{2}U(0, 2t),$$

where $U(a,t)$ is a parabolic cylinder function. For $t = 0$ we have

$$U_m(0) = \pi^{-1/2}2^{(2-m)/(2m)}\Gamma\left(\frac{1}{m}\right),$$

$$U_m'(0) = \pi^{-1/2}2^{-(2+m)/(2m)}\Gamma\left(-\frac{1}{m}\right).$$

As $t \to \infty$, $U_m(t)$ decreases exponentially and

$$U_m(t) \sim t^{(2-m)/4} e^{-t^{m/2}} .$$

For $t < 0$ we have

$$U_m(t) = \sqrt{\frac{2|t|}{\pi}} \left[\pi \operatorname{cosec} \frac{\pi}{m} I_{1/m}(|t|^{m/2}) + K_{1/m}(|t|^{m/2}) \right] , \tag{26}$$

where $I_{1/m}$ is a Bessel function with imaginary argument and $U_m(t)$ increases exponentially as $t \to -\infty$, with the asymptotic form

$$U_m(t) = \operatorname{cosec} \frac{\pi}{m} |t|^{(2-m)/4} \exp\{|t|^{m/2}\}[1 + O(t^{-m/2})] .$$

II. The standard equation (24) has F.S.S. $\{W_m(t), W_m(-t)\}$ where

$$\begin{aligned} W_m(t) &= -\sqrt{\frac{\pi t}{2}} \left[\tan \frac{\pi}{2m} J_{1/m}(t^{m/2}) + Y_{1/m}(t^{m/2}) \right] , \qquad t > 0 , \\ W_m(t) &= \sqrt{\frac{\pi |t|}{2}} \left[\cot \frac{\pi}{2m} J_{1/m}(|t|^{m/2}) - Y_{1/m}(|t| m/2) \right] , \quad t < 0 , \end{aligned} \tag{27}$$

and $Y_{1/m}$ is the Neumann function. The Wronskian of these solutions is

$$W(W_m(t), W_m(-t)) = \frac{m}{\cos \pi / m} .$$

Again, special cases are

$$W_2(t) = \sqrt{2} \cos(t + \pi/4) , \quad W_4(t) = 2^{3/4} W(0,\ 2t) ,$$

where $W(a,t)$ is a modified parabolic cylinder function. The solution $W_m(t)$ oscillates as $t \to \pm\infty$:

$$\begin{aligned} W_m(t) &= \frac{t^{(2-m)/4}}{\cos \pi/(2m)} \left[\cos\left(t^{m/2} + \frac{\pi}{4} \right) + O(t^{-m/2}) \right] , \quad t \to +\infty , \\ W_m(t) &= \frac{|t|^{(2-m)/4}}{\sin \pi/(2m)} \left[\cos\left(|t|^{m/2} - \frac{\pi}{4} \right) + O(t^{-m/2}) \right] , \quad t \to -\infty . \end{aligned}$$

We now give bounds for the F.S.S. Let $t = q_m$ be the least positive root of the equation

$$W_m(t) = \tan \frac{\pi}{2m} W_m(-t) .$$

We introduce a weight function $E_m(t)$ equal to

$$\begin{aligned} &\sqrt{\cot \pi/(2m)} , \quad t \geqslant q_m ; \quad \sqrt{\tan \pi/(2m)} , \quad t \leqslant -q_m ; \\ &\sqrt{\frac{W_m(-t)}{W_m(t)}} , \quad -q_m \leqslant t \leqslant q_m . \end{aligned} \tag{28}$$

Now $W_m(t)$ and $W_m(-t)$ have no zeros on the interval $[-q_m, q_m]$ and hence $E_m(t)$ is a non-decreasing function that is continuous on the whole line, with $E_m(-t) = E_m^{-1}(t)$. We introduce the positive even functions $M_m(t)$ and $N_m(t)$ which we can use to express the absolute values of the solutions in the F.S.S. and their derivatives:

$$W_m(t) = E_m^{-1}(t) M_m(t) \sin\theta_m(t), \quad W_m'(t) = E_m^{-1}(t) N_m(t) \sin\omega_m(t). \tag{29}$$

As $t \to \pm\infty$

$$M_m(t) \sim \sqrt{\frac{2}{\sin\frac{\pi}{m}}}|t|^{(2-m)/4}, \quad N_m(t) \sim m\sqrt{\frac{1}{2\sin\frac{\pi}{m}}}|t|^{(m-2)/4}. \tag{30}$$

III. The standard equation (23) has F.S.S. $V_m(t)$ and $\overline{V}_m(t)$, where

$$\begin{aligned} V_m(t) &= \sqrt{\frac{2t}{\pi}} K_{1/m}(t^{m/2}), \\ \overline{V}_m(t) &= \sqrt{\frac{2t}{\pi}}\left[\frac{\pi}{\sin\frac{\pi}{m}} I_{1/m}(t^{m/2}) + K_{1/m}(t^{m/2})\right], \quad t > 0. \end{aligned} \tag{31}$$

The Wronskian of these solutions is $m/(\sin\pi/m)$. We have $V_m(0) = \pi^{-1/2}2^{(2-m)/2m}\Gamma(1/m)$, $V_m'(0) = \pi^{-1/2}2^{-(2+m)(2m)}\Gamma(-1/m)$, $\overline{V}_m(0) = V_m(0)$ and $\overline{V}'_m(0) = -V_m'(0)$. For $m = 3$ these solutions can be expressed in terms of the Airy functions:

$$V_3(t) = 2^{5/6}3^{1/6}\pi^{1/2}\,\mathrm{Ai}\left(\left(\frac{3}{2}\right)^{2/3} t\right),$$

$$\overline{V}_3(t) = 2^{5/6}3^{-1/3}\pi^{1/2}\,\mathrm{Bi}\left(\left(\frac{3}{2}\right)^{2/3} t\right).$$

For $t \geqslant 0$ the functions $V_m(t)$ and $\overline{V}_m(t)$ are positive, $V_m(t)$ is monotonic decreasing, $\overline{V}_m(t)$ is monotonic increasing and

$$\begin{aligned} V_m(t) &= t^{(2-m)/4} e^{-t^{m/2}}[1 + O(t^{-m/2})], \\ \overline{V}_m(t) &= \frac{1}{\sin\frac{\pi}{m}} t^{(2-m)/4} e^{t^{m/2}}[1 + O(t^{-m/2})], \end{aligned}$$

as $t \to +\infty$. For $t < 0$ we have

$$\begin{aligned} V_m(t) &= \sqrt{\frac{\pi|t|}{2}}\left[\cot\frac{\pi}{2m} J_{1/m}(|t|^{m/2}) - Y_{1/m}(|t|^{m/2})\right], \\ \overline{V}_m(t) &= -\sqrt{\frac{\pi|t|}{2}}\left[\tan\frac{\pi}{2m} J_{1/m}(|t|^{m/2}) + Y_{1/m}(|t|^{m/2})\right]. \end{aligned} \tag{32}$$

Both solutions are oscillatory for $t \to -\infty$ with phase difference $\pi/2$:

$$V_m(t) = \frac{1}{\sin \pi/(2m)} |t|^{(2-m)/4} \left[\cos\left(|t|^{m/2} - \frac{\pi}{4}\right) + O(t^{-m/2})\right],$$
$$\overline{V}_m(t) = \frac{1}{\cos \pi/(2m)} |t|^{(2-m)/4} \left[\cos\left(|t|^{m/2} + \frac{\pi}{4}\right) + O(t^{-m/2})\right].$$

Let $t = -q_m$ be the root with smallest modulus of the equation

$$\overline{V}_m(t) = \tan \frac{\pi}{2m} V_m(t).$$

We introduce the weight function $E_m(t)$ equal to

$$\sqrt{\overline{V}_m(t)/V_m(t)}, \quad t \geqslant -q_m; \quad \sqrt{\tan \pi/(2m)}, \quad t \leqslant -q_m. \tag{33}$$

Then $E_m(t)$ is positive, continuous on the whole line and non-decreasing. We define $M_m(t)$ and $N_m(t)$ by formulae analogous to those in (29):

$$\begin{aligned} V_m(t) &= E_m^{-1}(t) M_m(t) \sin \theta_m(t), & \overline{V}_m(t) &= E_m(t) M_m(t) \cos \theta_m(t), \\ V_m'(t) &= E_m^{-1}(t) N_m(t) \sin \omega_m(t), & \overline{V}_m'(t) &= E_m(t) N_m(t) \cos \omega_m(t). \end{aligned} \tag{34}$$

There are the asymptotic formulae (30) for $M_n(t)$ and $N_n(t)$ as $|t| \to \infty$ with t replaced by $|t|$. Some approximate values for q_m are: $q_2 = 0.000$, $q_3 = 0.279$, $q_4 = 0.431$, $q_5 = 0.528$, $q_6 = 0.596$.

3.2 Auxiliary Functions. Let us consider the equation

$$w'' = [\lambda^2 f(x, \lambda) + g(x, \lambda)] w \tag{35}$$

on the finite or infinite interval $I = (a_1, a_2)$, $0 \in I$, where $\lambda > 0$ is a parameter. We introduce the conditions:

1) The function $x^{-2+m} f(x, \lambda)$ is real for real λ, non-vanishing on I, and $C^2(I)$ for each fixed $\lambda > 0$.

2) The function $g(x, \lambda)$ is continuous on I for each fixed $\lambda > 0$.

Let us put

$$\begin{aligned} \xi(x, \lambda) &= -\left(\int_x^0 \sqrt{|f(t, \lambda)|}\, dt\right)^{2/m}, & x \leqslant 0, \\ \xi(x, \lambda) &= \left(\int_0^x \sqrt{|f(t, \lambda)|}\, dt\right)^{2/m}, & x \geqslant 0. \end{aligned} \tag{36}$$

Then $\xi(x, \lambda)$ is monotonic increasing in x and is $C^2(I)$. We introduce the functions

$$\tilde{f}(x, \lambda) = \frac{4}{m^2} |f(x, \lambda)|\, |\xi|^{2-m}, \quad \Omega_m(t) = 1 + |t|^{(m-2)/2} \tag{37}$$

and the error control function

$$H_m(x,\lambda) = \int \left[\frac{1}{\tilde{f}^{1/4}(x,\lambda)} \frac{d^2}{dx^2} \frac{1}{\tilde{f}^{1/4}(x,\lambda)} - \frac{g(x,\lambda)}{\tilde{f}^{1/2}(x,\lambda)} \right] \frac{dx}{\Omega_m(\lambda^{2/m}\xi)}, \quad (38)$$

where the limits of integration are not important. For equation (23) we have

$$f(x,\lambda) = \frac{m^2}{4} x^{m-2}, \quad g(x,\lambda) \equiv 0, \quad \xi = x,$$

$$\tilde{f}(x,\lambda) \equiv 1, \quad H_m(x,\lambda) \equiv 0.$$

3.3 Asymptotic Behaviour of Solutions of Equation (35) in Case I. Let m be even and let the functions $f(x,\lambda)x^{2-m}$ be positive on I. Then equation (35) has solutions

$$\begin{aligned} w_1(x,\lambda) &= \tilde{f}^{-1/4}(x,\lambda)[U_m(-\lambda^{2/m}\xi) + \varepsilon_1(x,\lambda)], \\ w_2(x,\lambda) &= \tilde{f}^{-1/4}(x,\lambda)[U_m(\lambda^{2/m}\xi) + \varepsilon_2(x,\lambda)], \end{aligned} \quad (39)$$

The function U_m is defined by formulae (25) and (26). For $x \in I$ and $\lambda > 0$, we have the bounds

$$\frac{|\varepsilon_1|}{U_m'(-\lambda^{2/m}\xi)}, \frac{|\partial\varepsilon_1/\partial x|}{\mu_m \left|\frac{\partial}{\partial x} U_m(-\lambda^{2/m}\xi\right|} \leqslant \exp\left\{ \frac{\lambda_m}{\lambda^{2/m}} V_{q_1,x}(H_m) \right\} - 1, \quad (40)$$

where

$$\begin{aligned} \lambda_m &= \sup_{t\in R} \left[\frac{1}{m} \sin\frac{\pi}{m} \Omega_m(t) U_m(t) U_m(-t) \right], \\ \mu_m &= \sup_{t\in R} \left[\frac{m}{\sin\frac{\pi}{m}} \frac{1}{|U_m'(t)|U_m(-t)} \right], \end{aligned} \quad (41)$$

and $V_{a,b}(H_m)$ is the variation of H_m over the interval $[a,b]$, that is,

$$V_{a,b}(H_m) = \int_a^b |H_m'(x)| dx. \quad (42)$$

The bounds for $|\varepsilon_2|$ and $|\partial\varepsilon_2/\partial x|$ are given by (40) with $U_m(-t)$ replaced by $U_m(t)$ and $V_{a_1,x}$ by V_{x,a_2}.

3.4 Asymptotic Behaviour of Solutions of Equation (35) in Case II. Let m be even and let $f(x,\lambda)x^{2-m}$ be negative for $x \in I$. Equation (35) has solutions

$$w_{1,2}(x,\lambda) = \tilde{f}^{-1/4}(x,\lambda)[W_m(\mp\lambda^{2/m}\xi) + \varepsilon_{1,2}(x,\ \lambda)] \quad (43)$$

The function $\xi(x,\lambda)$ is determined by the formulae (27), and for $x \in I$ and $\lambda > 0$ we have

$$\frac{|\varepsilon_j(x,\lambda)|}{M_m(\lambda^{2/m}\xi)}, \frac{|\partial\varepsilon_j(x,\lambda)/\partial x|}{\lambda^{2/m}\tilde{f}^{1/2}(x,\lambda)N_m(\lambda^{2/m}\xi)} \leqslant \frac{\sigma_m}{\rho_m} E_m^{\pm 1}(\lambda^{2/m}\xi) \left[\exp\left\{ \frac{\rho_m}{\lambda^{2/m}} |V_{a_j,x}(H_m)| \right\} - 1 \right], \quad (44)$$

where the plus sign is taken for $j = 1$, and

$$\begin{aligned} \rho_m &= \sup_{t\in R}\left[\frac{1}{m}\sin\frac{\pi}{m}\Omega_m(t)M_m^2(t)\right], \\ \sigma_m &= \sup_{t\in R}\left[\frac{1}{m}\sin\frac{\pi}{m}\Omega_m(t)|W_m(t)|E_m(t)M_m(t)\right]. \end{aligned} \tag{45}$$

We note that $\sigma_m \leqslant \rho_m$ so that in (44) the ratio σ_m/ρ_m can be replaced by unity.

3.5 Asymptotic Behaviour of Solutions of Equation (35) in Case III. Let m be odd and let $f(x)x^{2-m}$ be positive for $x \in I$. Then equation (35) has solutions

$$\begin{aligned} w_1(x,\lambda) &= \tilde{f}^{-1/4}(x,\lambda)[\overline{V}_m(\lambda^{2/m}\xi) + \varepsilon_1(x,\lambda)], \\ w_2(x,\lambda) &= \tilde{f}^{-1/4}(x,\lambda)[V_m(\lambda^{2/m}\xi) + \varepsilon_2(x,\lambda)]. \end{aligned} \tag{46}$$

The functions V_m and $\overline{V}_m$ are determined by formulae (31) and (32), and for $x \in I$ and $\lambda > 0$ we have

$$\frac{|\varepsilon_j(x,\lambda)|}{M_m(\lambda^{2/m}\xi)}, \frac{|\partial\varepsilon_j(x,\lambda)/\partial x|}{\lambda^{2/m}\tilde{f}^{1/2}(x,\lambda)N_m(\lambda^{2/m}\xi)} \leqslant \frac{\sigma_{mj}}{\rho_m}E_m^{\pm 1}(\lambda^{2/m}\xi)\left[\exp\left\{\frac{\rho_m}{\lambda^{2/m}}|V_{a_j,x}(H_m)|\right\} - 1\right], \tag{47}$$

where ρ_m is as in (45) and

$$\begin{aligned} \sigma_{m1} &= \sup_{t\in R}\left[\frac{1}{m}\sin\frac{\pi}{m}\Omega_m(t)|\overline{V}_m(t)|E_m^{-1}(t)M_m(t)\right], \\ \sigma_{m2} &= \sup_{t\in R}\left[\frac{1}{m}\sin\frac{\pi}{m}\Omega_m(t)|V_m(t)|E_m(t)M_m(t)\right], \end{aligned} \tag{48}$$

where E_m and E_m^{-1} correspond to $\overline{V}_m$ and V_m respectively. Since $\sigma_{mj}/\rho_j \leqslant 1$ this ratio can be replaced by unity.

The above bounds for the remainder terms ε_j and $\partial\varepsilon_j/\partial x$ are applicable in very general situations but are rather unwieldly. They are generalizations of the WKB-bounds (Chap. 2, § 2). As in Chap. 2 we can use them to obtain simpler ones under additional assumptions concerning the dependence of f and g on λ.

3.6 Asymptotic Bounds for the Remainder Terms. We consider the equation

$$y'' - [\lambda^2 f(x) + g(x,\lambda)]y = 0 \tag{49}$$

under the following assumptions:

1) The function $f(x)/x^{m-2}$ is real, non-vanishing on I and of class $C^2(I)$; the function $g(x,\lambda)$ is of class $C(I)$ for $\lambda \geqslant \lambda_0 > 0$.

2) The integrals

$$\int^{a_j} (|f'|^2|f|^{-5/2} + |f''||f|^{-3/2})dx$$

converge.

This is the condition for applicability of the WKB-approximation (Chap. 2 § 3).

3) For $\lambda \geqslant \lambda_0 > 0$ and for some $\delta > 0$ we have

$$\int_{|x|>\delta} |g(x,\lambda)||f(x)|^{-1/2}\,dx \leqslant c\lambda^{\omega}\,.$$

If J is a finite interval and $J \subset I$, then

$$|g(x,\lambda)|\,|f(x)|^{-1/2} \leqslant c\lambda^{\omega}\,.$$

This condition holds, for instance, if $g(x,\lambda) = \lambda^{\omega}h(x)$ and the following integral converges:

$$\int_{|x|>\delta} |g(x)||f(x)|^{-1/2}\,dx\,.$$

Put $\omega_0 = \max(\omega, 0)$; then the variation $V_{a_1,a_2}(H_m)$ has the following orders as $\lambda \to \infty$:

$$\begin{aligned} &O(\lambda^{\omega_0-1+2/m}), && m = 2,\ 3,\\ &O(\lambda^{\omega_0-1/2}\ln\lambda), && m = 4,\\ &O(\lambda^{\omega_0-4/m}), && m > 4. \end{aligned} \tag{50}$$

Combined with the above bounds this allows us to refine the bounds for the remainder terms. For example in case I the relations

$$\varepsilon_1(x,\ \lambda)/U_m(-\lambda^{2/m}\xi), \quad \varepsilon_2(x,\ \lambda)/U_m(\lambda^{2/m}\xi)$$

have the following orders as $\lambda \to \infty$, uniform in $x \in I$:

$$\begin{aligned} &O(\lambda^{\omega_0-1}), && m = 2,\ 3,\\ &O(\lambda^{\omega_0-1}\ln\lambda), && m = 4,\\ &O(\lambda^{\omega_0-4/m}), && m > 4. \end{aligned} \tag{51}$$

3.7 Connection Formulae. Suppose that the conditions 1)–3) from paragraph 3.6 hold and let the integral

$$\int^{a_j} |f(x)|dx\,, \quad j = 1,2\,,$$

diverge. Then equation (49) has solutions of WKB-type on the intervals $J_1 = (a_1, -\delta)$ and $J_2 = (\delta, a_2)$ where $\delta > 0$. We state the connection formulae between them. It is assumed that $\lambda \geqslant \lambda_0 \gg 1$.

I. Equation (49) has solutions

$$\begin{aligned} \tilde{w}_1(x,\lambda) &\sim f^{-1/4}(x)\exp\left\{-\lambda\int_x^0 f^{1/2}(t)dt\right\}, \quad x \to a_1, \\ \tilde{w}_2(x,\lambda) &\sim f^{-1/4}(x)\exp\left\{-\lambda\int_0^x f^{1/2}(t)dt\right\}, \quad x \to a_2. \end{aligned} \tag{52}$$

These solutions decrease exponentially as $x \to a_1$ and $x \to a_2$, and are uniquely determined by their asymptotic behaviour. Comparison of the asymptotic behaviours as $x \to a_2$ of solutions w_2 and $\tilde{w}_2$ gives

$$\tilde{w}_2(x,\lambda) = \sqrt{\frac{2}{m}}\lambda^{(m-2)/m} w_2(x,\lambda),$$

and a similar identity connects the solutions w_1 and $\tilde{w}_1$. Since the asymptotic behaviour of the solutions w_1 and w_2 is known, we have

$$\begin{aligned} \tilde{w}_1(x,\lambda) &\sim \frac{1+k_1}{\sin\frac{\pi}{m}} f^{-1/4}(x)\exp\left\{\lambda\int_0^x f^{1/2}(t)dt\right\}, \quad x \to a_2, \\ \tilde{w}_2(x,\lambda) &\sim \frac{1+k_2}{\sin\frac{\pi}{m}} f^{-1/4}(x)\exp\left\{\lambda\int_x^0 f^{1/2}(t)dt\right\}, \quad x \to a_1. \end{aligned} \tag{53}$$

The error bounds (51) are true for the remainder terms k_1 and k_2.

II. Equation (49) has solutions which are uniquely determined by their asymptotic behaviour:

$$\begin{aligned} \tilde{w}_1(x,\lambda) &= |f(x)|^{-1/4}\left[\cos\left(\lambda\int_x^0 |f(t)|^{1/2}dt + \frac{\pi}{4}\right) + o(1)\right], \quad x \to a_1, \\ \tilde{w}_2(x,\lambda) &= |f(x)|^{-1/4}\left[\cos\left(\lambda\int_0^x |f(t)|^{1/2}dt + \frac{\pi}{4}\right) + o(1)\right], \quad x \to a_2. \end{aligned}$$

At the opposite ends of the interval we have

$$\begin{aligned} \tilde{w}_1(x,\lambda) &= (1+\gamma_1)\cot\frac{\pi}{2m}|f(x)|^{-1/4} \\ &\quad\times\left[\cos\left(\lambda\int_0^x |f(t)|^{1/2}dt - \frac{\pi}{4} + \delta_1\right) + o(1)\right], \quad x \to a_2, \\ \tilde{w}_2(x,\lambda) &= (1+\gamma_2)\cot\frac{\pi}{2m}|f(x)|^{-1/4} \\ &\quad\times\left[\cos\left(\lambda\int_x^0 |f(t)|^{1/2}dt - \frac{\pi}{4} + \delta_1\right) + o(1)\right], \quad x \to a_1. \end{aligned}$$

The bounds for the remainder terms are as before.

III. In this case we have a F.S.S. of the form

$$\tilde{w}_1(x,\lambda) = |f(x)|^{-1/4}\left[\cos\left(\lambda\int_x^0 |f(t)|^{1/2}dt + \frac{\pi}{4}\right) + o(1)\right], \quad x \to a_1,$$

$$\tilde{w}_1(x,\lambda) \sim \frac{1}{2}(1+k)\frac{f^{-1/4}(x)}{\sin\pi/(2m)}\exp\left\{\lambda\int_0^x f^{1/2}(t)dt\right\}, \quad x \to a_2,$$

$$\tilde{w}_2(x,\lambda) \sim f^{-1/4}(x)\exp\left\{-\lambda\int_0^x f^{1/2}(t)dt\right\}, \quad x \to a_2,$$

$$\tilde{w}_2(x,\lambda) = \frac{1+k}{\sin\pi/(2m)}|f(x)|^{-1/4} \times\left[\cos\left(\lambda\int_x^0 |f(t)|^{1/2}dt - \frac{\pi}{4} + \delta\right) + o(1)\right], \quad x \to a_1.$$

The bounds for the remainder terms are given by formula (51).

§ 5. The Fusion of a Turning Point and Regular Singular Point

1. Transformation of the Equation. We consider the equation

$$w'' - \left[\lambda^2 x^{\mu-2} f(x,\lambda) + \frac{g(x,\lambda)}{x^2}\right] w = 0 \tag{1}$$

on an interval $I = [0, a)$, finite or infinite, where $\lambda > 0$ is a large parameter. We introduce conditions for $x \in I$ and $\lambda \geqslant \lambda_0 > 0$:

1) $f(x,\lambda)$, $g(x,\lambda)$ are real and $f(x,\lambda)$ does not vanish.
2) $f''_{xx}(x,\lambda)$ and $g(x,\lambda)$ are continuous.

It will be assumed that $\mu > 0$. The number a and the parameter μ may depend on λ.

Here $x = 0$ is a turning point of order $\mu - 2$ (possibly not an integer) and is a second-order pole for the coefficient of ω, that is, a regular singular point of equation (1). The asymptotic behaviour of the F.S.S. is expressed in this case in terms of Bessel functions. We consider the following two cases I and II.

I . $f(x,\lambda) > 0$ for $x \in I$.
II. $f(x,\lambda) < 0$ for $x \in I$.

We apply the Liouville transform (Chap. 2, § 1)

$$w = \tilde{f}^{-1/4}(x,\lambda)W(\xi), \quad \tilde{f}(x,\lambda) = \frac{x^{\mu-2}|f(x,\lambda)|}{\zeta^{\mu-2}}, \tag{2}$$

$$\zeta = \left(\int_0^x t^{(\mu-2)/2}\sqrt{|(t,\lambda)|dt}\right)^{2/\mu}.$$

Then equation (1) becomes

$$\frac{d^2W}{d\zeta^2} = \left[\pm\frac{1}{4}\mu^2\lambda^2\zeta^{\mu-2} + \frac{\phi(\zeta,\lambda)}{\zeta^2}\right]W, \tag{3}$$

where the plus (minus) sign is taken for case I (case II), and

$$\phi(\zeta,\lambda) = \frac{\mu^2\zeta^2 g(x,\lambda)}{4x^2\tilde{f}(x,\lambda)} + \zeta^2\tilde{f}^{-1/4}(x,\lambda)\frac{d^2}{d\zeta^2}\tilde{f}^{1/4}(x,\lambda). \tag{4}$$

The interval I is mapped in a one-to-one way to the interval $J = [0,\beta)$, $\beta > 0$. Further ζ/x is positive and of class $C^2(J)$, and $\phi(\zeta,\lambda) \in C(J)$.

We introduce the parameter ν by the formula

$$\frac{1}{4}(\mu^2\nu^2 - 1) = \phi(0,\lambda) \tag{5}$$

and rewrite equation (3) as

$$\frac{d^2W}{d\zeta^2} = \left(\pm\frac{\mu^2\lambda^2}{4}\zeta^{\mu-2} + \frac{\mu^2\nu^2-1}{4\zeta^2} + \frac{\psi(\zeta,\lambda)}{\zeta}\right)W. \tag{6}$$

2. Case I. The plus sign is taken in equation (6). We take the standard equation as

$$\frac{d^2W}{d\zeta^2} = \left(\frac{\mu^2\lambda^2}{4}\zeta^{\mu-2} + \frac{\mu^2\nu^2-1}{4\zeta^2}\right)W. \tag{7}$$

The F.S.S. of this equation is formed by the Bessel functions $\zeta^{1/2}I_\nu(\lambda\zeta^{\mu/2})$, $\zeta^{1/2}K_\nu(\lambda\zeta^{\mu/2})$. It will be assumed that $\nu \geqslant 0$ (that is $\phi(0,\lambda) \geqslant -1/4$). We state the asymptotic formulae for these functions:

$$I_\nu(x) \sim \frac{1}{\sqrt{2\pi x}}e^x, \quad K_\nu(x) \sim \sqrt{\frac{\pi}{2x}}e^{-x}, \quad x \to +\infty,$$
$$I_\nu(x) \sim \frac{(x/2)^\nu}{\Gamma(\nu+1)}, \quad K_\nu(x) \sim \frac{\Gamma(\nu)}{2}\left(\frac{x}{2}\right)^{-\nu}, \quad x \to +0, \quad \nu > 0,$$
$$I_0(x) = 1 + \frac{x^3}{4}O(x^4), \quad K_0(x) \sim -\ln x, \quad x \to +0.$$

The Wronskian of these solutions is

$$W(K_\nu(x), I_\nu(x)) = 1/x.$$

There are asymptotic bounds for the F.S.S. of equation (6) of the same type as those in § 4, paragraph 3.3. However, we will restrict ourselves to giving bounds for the F.S.S. under stronger assumptions. Let $\nu \geqslant 0$, $\beta = +\infty$, let $\psi(\zeta)$ be independent of λ and suppose that the following conditions hold.

1) $\psi(\zeta)$ is continuous for $0 < \zeta < \infty$;
2) $\psi(\zeta) = 0(\zeta^{p-1})$, $\zeta \to +0$, $\rho > 0$:
3) $\int^{+\infty} \zeta^{-\mu/2}|\psi(\zeta)|d\zeta < \infty$.

Then equation (6) has F.S.S. of the form

$$\begin{aligned} W_1(\zeta,\lambda) &= \zeta^{1/2} I_\nu(\lambda\zeta^{\mu/2})(1+\varepsilon_1), \\ W_2(\zeta,\lambda) &= \zeta^{1/2} K_\nu(\lambda\zeta^{\mu/2})(1+\varepsilon_2), \end{aligned} \tag{8}$$

where ε_1 and ε_2 have orders

$$O(\lambda^{-1}), \quad \mu < 2\rho; \quad O(\lambda^{-1}\ln\lambda), \quad \mu = 2\rho; \quad O(\lambda^{-2\rho/\mu}), \quad \mu > 2\rho. \tag{9}$$

These results are also true for $\beta < \infty$.

3. Case II. Here we take the minus sign in equation (6) and the standard equation has the form

$$\frac{d^2W}{d\zeta^2} = \left(-\frac{\mu^2\lambda^2}{4}\zeta^{\mu-2} + \frac{\mu^2\nu^2-1}{4\zeta^2}\right)W. \tag{10}$$

With the same conditions on $\psi(\zeta)$ as above, equation (6) has F.S.S. of the form

$$\begin{aligned} W_1(\zeta,\lambda) &= \zeta^{1/2}[J_\nu(\lambda\zeta^{\mu/2}) + \varepsilon_1(\zeta,\lambda)], \\ W_2(\zeta,\lambda) &= \zeta^{1/2}[Y_\nu(\lambda\zeta^{\mu/2}) + \varepsilon_2(\zeta,\lambda)]. \end{aligned} \tag{11}$$

In this case the bounds are complicated since the functions $J_\nu(x)$ and $Y_\nu(x)$ have infinitely many zeros on the half-line $x > 0$. We use the representations

$$\begin{aligned} J_\nu(x) &= E_\nu^{-1}(x)M_\nu(x)\cos\theta_\nu(x), \\ Y_\nu(x) &= E_\nu(x)M_\nu(x)\sin\theta_\nu(x), \end{aligned}$$

where E and M are positive functions. Then

$$\frac{\varepsilon_1(\zeta,\lambda)}{E_\nu^{-1}(\lambda\zeta^{\mu/2})M_\nu(\lambda\zeta^{\mu/2})}, \quad \frac{\varepsilon_2(\zeta,\lambda)}{E_\nu(\lambda\zeta^{\mu/2})M_\nu(\lambda\zeta^{\mu/2})}$$

have the orders in (9) as $\lambda \to \infty$. Similar bounds in cases I and II occur for the derivatives of the solutions [Olver 7].

§ 6. Multiple Turning Points. The Complex Case

1. Statement of the Problem. We consider the equation

$$w'' - [\lambda^2 f(z) + g(z)]w = 0 \tag{1}$$

in a domain D (possibly unbounded) of the complex z-plane. Here $\lambda > 0$ is a large parameter and D contains the point $z = 0$. We introduce the conditions:

1) The functions $z^{2-m} f(z)$ and $g(z)$ are holomorphic in D and the first of them has no zeros;

2) $g(z) = O(z^{\gamma-1})$ as $z \to 0$, where $\gamma > 0$. The number $m \geqslant 0$ need not be an integer. Thus $z = 0$ may be a branch point of the functions f and g.

We introduce the functions

$$S(z) = \int_0^z \sqrt{f(t)}dt\,, \quad \zeta(z) = S^{2/m}(z)\,. \tag{2}$$

For small $|z|$

$$f(z) = f_0 + f_1 z + f_2 z^2 + \dots\,, \quad f_0 \neq 0\,,$$

$$\zeta(z) = \left(\frac{2}{m}\right)^{2/m} f_0^{1/m} z \left[1 + \frac{z}{m+2}\frac{f_1}{f_0} + \dots\right]\,,$$

so that we can single out a branch of $\zeta(z)$ which is holomorphic at $z = 0$. We need one more condition:

3) $\zeta(z)$ is single-sheeted in D.

Let S_j be the sectors of the complex z-plane introduced in § 3. We will denote by Δ the image of D under the mapping $\zeta = \zeta(z)$, and by D_j the inverse image of the domain $\Delta \cap S_j$. All the domains D_j are simply-connected.

Example. Let $f(z) = z^2(1-z)^6$, where D is the z-plane with a cut along the line $l = [1, +\infty)$. Then

$$\zeta(z) = \frac{1}{\sqrt{2}(1-z)}$$

and Δ is the ζ-plane with a cut along the line $\tilde{l} = (-\infty, -1/\sqrt{2}]$. In this case $m = 4$, S_0 is the sector $|\arg\zeta| < \pi/4$, S_1 is the sector $|\arg\zeta - \pi/2| < \pi/4$, and the sector S_2 (S_3) is symmetric with the sector S_0 (S_1) about $\zeta = 0$. Through $z = 0$ and $z = 1$ we draw two circles l_1 and $l_2 = l_1^*$, the centre of l_1 lying at the point $z = (1 + \sqrt{2}/2)$. Then D_0 is the crescent formed by their intersection, $D_3 = D_1^*$, the domain D_1 is bounded by the larger arc of the circle l_1 and the arc of l_2, and D_2 is the exterior of the union of the circles with a cut along l.

Suppose that $a_j \in [D_j]$, where a_j may possibly be infinite. We denote by $H_k(a_j)$, $k \neq j$, the set of all points $z \in [D_j] \cup [D_k]$, which can be joined to a_j by a canonical path γ. That is, as t moves along γ from a_j to z the function Re $S(z)$ is non-decreasing. The branch of this function is chosen so that Re $S(z) \leqslant 0$ in D_j and Re $S(z) \geqslant 0$ in D_k.

We will assume that k, j are integers with $k \neq j$. Also, if $m = m_1/m_2$ where m_1 and m_2 are mutually coprime integers with $m_1 \neq 1$, then we require that $|k-j| < m_1$. For other values of m no additional restrictions are placed on j and k. In particular we can always put $j = k \pm 1$; in this case S_j and S_k are adjacent sectors. We note that the union $D_k \cup D_j \cup \{0\}$ is connected but may not be a domain.

2. Auxiliary Functions. We put

$$e(t) = |\exp\{(-1)^j t^{m/2}\}|, \quad t \in [S_j],$$

where the branch of $t^{m/2}$ is chosen so that $t^{m/2} = |t|^{m/2} \exp\{(1/2)im \arg t\}$. Therefore

$$e(t) = 1, \quad t \in \partial S_j; \quad e(t) > 1, \quad t \in S_j,$$

and $e(t)$ is exponentially increasing inside the sector S_j. In the domain $[S_j] \cup [S_k]$ we introduce the functions

$$E_{jk}(t) = 1/e(t), \quad t \in [S_j]; \quad E_{jk}(t) = e(t), \quad t \in [S_k],$$

so that $E_{kj}(t) = E_{jk}^{-1}(t)$. Let $U_j(t)$ and $U_k(t)$ be the solutions of the standard equation introduced in § 3. We recall that $U_j(t) \to 0$ if $t \to \infty$ along any line in S_j and $U_j(t) \to \infty$ if $t \to \infty$ along any line lying in S_k. The solution $U_k(t)$ has similar properties. Let us bring in the functions

$$\begin{aligned}
M_{jk}(t) &= [|U_j(t)|^2 E_{jk}^{-2}(t) + |U_k(t)|^2 E_{jk}^2(t)]^{1/2}, \\
N_{jk}(t) &= [|U_j'(t)|^2 E_{jk}^{-2}(t) + |U_k'(t)|^2 E_{jk}^2(t)]^{1/2}, \\
\tilde{N}_{jk}(t) &= |t|^{(2-m)/4} \left[E_{jk}^{-2}(t) \left| \frac{d}{dt}(t^{(m-2)/4} U_j(t)) \right|^2 \right. \\
&\qquad \left. + E_{jk}^2(t) \left| \frac{d}{dt}(t^{(m-2)/4} U_k(t)) \right|^2 \right]^{1/2}, \qquad (3) \\
\Omega_m(t) &= (1 + |t|^{m/2})(1 + |t|)^{-1}.
\end{aligned}$$

If $t \to \infty$ along any line inside one of the sectors S_j, S_k then

$$M_{jk}(t) \sim (1 + \lambda_{jk}^2)^{1/2} |t|^{(2-m)/4},$$

$$N_{jk}(t) \sim \frac{m}{2}(1 + \lambda_{jk}^2)^{1/2} |t|^{(m-2)/4} \sim \tilde{N}_{jk}(t),$$

where the numbers λ_{jk} are as in § 3. We put

$$H(z_1, z_2, \lambda) = \int_{z_1}^{z_2} \left| \tilde{f}^{-1/4}(z) \frac{d^2}{dz^2} \tilde{f}^{-1/4}(z) - \frac{g(z)}{\tilde{f}^{1/2}(z)} \right| \frac{|dz|}{\Omega_m(\lambda^{2/m}\zeta(z))},$$

where

$$\tilde{f}(z) = z^{2-m} f(z).$$

3. Bounds for the Solutions [Olver 4]. Suppose that the above conditions are satisfied and that the quantities $H(z_1, z_2, \lambda)$ are finite for any $z \in H_k(a_j)$, $\lambda > 0$, where the integral is taken along a canonical path. Let A and B be arbitrary constants. Then the equation (1) has solution $w(z, \lambda)$ of the form

$$w(z,\ \lambda) = \tilde{f}^{-1/4}(z)[AU_j(\lambda^{2/m}\zeta(z)) + BU_k(\lambda^{2(m}\zeta(z)) + \varepsilon(\lambda,\ \zeta)], \tag{4}$$

where we have the bounds

$$\frac{|\varepsilon(\lambda,\zeta)|}{M_{jk}(\lambda^{2/m}\zeta)}, \frac{|\partial\varepsilon(\lambda,\zeta)/\partial\zeta|}{\lambda^{2/m}N_{jk}(\lambda^{2/m}\zeta)}, \frac{|\partial(\zeta^{(m-2)/4}\varepsilon)/\partial\zeta|}{\lambda^{2/m}|\zeta|^{(m-2)/4}\tilde{N}_{jk}(\lambda^{2/m}\zeta)}$$
$$\leqslant c_1\sigma_{jk}(\lambda)E_{jk}(\lambda^{2/m}\zeta)\left[\exp\left\{\frac{c_2}{\lambda^{2/m}}H(a_j, z, \lambda)\right\} - 1\right] \tag{5}$$

for $z \in H_k(a_j)$ and $\lambda \geqslant \lambda_0 > 0$. Here c_1 and c_2 are positive constants not depending on λ,

$$\sigma_{jk}(\lambda) = \sup_{t\in\gamma}[\Omega(\lambda^{2/m}t)E_{jk}^{-1}(\lambda^{2/m}t)M_{jk}(\lambda^{2/m}t)]$$
$$\times |AU_j(\lambda^{2/m}t) + BU_k(\lambda^{2/m}t)|, \tag{6}$$

where $\tilde{\gamma}$ is the image of the canonical path γ connecting the points z and a_j under the mapping $\zeta = \zeta(z)$, and the integral $H(a_j, z, \lambda)$ is taken along γ.

The bound (5) is very general but it contains rather cumbersome expressions for $\sigma_{jk}(\lambda)$ and $H(a_j, z, \lambda)$. Moreover is of interest for asymptotic behaviour of solutions only if the right-hand side of formula (6) (without the multiplier E_{jk}^{-1}) is small for $\lambda \gg 1$. The expression in the square brackets in formula (6) is bounded on each compact subset of $[S_j \cup S_k]$, and is bounded for $t \to \infty$, $t \in S_k$. If however $t \to \infty$ inside S_j then this expression is unbounded, except in the case $B = 0$. Consequently if $a_j = \infty$ is an interior point of S_j then the bound (5) is applicable only when $B = 0$.

4. Asymptotic Behaviour of the Solutions. We put $A = 1$ and $B = 0$ in formula (5). Then $|\sigma_{jk}(\lambda)| \leqslant c$ for $\lambda \geqslant \lambda_0 > 0$. Suppose that the points $a_j \in S_j$ and $a_k \in S_k$ can be connected by a canonical path γ, not passing through $z = 0$ (for instance if S_j and S_k are adjacent sectors). The points a_j and a_k may be finite or infinite. Suppose that the integrals

$$\int^a |\delta(z)||dz|, \quad \int^a \frac{|dz|}{|S(z)|}$$

converge along γ, where $a = a_j$ or $a = a_k$, and

$$\delta(z) = f^{-1/4}(z)\frac{d^2}{dz^2}f^{-1/4} - \frac{g(z)}{f^{1/2}(z)}.$$

Then there are the bounds

$$\lambda^{-2/m}H(a_j, a_k, \lambda) = O(\psi_m), \quad \lambda \to \infty,$$

where

$$\begin{aligned} \psi_m &= \lambda^{-1}, \quad 0 < m < 2(1+\gamma_1), \\ \psi_m &= \lambda^{-1}\ln\lambda, \quad m = 2(1+\gamma_1), \\ \psi_m &= \lambda^{-2(1+\gamma_1)/m}, \quad m > 2(1+\gamma_1), \quad \gamma_1 = \min(\gamma, 1). \end{aligned} \tag{7}$$

If in addition $g(z)$ is holomorphic at $z = 0$ then we can take for ψ_m the function

$$\lambda^{-1}, \ 0 < m < 4; \quad \lambda^{-1}\ln\lambda, \ m = 4; \quad \lambda^{-4/m}, \ m > 4. \tag{8}$$

All these conditions are satisfied if the functions $z^{2-m}f(z)$ and $g(z)$ are polynomials of degrees n_f and n_g respectively and $n_g < n_f/2 + 1$.

The final result is that equation (1) has a solution $w(z, \lambda)$ such that for $z \in H_j(a_j)$ and $\lambda \geqslant \lambda_0 > 0$

$$w(z, \lambda) = \tilde{f}^{-1/4}(z)U_j(\lambda^{2/m}\zeta(z)) + \varepsilon_{jk}(\zeta, \lambda), \tag{9}$$

where

$$|\varepsilon_{jk}(z, \lambda)| \leqslant cE_{jk}(\lambda^{2/m}\zeta)M_{jk}(\lambda^{2/m}\zeta)\psi_m(\lambda).$$

Similar bounds occur for the remaining functions from the left hand side of (5). We recall that $U_j(t) \to 0$ if $t \to \infty$ inside the sector S_j.

5. Connection Formulae. Let the above conditions be satisfied and let $S(a_j) = \infty$, so that either $a_j = \infty$ or a_j is a singular point of equation (1). We choose the branch of $S(z)$ such that Re $S(z) \geqslant 0$ in D_j. Then for each fixed $\lambda > 0$ equation (1) has a unique solution $w_0(z, \lambda)$ such that

$$w_0(z, \lambda) \sim f^{-1/4}(z)e^{-\lambda S(z)} \tag{10}$$

for $z \in D_j$ and $z \to a_j$. We state the asymptotic formulae for w_0 in D_k. We extend the functions $S(z)$ and $f^{1/4}(z)$ analytically from D_j into D_k along some canonical path which does not pass through $z = 0$; then Re $S(z) \leqslant 0$ in D_k. Let $b_k(\theta) \in D_k$, let $\tilde{\gamma}$ be a canonical path connecting $b_k(\theta)$ and $a_k \in D_k$, and let arg $S(z) = \theta$ on $\tilde{\gamma}$ where $|\theta| \leqslant \pi/2$. Then as $\lambda \to \infty$ and $z \in \tilde{\gamma}$

$$\begin{aligned} w_0(z, \lambda) = f^{-1/4}(z)[&i^{k-j-1}(\lambda_{jk} + O(\psi_m))e^{\lambda S(z)} \\ &\pm i^{k-j}(\lambda_{j,k\pm1} + O(\psi_m))e^{-\lambda S(z)}]. \end{aligned} \tag{11}$$

The function ψ_m is defined by formulae (7) and (8), and the sign + (−) is taken for $\theta \geqslant 0$ ($\theta \leqslant 0$). The bounds for the remainder terms are uniform in z and θ.

Similar results were obtained in [Olver 4] for the more general equation

$$w'' = [\lambda^2 f(z,\ \lambda) + g(z,\ \lambda)]w\,.$$

§ 7. Two Close Turning Points

1. Statement of the Problem. We consider the equation

$$y'' - \lambda^2 q(x,\ \alpha)y = 0 \tag{1}$$

on the interval $I = [a, b]$. Here $\lambda > 0$ is a large parameter, $\alpha \in J = [0, \alpha_0]$, $\alpha_0 > 0$, and the function $q(x, \alpha)$ is real of class $C^\infty(I \times J)$. The parameters α and λ are independent.

We will assume that for each fixed α, $0 < \alpha \leqslant \alpha_0$, equation (1) has precisely two simple turning points $x_1(\alpha)$ and $x_2(\alpha)$ which merge for $\alpha = 0$ into the double turning point x_0 where $a < x_0 < b$. The Weber equation is a typical example:

$$y'' \pm \lambda^2(x^2 - \alpha)y = 0\,.$$

We are required to construct the F.S.S. of equation (1) whose asymptotic behaviour is applicable as $\lambda \to \infty$ and uniform for $x \in I$ and $\alpha \in J$.

Another possibility is when the close turning points $x_1(\alpha)$ and $x_2(\alpha)$ are complex for $\alpha > 0$. In this case we assume additionally that $q(x, \alpha)$ is holomorphic in $x \in D$, where D is a domain of the complex x-plane, and $D \supset I$ for each fixed $\alpha \in J$. A typical example is the Weber equation

$$y'' \pm \lambda^2(x^2 + \alpha)y = 0\,.$$

This sort of problem arises, for instance, in quantum mechanics where the energy of a particle is close to the bottom of a potential well or to the top of a potential barrier. In what follows we will assume that $x_0 = 0$, $a < 0 < b$ and that $\alpha_0 > 0$ is sufficiently small and independent of λ.

2. Real Turning Points

2.1 The Structure of the Function $q(x, \alpha)$. Since the point $x = 0$ is a second order turning point for $\alpha = 0$ we have

$$q(0,\ 0) = q'_x(0,\ 0) = 0\,, \quad q''_{xx}(0,\ 0) \neq 0\,. \tag{2}$$

We require additionally that

$$q'_\alpha(0, 0) \neq 0\,. \tag{3}$$

For small x and α we have

$$q(x,\ \alpha) = q'_\alpha(0,\ 0)\alpha + \frac{1}{2}[q''_{xx}(0,\ 0)x^2 + 2q'_{x\alpha}(0,\ 0)x\alpha + q''_{\alpha\alpha}(0,\ 0)\alpha^2] + \dots$$

The fact that the turning points are real leads us to the condition

$$q'_\alpha(0,\ 0)q''_{xx}(0,\ 0) < 0\,. \tag{4}$$

We index the turning points so that $x_1(\alpha) < 0 < x_2(\alpha)$ for $\alpha > 0$. The function q can be represented as

$$q(x,\ \alpha) = (x^2 + a(\alpha)x + b(\alpha))\tilde{q}(x,\ \alpha)\,, \tag{5}$$

where $\tilde{q}(x,\alpha) \in C^\infty(I \times J)$ and $\tilde{q}(x,\alpha) \neq 0$ for $(x,\alpha) \in I \times J$. It follows from conditions (2)–(4) that

$$a(0) = b(0) = D(0) = 0\,, \quad D'(0) > 0\,, \quad D(\alpha) = a^2(\alpha) - 4b(\alpha)\,. \tag{6}$$

Since

$$x_{1,2}(\alpha) = \frac{1}{2}(-a(\alpha) \pm \sqrt{D(\alpha)})\,,$$

the $x_j(\alpha)$ are infinitely differentiable functions of $\sqrt{\alpha}$ for $\alpha \in J$.

2.2 Transformation of the Equation. The substitution

$$y = \sqrt{\phi'(S)}w\,, \quad x = \phi(S) \tag{7}$$

reduces equation (1) to the form

$$w''_{SS} - \left[\lambda^2 q(x,\alpha)[\phi'(S)]^2 - \frac{1}{2}\{\phi, S\}\right] w = 0\,, \tag{8}$$

where $\{\phi, S\}$ is the Schwarzian derivative (Chap. 2, § 1). The function $S(x,\alpha)$ is chosen so that the coefficient of λ^2 has the form $\pm(\beta(\alpha) - S^2)$. We must distinguish between two cases.

I. $q''_{xx}(0,0) < 0$

II. $q''_{xx}(0,0) > 0$.

In the first case $q(x,\alpha)$ is positive for $\alpha > 0$, $x_1(\alpha) < 0 < x_2(\alpha)$, in the second case negative. We obtain equations for S

$$\begin{aligned} S'^2(\beta - S^2) &= q(x,\ \alpha) \quad (I)\,,\\ S'^2(\beta - S^2) &= -q(x,\ \alpha) \quad (II)\,. \end{aligned} \tag{9}$$

We will consider case I; case II is studied in a similar fashion.

2.3 The Function $\beta(\alpha)$. It follows from the continuity of the function S that $S = \pm\sqrt{\beta}$ for $x = x_{1,2}(\alpha)$. We require that $S = -\sqrt{\beta} < 0$ for $x = x_1(\alpha)$. Then on $I(\alpha) = [x_1(\alpha),\ x_2(\alpha)]$ we have

$$\int_{\sqrt{\beta}}^{S} \sqrt{\beta - t^2}\,dt = \int_{x_1(\alpha)}^{x} \sqrt{q(t,\ \alpha)}\,dt\,, \tag{10}$$

where all roots are positive. From the condition $S = \sqrt{\beta}$ for $x = x_2(\alpha)$, we find that

$$\beta(\alpha) = \frac{2}{\pi} \int_{x_1(\alpha)}^{x_2(\alpha)} \sqrt{q(t,\alpha)}\,dt\,. \tag{11}$$

We can show that $\beta(\alpha) \in C^\infty(J)$. We suppose additionally that $q(x,\alpha)$ is holomorphic in x in a domain D containing the line segment I; this leads to a more convenient formula for $\beta(\alpha)$. We select the branch of the function $\sqrt{q(x,\alpha)}$ in $D\backslash I(\alpha)$ such that $\sqrt{q(x,\alpha)} = -i|\sqrt{q(x,\alpha)}|$ for $x \in I(\alpha)$, $x > x_2(\alpha)$; then $\sqrt{q(x,\alpha)} > 0$ on the upper side of the cut $I(\alpha)$. Let C be a simple closed contour, positively oriented and going around the interval $I(\alpha)$. Then

$$\beta(\alpha) = -\frac{1}{\pi} \oint_C \sqrt{q(x,\alpha)}\,dx\,. \tag{12}$$

Since the integrand is infinitely differentiable in α for $x \in C$, we have $\beta(\alpha) \in C^\infty(J)$. Further, if $q(x,\alpha)$ is holomorphic in α for $\alpha \in J$, then $\beta(\alpha)$ is holomorphic in J. If we expand $\sqrt{q(x,\alpha)}$ as a power series in α for $x \in C$, we obtain a series expansion for $\beta(\alpha)$. In particular

$$\beta'(0) = \sqrt{\frac{2}{|q''_{xx}(0,0)|}}\,|q'_\alpha(0,0)|\,. \tag{13}$$

2.4 The Function $S(x,\alpha)$. Let us fix $\alpha > 0$ and let $x_1(\alpha) \leqslant x \leqslant x_2(\alpha)$, $-\sqrt{\beta} \leqslant S \leqslant \sqrt{\beta}$. Then the left and right hand sides of (10) are strictly monotonic increasing functions of S and x. Hence $S(x,\alpha)$ is a strictly monotonic increasing continuous function. We extend the branch of $\sqrt{q(x,\alpha)}$ on the intervals $(x_2(\alpha), b]$ and $[a, x_1(\alpha))$ through the upper half-plane so that

$$\sqrt{q(x,\alpha)} = -i|\sqrt{q(x,\alpha)}|\,, \quad x > x_2(\alpha)\,,$$
$$\sqrt{q(x,\alpha)} = i|\sqrt{q(x,\alpha)}|\,, \quad x < x_1(\alpha)\,,$$

and extend $\sqrt{\beta - S^2}$ in a similar way. Then S belongs to $C(I)$, is strictly monotonic increasing and is infinitely differentiable everywhere on I except at the points $x_{1,2}(\alpha)$. For x near to $x_1(\alpha)$ the right-hand side of (11) can be represented as $(x - x_1(\alpha))^{3/2}Q(x,\alpha)$, and the left-hand side as $(S + \sqrt{\beta})^{3/2}\tilde{Q}(S,\beta)$ where Q and $\tilde{Q}$ are non-vanishing C^∞-functions. It follows from this that $S(x,\alpha) \in C^\infty(I)$ for $\alpha > 0$; for $\alpha = 0$ the proof is direct. Further $S(x,\alpha) \in C(I \times J)$.

3. Complex Turning Points. In this case conditions (2) and (3) stay as they are and (4) is changed to

$$q'_\alpha(0,\ 0)q''_{xx}(0,\ 0) > 0\,. \tag{14}$$

The representation (5) for q and condition (6) also stay the same. The turning points are complex conjugates: $x_2(\alpha) = \overline{x_1(\alpha)}$. We denote them by $x(\alpha)$ and $\overline{x(\alpha)}$, where $\operatorname{Im} x(\alpha) < 0$.

The substitution (7) reduces equation (1) to the form (8). As in paragraph 2 we need to distinguish between two cases.

III. $q''_{xx}(0,0) > 0$.

IV. $q''_{xx}(0,0) < 0$.

We obtain equations for S

$$\begin{aligned} &S'^2(\beta + S^2) = q(x,\alpha) \qquad (III)\,,\\ &S'^2(\beta + S^2) = -q(x,\alpha) \quad (IV)\,. \end{aligned} \tag{15}$$

We will consider case III; then

$$\int_{-i\sqrt{\beta}}^{S} \sqrt{\beta + t^2}\,dt = \int_{x_1(\alpha)}^{x} \sqrt{q(t,\alpha)}\,dt\,. \tag{16}$$

We select the branches of the roots $\sqrt{q(t,\alpha)}$ and $\sqrt{t^2+\beta}$ to be positive for $t \in I$. These roots are then positive on the intervals $[x(\alpha),\ \overline{x(\alpha)}]$ and $[-i\sqrt{\beta},\ i\sqrt{\beta}]$. Therefore

$$\int_{-i\sqrt{\beta}}^{0} \sqrt{\beta + t^2}\,dt = \frac{i\pi\beta}{4}\,, \qquad \int_{x(\alpha)}^{\operatorname{Re}\, x(\alpha)} \sqrt{q(t,\alpha)}\,dt = iA\,, \quad A > 0\,.$$

From the condition $S = -i\sqrt{\beta}$ for $x = \overline{x(\alpha)}$, we find that

$$\beta(\alpha) = -\frac{2i}{\pi} \int_{x(\alpha)}^{\overline{x}(\alpha)} \sqrt{q(t,\alpha)}\,dt\,.$$

Let C be a simple closed contour in the complex x-plane enclosing the segment $[x(\alpha), \overline{x(\alpha)}]$ and positively oriented. Then

$$\beta(\alpha) = -\frac{i}{\pi} \oint_C \sqrt{q(x,\alpha)}\,dx\,, \tag{17}$$

where we choose the branch of the root so that $\sqrt{q(x,\alpha)} > 0$ for $x \in C$ and $x > \operatorname{Re} x(\alpha)$. The function $\beta(\alpha)$ is infinitely differentiable for $\alpha \in J$ and formula (13) is true here. For each fixed $\alpha \in J$ the function $S(x,\alpha)$ is holomorphic in x in a complex neighbourhood of I and continuous in (x,α), as is proved the same way as in paragraph 2.4. For $x \in I$ the function $S(x,\alpha)$ is real and monotonic increasing. This follows from the choice of the branches of the roots and from the fact that for $x \in I$

$$\int_0^S \sqrt{t^2+\beta}\,dt = \int_{\mathrm{Re}\ x(\alpha)}^{x} \sqrt{q(t,\alpha)}\,dt\,.$$

4. The Principal Asymptotic Term of the Solutions. Using the transformation (7) equation (1) can be reduced to the form (8), and so it is sufficient to obtain asymptotic formulae for the solution of the reduced equation. Again we denote the independent variable by x and take β as α^2 (recall that $\beta(\alpha) \geqslant 0$ for $\alpha \geqslant 0$), where $\alpha \geqslant 0$.

4.1 Case I. Let us consider the equation

$$y'' - \lambda^2\left(\frac{x^2}{4} - \alpha^2\right)y + f(x,\alpha)y = 0 \tag{18}$$

on the interval $I = [-a, a]$, $a > 0$, where the function f is continuous in x and α. The standard equation has the form

$$w'' - \left(\frac{x^2}{4} - \alpha^2\right)w = 0\,. \tag{19}$$

Its F.S.S. is formed by the parabolic cylinder functions (or Weber functions) $U(-\alpha^2, x)$ and $V(-\alpha^2, x)$ [Abramowitz]. Using the notation of Whittaker we have

$$U(a,\ x) = D_{a-1/2}(x)\,,$$

$$V(a,\ x) = \frac{1}{\pi}\Gamma\left(a+\frac{1}{2}\right)[\sin \pi a D_{-a-1/2}(x) + D_{-a-1/2}(-x)]\,.$$

The Wronskian of these solutions is

$$W(U,V) = \sqrt{2/\pi}\,.$$

As $x \to \infty$ with a fixed, we have

$$U(a,\ x) \sim x^{-a-1/2}e^{-x^2/4}\,,$$

$$V(a,\ x) \sim \sqrt{2/\pi}x^{a-1/2}e^{x^2/4}\,.$$

The asymptotic expansion of these solutions as $x \to -\infty$ seems to be complicated [Abramowitz]. Equation (18) has F.S.S. of the form

$$y_1(x,\ \lambda,\ \alpha) = U(-\lambda\alpha^2, x\sqrt{\lambda}) + O(\lambda^{-2/3})|U(-\lambda\alpha^2,\ x\sqrt{\lambda})|\,. \tag{20}$$

The bounds for the remainder terms are uniform in x and α for $x \in I$ and $\alpha \in J = [0, \alpha_0]$, $\alpha_0 > 0$. This is also true for the latter bounds.

4.2 Case III. The reduced equation has the form

$$y'' - \lambda^2\left(\frac{x^2}{4} + \alpha^2\right)y + f(x,\alpha)y = 0\,. \tag{21}$$

In this case the F.S.S. of the standard equation

$$w'' - \left(\frac{x^2}{4} + \alpha^2\right) w = 0 \tag{22}$$

is formed by the functions $U(\alpha^2, x)$ and $U(\alpha^2, -x)$. Equation (21) has a F.S.S. of the form

$$y_{1,2}(x, \lambda, \alpha) = U(\lambda\alpha^2, \pm x\sqrt{\lambda})[1 + O(\lambda^{-1} \ln \lambda)]\,. \tag{23}$$

From the point of view of quantum mechanics cases I and III correspond to the case where the particle energy is close to the bottom of a potential well. The next possibilities correspond to the case where the particle energy is close to the top of a potential barrier.

4.3 Cases II and IV. We consider the equation

$$y'' + \lambda^2 \left(\frac{x^2}{4} \pm \alpha^2\right) y + f(x, \alpha) y = 0\,, \tag{24}$$

where the function $f(x, \alpha)$ has the same properties as those in paragraph 4.1. The standard equation takes the form

$$w'' + \left(\frac{x^2}{4} \pm \alpha^2\right) w = 0$$

and has F.S.S.

$$\{W(\alpha^2, x),\, W(\alpha^2, -x)\}\,, \quad \{W(-\alpha^2, x),\, W(-\alpha^2, -x)\}\,.$$

The first (second) F.S.S. corresponds to α^2 $(-\alpha^2)$. The Wronskian of these solutions is

$$W(W(a, x),\, W(a - x)) = 1\,.$$

As $x \to \infty$ we have

$$W(a, x) = \sqrt{\frac{2k}{x}} \cos\left(\frac{x^2}{4} - a \ln x + \frac{\pi}{4} + \frac{\Phi}{2}\right) + O\left(\frac{1}{x^{5/2}}\right),$$

$$W(a, -x) = \sqrt{\frac{2}{kx}} \sin\left(\frac{x^2}{4} - a \ln x + \frac{\pi}{2} + \frac{\Phi}{2}\right) + O\left(\frac{1}{x^{5/2}}\right),$$

where

$$k = \sqrt{1 - e^{2\pi a}} - e^{\pi a}\,, \quad \Phi = \arg \Gamma(1/2 + ia)\,.$$

Equation (24) has F.S.S. [Olver 8, Olver 5]

$$\begin{aligned} w^{+}_{1,2}(x,\ \lambda,\ \alpha) &= W(\pm\lambda\alpha^2,\ -x\sqrt{\lambda}) \\ &\quad + O(\lambda^{-1}\ln\lambda|W(\pm\lambda\alpha^2,\ -x\sqrt{\lambda})|)\,, \\ w^{-}_{1,2}(x,\ \lambda,\ \alpha) &= W(\pm\lambda\alpha^2,\ x\sqrt{\lambda}) \\ &\quad + O(\lambda^{-2/3}\ln\lambda|W(\pm\lambda\alpha^2,\ x\sqrt{\lambda})|)\,. \end{aligned} \tag{25}$$

The bounds for the remainder terms are uniform in x and α.

§ 8. Fusion of Several Turning Points

1. The Characteristic Polygon. We consider the equation

$$\varepsilon^2 w'' - q(z,\ \varepsilon)w = 0 \tag{1}$$

under the following conditions:

1) The function $q(z,\varepsilon)$ is holomorphic in both the variables in the domain

$$D : |z| \leqslant r_0\,, \quad |\varepsilon| \leqslant \varepsilon_0\,, \quad |\arg\varepsilon| \leqslant \theta_0\,.$$

2) As $\varepsilon \to 0$ with $|\arg\varepsilon| \leqslant \theta_0$ there is the asymptotic expansion

$$q(z,\varepsilon) = \sum_{r=0}^{\infty} q_r(z)\varepsilon^r\,,$$

uniform in z, $|z| \leqslant r_0$, where the $q_r(z)$ are holomorphic for $|z| \leqslant r$.

3) $q_0(z) = z^n\tilde{q}_0(z)$, where $\tilde{q}_0(0) \neq 0$ and the function $\tilde{q}_0(z)$ is holomorphic for $|z| \leqslant r_0$.

The point $z = 0$ is a turning point when $n \geqslant 1$. The case $n = 1$ was considered in §§ 1,2, and the case $n = 2$ in § 7. Hence we consider $n \geqslant 3$. The numbers r_0 and ε_0 are assumed to be positive and sufficiently small. Without loss of generality we can assume that $q(z,0) = z^n$ since we can reduce equation (1) to this form by a change of variable and function (Chap. 2 § 1; Chap. 4, § 4). Then equation (1) becomes

$$\varepsilon^2 w'' - [z^n + \varepsilon\tilde{q}(z,\ \varepsilon)]w = 0\,.$$

The solutions of the equation $q(z,\varepsilon) = 0$ are called *turning points* $z(\varepsilon)$ for equation (1). If $\tilde{q}(z,\varepsilon) \not\equiv 0$ then for $\varepsilon \neq 0$ there are n turning points $z_1(\varepsilon), \ldots, z_n(\varepsilon)$ which merge into the one turning point $z = 0$ as $\varepsilon \to 0$. They are called *secondary turning points*.

Suppose that $q_r(z) \not\equiv 0$; then

$$q_r(z) = \sum_{k=m_r}^{\infty} q_{rk} z^k , \quad q_{r,m_r} \neq 0 . \tag{2}$$

In the plane with Cartesian coordinates X, Y we mark the points

$$Q = (1,\ -1), \quad P_0 = (0,\ n/2), \quad P_r = (r/2,\ m_r/2), \quad r = 1,\ 2,\ \ldots$$

If $q_r(z) \equiv 0$ then P_r is absent.

Let Γ be the piecewise linear curve connecting P_0 and Q. The vertices of Γ are some of the points P_r. Let Γ also be convex downwards with all points P_r lying on or above Γ. Then Γ is called the *characteristic polygon* of the turning point $z = 0$ [Fedoryuk 13]. The piecewise linear curve Γ consists of one or two components. There are three possibilities.

1. Γ is the line segment P_0Q. In this case either $2m_1 > n-2$ or $q_r(z) \equiv 0$, giving $q(z,\varepsilon) = z^n + \varepsilon^2 q(x,\varepsilon)$ (equations of this type were considered in the previous paragraphs).
2. Γ is the line segment P_0Q and $P_1 \in \Gamma$. Here $2m_1 = n-2$.
3. Γ is the polygonal line with vertices P_0, P_1 and Q. Here P_1 lies below the segment P_0Q, and $2m_1 < n-2$.

The asymptotic behaviour of the solutions in a neighbourhood of $z = 0$ in cases 2 and 3 is considerably more complicated than that in case 1.

2. Reduction of Equation (1) when Γ is a Line Segment [Sibuya 1]

2.1 Reduction in the Whole Neighbourhood of a Turning Point. We replace equation (1) by the equivalent system

$$\varepsilon u' = \begin{bmatrix} 0 & 1 \\ q(z,\varepsilon) & 0 \end{bmatrix} u , \quad u = \begin{bmatrix} w \\ \varepsilon w' \end{bmatrix} . \tag{3}$$

Suppose that the following condition is satisfied as $z \to 0$:

$$q_1(z) = O(z^{(n-1)/2}), \quad n \text{ odd},$$
$$q_1(z) = O(z^{n/2}), \qquad n \text{ even}.$$

Then there is a matrix function $T(z,\varepsilon)$ such that the transformation

$$u = T(z,\ \varepsilon)v \tag{4}$$

reduces system (3) to the form

$$\varepsilon v' = \begin{bmatrix} 0 & 1 \\ z_n + \varepsilon \sum_{j=2}^{n} \beta_j(\varepsilon) z^{n-j} & 0 \end{bmatrix} v . \tag{5}$$

Also, $T(z,\varepsilon)$ is holomorphic in z and ε in the domain D (see condition (1)) and has asymptotic expansion

$$T(z,\varepsilon) = \sum_{r=0}^{\infty} T_r(z)\varepsilon^r\,, \quad \varepsilon \to 0\,, \quad |\arg \varepsilon| \leqslant \theta_0\,, \tag{6}$$

uniformly in z, $|z| \leqslant r$. Further $\det T_0(z) = 1$. The functions $\beta_j(\varepsilon)$ are holomorphic for $|\varepsilon| \leqslant \varepsilon_0$ and $|\arg \varepsilon| \leqslant \theta_1$, and have asymptotic expansions in this sector

$$\beta_j(\varepsilon) = \sum_{k=0}^{\infty} \beta_{jk}\varepsilon^k\,, \quad \varepsilon \to 0\,.$$

System (5) is equivalent to the equation

$$\varepsilon^2 W'' - [z^n + \varepsilon \sum_{j=2}^{n} \beta_j(\varepsilon) z^{n-j}]W = 0\,, \tag{7}$$

in which the coefficients of W are polynomials in z of degree n.

In this way the study of the asymptotic solution of equation (1) in the neighbourhood of the turning point $z = 0$ leads to the study of the solutions of an equation of the form

$$w'' - Q(z)w = 0\,, \quad Q(z) = z^n + a_1 z^{n-1} + \ldots + a_n\,.$$

The conditions on $q_1(z)$ are equivalent to requiring that the characteristic polygon Γ consists of one segment.

2.2 Reduction of Equation (1) in a Sector. Suppose that as $z \to 0$

$$q_1(z) = O(z^{(n-1)/2})\,, \quad n \text{ odd}\,,$$
$$q_1(z) = O(z^{(n-2)/2})\,, \quad n \text{ even}\,.$$

We remark that if n is even and $q_1(z) = q_0 z^{(n-2)/2} + \ldots$, then Γ consists of two segments. Let $\delta > 0$ be sufficiently small, let D be as in paragraph 1, and let $\theta_0 = (m+2)\delta/4$. Then the transformation (4) exists, reducing the system (3) to the form (5) and having the same properties. The only difference is that the reduction is effected for z lying in the sector

$$|z| \leqslant r_0\,, \quad -\frac{\pi}{n+2} + \delta \leqslant \arg z - \frac{2k\pi}{n+2} \leqslant \frac{3\pi}{n+2} + \delta\,,$$

where k is a fixed integer.

3. Asymptotic Behaviour of the Solutions when Γ Consists of One Component [Nakano 2]. In this case equation (1) has the form

$$\varepsilon^2 w'' + [z^n + \varepsilon z^m q_1(z) + \varepsilon^2 q_2(z,\, \varepsilon)]w = 0\,, \tag{8}$$

where either $q_1(0) \neq 0$ and $2m > n-2$ or $q_1(z) \equiv 0$.

The neighbourhood $|z| \leqslant r$ of the turning point decomposes into two domains: the *exterior*

$$D_e : c_1\varepsilon^{2/(n+2)} \leqslant |z| \leqslant r$$

and the *interior*

$$D_i : |z| \leqslant c_2\varepsilon^{2/(n+2)},$$

where c_1, c_2 are positive constants independent of ε, and c_2 is sufficiently small. We construct the asymptotic expansions of the solution in these domains, called respectively the *exterior* and *interior expansions* (or *solutions*). The exterior expansion is simply a modification of the WKB-approximation (Chap. 2, § 3). The interior expansion is expressed in terms of the solution of the standard equation

$$\varepsilon^2 w'' + z^n w = 0,$$

that is, in terms of Bessel functions (§ 3). Since the intersection $D_i \cap D_e$ is non-empty both asymptotic expansions are applicable in this domain, which allows us to combine them (or piece them together) and obtain the asymptotic behaviour of the F.S.S. for equation (1) in the whole neighbourhood of a turning point. In actual fact the situation is rather more complicated, since one must decompose D_e and D_i into sectors in each of which are constructed the asymptotic behaviours of the corresponding F.S.S., and find the connection formulae between them.

3.1 The Exterior Expansion. We make the transformation

$$\begin{bmatrix} w \\ \varepsilon w' \end{bmatrix} = \begin{bmatrix} 1 & 1 \\ iz^{n/2} & -iz^{n/2} \end{bmatrix} \begin{bmatrix} u_1 \\ u_2 \end{bmatrix}, \quad z = \zeta^2.$$

We then obtain the system

$$(\zeta^{-(n+2)}\varepsilon)\zeta\frac{du}{d\zeta} = B(\zeta,\varepsilon)u,$$

$$B(\zeta,\varepsilon) = \sum_{j=0}^{\infty} B_j(\zeta)[\zeta^{-(q+2)}\varepsilon]^j.$$

We remark that $\rho = \zeta^{-(n+2)}\varepsilon \to 0$ if $|z| \geqslant \varepsilon^{1/(n+2)-\alpha}$, $\alpha > 0$, so that there is a F.A.S. of solutions u as a series in powers of ρ. We have

$$B_0(\zeta) = \begin{bmatrix} -2i & 0 \\ 0 & 2i \end{bmatrix}, \quad B_1(\zeta) = \begin{bmatrix} -ib_1(\zeta) - n/2 & -ib_1(\zeta) + n/2 \\ ib_1(\zeta) + n/2 & ib_1(\zeta) - n/2 \end{bmatrix},$$

$$B_j(\zeta) = \begin{bmatrix} -ib_j(\zeta) & -ib_j(\zeta) \\ ib_j(\zeta) & ib_j(\zeta) \end{bmatrix}, \quad j \geqslant 2,$$

where the $b_j(\zeta)$ are series in powers of ζ. In particular,

$$b_1(\zeta) = \begin{cases} q_{1k_2}\zeta^2 + q_{1k_4}\zeta^4 + \ldots + q_{1k_{2m}}\zeta^{2m} + \ldots, & n \text{ even}, \\ q_{1k_2}\zeta^2 + q_{1k_3}\zeta^3 + \ldots + q_{1k_{2m}}\zeta^{2m-1} + \ldots, & n \text{ odd}, \end{cases}$$

where $k_{2m} = n/2 + m - 1$ and $k_{2m-1} = (n-1)/2 + m - 1$.

Let $D_k = S_k \cap D_e$ where S_k is the sector

$$\frac{2(k-1)}{n+2}\pi < \arg z < \frac{2(k+1)}{n+2}\pi .$$

Equation (8) has fundamental matrix

$$\begin{aligned} Y &= \begin{bmatrix} w_1 & w_2 \\ \varepsilon w_1' & \varepsilon w_2' \end{bmatrix} \\ &= z^{-n/4} \begin{bmatrix} 1 & 1 \\ iz^{n/2} & -iz^{n/2} \end{bmatrix} \left[I + \sum_{j=0}^{m} U_j(\zeta)(\varepsilon\zeta^{-n-2})^j + R_m(\zeta,\varepsilon) \right] \\ &\quad \times \exp\{(\varepsilon\zeta^{-n-2})^{-1}\Lambda_0 + \Lambda_1\}, \end{aligned} \tag{9}$$

where

$$\Lambda_0 = \frac{2i}{n+2}\begin{bmatrix} 1 & 0 \\ 0 & -1 \end{bmatrix}, \quad \Lambda_1 = a(\zeta)\begin{bmatrix} -1 & 0 \\ 0 & 1 \end{bmatrix}, \tag{10}$$

$$a(\zeta) = i\int_0^\zeta t^{-1} b_1(t)dt .$$

The bounds for the remainder term in D_k are

$$|R_m(\zeta, \varepsilon)| \leqslant c_m(\varepsilon\zeta^{-n-2})^{m+1} . \tag{11}$$

As $\rho \to \infty$ the exterior expansion is the standard WKB-expansion (Chap. 2, § 3), and the matrices $U_j(\zeta)$ are determined from recurrence relations. In these formulae $z^{1/2} > 0$ and $z^{1/4} > 0$ for $z > 0$.

8.2 The Interior Expansion. In the interior domain D_i equation (1) can be considered as a perturbation of the equation $\varepsilon^2 w'' + z^n w = 0$, the solutions of which can be expressed in terms of Bessel functions. We make the transformation

$$\tilde{z} = \varepsilon^{-2/(n+2)} z, \quad \begin{bmatrix} w \\ \varepsilon w' \end{bmatrix} = \begin{bmatrix} 1 & 0 \\ 0 & \rho^n \end{bmatrix} v, \quad \rho = \varepsilon^{1/(n+2)}$$

to obtain the system

$$\frac{dv}{d\tilde{z}} = \begin{bmatrix} 0 & 1 \\ -\tilde{z}^n + \psi(\tilde{z},\rho) & 0 \end{bmatrix} v, \tag{12}$$

where

$$\psi(\tilde{z},\rho)=\sum_{j=1}^{\infty} c_j(\tilde{z})\rho^j .$$

The new variable $\tilde{z}$ is called the *interior variable*. We seek the F.A.S. in the form

$$v=\sum_{j=0}^{\infty} v_j(\tilde{z})\rho^j .$$

Then we obtain the recurrence system of equations for v_j

$$\frac{dv_0}{d\tilde{z}}=C_0(\tilde{z})v_0\,,\quad \frac{dv_j}{d\tilde{z}}=\sum_{k=0}^{j} C_k(\tilde{z})v_{j-k}\,,\quad j\geqslant 1\,, \tag{13}$$

$$C_0(\tilde{z})=\begin{bmatrix}0 & 1\\ -\tilde{z}^n & 0\end{bmatrix},\quad C_j(\tilde{z})=\begin{bmatrix}0 & 0\\ c_j(\tilde{z}) & 0\end{bmatrix}.$$

We consider equation (13) as a matrix equation; that is, v is a 2×2 matrix. The first equation in (13) has fundamental matrix

$$v_0(\tilde{z})=\xi^{\nu}\begin{bmatrix}0 & 0\\ 1 & \tilde{z}^{n/2}\end{bmatrix}\begin{bmatrix}H_{\nu}^{(1)}(\xi) & H_{\nu}^{(2)}(\xi)\\ H_{\nu-1}^{(1)}(\xi) & H_{\nu-1}^{(2)}(\xi)\end{bmatrix}, \tag{14}$$

$$\xi=\frac{2}{n+2}\tilde{z}^{(n+2)/2},\quad \nu=\frac{1}{n+2}.$$

The solutions of the next equations are found using the method of variation of constants. With an appropriate choice for the contours of integration we obtain

$$v_j(\tilde{z})=\begin{bmatrix}1 & 0\\ 0 & \tilde{z}^{n/2}\end{bmatrix}\xi^{1/(n+2)}V_j(\tilde{z})\begin{bmatrix}g_1(\xi) & 0\\ 0 & g_2(\xi)\end{bmatrix}, \tag{15}$$

where

$$g_{1,2}(\xi)=\sqrt{\frac{2}{\pi\xi}}\exp\left\{\pm i\left(\xi-\frac{\nu\pi}{2}-\frac{\pi}{4}\right)\right\}.$$

All the subsequent asymptotic expansions and bounds are valid in the domain $D_0(\tilde{z})=D_i\cap S_0$. Equation (9) has fundamental matrix of the form

$$Y_0(\tilde{z},\rho)=\begin{bmatrix}1 & 0\\ 0 & \rho^n\end{bmatrix}\left[\sum_{j=0}^{m} v_j(\tilde{z})\rho^j+O(\rho^{m+1})\right],\quad |\tilde{z}|\leqslant a_0\,, \tag{16}$$

$$Y_0(\tilde{z},\rho) = \begin{bmatrix}1 & 0\\ 0 & \rho^n\end{bmatrix} \xi^\nu \begin{bmatrix}1 & 0\\ 0 & \tilde{z}^{n/2}\end{bmatrix} \Bigg[\sum_{j=0}^{m} w_j(\tilde{z})(\tilde{z}^{1/2}\rho)^j$$
$$+ O((\tilde{z}^{1/2}\rho)^{m+1})\Bigg] \begin{bmatrix}g_1(\xi) & 0\\ 0 & g_2(\xi)\end{bmatrix}, \quad |\tilde{z}| \geqslant a_0 .$$

Here $w_j(\tilde{z}) = \tilde{z}^{-j/2}V_j(\tilde{z})$, the matrix functions $w_j(\tilde{z})$ are bounded for $|\tilde{z}| \geqslant a_0$, and $\xi > 0$ for $\tilde{z} > 0$. The matrices $V_j(\xi)$ have the following asymptotic expansions as $\xi \to \infty$.

A. n even. Then

$$V_{2m-1}(\xi) = 0, \quad 2m-1 < n+2,$$

$$V_{2m}(\xi) = \xi^{2m/(n+2)} \sum_{j=0}^{\infty} V_{2m,j}\xi^{-j}, \quad 2m < n+2,$$

$$V_k(\xi) = \xi^{k/(n+2)}\left[V_{k^0} + \sum_{j=1}^{\infty} V_{kj}\xi^{-j}\ln\xi\right], \quad k \geqslant n+2.$$

In particular

$$V_2(\xi) = \frac{1}{2}\left(\frac{n+2}{2}\right)^{2/(n+2)} q_{1k_2}\xi^{2/(n+2)}V_0(\xi)\left\{\begin{bmatrix}-i & 0\\ 0 & i\end{bmatrix} + O(\xi^{-1})\right\}.$$

B. n odd. Then

$$V_k(\xi) = \xi^{k/(n+2)} \sum_{j=1}^{\infty} V_{kj}\xi^{-j}, \quad k < n+2,$$

$$V_k(\xi) = \xi^{k/(n+2)}\left[V_{k^0} + \sum_{j=1}^{\infty} V_{kj}\xi^{-j}\ln\xi\right], \quad k \geqslant n+2,$$

In particular

$$V_1(\xi) = \left(\frac{n+2}{2}\right)^{1/(n+2)} q_{1k_1}\xi^{1/(n+2)}V_0(\xi)\left\{\begin{bmatrix}-i & 0\\ 0 & i\end{bmatrix} + O(\xi^{-1})\right\}.$$

3.3 Connection Formulae. We have

$$W_0(z,\varepsilon) = W_0(\tilde{z},\ \rho)\Omega_0(\rho).$$

The connection matrix has the form

$$\Omega_0(\rho) = \sqrt{\frac{\pi}{2}}\left(\frac{2}{n+2}\right)^{n/2(n+2)} \rho^{-n/2}$$
$$\times \left\{\begin{bmatrix}\exp\left\{i\left(\frac{\nu\pi}{2}+\frac{\pi}{4}\right)\right\} & 0\\ 0 & \exp\left\{-i\left(\frac{\nu\pi}{2}+\frac{\pi}{4}\right)\right\}\end{bmatrix} + O(\varepsilon)\right\}, \qquad (17)$$

$$\nu = \frac{1}{n+2}.$$

This formula allows us to find the value of the exterior series expansion at the turning point $z = 0$. We have

$$Y_0(0,\varepsilon) = \begin{cases} v_0(0)[I + A\rho^2 + O(\rho^4)]\Omega_0(\rho), & n \quad \text{even}, \\ v_0(0)[I + B\rho^2 + O(\rho^4)]\Omega_0(\rho), & n \quad \text{odd}, \end{cases}$$

$$v_0(0) = \begin{bmatrix} \frac{i2^\nu}{\Gamma(1-\nu)\sin\nu\pi} & 0 \\ 0 & \left(\frac{n+2}{2}\right)^{n/(n+2)} \frac{i2^{1-\nu}}{\Gamma(\nu)\sin\nu\pi} \end{bmatrix} \begin{bmatrix} -1 & 1 \\ e^{-i\nu\pi} & -e^{i\nu\pi} \end{bmatrix},$$

where A and B are constant matrices, and $\nu = 1/(n+2)$.

The analogous interior F.S.S. in $D_k(\tilde{z}) \cap D_i$ and connection formulae were obtained in [Nakano 2].

4. Asymptotic Behaviour of the Solutions when Γ has Two Components [Nakano 1]. In this case the neighbourhood of the turning point must be decomposed into several zones. We restrict ourselves to the example [Roos 2]

$$\varepsilon^2 w'' - (z^n - \varepsilon)w = 0, \quad n \geqslant 3. \tag{18}$$

4.1 Exterior Expansion. The exterior domain D_e is the annulus

$$M\varepsilon^{1/n} \leqslant |z| \leqslant r,$$

where $M > 0$ is a sufficiently large number, and the exterior expansion for the fundamental matrix W of equation (18) has the form

$$W(z,\varepsilon) = \begin{bmatrix} z^{-n/4} & 0 \\ 0 & z^{n/4} \end{bmatrix} \tilde{W}(z,\varepsilon) \times \exp\left\{ \left[\frac{1}{\varepsilon} \frac{2}{n+2} z^{(n+2)/2} + \frac{1}{n-2} z^{(2-n)/2} \right] \begin{bmatrix} 1 & 0 \\ 0 & -1 \end{bmatrix} \right\}, \tag{19}$$

where

$$\tilde{W}(z,\varepsilon) = \begin{bmatrix} 1 & -1 \\ 1 & 1 \end{bmatrix} + O(\varepsilon).$$

The domain D_e can be covered by domains $D_{ek}, k = 0, 1, \dots, n+1$, each of which is contained in the sector

$$G_{k+3} : \delta + \frac{-\pi + 2\pi k}{n+2} \leqslant \arg z \leqslant \frac{3\pi + 2\pi k}{n+2} - \delta,$$

$$M\varepsilon^{1/n} \leqslant |z| \leqslant r,$$

and is constructed in the following manner. Let $\tilde{G}_{k+3}$ be the image of the domain G_{k+3} under the mapping $\zeta = z^{(n+2)/2}$, so that

$$-\frac{\pi}{2} + \pi k + \frac{n+2}{2}\delta \leqslant \arg \zeta \leqslant \frac{3\pi}{2} + \pi k - \frac{n+2}{2}\delta,$$

$$\zeta_0 = r^{(n+2)/2}.$$

Let $\tilde{D}_{e,k} \subset \tilde{G}_{k+3}$ be a domain such that each point $\zeta \in \tilde{D}_{e,k}$ can be connected to the point $-\zeta_0$ by a curve along which Re ζ is non-decreasing, and to the point ζ_0 by a curve along which Re ζ is non-increasing. The domain $D_{e,k}$ is the inverse-image of the domain $\tilde{D}_{e,k}$. In particular if $k = 0$ then $\tilde{G}_3$ is the sector

$$M^{(n+2)/2}\varepsilon^{1/n} \leqslant |\zeta| \leqslant \zeta_0, \quad -\frac{\pi}{2} + \delta \leqslant \arg \zeta \leqslant \frac{3\pi}{2} - \delta.$$

We draw the tangents to the circle $C : |\zeta| = M^{(n+2)/2}\varepsilon^{1/n}$ down as far as the intersection with the bounding rays of the sector $\tilde{G}_3$ and remove from $\tilde{G}_3$ the two small domains cut off by these tangents; we then obtain the domain $\tilde{D}_{e,3}$. In each of the domains D_e there exists a fundamental matrix $W_k(z, \varepsilon)$ of the form (19). We can obtain an asymptotic expansion for the matrix $\tilde{W}$ in powers of ε.

4.2 Interior Expansion. Let us make the change of variable

$$z = \varepsilon^{1/n} t, \quad \rho = \varepsilon^{(n-2)/2n}. \tag{20}$$

Then equation (18) has the form

$$\rho^2 \frac{d^2 u}{dt^2} - (t^n - 1)u = 0. \tag{21}$$

For the interior domain D_e we take the disk $|t| \leqslant m$ (that is $|x| \leqslant M\varepsilon^{1/n}$), from which are removed small fixed neighbourhoods of all the turning points $t_k = \exp\{2\pi i k/n\}$, $k = 0, 1, \ldots, n-1$. These points are called ***secondary turning points***.

The standard equation (21) is not integrable, unlike the case considered in Sect. 3, but the asymptotic behaviour of its F.S.S. as $\rho \to 0$ can be obtained everywhere in the domain D_e using the methods described in Chap. 3, § 3. From each of the turning points there arise three Stokes lines, partitioning the t-plane into domains of half-plane type (precisely $n+2$ of them) and domains of band type. We will construct the asymptotic behaviour of one of the interior fundamental matrices.

From the turning point $t = 1$ there emerge two Stokes lines l_1 and $l_2 = l_1^*$, where Im $t > 0$ for $t \in l_1$ with asymptotes $\arg t = \pm\pi/(n+2)$. They bound the domain D_1 of half-plane type containing the half-line $(1, +\infty)$. For n odd the third Stokes line l_0 is the half-line $(-\infty, 1)$, while for n even it is the segment $[-1, 1]$. There is attached to the Stokes line l_1 a domain D_2 of band type, and adjacent to D_2 is a domain D_3 of half-plane type such that one of its bounding Stokes lines l_3 has the same asymptote $\arg t = \pi/(n+2)$ as the line l_1.

We put

$$D_0 = D_1 \cup l_1 \cup D_2 \cup l_3 \cup D_3 .$$

The domain D_0 is canonical (Chap. 3, § 3, paragraph 2.1). We remove from D_0 a small neighbourhood of the bounding Stokes lines and the domain so obtained is denoted by $\tilde{D}_0$. Suppose that the F.S.S. is defined by $(1, l_1, D_0)$. Then equation (18) has a fundamental matrix of the form

$$W_i(t,\rho) = e^{\frac{n-1}{12}\pi i} \begin{bmatrix} p^{-1/4}(t) & 0 \\ 0 & \varepsilon^{1/2} p^{1/4}(t) \end{bmatrix} \tilde{W}(t,\rho) \begin{bmatrix} e^{\rho^{-1}S} & 0 \\ 0 & e^{-\rho^{-1}S} \end{bmatrix}, \quad (22)$$

where

$$S(t) = \int_1^t \sqrt{p(\tau)}d\tau , \quad p(t) = t^n - 1 ,$$

$$\tilde{W}(t,\rho) = \begin{bmatrix} 1 & -1 \\ 1 & 1 \end{bmatrix} + O(\rho) .$$

We denote by $W_e(z, e)$ the exterior fundamental matrix corresponding to the domain $D_{e,3}$ (paragraph 4.1). Then

$$W_i(t\ \rho) = W_e(z,\ e)\Omega(e) .$$

Since the transition matrix $\Omega(\varepsilon)$ does not depend on z we can take an arbitrary $z \in D_i \cap D_e$ in this formula and replace the fundamental matrices W_i and W_e by their asymptotic formulae (19) and (20). We obtain

$$x_0 = \alpha\rho^{1/(n-2)} , \quad t_0 = \alpha\rho^{-1/(n-2)} , \quad |\alpha| = 1 ,$$

so that $x_0 \in D_{e,3}$. Then

$$\begin{aligned} \frac{1}{\varepsilon}\frac{2}{n+2}x_0^{(n+2)/2} + \frac{1}{n-2}x_0^{-n/2+1} - \frac{1}{\rho}\int_1^{t_0} \sqrt{p(t)}dt \\ = c_1\rho^{-1} + O(\rho^{(n+2)/(2(n-2))}) . \end{aligned}$$

For $\Omega(\varepsilon)$ we have the asymptotic formula

$$\Omega(\varepsilon) = e^{\frac{n-1}{12}\pi i}\varepsilon^{1/4}[I + O(\rho)] \begin{bmatrix} e^{c_1\rho^{-1}} & 0 \\ 0 & e^{-c_1\rho^{-1}} \end{bmatrix} ,$$

where c_1 is a constant.

In [Roos 1] the equation

$$\varepsilon^2 w'' - (z^5 - \varepsilon z)w = 0 .$$

is studied in detail.

Chapter 5. n^{th}-Order Equations and Systems

In this chapter scalar equations

$$y^{(n)} + \lambda q_1(x,\ \lambda^{-1})y^{(n-1)} + \ldots + \lambda^n q_n(x)y = 0$$

and systems

$$y' = \lambda A(x,\ \lambda^{-1})y\,, \quad \lambda \to \infty$$

are considered. Asymptotic formulae are given for solutions on a finite interval, on a half-line and in the complex plane of x.

§ 1. Equations and Systems on a Finite Interval

1. n^{th}-Order Equations. We consider the equation

$$ly \equiv y^{(n)} + \sum_{k=1}^{n} \lambda^k q_k(x) y^{(n-k)} = 0 \tag{1}$$

on the segment $I = [a, b]$. Here $\lambda > 0$ is a large parameter, and the coefficients $q_k(x)$ are complex-valued and $C^\infty(I)$. The asymptotic behaviour of the solutions of equation (1) as $\lambda \to \infty$ in the simplest case can be expressed in terms of the roots of the characteristic equation

$$l(x, p) \equiv p^n + \sum_{k=0}^{n-1} q_k(x) p^{n-k} = 0\,. \tag{2}$$

The function $l(x, p)$ is called the *λ-symbol* of the operator l. We note that equation (1) can be rewritten as

$$l(x,\ \lambda^{-1}D)y = 0\,, \quad D = d/dx\,.$$

1.1 Formal Asymptotic Solutions. We look for the F.A.S. of equation (1) in the form

$$y = e^{\lambda S(x)} \sum_{j=0}^{\infty} \lambda^{-j} a_j(x)\,. \tag{3}$$

We substitute this series into (1), divide by $\exp\{\lambda S(x)\}$ and equate the coefficients of powers of λ^{-1} to zero. We then obtain a recurrence system for the unknown functions $S(x)$, $a_0(x)$, $a_1(x), \ldots$.

We write out the first two equations. Applying the Leibniz formula

$$l(f(x)g(x)) = \sum_{j=0}^{n} \frac{1}{j!} D^j f(x) \left(\frac{\partial}{\partial p}\right)^j l(x,p)|_{p=D} g(x),$$

we obtain

$$l[a_0(x)e^{\lambda S(x)}] = e^{\lambda S(x)}[l(x, S'(x))a_0(x) + \lambda^{-1}(l_p(x, S'(x))a_0'(x) + \frac{1}{2}l_{pp}(x, S'(x))S''(x)a_0(x)) + O(\lambda^{-2})]. \quad (4)$$

For $S'(x)$ we obtain the equation $l(x, S'(x)) = 0$; that is, $S'(x)$ is a root of the characteristic equation (2). Let $p_j(x)$ be one of the roots of this equation. Putting $S'(x) = p_j(x)$ we obtain

$$S(x) = \int_{x_0}^{x} p_j(t)dt, \quad x_0 \in I.$$

The function $a_0(x)$ is determined from the equation

$$l_p(x, p_j(x))a_0'(x) + \frac{1}{2}p_j'(x)l_{pp}(x, p_j(x))a_0(x) = 0,$$

and finally we obtain the F.A.S. as

$$y_j(x, \lambda) = \tilde{y}_j(x, \lambda; x_0)[1 + O(\lambda^{-1})],$$

where

$$\tilde{y}_j(x, \lambda; x_0) = \exp\left\{\lambda \int_{x_0}^{x} p_j(t)dt - \frac{1}{2}\int_{x_0}^{x} p_j'(t)\frac{l_{pp}(t, p_j(t))}{l_p(t, p_j(t))}dt\right\}. \quad (5)$$

We have restricted ourselves to finding the functions $S(x)$ and $a_0(x)$. Using the same method we can obtain the other coefficients $a_1(x)$, $a_2(x), \ldots$ of the series (3). But the formulae for them turn out to be unwieldly and therefore in applications one generally uses only the principal term of the asymptotic behaviour.

1.2 Sufficient Conditions for the Existence of the Asymptotic Behaviour of the Solutions. If the characteristic equation (2) has a multiple root when $x = x_0$ then x_0 is called a *turning point* of equation (1). Turning points are determined by eliminating p from the system

$$l(x, p) = 0, \quad l_p(x, p) = 0. \quad (6)$$

Let $p_1(x), \ldots, p_n(x)$ be the roots of equation (2). We introduce the conditions:

1) Equation (1) has no turning points for $x \in I$; that is

$$p_j(x) \neq p_k(x), \quad j \neq k, \quad x \in I.$$

2) The differences Re $(p_j(x) - p_k(x))$ do not change sign for j fixed, $x \in I$, $k = 1, \dots, n$.

If condition (1) is satisfied then $p_k(x) \in C^\infty(I)$, $k = 1, \dots, n$.

Equation (1) has a solution of the form

$$y_j(x,\lambda) = \tilde{y}_j(x,\lambda; x_0)\left[1 + \sum_{k=1}^{N-1} \lambda^{-j} a_{jk}(x) + O(\lambda^{-N})\right], \quad \lambda \to \infty. \tag{7}$$

Here $N \geqslant 1$ is arbitrary, all the $a_{jk}(x) \in C^\infty(I)$, and the bound for the remainder term is uniform in $x \in I$.

The asymptotic formula (7) can be differentiated in x and λ an arbitrary number of times preserving the uniformity in x of the bound for the remainder term. In particular, for the principal asymptotic term, we have

$$y_j^{(m)}(x,\ \lambda) \sim \lambda^m p_j^m(x) \tilde{y}_j(x,\lambda; x_0), \quad \lambda \to \infty,$$

uniformly in $x \in I$.

If condition (2) is satisfied for all j then the solutions of the form (7) form a F.S.S. of equation (1). The reasoning behind conditions (1) and (2) was discussed in Chap. 2, § 7.

The principal asymptotic term $\tilde{y}_j$ can be written in another way. Using the identities (here $p = p(x)$)

$$\frac{d}{dx} l_p(x,p) = p'(x) l_{pp}(x,p) + l_{px}(x,p),$$

$$\frac{p'(x) l_{pp}(x,p)}{l_p(x,p)} = \frac{d}{dx} \ln l(x,p) - \frac{l_{px}(x,p)}{l_p(x,p)},$$

we obtain from (5)

$$\tilde{y}_j(x,\lambda; x_0) = [l_p(x, p_j(x))]^{-1/2} \times \exp\left\{\lambda \int_{x_0}^x p_j(t)dt + \frac{1}{2}\int_{x_0}^x \frac{l_{px}(t,p_j(t))}{l_p(t,p_j(t))} dt\right\}. \tag{8}$$

We can also represent $\tilde{y}_j$ in the form

$$\tilde{y}_j(x,\lambda; x_0) = \exp\left\{\lambda \int_{x_0}^x p_j(t)dt - \int_{x_0}^x {\sum_{k=1}^{n}}' \frac{p_j'(t)dt}{p_j(t) - p_k(t)}\right\}, \tag{9}$$

where the prime denotes $k \neq j$.

Examples. 1. Let us consider the two-term equation

$$y^{(n)} - \lambda^n q(x) y = 0 .$$

It has no turning points if $q(x) \neq 0$, $x \in I$. Let $q^{1/n}(x)$ be a fixed branch of the root and let $\omega_1, \dots, \omega_n$ be the distinct n-th roots of unity. The asymptotic behaviour of the F.S.S. has the form

$$y_j(x, \lambda) \sim [q(x)]^{-1/2+1/2n} \exp\left\{ \lambda \omega_j \int_{x_0}^{x} q^{1/n}(t) dt \right\}$$

under the condition that all the differences Re $[(\omega_j - \omega_k) q^{1/n}(x)]$ do not change sign for $x \in I$. If $q(x)$ is real this condition is certainly satisfied.

2. We next consider the "biquadratic" equation

$$y^{(4)} - 2\lambda^2 a(x) y'' + \lambda^4 b(x) y = 0 .$$

The roots of the characteristic equation are

$$p_j(x) = \pm\sqrt{a(x) \pm \sqrt{D(x)}}, \quad D(x) = a^2(x) - b(x) .$$

A turning point is a root of one of the equations

$$b(x) = 0 , \quad D(x) = 0 .$$

We find the asymptotic behaviour of the F.S.S. from (7)

$$y_j(x, \lambda) \sim p_j^{-1/2}(x) D^{-1/4}(x) \exp\left\{ \lambda \int_{x_0}^{x} p_j(t) dt - \frac{1}{2} \int_{x_0}^{x} p_j'(t) D^{-1/2}(t) dt \right\} .$$

Remark. To construct a finite number of terms of the asymptotic expansion of the solution it is sufficient for the coefficients of equation (1) to be in $C^k(I)$, $k < \infty$. For example, a solution of the form (7), for $N \geqslant 1$ fixed, exists if all $q_j(x) \in C^{N+1}(I)$. Further, under conditions (1) and (2), there is a solution of equation (1) which has asymptotic series expansion

$$y_j(x, \lambda) = \tilde{y}_j(x, \lambda; x_0) \left[1 + \sum_{k=1}^{\infty} \lambda^{-k} a_{jk}(x) \right], \quad \lambda \to \infty .$$

However this solution can not be constructed explicitly. (See Chap. 2, § 3).

1.3 Reduction of Equation (1) to a System. Another method of constructing the F.S.S. of equation (1) is as follows. First of all the equation is reduced by standard methods to a first-order system. Then this system is transformed to almost diagonal form

$$z' = [\lambda \Lambda_0(x) + \Lambda_1(x) + \dots + \lambda^{-N+1} \Lambda_N(x) + O(\lambda^{-N})] z ,$$

where the $\Lambda_j(x)$ are diagonal matrices. Discarding $O(\lambda^{-N})$ we have a splitting system which is integrable, and hence we obtain the F.S.S.

We give the explicit formulae for the case $N = 1$. The substitution

$$y = y_1 , \quad y_1' = \lambda y_2, \dots, \quad y_{n-1}' = \lambda y_{n-1}$$

reduces equation (1) to the first-order system

$$y = \begin{bmatrix} y_1 \\ \cdots \\ y_n \end{bmatrix}, \quad A(x) = \begin{bmatrix} 0 & 1 & 0 & \dots & 0 \\ 0 & 0 & 1 & \dots & 0 \\ \dots & \dots & \dots & \dots & \dots \\ 0 & 0 & 0 & \dots & 1 \\ -q_n & -q_{n-1} & -q_{n-2} & \dots & -q_1 \end{bmatrix}. \tag{10}$$

The matrix $A(x)$ has eigenvalues $p_j(x)$ and eigenvectors $(1, p_j(x), \dots, p_j^{n-1}(x))^T$, $1 \leqslant j \leqslant n$. The matrix

$$T_0(x) = \begin{bmatrix} 1 & 1 & \dots & 1 \\ p_1(x) & p_2(x) & \dots & p_n(x) \\ \dots & \dots & \dots & \dots \\ p_1^{n-1}(x) & p_2^{n-1}(x) & \dots & p_n^{n-1}(x) \end{bmatrix}$$

reduces $A(x)$ to diagonal form; that is

$$T_0^{-1}(x)A(x)T_0(x) = \Lambda_0(x) = \operatorname{diag}(p_1(x), \dots, p_n(x)).$$

The substitution $y = T_0(x)z$ transforms system (10) to the form

$$z' = \left[\lambda \Lambda_0(x) - T_0^{-1}(x)\frac{dT_0(x)}{ds}\right] z, \tag{11}$$

that is, it diagonalizes system (10) to within $O(1)$. We have

$$\begin{aligned} \left(T_0^{-1}(x)\frac{dT_0(x)}{dx}\right)_{jj} &= -\frac{p_j'(x)}{2}\frac{l_{pp}(x, p_j(x))}{l_p(x, p_j(x))} \equiv -p_j^{(1)}(x), \\ \left(T_0^{-1}(x)\frac{dT_0(x)}{dx}\right)_{jk} &= \frac{p_k'(x)}{p_k(x) - p_j(x)}\frac{l_p(x, p_k(x))}{l_p(x, p_j(x))}, \quad k \neq j. \end{aligned} \tag{12}$$

We make the transformation $z = (I + \lambda^{-1}T_1)w$, that is,

$$y = T_0(x)[I + \lambda^{-1}T_1(x)]w; \tag{13}$$

we then obtain the system

$$\begin{aligned} w' = \Bigg[&(I + \lambda^{-1}T_1)^{-1}\left(\lambda\Lambda_0 - T^{-1}\frac{dT_0}{dx}\right)(I + \lambda^{-1}T_1) \\ &- \lambda^{-1}(I + \lambda^{-1}T_1)^{-1}\frac{dT_1}{dx}\Bigg] w. \end{aligned} \tag{14}$$

The matrix of this system is

$$\lambda\Lambda_0 + \left\{[\Lambda_0, T_1] - T_0^{-1}\frac{dT_0}{dx}\right\} + O(\lambda^{-1}),$$

where $[\Lambda_0, T_1] = \Lambda_0 T_1 - T_1 \Lambda_0$ (the commutator of the matrices Λ_0 and T_1). We choose T_1 so that the matrix in parentheses is diagonal, that is,

$$[\Lambda_0, T_1] - T_0^{-1}\frac{dT_0}{dx} = \Lambda_1 \,.$$

This is not a unique choice since the diagonal elements of $[\Lambda_0, T_1]$ are zero. For definiteness we put

$$(T_1)_{jj} = 0\,, \quad (T_1)_{jk} = (p_j - p_k)^{-1}\left(T_0^{-1}\frac{dT_0}{dx}\right)_{jk}\,; \tag{15}$$

then for ω we obtain the system

$$w' = [\lambda\Lambda_0(x) + \Lambda_1(x) + O(\lambda^{-1})]w\,.$$

Here $\Lambda_1(x) = \operatorname{diag}\,(p_1^{(1)}(x), \ldots, p_n^{(1)}(x))$, and the functions $p_j^{(1)}(x)$ are as in (12). Discarding $O(\lambda^{-1})$ we obtain a splitting system, the solutions of which are

$$w_k(x,\ \lambda) = c_k\tilde{y}_k(x, \lambda; x_0)\,, \quad 1 \leqslant k \leqslant n\,.$$

Putting $c_k = 0$ for $k \neq j$ and $c_j = 1$, and taking into account that $y = T_0(x)[I + O(\lambda^{-1})]w$, we obtain the F.A.S. $y_j(x, \lambda)$ for equation (1).

We can make a transformation of the form

$$y = T_0(x)[I + \lambda^{-1}T_1(x) + \ldots + \lambda^{-N+1}T_{N-1}(x)]z$$

and choose the matrices $T_2(x), \ldots, T_{N-1}(x)$ so that the matrix of the system so obtained is diagonal to within $O(\lambda^{-N})$.

This method of constructing the asymptotic behaviour of the solutions appears more involved than that given in paragraph 1.1. But it has a series of advantages. First of all this method is applicable to first-order systems. Also it enables us to prove the asymptotic formulae (see § 4). Finally even the simplest transformation $y = T_0(x)z$ is useful in numerical calculations.

1.4 Additional Parameters and Complex λ. We consider the equation

$$y^{(n)} + \sum_{k=1}^{n} \lambda^k q_k(x, \mu) y^{(n-k)} = 0\,, \tag{16}$$

where μ is a parameter and $\mu \in D$. Let $p_1(x, \mu), \ldots, p_n(x, \mu)$ be the roots of the characteristic equation

$$l(x, p, \mu) \equiv p^n + \sum_{k=1}^{n} q_k(x, \mu) p^{n-k} = 0\,. \tag{17}$$

We introduce conditions similar to (1) and (2) of paragraph 1.1:

1) $|p_j(x, \mu) - p_k(x, \mu)| \geqslant \delta > 0$ for $x \in I$, $\mu \in D$ and all j, k, $j \neq k$, where δ does not depend on x, μ.

2) The differences $\operatorname{Re}\,(p_j(x, \mu) - p_k(x, \mu))$ do not change sign for j fixed, $(x, \mu) \in I \times D$, $k = 1, \ldots, n$.

The parameter μ can be real or complex.

A. If μ is real, D is an interval of the real line.

In this case we assume that $q_k(x,\mu) \in C^\infty(I \times D)$ for all k.

B. If μ is complex, D is a domain of the complex plane.

In this case we assume that $q_k(x,\mu) \in C^\infty(I \times D)$ and that $q_k(x,\mu)$ is holomorphic in μ in the domain D for each fixed $x \in I$ and for all k.

Then equation (16) has solution $y_j(x,\lambda,\mu)$ of the form (5) (in this formula we must of course replace $p_j(x)$ by $p_j(x,\mu)$ and $l(x,p)$ by $l(x,p,\mu)$). The bound for the remainder term is uniform in $(x,\mu) \in I \times D$ and formula (5) can be differentiated in x, λ and μ an arbitrary number of times, preserving the uniformity in x and μ of the bound for the remainder term. The coefficients $a_{jk}(x,\mu)$ of expansion (5) satisfy the same conditions as the coefficients of equation (16) (see A and B).

In case B the solution $y_j(x,\lambda,\mu)$ is holomorphic in μ in the domain D for each fixed $x \in I$ and $\lambda \geqslant \lambda_0 \gg 1$. All these statements carry over to the case where the coefficients of the equation depend on several parameters: $\mu = (\mu_1, \dots, \mu_m)$.

We return to equation (1). Let D be an unbounded domain of the complex λ-plane and let $|\lambda| \geqslant \lambda_0 > 0$ in D. As a rule we can take D as a sector of the form $|\lambda| > \lambda_0$, $\alpha < \arg \lambda > \beta$. Suppose that conditions 1) and 2) are satisfied for the functions $\lambda p_1(x), \dots, \lambda p_n(x)$ for all $\lambda \in D$. Then condition 1) is unchanged but condition 2) takes a different form:

2′) The differences Re $[\lambda(p_j(x) - p_k(x))]$ do not change sign for j fixed, $k = 1, \dots, n$, $(x,\lambda) \in I \times D$.

Equation (1) has solution $y_j(x,\lambda)$ for which the asymptotic formula (5) holds as $|\lambda| \to \infty$, $\lambda \in D$, uniformly in $x \in I$. This solution is holomorphic in D for each fixed $x \in I$.

Examples. 1. Let $q(x) > 0$ in I. Then as $\lambda \to \infty$ the equation

$$y^{(n)} - \lambda^n q(x) y = 0$$

has a solution y_1 such that

$$y_1(x,\lambda) \sim [q(x)]^{-1/2+1/2n} \exp\left\{\lambda \int_{x_0}^{x} q^{1/n}(t)\,dt\right\}$$

(see example 1 of paragraph 1.2). Here $q^{1/n}(x) > 0$. We will investigate in which sector of the form $0 \leqslant \arg \lambda < \alpha$, $|\lambda| \geqslant \lambda_0 \gg 1$ this asymptotic behaviour is true. The condition for applicability of the asymptotic behaviour is that the differences Re $(e^{i\phi}(1 - \omega_k))$ do not change sign. Here $\phi = \arg \lambda$ and $\omega_k = \exp\{2\pi k i/n\}$. We connect the point $\omega_0 = 1$ to all the points ω_k by segments. Under a rotation about the point ω_0 by an angle $\phi < 2\pi/n$ all the differences preserve their signs. Under a rotation by an angle slightly more than $2\pi/n$, the difference Re $[e^{i\phi}(i - \omega_{n-1})]$ changes sign. Thus the asymptotic formula is true for $|\lambda| \to \infty$ and $0 \leqslant \arg\lambda \leqslant 2\pi/n$. We can show that this formula ceases to be valid in a larger sector.

2. Let all the roots of the characteristic equation (2) be purely imaginary: $p_j(x) = i\tilde{p}_j(x)$, and suppose they are distinct for $x \in I$. Then condition 2) is satisfied for each j, and equation (1) has F.S.S. of the form (5) which are rapidly oscillating as $\lambda \to \infty$.

If $0 \leqslant \arg\lambda \leqslant \pi$ then all the differences $\operatorname{Re}[\lambda(p_j(x) - p_k(x))]$, $1 \leqslant j, k \leqslant n$, do not change sign, and equation (1) has F.S.S. $\{y_1^+(x,\lambda), \dots, y_n^+(x,\lambda)\}$ for which the asymptotic behaviour (5) holds true as $|\lambda| \to \infty$ with $\operatorname{Re}\lambda \geqslant 0$. In a similar way equation (1) has F.S.S. $\{y_1^-(x,\lambda), \dots, y_n^-(x,\lambda)\}$ with the same asymptotic behaviour but as $|\lambda| \to \infty$ with $\operatorname{Re}\lambda \leqslant 0$. These two F.S.S. are in general different.

In many problems one encounters an equation of the form

$$y^{(n)} + \sum_{k=1}^{n} \lambda^k q_k(x, \lambda^{-1}) y^{(n-k)} = 0, \tag{18}$$

where the coefficients $q_k(x, \lambda^{-1})$ are polynomials in λ^{-1} or have an asymptotic series expansion in powers of λ^{-1}. This equation is of the form (15) where $\mu = \lambda^{-1}$ and the above results can be applied. The characteristic equation has the form

$$p^n + \sum_{k=1}^{n} q_k(x, \lambda^{-1}) p^{n-k} = 0. \tag{19}$$

Equation (18) can be investigated as follows. The coefficients of the equation are expanded in asymptotic series

$$q_k(x, \lambda^{-1}) = \sum_{m=0}^{\infty} \lambda^{-m} q_{km}(x).$$

We look for the F.A.S. in the form (3) and then we reiterate the procedure described in paragraph 1.1. The function $S'(x)$ must be a root of the characteristic equation

$$p^n + \sum_{k=1}^{n} q_k(x, 0) p^{n-k} = 0,$$

which is rather different from (19). This method leads to asymptotic expansions of the form (5), while the former method leads to asymptotic expansions of the form

$$y_j(x,\lambda) = \exp\left\{\lambda \int_{x_0}^{x} p_j(t, \lambda^{-1})\,dt\right\} \left[\sum_{k=0}^{N-1} a_{jk}(x, \lambda^{-1}) \lambda^{-k} + O(\lambda^{-N})\right].$$

All the entries in this formula can be expanded in asymptotic series in powers of λ^{-1}; if we bring together all the terms of the same power of λ^{-1} then we obtain an asymptotic series of the form (5).

2. Equation of Self-adjoint Form

2.1 Equations of Even Order. We consider the equation

$$\begin{aligned} ly = (-1)^n \frac{d^n}{dx^n}\left(q_0(x)\frac{d^n y}{dx^n}\right) \\ +(-1)^{n-1}\lambda^2 \frac{d^{n-1}}{dx^{n-1}}\left(q_1(x)\frac{d^{n-1}y}{dx^{n-1}}\right) + \ldots + \lambda^{2n} q_n(x) y = 0 \end{aligned} \tag{20}$$

on the interval $I = [a, b]$. Here $\lambda > 0$ is a large parameter, the coefficients $q_k(x)$ are complex-valued and all $q_k(x) \in C^\infty(I)$. We will assume also that $q_0(x) \neq 0$ for $x \in I$. If all the $q_k(x)$ are real and $y(x)$ and $z(x) \in C_0^\infty(I)$ then the Lagrange identity

$$\int_a^b ly\bar{z}dx = \int_a^b yl\bar{z}dx$$

is true here. This means that the differential operator L with domain of definition $D(L) = C_0^\infty(I)$, and defined by the formula $Ly = ly$, is symmetric.

Equation (20) is a special case of equation (16) but, in view of its special form, a direct study is useful.

We call the equation

$$l(x,p) \equiv q_0(x)p^{2n} - q_1(x)p^{2n-2} + \ldots + (-1)^n q_n(x) = 0 \tag{21}$$

the *characteristic equation* of (20). Let $p_1(x), \ldots, p_{2n}(x)$ be the roots of (21), and introduce the notation

$$\tilde{y}_j(x,\lambda;x_0) = [l_p(x,p_j(x))]^{-1/2} \exp\left\{\lambda \int_{x_0}^x p_j(t)dt\right\}. \tag{22}$$

Suppose that conditions 1) and 2) of paragraph 1.1 are satisfied. Then equation (20) has a solution of the form

$$y_j(x,\lambda) = \tilde{y}_j(x,\lambda;x_0)\left[1 + \sum_{k=1}^{N-1} \lambda^{-k} a_{jk}(x) + O(\lambda^{-n})\right], \tag{23}$$

where $N \geqslant 1$ is arbitrary. We note that this formula is simpler than formula (5).

All that we showed in paragraph 1 for the solutions to equation (1) is true for the solutions of equation (20): differentiation of the asymptotic behaviour, the asymptotic behaviour for complex λ, asymptotic behaviour of the solutions when the coefficients of the equation depend on a parameter, and so on. The only difference is that there is an alternative and more convenient first-order system which is equivalent to (20). To obtain this system, we put

$$\begin{aligned} y_1 &= y\,, \quad y_1' = \lambda y_2, \ldots, \quad y_{n-1}' = \lambda y_n\,, \\ y_n' &= \lambda q_0^{-1}(x) y_{n+1}\,, \quad y_{n+1}' = \lambda(q_1(x) y_n - y_{n+2})\,, \\ y_{n+2}' &= \lambda(q_2(x) y_{n-1} - y_{n+3}), \ldots, \quad y_{2n}' = \lambda q_n(x) y_1\,. \end{aligned} \tag{24}$$

The functions $\lambda^{-k} y_k(x)$ are called *quasi-derivatives* and denoted by $y^{[k]}(x)$, so that $y_k = \lambda^k y^{[k]}$.

The substitution(24) reduces equation (20) to the first-order system

$$y' = \lambda A(x) y\,,$$

$$A(x) = \begin{bmatrix} 0 & 1 & & & & 0 \\ & \ddots & \ddots & & \iddots & \\ & & 1 & & & \\ & & 0 & q_0^{-1} & & \\ & & q_1 & 0 & -1 & \\ & & & & \ddots & \\ & \iddots & & & \ddots & \\ & & & & & -1 \\ q_n & & & & & 0 \end{bmatrix}, \quad y = \begin{bmatrix} y_1 \\ y_2 \\ \cdots \\ y_{2n} \end{bmatrix}. \tag{25}$$

The element $q_0^{-1}(x)$ in $A(x)$ lies in the nth row and the $(n+1)^{\text{th}}$ column, and all the elements not shown are zero. Note that $A(x)$ does not contain the derivatives of the functions $q_0(x), \ldots, q_n(x)$; in the usual way of reducing the equation to a system (see paragraph 1.3) the matrix would also contain these derivatives.

We diagonalize the system (25) to within $O(\lambda^{-1})$ using the same method as in paragraph 1.3. The eigenvalues of $A(x)$ are $p_1(x), \ldots, p_{2n}(x)$. The matrix $T_0(x)$ with elements

$$(T_0)_{jk} = p_k^{j-1}\,, \quad (T_0)_{n+j,k} = p_k^{n-j+1} \sum_{m=0}^{j-1} (-1)^m q_{j-m-1} p_k^{2m}\,,$$

where $1 \leqslant j \leqslant n$, $1 \leqslant k \leqslant 2n$, reduces $A(x)$ to diagonal form. The elements of the inverse matrix are

$$\begin{aligned} (T_0^{-1})_{j,n-k} &= (-1)^k p_j^{n-k} (l_p(x, p_j))^{-1} \sum_{m=0}^{k} (-1)^m q_{k-m} p_j^{2m}\,, \\ &\qquad 0 \leqslant k \leqslant n-1\,, \\ (T_0^{-1})_{j,n+k} &= (-1)^{k+1} p_j^{n-k} (l_p(x, p_j))^{-1}\,, \quad 1 \leqslant k \leqslant n\,, \quad 1 \leqslant j \leqslant 2n\,. \end{aligned}$$

Further

$$\left(T_0^{-1} \frac{dT_0}{dx} \right)_{jj} = \frac{1}{2} \frac{d}{dx} \ln l_p(x, p_j)\,,$$

$$\left(T_0^{-1}\frac{dT_0}{dx}\right)_{jk} = (p_j - p_k)^{-1} l_p^{-1}(x, p_j) \times \sum_{m=0}^{n} (-1)^m q_m'(p_j p_k)^{n-m}, \quad j \neq k. \tag{26}$$

We construct $T_1(x)$ by formula (15); then the transformation (13) reduces the system (24) to the form

$$w' = [\lambda \Lambda_0 + \Lambda_1 + O(\lambda^{-1})]w.$$

Here

$$\Lambda_0(x) = \operatorname{diag}(p_1(x), \dots, p_{2n}(x)),$$
$$\Lambda_1(x) = -\operatorname{diag}\left(T_0^{-1}(x)\frac{dT_0(x)}{dx}\right).$$

If we discard $O(\lambda^{-1})$ then this system can be integrated and we obtain

$$w_j = c_j \tilde{y}_j(x, \ \lambda; \ x_0).$$

Thus we have constructed the F.A.S. of the form (23).

Examples. 1. Let us consider the two-term equation

$$(-1)^n (p(x) y^{(n)})^{(n)} + \lambda^{2n} q(x) y = 0.$$

Suppose that $p(x) \neq 0$ and $q(x) \neq 0$ in I. Then equation (20) has no turning points. We have

$$l(x, \ p) = p(x)p^{2n} + (-1)^n q(x), \quad l_p(x, \ p) = 2np(x)p^{2n-1},$$

$$p_j(x) = \sqrt[2n]{(-1)^{n+1}\frac{q(x)}{p(x)}},$$

where the roots take all the possible values. Under condition 2) of paragraph 1.4 there exists a F.S.S. of the form

$$y_j(x, \lambda) \sim p^{-1/2}(x)\left(\frac{p(x)}{q(x)}\right)^{n-1/2} \exp\left\{\lambda \int_{x_0}^{x} \sqrt[2n]{(-1)^{n+1}\frac{q(t)}{p(t)}}dt\right\}.$$

If $p(x)$ and $q(x)$ are real then condition 2) is satisfied.

2. Let us consider the fourth-order self-adjoint equation

$$(q_0(x)y'')'' - \lambda^2 (q_1(x)y')' + \lambda^4 q_2(x) y = 0,$$

where $q_0(x) \neq 0$. We have

$$l(x, \ p) = q_0(x)p^4 - q_1(x)p^2 + q_2(x),$$
$$l_p(x, \ p) = 4q_0(x)p^3 - 2q_1(x)p.$$

Eliminating p from the system $l = 0$, $l_p = 0$ we find that the turning points are the roots of one of the equations

$$q_2(x) = 0, \quad q_1^2(x) - 4q_0(x)q_2(x) = 0.$$

The roots of the characteristic equation are

$$p_j(x) = \pm\sqrt{\frac{1}{2q_0(x)}(q_1(x) \pm \sqrt{D(x)})},$$

$$D(x) = q_1^2(x) - 4q_0(x)q_2(x).$$

If $q_2(x) \neq 0$ and $D(x) \neq 0$ in I and condition (2) is satisfied, there is a F.S.S. of the form

$$y_1(x, \lambda) \sim p_j^{-1/2}(x)D^{-1/4}(x)\exp\left\{\lambda \int_{x_0}^{x} p_j(t)dt\right\}.$$

2.2 Equations of Odd Order. We consider the equation

$$ly \equiv \sum_{k=0}^{n}(-1)^n\lambda^{2(n-k)}i\left[\left(\frac{d}{dx}\right)^{k+1}(q_{k+1}(x)y^{(k)}) + \left(\frac{d}{dx}\right)^{k}(q_{k+1}(x)y^{(k+1)})\right] = 0 \tag{27}$$

on the interval I under the same conditions on the coefficients as in paragraph 2.1. We assume also that $q_{n+1}(x) \neq 0$ in I. If all the coefficients of equation (27) are real, then the operator L, with domain of definition $D(L) = C_0^\infty(I)$ and defined by the formula $Ly = ly$, is symmetric. The equation

$$l(x,p) \equiv \sum_{k=0}^{n}(-1)^{n+k+1}[2q_{k+1}(x)p^{2k+1} + q'_{k+1}(x)p^{2k}] = 0 \tag{28}$$

is called the *characteristic equation*. We introduce the notation

$$p_j^{(1)}(x) = -\frac{1}{2}\frac{d}{dx}\ln l_p(p_j(x),x) + l_p^{-1}(p_j(x),x)\sum_{l=0}^{n}(-1)^n q'_{l+1}(x)p_j^{2l}(x), \tag{29}$$

$$\tilde{y}_j(x,\lambda;x_0) = \exp\left\{\lambda\int_{x_0}^{x} p_j(t)dt + \int_{x_0}^{x} p_j^{(1)}(t)dt\right\},$$

where $p_1(x), \dots, p_{2n+1}(x)$ are the roots of equation (28).

Suppose that conditions 1) and 2) from paragaph 1.1. are satisfied. Then equation (27) has solution

$$y_j(x,\lambda) = \tilde{y}_j(x,\lambda,x_0)\left[1 + \sum_{k=1}^{N} \lambda^{-k} a_{jk}(x) + O(\lambda^{-N-1})\right], \tag{30}$$

where $N \geqslant 1$ is arbitrary, $a_{jk}(x) \in C^\infty(I)$, and the bound for the remainder term is uniform in $x \in I$. Everything mentioned in paragraph 1 concerning the solutions of equation (1) holds true for the solutions of equation (27).

Example. Let $n = 1$ and $q_2(x) \equiv 1$. Then equation (27) and the characteristic equation have the form

$$2y''' - \lambda^2(2q_1(x)y' + q_1'(x)y) = 0,$$
$$l(x,\ p) \equiv 2p^2 - 2q_1(x)p - q_1'(x) = 0.$$

The principal asymptotic term for the solution y_j is

$$y_j(x,\lambda) = \sqrt{\frac{p_j(x)}{4q_1(x)p_j(x) + 3q_1'(x)}}$$
$$\times \exp\left\{\lambda \int_{x_0}^{x} p_j(t)dt - \int_{x_0}^{x} \frac{p_j(t)q_1'(t)}{4q_1(t)p_j(t) + 3q_1'(t)} dt\right\} [1 + O(\lambda^{-1})].$$

§ 2. Systems of Equations on a Finite Interval

1. Systems of First-Order Equations. We consider the system of n equations

$$A(x)y' = \lambda B(x)y \tag{1}$$

on the interval $I = [a, b]$. Here $\lambda > 0$ is a large parameter, $A(x)$ and $B(x)$ are $n \times n$ matrices, the elements of which are complex-valued functions of class $C^\infty(I)$. We will assume that $A(x)$ is non-singular, that is,

$$\det A(x) \neq 0, \quad x \in I.$$

System (1) corresponds to the characteristic equation

$$\det \|B(x) - pA(x)\| = 0. \tag{2}$$

The point x_0 is called a *turning point* of system (1) if equation (2) has a multiple root at $x = x_0$. In this paragraph we will assume that system (1) does not have a turning point for $x \in I$, that is, the roots $p_1(x), \ldots, p_n(x)$ of the characteristic equation are distinct for all $x \in I$.

1.1 Some Results from Linear Algebra. Let A and B be constant matrices of order $n \times n$ with complex elements, let p be a complex number, and let A

be non-singular, that is, $\det A \neq 0$. We will consider the family of matrices $B - pA$ depending on the parameter p— a *linear bundle* of matrices. If

$$(B - p_0 A)e = 0,$$

where e is a non-zero vector, then p_0 is called an *eigenvalue*, and the vector e an *eigenvector*, of the bundle. In particular if $A = I$ then p_0 is an eigenvalue and e is an eigenvector of B.

Eigenvalues of the bundle are roots of the characteristic equation

$$\det (B - pA) = 0.$$

We will assume that the eigenvalues $p_1, \ldots, p_n$ of the bundle are distinct. Then the eigenvectors $e_1, \ldots, e_n$ are linearly independent. We recall that e_k is a column vector.

The row vector $e^\star \neq 0$ is called a *left eigenvector* of the bundle if

$$e^\star(B - p_0 A) = 0$$

for some p_0. It is clear that $e^{\star T}$ is an eigenvector of the adjoint bundle $B^T - pA^T$. The eigenvalues of the original and of the adjoint bundles are the same and so the left eigenvectors $e_1^\star, \ldots, e_n^\star$ are linearly independent. We call the vectors $e_1, \ldots, e_n$ the *right eigenvectors* of the bundle $B - pA$. We have

$$e_j^\star A e_k = 0, \quad j \neq k$$

which follows when we multiply the identity $(B - p_k A)e_k = 0$ on the left by $e_j^\star$ to obtain $(p_j - p_k)e_j^\star A e_k = 0$. Further

$$e_j^\star A e_j \neq 0, \quad j = 1, \ldots, n, \tag{3}$$

for otherwise the vectors $e_1, \ldots, e_n$ would be linearly dependent.

We consider next the system of equations

$$(B - p_j A)f = g,$$

where f is an unknown vector. Its determinant is zero and therefore it is not solvable for all g. A necessary and sufficient condition for solvability is

$$e_j^\star g = 0. \tag{4}$$

To prove the necessity of this condition we have only to multiply both sides of the system on the left by $e_j^\star$. Sufficiency follows, for instance, from the third Fredholm theorem.

We now consider the bundle $B(x) - pA(x)$, the eigenvalues $p_1(x), \ldots, p_n(x)$ of which are distinct for $x \in I$. Then

1) the functions $p_1(x), \ldots, p_n(x)$ are infinitely differentiable for $x \in I$;

2) the right and left eigenvectors $e_1(x), \ldots, e_n(x)$, $e_1^\star(x), \ldots, e_n^\star(x)$ can be chosen so that they are infinitely differentiable for $x \in I$.

In what follows we shall use such eigenvectors.

We consider the polynomial matrix bundle $L(p) = p^m A_0 + p^{m-1} A_1 + \ldots + A_m$, where the A_j are constant $n \times n$ matrices and $\det A_0 \neq 0$. The left and right eigenvectors are defined respectively by the condition

$$L(p)e = 0\,, \quad e^\star L(p) = 0\,.$$

The eigenvalues of the bundle are the roots of the characteristic equation

$$\det(p^m A_0 + p^{m-1} A_1 + \ldots + A_m) = 0\,,$$

and the eigenvalues of the bundle and of the adjoint bundle $L^T(p)$ are the same. Suppose that the eigenvalues $p_1, \ldots, p_{nm}$ of the bundle $L(p)$ are distinct. Then the right eigenvectors $\{e_1, \ldots, e_{nm}\}$ are linearly independent; this is true also for the left eigenvectors. Necessary and sufficient conditions for the solvability of the system $L(p_j)f = g$ are the same as (4). Moreover in this case

$$e_j^\star \frac{\partial L}{\partial p}(p_j)e_j \neq 0\,, \quad j = 1, 2, \ldots, nm\,. \tag{5}$$

Let us consider the bundle $L(x,p) = p^m A_0(x) + \ldots + A_m(x)$ where $A_j \in C^\infty(I)$ and $\det A_0(x) \neq 0$. If the eigenvalues of the bundle are all distinct for $x \in I$ then the above statements 1) and 2) for the linear bundle are true.

1.2 Formal Asymptotic Solutions. We seek a F.A.S. of system (1) in the form

$$y = \exp\{\lambda S(x)\} \sum_{k=0}^{\infty} \lambda^{-k} f_k(x)\,. \tag{6}$$

Substituting this series into (1) we arrive at

$$\sum_{k=0}^{\infty} \lambda^{-k} [\lambda(A(x)S'(x) - B(x))f_k(x) + A(x)f_k'(x)] = 0\,.$$

Equating the coefficients of powers of λ^{-1} to zero we obtain the recurrence system of equations

$$\begin{aligned} &[B(x) - S'(x)A(x)]f_0(x) = 0\,, \\ &[B(x) - S'(x)A(x)]f_{k+1}(x) = -A(x)f_k'(x)\,, \quad k \geqslant 0\,. \end{aligned} \tag{7}$$

It follows from the first equation that $S'(x)$ is an eigenvalue, and $f_0(x)$ a (right) eigenvector, of the bundle $B(x) - pA(x)$. We recall that the eigenvalues $p_1(x), \ldots, p_n(x)$ are assumed to be distinct for $x \in I$. We put $S'(x) = p_j(x)$. Then $f_0(x) = \alpha(x)e_j(x)$ where $e_j(x)$ is an eigenvector of the bundle and $\alpha(x)$ is an unknown function with $\alpha(x) \neq 0$ in I.

We find $\alpha(x)$ from the second equation of (7):

$$[B(x) - p_j(x)A(x)]f_0(x) = -A(x)f_0'(x)\,.$$

Since the matrix $B(x) - p_j(x)A(x)$ is singular this system is not solvable for all right-hand sides. We use the necessary and sufficient conditions (4) for solvability of this system:

$$e_j^\star(x)A(x)f_0'(x) = 0\,.$$

Thus we obtain

$$\alpha(x) = \exp\left\{-\int_{x_0}^{x} \frac{e_j^\star(t)A(t)e_j'(t)}{e_j^\star(t)A(t)e_j(t)}dt\right\},$$

where $e_j'(t) = de_j(t)/dt$. Note that the denominator does not vanish because of (3).

We have therefore found the function $S(x)$ and the vector function $f_0(x)$, and hence the principal asymptotic term $\tilde{y}_j$ is

$$\tilde{y}_j(x,\lambda) = \exp\left\{\lambda\int_{x_0}^{x} p_j(t)dt - \int_{x_0}^{x} \frac{e_j^\star(t)A(t)e_j'(t)}{e_j^\star(t)A(t)e_j(t)}dt\right\} \times [e_j(x) + O(\lambda^{-1})]\,. \tag{8}$$

It appears that $\tilde{y}_j$ depends on the choice of eigenvectors $e_j(x)$, $e_j^\star(x)$. However, we now show that the vector function

$$h(x) = \alpha(x)e_j(x)$$

is invariant in the following sense: if we replace e_j, $e_j^\star$ by $\tilde{e}_j$, $\tilde{e}_j^\star$ then $\tilde{h}(x) =$ (const.) $h(x)$. In fact if $\tilde{e}_j$, $\tilde{e}_j^\star$ are another pair of eigenvectors then $\tilde{e}(x) = \beta(x)e(x)$ and $\tilde{e}^\star(x) = \gamma(x)e^\star(x)$ (the index j is omitted for brevity), where β and $\gamma \in C^\infty(I)$, while $\beta(x) \neq 0$ and $\gamma(x) \neq 0$ in I. We have

$$\int_{x_0}^{x} \frac{\tilde{e}^\star A\tilde{e}'}{\tilde{e}A\tilde{e}'}dt = \int_{x_0}^{x} \frac{e^\star Ae'}{e^\star Ae}dt + \ln\beta(x) + c\,,$$

so that

$$\tilde{h}(x) = \alpha(x)\beta^{-1}(x)e^{c}\tilde{e}(x) = (\text{const.})\; h(x)\,.$$

However, a detailed analysis is not necessary. In fact, because of the uniqueness of the F.A.S. each term $\exp\{\lambda S(x)\}f_k(x)$ of the series (6) is invariant in the above sense. Moreover, under a change of bases $\{e_1,\dots,e_n\}$ and $\{e_1^\star,\dots,e_n^\star\}$ they are all multiplied by the same constant.

Next we find $f_1(x)$. We have

$$f_1(x) = \sum_{k=1}^{n} \alpha_k(x)e_k(x)\,,$$

where the $a_k(x)$ are unknown functions. Substituting this expansion into the second equation in (7) we get

$$\sum_{k=1}^{n}(p_k - p_j)\alpha_k A e_k = -Af_0' .$$

Multiplying this identity on the left by $e_l^\star$ and remembering that $e_i^\star A e_k = 0$ for $k \neq l$ (see paragraph 1.1), we arrive at

$$\alpha_k(x) = \frac{e_k^\star(x)A(x)f_0'(x)}{p_j(x) - p_k(x)}, \quad k \neq j .$$

The function $\alpha_1(x)$ is still undetermined. However, $\alpha(x)$ is found from the condition of solvability for the equation with $k = 1$ in (7), that is, from the condition $e_j^\star A f_1' = 0$.

Continuing this procedure we can successively find the vector functions $f_2(x)$, $f_3(x)$.... All the coordinates of the vector function $f_k(x)$ in the basis $\{e_1(x), \ldots, e_n(x)\}$, except the jth, are found from the kth equation. The coordinate with index j is found from the condition for solvability of the $(k+1)^{\text{st}}$ equation. Thus we have constructed a F.A.S. of system (1) having the form (2).

We can also construct the F.A.S. in a different way. We look for a matrix

$$T(x, \lambda) = \sum_{k=0}^{\infty} T_k(x)\lambda^{-k}$$

such that the transformation

$$y = T(x, \lambda)z$$

diagonalizes (1); that is, reduces it to the form

$$z' = \lambda \Lambda z , \quad \Lambda = \sum_{k=0}^{\infty} \lambda^{-k} \Lambda_k(x) , \tag{9}$$

where the $\Lambda_k(x)$ are diagonal matrices. System (9) is decomposable and is therefore integrable.

Let us consider a system of a form more general than (1):

$$A(x, \lambda^{-1})y' = \lambda B(x, \lambda^{-1})y , \tag{10}$$

where the matrices A and B have asymptotic series:

$$A(x, \lambda^{-1}) = \sum_{k=0}^{\infty} \lambda^{-k} A_k(x) , \quad B(x, \lambda^{-1}) = \sum_{k=0}^{\infty} \lambda^{-k} B_k(x) .$$

We will assume that the matrix $A_0(x)$ is non-singular, that the eigenvalues $p_1(x), \ldots, p_n(x)$ of the bundle $B_0(x) - pA_0(x)$ are distinct and that the usual smoothness conditions hold for A and B. We consider first the case when $A(x) = I$ and $B_0(x)$ is diagonal with diagonal elements $p_1(x), \ldots, p_m(x)$. Then $\Lambda_0(x) = B_0(x)$ in formula (9) and we can put $T_0(x) = I$. With the substitution $y = Tz$ system (10) becomes

$$T\frac{dz}{dx} + \frac{dT}{dx}z = \lambda Bz\,,$$

and since it must be the same as system (9) the matrix T must satisfy the equation

$$\frac{dT}{dx} = \lambda(BT - T\Lambda)\,.$$

Substituting the asymptotic series for T and Λ into this formula, we obtain the recurrence system of equations

$$B_0 - \Lambda_0 = 0\,, \quad \sum_{j=0}^{k}(B_{k-j}T_j - T_j\Lambda_{k-j}) = \frac{dT_{k-1}}{dx}\,, \quad k \geqslant 1\,.$$

Here not only the matrices T_k are unknown but also the matrices Λ_k, $k \geqslant 1$. The first equation is satisfied identically, and the second has the form

$$\Lambda_0 T_1 - T_1\Lambda_0 = \Lambda_1 - B_1\,.$$

Since Λ_0 is diagonal, all diagonal elements of $\Lambda_0 T_1 - T_1\Lambda_0$ are zero. Therefore

$$\Lambda_1 = \text{diag}\, B_1\,, \quad (T_1)_{jk} = \frac{(B_1)_{jk}}{p_k - p_j}\,, \quad j \neq k\,.$$

The elements $(T_1)_{jj}$ are still undetermined; we put them equal to zero. The equation with index k has the form

$$\Lambda_0 T_k - T_k\Lambda_0 = \Lambda_k + C_k\,,$$

where C_k is an unknown matrix. Thus we find

$$\Lambda_k = -\text{diag}\, C_k\,, \quad (T_k)_{jl} = \frac{(C_k)_{jl}}{p_j - p_l}\,, \quad j \neq l\,,$$

and we put the elements $(T_k)_{jj}$ equal to zero. Thus we have constructed the required matrix $T(x, \lambda)$.

Let us return to the system (10). We make the substitution $y = T_0(x)\tilde{z}$, where the matrix $T_0(x)$ reduces $A_0^{-1}(x)B_0(x)$ to diagonal form; that is, $T_0^{-1}A_0^{-1}B_0T_0 = \Lambda_0 = \text{diag}\,(p_1, \ldots, p_n)$. Then system (10) becomes

$$\tilde{z}' = C(x, \lambda)\tilde{z}\,, \quad C = \lambda T_0^{-1}A^{-1}BT_0 - T^{-1}\frac{dT_0}{dx}\,.$$

Because of the choice of $T_0(x)$ the matrix C has asymptotic expansion

$$C(x,\lambda) = \Lambda_0(x) + \sum_{k=1}^{\infty} \lambda^{-k} C_k(x),$$

and we arrive at the case studied above.

1.3 Sufficient Conditions for the Existence of the Asymptotic Behaviour of the Solutions. We introduce the conditions:

1) the system (1) has no turning points for $x \in I$; that is, $p_j(x) \neq p_k(x)$, $j \neq k$, $x \in I$;

2) the differences Re $[p_j(x) - p_k(x)]$ do not change sign for j fixed, $k = 1, \ldots, n$, $x \in I$.

We introduce the notation

$$y_{j0}(x,\lambda) = \exp\left\{\lambda \int_{x_0}^{x} p_j(t)dt - \int_{x_0}^{x} \frac{e_j^*(t)A(t)e_j'(t)}{e_j^*(t)A(t)e_j(t)} dt\right\}.$$

Recall that $e_j^*(x)$ and $e_j(x)$ are the left and right eigenvectors of the matrix bundle $B(x) - pA(x)$.

System (1) has a solution of the form

$$y_j(x,\lambda) = y_{j0}(x,\lambda)\left[e_j(x) + \sum_{k=1}^{N-1} \lambda^{-j} f_{jk}(x) + O(\lambda^{-N})\right], \quad \lambda \to \infty. \tag{11}$$

Here $N \geqslant 1$ is arbitrary, the vector-functions $f_{jk}(x)$ are of class $C^\infty(I)$ for all k, and the bound for the remainder term is uniform in $x \in I$. The asymptotic formula (11) can be differentiated in x and λ an arbitrary number of times preserving the uniformity in x of the bound for the remainder term. In particular, for the principal asymptotic term we have

$$y_j^{(m)}(x,\lambda) = \lambda^m p_j^m(x) y_{j0}(x,\lambda)[e_j(x) + O(\lambda^{-1})].$$

If conditions 1), 2) are satisfied for all j then the solutions $y_1(x,\lambda), \ldots, y_n(x,\lambda)$ are linearly independent and form the F.S.S. for $\lambda \gg 1$.

Examples. 1. Let $B(x)$ be a real symmetric matrix with distinct eigenvalues $p_1(x), \ldots, p_n(x)$, and let $A(x) = I$. Since the matrices are real condition 2) is satisfied. Then formula (7) can be simplified if the eigenvectors $e_1(x), \ldots, e_n(x)$ are chosen to be real and unitary (and of class $C^\infty(I)$). In fact, in this case $e_k^*(x) = e_k^T(x)$ for all k, as follows from the identities

$$A^T(x)e_k(x) = A(x)e_k(x) = p_k(x)e_k(x),$$

$$e_k^T(x)A(x) = p(x)e_k^T(x), \quad e_k^T(x)e_k(x) = \delta_{jk}.$$

Differentiating the identity $e_k^T(x)e_k(x) = 1$, we obtain $e_k^*(x)e_k'(x) = 0$ and formula (8) becomes

$$y_j(x,\lambda) = \exp\left\{\int_{x_0}^{x} p_j(t)dt\right\}[e_j(x) + O(\lambda^{-1})]\,. \tag{12}$$

2. Let $A(x) = I$, and let $B(x)$ be a Hermitian matrix with distinct eigenvalues; then condition 2) is satisfied. We will demonstrate the existence of eigenvectors $e_1(x), \dots, e_n(x)$, $e_1^\star(x), \dots e_n^\star(x)$ such that

$$e_k^\star(x)e_k'(x) = 0\,. \tag{13}$$

Then the asymptotic formula (12) is valid for the solutions of system (1).

We select an orthonormal basis $g_1(x), \dots, g_n(x)$ from the $C^\infty(I)$ eigenvectors such that

$$\overline{g}_k^T(x)g_j(x) = \delta_{kj}\,, \quad x \in I\,.$$

Differentiating this identity we obtain

$$\operatorname{Re}\,(\overline{g_k^T(x)}, g_k(x)) = 0\,.$$

Since $B(x)$ is Hermitian, we can take $\overline{g_k^T(x)}$ as $g_k^\star(x)$. Then condition (4) is satisfied. We put

$$e_k(x) = e^{i\alpha_k(x)}g_k(x)\,, \quad e_k^\star(x) = \overline{e_k^T(x)}\,.$$

We have

$$\begin{aligned}\frac{d}{dx}(e_k^\star(x)e_k(x)) &= \overline{g_k^T(x)}(i\alpha_k'(x)g_k(x) + g_k'(x)) \\ &= i\alpha_k'(x) + \overline{g_k^T(x)}g_k'(x) = 0\,,\end{aligned}$$

so that

$$\alpha_k'(x) = -i\overline{g_k^T(x)}g_k'(x)\,.$$

The right-hand side of this equation is real and the function $\alpha_k(x)$ can be chosen to be real.

3. We consider the system

$$y' = i\lambda B(x)y\,,$$

where $B(x)$ is the matrix from Example 2. As in § 1, paragraph 1.4, Example 2, we can show that there exist two F.S.S.

$$\{y_1^+(x,\lambda), \dots, y_n^+(x,\lambda)\}\,, \quad \{y_1^-(x,\lambda), \dots, y_n^-(x,\lambda)\}\,.$$

There is an asymptotic series expansion for the first of these of the form (8) as $|\lambda| \to \infty$, $\operatorname{Im}\lambda \geqslant 0$, and for the second as $|\lambda| \to \infty$, $\operatorname{Im}\lambda \leqslant 0$.

1.4 Additional Parameters and Complex λ. We consider the system

$$A(x,\mu)y' = \lambda B(x,\mu)y\,,$$

where μ is a parameter. The results established above carry over to this system; the corresponding formulations and conditions on the matrices A and B are of the same type as those in § 1, paragraph 1.4.

2. Second-Order Systems

2.1 Two-Term Systems. We consider the system of n equations

$$y'' - \lambda^2 A(x)y = 0 \tag{14}$$

on an interval I, where the elements of $A(x)$ are in $C^\infty(I)$. The characteristic equation has the form

$$\det(p^2 I - A(x)) = 0\,.$$

The roots of this equation are $\pm\sqrt{\mu_1(x)}, \dots, \pm\sqrt{\mu_n(x)}$, where $\mu_1(x), \dots, \mu_n(x)$ are the eigenvalues of $A(x)$. The system (14) has turning points of two types.

1. x_0 is a turning point if $\mu_j(x_0) = 0$ for some j. At this point the roots $\sqrt{\mu_j(x)}$ and $-\sqrt{\mu_j(x)}$ coalesce.

2. x_0 is a turning point if $\mu_j(x_0) = \mu_k(x_0)$, $j \neq k$. At this turning point the roots $\sqrt{\mu_j(x)}$ and $\sqrt{\mu_k(x)}$ coalesce, as do $-\sqrt{\mu_j(x)}$ and $-\sqrt{\mu_k(x)}$.

We will put $p_j(x) = \sqrt{\mu_j(x)}$ and $p_{n+j}(x) = -\sqrt{\mu_j(x)}$, $1 \leqslant j \leqslant n$. We suppose that system (15) has no turning points for $x \in I$. Let us fix the bases $\{e_1(x), \dots, e_n(x)\}$ and $\{e_i^*(x), \dots, e_n^*(x)\}$ formed by the right and left eigenvectors, chosen so that

$$e_j^*(x)e_k(x) = \delta_{jk}\,. \tag{15}$$

We denote

$$\tilde{y}_j^\pm(x,\lambda;x_0) = \mu_j^{-1/4}(x)\exp\left\{\pm\lambda\int_{x_0}^x \sqrt{\mu_j(t)}dt - \int_{x_0}^x e_j^*(t)\frac{de_j(t)}{dt}dt\right\}. \tag{16}$$

Suppose that none of the differences $\operatorname{Re}(\sqrt{\mu_j(x)} \pm \sqrt{\mu_k(x)})$ changes sign in I. Then the system (14) has $2n$ solutions

$$y_j^\pm(x,\lambda) = \tilde{y}_j^\pm(x,\lambda;x_0)\left[e_j(x) + \sum_{k=1}^{N-1}\lambda^{-k}e_{jk}(x) + O(\lambda^{-N})\right], \tag{17}$$

$$1 \leqslant j \leqslant n\,, \quad \lambda \to \infty\,,$$

where $N \geqslant 1$ is arbitrary and $e_{jk}(x) \in C^\infty(I)$. For $\lambda \gg 1$ these solutions form the F.S.S. If $A(x)$ is a real symmetric matrix then formula (16) simplifies to

$$\tilde{y}_j^\pm(x,\lambda;x_0) = \mu_j^{-1/4}(x)\exp\left\{\pm\lambda\int_{x_0}^x \sqrt{\mu_j(t)}dt\right\}. \tag{18}$$

The asymptotic expansion (17) can be differentiated in x and λ an arbitrary number of times. Analogous results are true when λ is complex and when the matrix A depends on a parameter ν: $A = A(x,\nu)$. We can formulate results corresponding to those in § 1, paragraph 1.4. This remark relates to all the systems considered below.

2.2 Systems of Self-adjoint Form. Let us consider the system of equations

$$(A(x)y')' - \lambda^2 B(x)y = 0\,, \tag{19}$$

where $A(x)$, $B(x) \in C^\infty(I)$ and $\det A(x) \neq 0$ for $x \in I$. The characteristic equation has the form

$$\det(p^2 A(x) - B(x)) = 0\,,$$

and its roots are

$$p_j(x) = \sqrt{\mu_j(x)}\,, \quad p_{n+j}(x) = -\sqrt{\mu_j(x)}\,, \quad 1 \leqslant j \leqslant n\,,$$

where the $\mu_j(x)$ are the eigenvalues of the bundle $\mu A(x) - B(x)$. Suppose that the $\mu_j(x)$ satisfy the conditions given in paragraph 2.1, and that $\{e_1(x), \dots, e_n(x)\}$ and $\{e_1^\star(x), \dots, e_k^\star(x)\}$ are the bases formed by the right and left eigenvectors of this bundle, normalized by the condition

$$e_j^\star(x)A(x)e_k(x) = \delta_{jk}\,. \tag{20}$$

Let

$$\begin{aligned}\tilde{y}_j^\pm(x, \lambda; x_0) = \mu_j^{-1/4}(x)\exp\Big\{\pm\lambda\int_{x_0}^x \sqrt{\mu_j(t)}dt \\ \mp\frac{1}{2}\int_{x_0}^x [e^\star(t)A(t)De(t) - De^\star(t)A(t)e(t)]dt\Big\}\,,\end{aligned} \tag{21}$$

where $D = d/dx$. The system (19) has $2n$ solutions of the form (17) where the $\tilde{y}_j^\pm$ have the form (21).

2.3 Systems of General Form. We consider the system

$$A(x)y'' + \lambda B(x)y' + \lambda^2 C(x)y = 0 \tag{22}$$

on an interval I, where $A(x)$, $B(x)$ and $C(x) \in C^\infty(I)$, and $\det A(x) \neq 0$ in I. The characteristic equation has the form

$$\det(p^2 A(x) + pB(x) + C(x)) = 0\,.$$

We assume that the eigenvalues $p_1(x), \dots, p_n(x)$ of the bundle $L(p) = p^2 A(x) + pB(x) + C(x)$ are distinct for $x \in I$ and that all the differences $\operatorname{Re}\,(p_j(x) - p_k(x))$, $1 \leqslant j,\ k \leqslant 2n$, do not change sign for $x \in I$. Let $\{e_1(x), \dots, e_n(x)\}$ and $\{e_1^\star(x), \dots, e_n^\star(x)\}$ be the right and left eigenvectors of the bundle, normalized by the conditions

$$e_j^\star(x)[2p_j(x)A(x) + B(x)]e_k(x) = \delta_{jk}\,. \tag{23}$$

Let

$$\tilde{y}_j(x,\lambda;x_0) = \exp\left\{\lambda\int_{x_0}^{x} p_j(t) + \int_{x_0}^{x} p_j^{(1)}(t)dt\right\}, \tag{24}$$

$$p_j^{(1)}(x) = -e_j^*[p_j'(x)A(x)e_j(x) + 2p_j(x)A(x)e_j'(x) + B(x)e_j'(x)].$$

Then system (22) has $2n$ solutions of the form

$$y_j(x,\lambda) = \tilde{y}_j(x,\lambda;x_0)\left[e_j(x) + \sum_{k=1}^{N-1}\lambda^{-k}e_{jk}(x) + O(\lambda^{-N})\right], \tag{25}$$

$$1 \leqslant j \leqslant 2n, \quad \lambda \to \infty,$$

where $N \geqslant 1$ is arbitrary and $e_{jk} \in C^\infty(I)$. These solutions form the F.S.S. for $\lambda \gg 1$.

3. Systems of n Equations of mth-Order. Let us consider the system

$$y^{(m)} + \sum_{k=0}^{m-1}\lambda^{m-k}A_k(x)y^{(k)} = 0 \tag{26}$$

on an interval I, where $y(x)$ is an n-vector, and the $A_k(x)$ are matrices of order $n\times n$ with complex elements of class $C^\infty(I)$. The characteristic equation has the form

$$\det L(x,p) = 0, \quad L(x,p) = p^m + \sum_{k=0}^{m-1}A_k(x)p^k.$$

We assume that the eigenvalues $p_1(x),\ldots,p_{nm}(x)$ of the bundle L are distinct for all $x \in I$ and that all the differences Re $(p_j(x) - p_k(x))$, $1 \leqslant j,k \leqslant mn$, do not change sign on the interval I. Let $e_j(x)$ and $e_j^*(x)$, $1 \leqslant j,k \leqslant mn$, be the right and left eigenvectors of L. We denote

$$\tilde{y}_j(x,\lambda;x_0) = \exp\left\{\lambda\int_{x_0}^{x} p_j(t)dt + \int_{x_0}^{x} p_j^{(1)}dt\right\},$$

$$p_j^{(1)}(x) = -\left[e_j^*\frac{\partial L}{\partial p}De_j + \frac{1}{2}Dp_j\frac{\partial^2 L}{\partial p^2}e_j\right]\left[e_j^*\frac{\partial L}{\partial p}e_j\right]^{-1}, \tag{27}$$

where the values of the eigenvectors e_j and e_j^* are taken at the point x, the values L_p, L_{pp} at the point $(x, p_j(x))$, and $D = d/dx$. The system (26) has mn solutions of the form (25) where $p_j^{(1)}$ has the form (27), $N \geqslant 1$ is arbitrary, and these solutions form the F.S.S. for $\lambda \gg 1$.

§ 3. Equations on an Infinite Interval

1. Equation with Almost Constant Coefficients. We consider the equation

$$ly \equiv y^{(n)} + q_{n-1}(x)y^{(n-1)} + \ldots + q_0(x)y = 0 \tag{1}$$

on the half-line $\mathbb{R}^+ = [0, \infty)$ with complex continuous coefficients. The characteristic equation is

$$l(x,p) \equiv p^n + q_{n-1}(x)p^{n-1} + \ldots + q_0(x) = 0\,. \tag{2}$$

The asymptotic behaviour of solutions of equation (1) as $x \to \infty$ depends essentially on the behaviour of the roots $p_1(x), \ldots, p_n(x)$ of the characteristic equation at infinity.

Suppose that the coefficients of equation (1) have the form

$$q_k(x) = q_k^0 + r_k(x)\,, \tag{3}$$

where q_k^0 is constant and the following conditions are satisfied:

1) $\int_0^\infty |r_k(x)|dx < \infty\,, \quad 1 \leqslant k \leqslant n\,;$
2) the roots $\rho_1, \ldots, \rho_n$ of the equation

$$\rho^n + q_{n-1}^0 \rho^{n-1} + \ldots + q_0^0 = 0 \tag{4}$$

are distinct.

Then equation (1) has F.S.S. $\{y_1, \ldots, y_n\}$ such that

$$y_j(x) = e^{\rho_j x}[1 + o(1)]\,, \quad x \to \infty\,. \tag{5}$$

Further, this asymptotic formula can be differentiated n times, that is,

$$y_j^{(k)}(x) = \rho_j^k e^{\rho_j x}[1 + o(1)]\,, \quad 0 \leqslant k \leqslant n\,.$$

This result is a corollary of Levinson's theorem on L-diagonal systems (Chap. 2, § 5). The next result arises out of the Perron theorem (Chap. 2, § 5). Suppose that the coefficients of equation (1) have the form (3) and suppose that the following conditions are satisfied:

3) $\lim_{x\to\infty} r_k(x) = 0, \quad 1 \leqslant k \leqslant n;$
4) $\operatorname{Re} \rho_j \neq \operatorname{Re} \rho_k$ for all $j \neq k$ where the ρ_j are the roots of equation (4).

Then equation (1) has F.S.S. $\{y_1, \ldots, y_n\}$ such that

$$\lim_{x\to\infty} \frac{y_k'(x)}{y_k(x)} = \rho_k\,, \quad 1 \leqslant k \leqslant n\,. \tag{6}$$

This formula can be written as

$$y_k(x) = \exp\{\rho_k x + o(x)\}\,, \quad x \to \infty\,,$$

so that the asymptotic behaviour in (6) is rather coarse.

Analogous results are true for n^{th}-order equations of self-adjoint form

$$(-1)^n \frac{d^n}{dx^n}\left[q_0(x)\frac{d^n y}{dx^n}\right] + (-1)^{n-1}\frac{d^{n-1}}{dx^{n-1}}\left[q_1(x)\frac{d^{n-1}y}{dx^{n-1}}\right] + \dots \\ + q_n(x)y = 0 . \tag{7}$$

Suppose that the coefficients $q_k(x)$ have the form (3) for $1 \leqslant k \leqslant n$, and

$$\frac{1}{q_0(x)} = q_0^0 + r_0(x) , \quad q_0^0 \neq 0 ,$$

where condition 1) is satisfied for $k = 0, 1, \dots, n$ and condition 2) holds for the roots of the equation

$$q_0^0 \rho^{2n} - q_1^0 \rho^{2n-2} + \dots + (-1)^n q_n^0 = 0 ,$$

i.e. all the roots have the same growth order as $x \to \infty$. Then equation (7) has F.S.S. $\{y_1, \dots, y_n\}$ of the form (5), and similar formulae are true for the quasi-derivatives $y^{[k]}(x)$, $1 \leqslant k \leqslant 2n$.

2. Equation (1) with Asymptotically Simple Roots. Suppose that for some j there exist finite or infinite limits

$$\lim_{x\to\infty} p_j(x)/p_k(x) = c_{jk} , \quad 1 \leqslant j , \ k \leqslant n . \tag{8}$$

The root $p_j(x)$ of the characteristic equation (2) is called *asymptotically simple* if $c_{jk} \neq 1$ for all $k \neq j$.

Throughout what follows we will assume that $q_k(x) \in C^2(\mathbb{R}^+)$, $0 \leqslant k \leqslant n-1$.

2.1 Roots of the Same Order. Let $q_0(x) \neq 0$ for $x \gg 1$, let all the roots of equation (2) be asymptotically simple, and suppose that $c_{jk} \neq 0, 1$ for all $j \neq k$. Then there are the finite limits

$$\lim_{x\to\infty} q_k(x) q_0^{-1+k/n}(x) = a_k , \quad 1 \leqslant k \leqslant n-1 , \tag{9}$$

and the asymptotic behaviour of the roots has the form

$$p_j(x) = [\rho_j + o(1)] q_0^{1/n}(x) , \quad x \to \infty , \tag{10}$$

where $\rho_1, \dots, \rho_n$ are the roots of the equation

$$\rho^n + a_{n-1}\rho^{n-1} + \dots + a_1 p + 1 = 0 . \tag{11}$$

The converse is true: if the asymptotic behaviour of all the roots of equation (2) have the form (10) where $\rho_1, \dots, \rho_n$ are distinct non-zero numbers then the limits (9) exist and are finite.

Condition (9) is satisfied by the equation

$$y^{(n)} + a_{n-1}q(x)y^{(n-1)} + \ldots + a_1 q^{n-1}(x)y' + q^n(x)y = 0\,, \tag{12}$$

where $a_1, \ldots, a_{n-1}$ are constants and $q(x) \neq 0$ for $x \gg 1$.

If all the coefficients $q_0(x), \ldots, q_{n-1}(x)$ of equation (1) are polynomials in x then under condition (9) the point $x = \infty$ is an irregular singular point for equation (1) (Chap. 1, § 3, paragraph 1.1). In the general case condition (9) means that all the terms $q_k(x)y^{(k)}$ of the operator l have "equal weight" for $x \gg 1$; that is, they all, generally speaking, affect the asymptotic behaviour of the solution.

Let us put

$$p_j^{(1)}(x) = -\frac{p_j'(c)}{2}\,\frac{l_{pp}(x, p_j(x))}{l_p(x, p_j(x))}\,,$$
$$\tilde{y}_j(x) = \exp\left\{\int_{x_0}^{x} [p_j(t) + p_j^{(1)}(t)]dt\right\}. \tag{13}$$

We introduce the conditions:

1) For some j and for all $k \neq j$ the functions Re $\phi_{jk}(x)$, where

$$\phi_{jk}(x) = p_j(x) - p_k(x) + p_j^{(1)}(x) - p_k^{(1)}(x)\,, \tag{14}$$

do not change sign for $x \gg 1$;

2) $$\int^{\infty} \alpha(x)dx < \infty\,,$$

where

$$\alpha(x) = \sum_{k=0}^{n-1}(|q_k'|^2|q_0|^{-2+(2k-1)/n} + |q_k''|\,|q_0|^{-1+(k-1)/n})\,.$$

Suppose that conditions 1), 2) and (9) hold and suppose that equation (11) has no repeated roots. Then equation (1) has solutions $y_j(x)$ such that

$$y_j(x) = \tilde{y}_j(x)[1 + o(1)]\,, \quad x \to \infty\,. \tag{15}$$

If further for all j the following condition is satisfied

3) $$q_k'(x) = 0\,(q_0^{1-(k-1)/n}(x))\,, \quad x \to \infty\,,$$

then $p_l^{(1)}(x) = o\,(p_l(x))$ as $x \to \infty$ for all l. Then for $0 \leqslant m \leqslant n$ we have

$$y_j^{(m)}(x) = p_j^m(x)\tilde{y}_j(x)[1 + o(1)]\,, \quad x \to \infty\,. \tag{16}$$

If condition 1) is satisfied for all j then equation (1) has an F.S.S. $\{y_1(x), \ldots, y_n(x)\}$ of the form (15).

The proofs are based on the following fact. Equation (1) is equivalent to the system (10) of § 1:

$$z' = A(x)z\,, \quad z = (y, y', \ldots, y^{n-1})^T\,.$$

Suppose that $T_0(x)$ and $T_1(x)$ are the matrices constructed in § 1, paragraph 1. Then the transformation $z = T_0(x)(I + T_1(x))w$ reduces the system to L-diagonal form, after which it remains only to apply Levinson's theorem (Chap. 2, § 5). This remark also relates to paragraphs 2.2, 3.1 of the present section.

Condition 1) was discussed in § 8 of Chap. 8. We now discuss conditions 2) and 3) in the case of equation (12). Here $q_0(x) = q^n(x)$, and conditions 2) and 3) become

$$\int^{\infty} \left(\left| \frac{q_0'^2(x)}{q_0^{2+1/n}(x)} \right| + \left| \frac{q_0''(x)}{q_0^{1+1/n}(x)} \right| \right) dx < \infty\,.$$

For the equation $y'' + q(x)y = 0$ these conditions are precisely the conditions for the applicability of the WKB-approximation which was given in § 6 of Chap. 2. If $q_0(x) = x^\alpha$ then conditions 2) and 3) are satisfied for $\alpha > n$. Suppose that all the coeffcients $q_0(x), \ldots, q_{n-1}(x)$ of equation (1) are analytic functions which are either holomorphic or have a pole at the point at infinity $z = \infty$. Then it follows from conditions (9) and 2) that $z = \infty$ is an irregular singular point of equation (1) and that condition 3) is satisfied.

The conditions 2) and 3) for equation (12) mean that the function $q(x)$ behaves sufficiently regularly at infinity and can decrease, but more slowly than x^{-1}. These conditions are satisfied, for instance, for the functions $q(x) = e^{ax^\alpha}$, $q(x) = x^\beta(\ln x)^\gamma$ where $a \neq 0$, $\operatorname{Re} a \geqslant 0$, $\alpha > 0$ and either $\beta > -1$ with γ arbitrary, or $\beta = -1$ with $\gamma > 1$. In terms of the roots of the characteristic equation (2) conditions 2) and 3) can be written as

$$\int^{\infty} \left[\frac{|p_j'|^2}{|p_j|^3} + \frac{|p_j''|}{|p_j|^2} \right] dx < \infty\,, \tag{17}$$

$$p_j'(x) = o(p_j^2(x))\,, \quad x \to \infty\,, \quad j = 1, \ldots, n\,.$$

2.2 One of the Roots of Least Growth Order. Let

$$\lim_{x \to \infty} p_n(x)/p_k(x) = 0\,, \quad 1 \leqslant k \leqslant n-1\,, \tag{18}$$

and let all the other limits (8) be finite if $k \neq n$. Suppose that $q_1(x) \neq 0$ for $x \gg 1$. Then there exist finite limits

$$\lim_{x \to \infty} q_k(x)[q_1(x)]^{(k-n)/(n-1)} = a_k\,, \quad 0 \leqslant k \leqslant n-1\,, \tag{19}$$

where $a_0 = 0$. The asymptotic behaviour of the roots of equation (2) as $x \to \infty$ is

$$p_k(x) = [\rho_k + o(1)]q_1^{1/(n-1)}(x)\,, \quad 1 \leqslant k \leqslant n-1\,,$$
$$p_n(x) = \frac{-q_0(x) + o(q_0(x))}{q_1(x)}\,, \tag{20}$$

where $\rho_1, \dots, \rho_{n-1}$ are the roots of the equation

$$\rho^{n-1} + a_{n-1}\rho^{n-2} + \ldots + a_2\rho + a_1 = 0\,. \tag{21}$$

We replace conditions 2) and 3) by the following:

2′) $\int^\infty (|q_k'|^2|q_1|^{-2+(2k+1)/(n-1)} + |q_k''|\,|q_1|^{-1+k/(n-1)})dx < \infty$ for all k;

3′) $q_k'(x) = o(q_1^{(n+1-k)/(n-1)}(x))$ as $x \to \infty$ and for all k.

If the conditions 1) for all j, (19) and 2′) are satisfied, and if equation (21) has no repeated roots, then equation (1) has F.S.S. $\{y_1(x), \dots, y_n(x)\}$ for which the asymptotic formula (15) holds. If in addition condition 3′) is satisfied then $p_k^{(1)}(x) = o(p_k(x))$ as $x \to \infty$, $1 \leqslant k \leqslant n-1$, and formula (16) is true for the solutions $y_1(x), \dots, y_{n-1}(x)$. We note that the condition 2′) is equivalent to the first condition in (17) for all the roots $p_k(x)$, and condition 3′) is equivalent to the second one for $p_1(x), \dots, p_{n-1}(x)$. Formula (16) is true for the solution $y_n(x)$ if the above conditions hold with the only difference being that condition 3′) is replaced by the following:

$$\sum_{k=2}^{n} |q_k| \left|\frac{q_0}{q_1}\right|^{k-2} \sum_{j=0}^{n-1} |q_j'| \left|\frac{q_0}{q_1}\right|^{j} = o(q_0)\,, \quad x \to \infty\,.$$

Here $p_n^{(1)}(x) = o(p_n(x))$ as $x \to \infty$.

We note also that the principal asymptotic term of the solution as $x \to \infty$ of equation (1) has precisely the same form as the principal asymptotic term of the solution as $\lambda \to \infty$ of equation (1) of § 1, as is clear by comparing formulae (15) and (5) of § 1.

2.3 Bounds for the Remainder Terms. Suppose that conditions (9), 1), and 2) from paragraph 2.2 are satisfied and suppose that equation (11) has no repeated roots. Formula (15) can be written as

$$y_j(x) = \tilde{y}_j(x)[1 + \varepsilon_j(x)]\,, \quad \lim_{x\to\infty} \varepsilon_j(x) = 0\,.$$

We give bounds for the remainder term $\varepsilon_j(x)$. Let us separate the functions $\phi_{j1}(x), \dots, \phi_{jn}(x)$ (see (14)) into two classes.

1) $k \in H_1(j)$, if $\operatorname{Re} \phi_{jk}(x) \leqslant 0$, for $x \gg 1$.

2) $k \in H_2(j)$, if $\operatorname{Re} \phi_{jk}(x) \geqslant 0$, for $x \gg 1$.

Then for $x \gg 1$

$$|\varepsilon_j(x)| \leqslant c\left[\int_x^\infty \alpha(t)dt + \sum_{k\in H_2(j)x} \int_x^\infty \exp\{\operatorname{Re}\, \phi_{kj}(t)\}\alpha(t)dt\right]. \tag{22}$$

The function $\alpha(x)$ is as in condition 2). In particular, if Re $\phi_{jk}(x) \leqslant 0$ for $x \gg 1$ and all k, then

$$|\varepsilon_j(x)| \leqslant c \int_x^\infty \alpha(t)dt\,.$$

This case occurs, for instance, when all the coefficients of equation (1) are real and all the roots of equation (2) are purely imaginary. If more exact information about the behaviour of the coefficients is known then we can refine the bound (22) by applying the Laplace method for the asymptotic bounds of the integrals.

Example. In the case of equation (12) with $q(x) = x^\beta$, $\beta > -1$, we have

$$\varepsilon_j(x) = O(x^{-\beta-1})\,, \quad x \to \infty\,.$$

Similar bounds are true for the case considered in paragraph 2.2.

2.4 Dual Asymptotic Behaviour. We consider the equation

$$ly \equiv y^{(n)} + \lambda q_{n-1}(x)y^{(n-1)} + \ldots + \lambda^n q_0(x)y = 0\,, \tag{23}$$

where λ is a large parameter. We will restrict ourselves to the case where conditions (9), 1), and 2) from paragraph 2.1 hold, where equation (11) has no repeated roots. Let $\tilde{y}_j(x, \lambda; x_0)$ be the function defined by formula (5) of § 1. We state sufficient conditions for the existence of a solution of equation (23) such that

$$y_j(x, \lambda) = \tilde{y}_j(x, \lambda; x_0)[1 + \lambda^{-1}\varepsilon_j(x, \lambda)] \tag{24}$$

where the bounds for the remainder term when $x \geqslant x_0$ and $\lambda \geqslant \lambda_0 > 0$ are

$$|\varepsilon_j(x, \lambda)| \leqslant k(x)\,, \quad \lim_{x\to\infty} k(x) = 0\,. \tag{25}$$

The asymptotic behaviour (24) is dual: the remainder term is small for $\lambda \gg 1$ with $x \geqslant x_0$ fixed, and for $x \to \infty$ with $\lambda \geqslant \lambda_0$ fixed.

Equation (23) has solution $y_j(x, \lambda)$ of the form (24), (25) if for all $k \neq j$ the following conditions are satisfied:

4) $$\int^\infty |\text{Re}\,(p_j(x) - p_k(x))|dx = \infty\,;$$

5) $$|p_j^{(1)}(x) - p_k^{(1)}(x)| \leqslant c|p_j(x) - p_k(x)|\,, \quad x \gg 1\,.$$

If these conditions are satisfied for all j and k, $j \neq k$, then for $\lambda \geqslant \lambda_0 \gg 1$ the solutions $y_1, \ldots, y_n$ of the form (24) form the F.S.S. In this case the solutions $y_1, \ldots, y_n$ differ in their growths as $x \to \infty$, that is, the limits

$$\lim_{x\to\infty} y_j(x,\lambda)[y_k(x,\lambda)]^{-1}$$

are either zero of infinity when $j \neq k$.

We consider next the case where several solutions can have equal orders of growth at infinity. Suppose that for some j and all $k \neq j$ either conditions 4) and 5) are satisfied or the following condition holds:

6) Re $p_j(x) =$ Re $p_k(x)$, $x \gg 1$, and Re $[p_j^{(1)}(x) - p_k^{(1)}(x)]$ does not change sign for $x \gg 1$.

Then all the above assertions remain in force. In both cases considered the asymptotic behaviour can be differentiated m times to give

$$y_j^{(m)}(x,\lambda) = \lambda^m p_j^m(x)\tilde{y}_j(x,\lambda;x_0)[1 + \lambda^{-1}\varepsilon_{jm}(x,\lambda)]\,,$$

and the bound (25) holds for the functions ε_{jm}.

These statements are true for complex λ where $\lambda \to \infty$ in some sector of the form S: $\alpha \leqslant \arg\lambda \leqslant \beta$, $0 < \beta - \alpha < 2\pi$, if condition 5) is satisfied together with the condition

4′) $\int^\infty |\mathrm{Re}[e^{i\phi}(p_j(x) - p_k(x))]|dx < \infty$ for all $k \neq j$, $\alpha \leqslant \phi \leqslant \beta$, $\phi = \arg\lambda$.

Example. Suppose that the limit $\lim_{x\to\infty} \arg q_0(x) = \phi_0$ exists and S is the sector $\alpha \leqslant \phi \leqslant \beta$, $\lambda \neq 0$, such that

$$\mathrm{Re}\,[\exp\{i(\phi + \phi_0/n)\}(\rho_j - \rho_k)] \neq 0\,, \quad \lambda \in S\,,$$

for all $k \neq j$. Here $\rho_1, \ldots, \rho_n$ are the roots of equation (11). Then equation (23) has a solution of the form (24) and the bound (25) holds for $\lambda \in S$, $|\lambda| \geqslant \lambda_0 \gg 1$, $x \geqslant x_0$.

2.5 Roots with a Variety of Growth Orders. Suppose that the roots of equation (2) are asymptotically non-repeated and have different growth orders, that is, some of the limits c_{jk} (see (8)) are 0 or ∞. We index them by their growth orders: the number c_{jk} is finite for $j \leqslant k$ and at least two of the numbers c_{nj}, $j < n$, are zero. The case where only one of these numbers is zero is considered in paragraph 2.9.

The asymptotic formulae (15) are preserved, under certain conditions on the coefficients of equation (1) and the roots of equation (2), but these conditions turn out to be highly complex. We will restrict ourselves to the case where the coefficients of equation (1) are $C^\infty(\mathbb{R}^+)$ and as $x \to \infty$ have asymptotic expansions

$$q_k(x) = \sum_{j=0}^{\infty} a_{kj} x^{-\alpha_{kj}}\,, \tag{26}$$

where the α_{kj} are real, and the a_{kj} are complex constants with $\alpha_{k0} < \alpha_{k1} < \ldots < \alpha_{kj} < \ldots$, and $\alpha_{kj} \to \infty$. We assume also that the asymptotic expansion (26) can be differentiated in x an arbitrary number of times. If $a_{kj} = 0$

for fixed k and for all j then $q_k^{(m)}(x) = O(x^{-\infty})$ as $x \to \infty$ and for all $m = 0, 1, \ldots$. A typical example is the equation

$$y^{(n)} + a_{n-1}x^{\alpha_{n-1}}y^{(n-1)} + \ldots + a_0 x^{\alpha_0} y = 0 ,$$

where $a_0, a_1, \ldots, a_{n-1}$ are constants. Here all the roots of the characteristic equation (2) have asymptotic expansions of the form (26):

$$p_k(x) = \sum_{j=0}^{\infty} p_{kj} x^{\beta_{kj}} , \quad x \to \infty ,$$

which can be differentiated term by term an arbitrary number of times.

Suppose that all the roots of equation (2) are asymptotically simple, with $\beta_{k0} \neq 0$ and $\beta_{k0} > -1$ for all k, and suppose that condition (1) of paragraph 2.1 is satisfied. Then equation (1) has F.S.S. $y_1(x), \ldots, y_n(x)$ for which formula (15) is true. For these solutions there are asymptotic expansions of the form

$$y_k(x) = x^{\gamma_k} \exp\left\{ \sum_{j=0}^{N} c_{kj} x^{\delta_{kj}} \right\} \left[1 + \sum_{j=1}^{\infty} d_{kj} x^{\varepsilon_{kj}} \right] , \quad x \to \infty .$$

Here

$$\delta_{k0} > \delta_{k1} > \ldots > \delta_{kN} > -1 , \quad \delta_{k0} = \beta_{k0} + 1 ,$$

$$c_{k0} = p_{k0}/(\beta_{k0} + 1) , \quad 0 > \varepsilon_{k0} > \varepsilon_{k1} > \ldots > \varepsilon_{kj} > \ldots , \quad \varepsilon_{kj} \to -\infty .$$

2.6 Equations with Three Terms. We now consider the equation

$$y^{(n)} + a_k x^{\alpha_k} y^{(k)} + a_0 x^{\alpha_0} y = 0 , \tag{27}$$

where α_0, α_k are real, and a_0, a_k are complex constants with $a_0 \neq 0$, $a_k \neq 0$, and $\alpha_0 > -n$. Let us determine the asymptotic behaviour of the roots of the characteristic equation

$$p^n + a_k x^{\alpha_k} p^k + a_0 x^{\alpha_0} = 0 ,$$

using Newton's diagram. We pick out three points in the (ρ, α)-plane: $M_1 = (n, 0)$, $M_2 = (k, \alpha_k)$, $M_3 = (0, \alpha_0)$, and let I be the line segment $M_1 M_3$. There are three possibilities:

1. M_2 does not lie above I so that $\alpha_k \leqslant (n-k)\alpha_0/n$, and the Newton diagram is the line segment I. The asymptotic behaviour of the roots $p_j(x)$ has the form $p_j(x) \sim p_{j_0} x^{\alpha_0/n}$ as $x \to \infty$, where p_{j_0} is a root of the equation $q^n + a_k q^k + a_0 = 0$ if $\alpha_k = (n-k)\alpha_0/n$, and a root of the equation $q^n + a_0 = 0$ if $\alpha_k < (n-k)\alpha_0/n$. This case is the same as that considered in paragraph 2.1, and equation (27) has F.S.S. consisting of solutions of the form (15). If we compute the asymptotic behaviour of the roots $p_j(x)$ up to

the order $O(x^{-1-\delta})$, $\delta > 0$, then we can obtain the asymptotic expansion of the solutions in the form

$$y(x) = x^{\alpha} e^{Q(x)} [1 + o(1)] ,$$

where $Q(x)$ is a finite sum of powers.

2. The point M_2 lies above the line segment I so that $\alpha_k > (n-k)\alpha_0/n$, and the Newton diagram is the section of a polygon consisting of $M_1 M_2$ and $M_2 M_3$. In this case we have two sets of roots with different asymptotic behaviours

$$p_j(x) \sim (-a_k)^{1/(n-k)} x^{\alpha_k/(n-k)} , \quad j = 1, \ldots, n-k ,$$

$$p_j(x) \sim \left(-\frac{a_0}{a_k}\right)^{1/k} x^{(\alpha_0 - \alpha_k)/k} , \quad j = n-k+1, \ldots, n .$$

We obtain the principal asymptotic term for the roots of the first or second set if we ignore the last or first term respectively in the characteristic equation. The roots of the first set are such that $p_j(x) \gg x^{-1}$ as $x \to \infty$ and are large in comparison with the roots of the second: $p_j(x) \gg p_l(x)$ for $x \gg 1$ if $1 \leqslant j \leqslant n-k$, $n-k+1 \leqslant l \leqslant n$. Equation (27) has $n-k$ solutions $y_1(x), \ldots, y_{n-k}(x)$ solutions for which the asymptotic expansion (15) holds true. The situation is more complex for solutions corresponding to the second series of roots.

2a. Suppose that $\alpha_0 - \alpha_k > -k$, so that $p_j(x) \gg x^{-1}$ as $x \to \infty$. Then equation (27) has solutions $y_{n-k+1}(x), \ldots, y_n(x)$ which have asymptotic expansions of the form (15).

2b. Suppose that $\alpha_0 - \alpha_k \leqslant -k$, so that $p_j(x) = O(x^{-1})$ as $x \to \infty$. Then equation (27) has k solutions $y_{n-k+1}(x), \ldots, y_n(x)$ for each of which there is an asymptotic formula of the form

$$y(x) \sim x^{\alpha} (\ln x)^m , \quad x \to \infty ,$$

where $m \geqslant 0$ is an integer. These solutions behave in the same way as the solutions of equation (1) in the neighbourhood of the regular singular point $z = \infty$ (Chap. 1, § 2).

Examples. 1. Let us consider the equation

$$y^{(4)} - a x^{\alpha} y'' + b x^{\beta} y = 0 , \tag{28}$$

where $a \neq 0$, $b \neq 0$, $\alpha > -2$, and $2\alpha > \beta$. The characteristic roots split into two groups as $x \to \infty$:

$$p_{1,2}(x) \sim \pm \sqrt{a} x^{\alpha/2} , \quad p_{3,4}(x) \sim \pm \sqrt{\frac{b}{a}} x^{(\beta-\alpha)/2} .$$

Equation (28) has two solutions of the form

$$y_j(x) \sim x^{-5\alpha/4} \exp\left\{\int_{x_0}^{x} p_j(t)dt\right\}, \quad j = 1, 2.$$

If $3\alpha > 2(1+\beta)$ then these formulae simplify to

$$y_{1,2}(x) \sim x^{-5\alpha/4} \exp\left\{\pm\frac{2\sqrt{a}}{\alpha+2} x^{\alpha/2+1}\right\}.$$

If in addition $\alpha < \beta + 2$ then equation (28) has two solutions of the form

$$y_j(x) \sim x^{(\alpha-\beta)/4} \exp\left\{\int_{x_0}^{x} p_j(t)dt\right\}, \quad j = 3, 4.$$

Under the condition $3\alpha > 2(1+\beta)$ these formulae simplify to

$$y_{3,4}(x) \sim x^{(\alpha-\beta)/4} \exp\left\{\pm\sqrt{\frac{a}{b}}\frac{2}{\beta-\alpha+2} x^{(\beta-\alpha+2)/2}\right\}.$$

2. Let us consider equation (27) where $n = 2k$ and $\alpha_0 < 2\alpha_k$. Then the characteristic equation is solvable exactly and gives

$$p^k = -\frac{a_k}{2} x^{\alpha_k}\left[1 \pm \sqrt{1 - 4a_0 a_k^{-2} x^{\alpha_0 - 2\alpha_k}}\right],$$

where the value of the root is unity at $x = \infty$. The first (second) set of roots corresponds to the $+$ $(-)$ square root. The first set of roots corresponds to k solutions of the form

$$y_j(x) = x^{-\delta_1} \exp\left\{\int_{x_0}^{x} p_j(t)dt\right\}, \quad \delta_1 = \frac{(n-1)\alpha_k}{n},$$

and the second set to k solutions of the form

$$y_{j+k}(x) = x^{-\delta_2} \exp\left\{\int_{x_0}^{x} p_{j+k}(t)dt\right\}, \quad \delta_2 = \frac{(k-1)(\alpha_0 - \alpha_k)}{n}.$$

The asymptotic behaviour of the integrals $\int_{x_0}^{x} p_j(t)dt$ as $x \to \infty$ is easily calculated.

3. Equations of Self-Adjoint Form with Asymptotically Simple Roots

3.1 Equations of Even Order. Let us consider the equation

$$ly \equiv \sum_{k=0}^{n}(-1)^k [q_{n-k}(x)y^{(k)}]^{(k)} = 0 \tag{29}$$

on the half-line $\mathbb{R}^+ = [0, \infty)$, where $q_k(x)$ is a complex-valued function of class $C^2(\mathbb{R}^+)$ and $q_0(x) \neq 0$ for $x \gg 1$. If all the functions $q_k(x)$ are real then the operator L, with domain of definition $D(L) = C_0^\infty(\mathbb{R}^+)$ and $Ly = ly$, is symmetric in $L_2(\mathbb{R}^+)$. We introduce conditions of the same type as those in paragraph 2.1.

1) The limits $\lim_{x\to\infty} q_k(x)q_0^{-1}(x)\tau^{-2k}(x) = c_k$ exist and are finite where

$$\tau(x) = [q_n(x)/q_0(x)]^{1/2n}\,. \tag{30}$$

2) The equation

$$g(\xi) = \sum_{k=0}^{n}(-1)^k c_k \xi^{2n-2k} = 0 \tag{31}$$

has no repeated roots.

Equation (29) corresponds to the characteristic equation

$$l(x,p) \equiv \sum_{k=0}^{n}(-1)^k q_{n-k}(x)p^{2n-2k} = 0\,. \tag{32}$$

It follows from conditions 1) and 2) that the asymptotic formulae for the roots of equation (30) are

$$p_j(x) = [\xi_j + o(1)]\tau(x)\,, \quad x \to \infty\,, \tag{33}$$

where the ξ_j are the roots of equation (31). In particular all roots of equation (32) are asymptotically simple. A typical example of an equation of the type (29), for which conditions 1) and 2) are satisfied, is the equation

$$\sum_{k=0}^{n}(-1)^k c_k [q^{2(n-k)}y^{(k)}]^{(k)} = 0\,, \tag{34}$$

where the c_k are constants, $q(x) \neq 0$ for $x \gg 1$, and $c_0 = 1$.

The remaining conditions are (paragraph 2.1):

3) for some j and all k, Re $[(\xi_j - \xi_k)\tau(x)]$ does not change sign for $x \gg 1$;

4) for all k

$$\int^\infty (|q_k'|^2|q_0|^{-2}|\tau|^{-4k-1} + |q_k''|\,|q_0|^{-1}|\tau|^{-2k-1})dx < \infty\,.$$

5) $\lim_{x\to\infty} q_k'(x)q_0^{-1}(x)\tau^{-2k-1}(x) = 0$ for all k.

We introduce the notation

$$\tilde{y}_j(x) = \left[\left.\frac{\partial l(x,p)}{\partial p}\right|_{p=p_j(x)}\right]^{-1/2} \exp\left\{\int_{x_0}^{x} p_j(t)dt\right\}\,. \tag{35}$$

If conditions 1) – 5) are satisfied then equation (29) has solution $y_j(x)$ for which

$$y_j^{(k)}(x) = p_j^k(x)\tilde{y}_j(x)[1 + o(1)], \quad 0 \leqslant k \leqslant n-1,$$

and

$$y_j^{[k]}(x) = (-1)^{n-k} p_j^k(x) q_0(x) \tilde{y}_j(x) \left[\sum_{m=0}^{k-1} (-1)^m c_m \xi_j^{-2m} + o(1) \right], \tag{36}$$

$$n \leqslant k \leqslant 2n-1,$$

as $x \to \infty$, where $y^{[k]}$ is the quasi-derivative of order k (§ 1). These formulae can be simplified for $q_0(x) \equiv 1$ to give

$$y_j(x) \sim [q_n(x)]^{-1/2+1/4n} \exp\left\{ \int^x p_j(t)dt \right\}, \quad x \to \infty.$$

The admissible class of functions $q(x)$ for equation (34) is the same as for equation (12). If condition (3) is satisfied for all j then the solutions $y_1(x), \ldots, y_{2n}(x)$ form the F.S.S.

Examples. 1. We consider the two-term equation

$$(-1)^n [q_0(x) y^{(n)}]^{(n)} + q_n(x) y = 0.$$

With the above conditions it has solutions $y_j(x)$ such that, as $x \to \infty$,

$$y_j(x) = \rho^{-(2n-1)/2} q_0^{-1/2} \exp\left\{ i\omega_j \int_{x_0}^x \rho(t)dt \right\} [1 + o(1)],$$

where

$$\rho(x) = \sqrt[2n]{q_n(x)/q_0(x)}$$

and ω_j is a root of the equation $\omega^{2n} = -1$.

The same asymptotic formulae for the solutions of equation (29) are true in the case where the principal terms of the equation are $(-1)^n [q_0(x) y^{(n)}]^{(n)}$ and $q_n(x)y$, but the intermediate terms are subordinate to them for $x \gg 1$; for sufficient conditions see [Naimark].

2. We consider the fourth order equation of the form (29). Then

$$l(x,p) = q_0(x)p^4 - q_1(x)p^2 + q_2(x),$$

$$p_j(x) = \pm\sqrt{\frac{q_1(x) \pm \sqrt{D(x)}}{2q_0(x)}}, \quad D = q_1^2 - 4q_0q_2,$$

and if $p(x)$ is a root of the characteristic equation then

$$l_p(x, p(x)) = \pm 2p(x)\sqrt{D(x)}.$$

Equation (29) has F.S.S. of the form

$$y_j(x) \sim p_j^{-1/2}(x) D^{-1/4}(x) \exp\left\{\int_{x_0}^{x} p_j(t)dt\right\}, \quad x \to \infty .$$

3.2 Dual Asymptotic Behaviour. We consider the equation

$$ly \equiv \sum_{k=0}^{n} \varepsilon^{2k}(-1)^k [q_{n-k}(x) y^{(k)}]^{(k)} = 0, \tag{37}$$

where $\varepsilon > 0$ is a parameter. Suppose that conditions 1), 2), 4) and 5) from paragraph 3.1 are satisfied together with the following condition.

3′) For some j and for all $k \neq j$

$$f_{jk}(x) \equiv \mathrm{Re}[(\xi_j - \xi_k)\tau(x)] \neq 0, \quad x \gg 1, \quad \int^{\infty} f_{jk}(x)dx = \infty .$$

We introduce the functions $y^{[k;\varepsilon]}(x)$ which for $\varepsilon = 1$ are just the quasi-derivatives:

$$\begin{aligned} y^{[k;\varepsilon]}(x) &= \varepsilon^{-k} y^{(k)}(x), \quad 0 \leqslant k \leqslant n-1, \\ y^{[n;\varepsilon]}(x) &= \varepsilon^{-n} q_0(x) y^{(n)}(x), \\ y^{[n+k;\varepsilon]}(x) &= \varepsilon^{-2k} q_k(x) y^{[n-k;\varepsilon]}(x) - (y^{[n+k-1;\varepsilon]}(x))', \quad 1 \leqslant k \leqslant n-1, \end{aligned}$$

and put

$$\tilde{y}_j(x,\varepsilon) = l_p^{-1/2}(x, p_j(x)) \exp\left\{\varepsilon^{-1} \int_{x_0}^{x} p_j(t)dt\right\} .$$

Then for any $\varepsilon_0 > 0$ there exists $x(\varepsilon_0) < 0$ such that equation (37) has a solution $y_j(x,\varepsilon)$ for which

$$y_j^{(k)}(x,\varepsilon) = \varepsilon^{-k} p_j^k(x) \tilde{y}_j(x,\varepsilon)[1 + \varepsilon\psi_{jk}(x,\varepsilon)], \quad 0 \leqslant k \leqslant n-1, \tag{38}$$

$$\begin{aligned} y_j^{[k;\varepsilon]}(x,\varepsilon) = (-1)^{n-k} p_j^k(x) q_0(x) \tilde{y}_j(x,\varepsilon) &\left[\sum_{m=0}^{k-1} (-1)^m c_m \xi_j^{-2m} + \phi_{jk}(x)\right] \\ &\times [1 + \varepsilon\psi_{jk}(x,\varepsilon)], \quad n \leqslant k \leqslant 2n-1 . \end{aligned}$$

Here $\phi_{jk}(x) \to 0$ as $x \to \infty$ and, for $x \geqslant x(\varepsilon_0)$ and $0 < \varepsilon \leqslant \varepsilon_0$, there are the uniform bounds

$$|\psi_{jk}(x,\varepsilon)| \leqslant \phi(x), \quad \lim_{x\to\infty} \phi(x) = 0 .$$

The asymptotic fomulae (38) are dual; they are applicable both for $x \to \infty$ uniformly in ε, $0 < \varepsilon < \varepsilon_0$, and as $\varepsilon \to +0$ uniformly in $x \geqslant x(\varepsilon_0)$. If condition 3′) is satisfied for all j then the solutions $y_1, \dots, y_{2n}$ form the F.S.S.

This result is true with even weaker restrictions on the coefficients of equation (29). It is sufficient that conditions 4), 5) and 3′) are satisfied for all j together with the following condition.

1′) For $x \gg 1$ and for any pair j and k, $j \neq k$, there holds one of the bounds

$$0 < a_1 \leqslant |p_j(x)/p_k(x)| \leqslant a_2 < 1\,,$$
$$0 < a_1 \leqslant |p_k(x)/p_j(x)| \leqslant a_2 < 1\,.$$

Condition 1′) replaces conditions 1) and 2), and means that the roots $\xi_j(x)$ of the equation

$$g(x,\xi) \equiv \sum_{k=0}^{n} (-1)^k c_k(x) \xi^{2n-2k} = 0 \tag{39}$$

have the property that for all j and k, $j \neq k$, and for $x \gg 1$

$$0 < \delta_1 \leqslant |\xi_j(x) + \xi_k(x)| \leqslant \delta_2 < \infty\,.$$

3.3 The Equation $ly = \mu y$ with Real Coefficients. We consider the equation

$$ly = \mu y\,, \tag{40}$$

where l is the operator from (29) and all the functions $q_k(x)$ are real. Equations of the form (39) arise in various problems of spectral analysis where μ is the spectral parameter. We assume that

$$\lim_{x\to\infty} q_0(x) = 1\,, \quad \lim_{x\to\infty} |q_n(x)| = \infty$$

and that conditions 1), 2), 4) and 5) of paragraph 2.1 are satisfied. The characteristic equation has the form

$$l(x,p) = (-1)^n \mu\,.$$

The asymptotic formula as $x \to \infty$ for the roots $p_j(x,\mu)$ of this equation, with μ fixed, is

$$p_j(x,\mu) = q_n^{1/2n}(x)[\xi_j(x) + (-1)^n \mu q_n^{-1}(x)[g_\xi(x,\xi_j(x))]^{-1} + O(q_n^{-2}(x))][1+o(1)]\,.$$

We put $\xi'_j = \xi_j$ if $q_n(\infty) = +\infty$ and $\xi'_j = e^{i\pi/2n}\xi_j$ if $q_n(\infty) = -\infty$. Then the points $\xi'_j, \ldots, \xi'_{2n}$ lie symmetrically about the real line in the complex ξ-plane. We index them so that Re $\xi'_j \leqslant$ Re ξ'_{j+1} and take the F.S.S. $\{y_1, \ldots, y_{2n}\}$ consisting of solutions of the form (36). We also fix μ.

1. Suppose that Re $\xi'_j \neq 0$ for all j and, if ξ'_j is not real, then Im $g'(\xi'_j) \neq 0$. Then $y_1, \ldots, y_n \in L_2(\mathbb{R}^+)$ and no non-trivial linear combination of the solutions $y_{n+1}, \ldots, y_{2n}$ belongs to $L_2(\mathbb{R}^+)$.

2. Suppose that Re $\xi_j' = 0$ for $1 \leqslant j \leqslant 2k$ with Re $\xi' \neq 0$ for the remaining l. Suppose also that $g'(\xi_j') \neq g'(\xi_l')$ for $j \neq l$, $1 \leqslant j, l \leqslant 2k$ and $g'(\xi_j') \neq 0$ if ξ_j' is neither real nor purely imaginary. If Im $\mu \neq 0$ then the maximal number of linearly independent solutions of equation (40) is n or $n+k$ depending on whether the integral

$$\int^{\infty} |q_n(x)|^{-1+1/2n} dx$$

diverges or converges.

These results follow from formula (26) and from the asymptotic behaviour of the roots $p_j(x,\mu)$, and they allow us to calculate the deficiency index of the symmetric operator L acting in $L_2(\mathbb{R}^+)$ corresponding to the differential operator l. The deficiency index of L is found in [Fedoryuk 8] for the case where the characteristic equation (32) has asymptotically repeated roots.

3.4 Equations of Odd Order. We consider the equation

$$\begin{aligned} ly \equiv i \sum_{k=0}^{n} (-1)^k \varepsilon^{2k+1} \Bigg\{ \Bigg[\left(\frac{d}{dx}\right)^{k+1} \left(q_{k+1}(x) \left(\frac{d}{dx}\right)^{k} y \right) \Bigg] \\ + \left(\frac{d}{dx}\right)^{k} \left(q_{k+1}(x) \left(\frac{d}{dx}\right)^{k+1} y \right) \Bigg\} = 0 \end{aligned} \tag{41}$$

with complex-valued coefficients of class $C^3(\mathbb{R}^+)$, where $\varepsilon > 0$ is a parameter, $q_1(x) \neq 0$, and $q_{n+1}(x) \neq 0$ for $x \geqslant 0$. The characteristic equation has the form

$$l(x,p) \equiv \sum_{k=0}^{n} (-1)^{k+1} [2q_{k+1}(x) p^{2k+1} + q_{k+1}'(x) p^{2k}] = 0 \,. \tag{42}$$

Let us put $\tau(x) = [q_1(x)/q_{n+1}(x)]^{1/2n}$ and introduce conditions of the same type as in paragraph 2.1. A typical example is equation (41) in which

$$q_{n+1}(x) = 1\,, \quad q_{k+1}(x) = [q(x)]^{2(n-k)}\,, \quad k < n\,, \tag{43}$$

where the function $q(x)$ lies in any of the classes described in paragraph 2.1. The following two conditions mean that the roots $p_j(x)$ of equation (42) are asymptotically simple:

1) $$\lim_{x\to\infty} q_{n-k+1} q_{n+1}^{-1} \tau^{-2k} = c_{2k+1}\,,$$

$$\lim_{x\to\infty} q_{n-k+1}' q_{n+1}^{-1} \tau^{-2k-1} = 0\,.$$

2) The equation $g(\xi) \equiv \sum_{k=0}^{n} (-1)^k c_{2k+1} \xi^{2n-2k+1} = 0$

has no repeated roots.

We note that $c_1 = 1$. As $x \to \infty$ the roots $p_j(x)$ have the asymptotic behaviour

$$p_j(x) = [\xi_j + o(1)]\tau(x), \quad 1 \leqslant j \leqslant 2n,$$

$$p_0(x) = \frac{q_1'(x) + o(q_1'(x))}{q_1(x)},$$

where ξ_j are non-zero roots of the equation $g(\xi) = 0$. The root $p_0(x)$ has smallest growth, that is,

$$\lim_{x\to\infty} p_0(x)/p_j(x) = 0, \quad j \neq 0,$$

which follows from condition 1). The next two conditions, as in paragraph 2.1, guarantee some regularity in the behaviour of the coefficients of equation (41) at infinity:

3) $\displaystyle\int^{\infty} F(x)dx < \infty,$ where

$$F(x) = \sum_{k=0}^{\infty}(|q'_{n-k+1}q_{n+1}^{-1}|^2|\tau|^{-4k-1} + |q''_{n-k+1}q_{n+1}^{-1}|^2|\tau|^{-4k-3}$$
$$+ |q''_{n-k+1}q_{n+1}^{-1}|\,|\tau|^{-2k-1} + |q'''_{n-k+1}q_{n+1}^{-1}|\,|\tau|^{-2k-2});$$

4) $\displaystyle\lim_{x\to\infty} q''_{n-k+1}q_{n+1}^{-1}\tau^{2k-2} = 0.$

These conditions are satisfied, for instance, when the coefficients of equation (41) have the form (43) and $q(x) = ax^{\alpha}$, with $\alpha > -1$ and $a \neq 0$.

We introduce the notation

$$p_j^{(1)}(x) = -\frac{1}{2}\frac{d}{dx}\ln l_p(x, p_j(x)) - [l_p(x, p_j(x))]^{-1}\sum_{k=0}^{n}(-1)^k q'_{k+1}(x)p_j^{2k}(x), \tag{44}$$

$$\tilde{y}_j(x,\varepsilon) = \exp\left\{\varepsilon^{-1}\int_{x_0}^{x} p_j(t)dt + \int_{x_0}^{x} p_j^{(1)}(t)dt\right\}.$$

It follows from conditions 1), 2), and 4) that

$$p_j^{(1)}(x) = o(p_j(x)), \quad x \to \infty,$$

for all $j \neq 0$. Moreover as $x \to \infty$

$$p_j^{(1)}(x) = -\frac{2n+1}{4n}\frac{q_1'(x) + o(q_1'(x))}{q_1(x)}, \quad j \neq 0,$$

$$p_0^{(1)}(x) = o(q_1'(x)/q_1(x)).$$

If the functions Re $(p_j(x) - p_k(x))$ do not change sign for $x \gg 1$ for all j and k then equation (41) has a F.S.S. of the form

$$y_j(x,\varepsilon) = \tilde{y}_j(x,\varepsilon)[1+\varepsilon\psi_j(x,\varepsilon)], \quad 0 \leqslant j \leqslant 2n. \tag{45}$$

There are the uniform bounds

$$|\psi_j(x,\varepsilon)| \leqslant k(x), \quad \lim_{x\to\infty} k(x) = 0,$$

for the remainder terms for $0 < \varepsilon \leqslant \varepsilon_0$ and $x \geqslant x(\varepsilon_0)$, where ε_0 is sufficiently small. The asymptotic behaviour (45) is dual.

We put $\varepsilon = 1$ and suppose that $q_{n+1}(x) \to 1$ as $x \to \infty$. We can simplify (45) to

$$y_j(x) = [q_1(x)]^{-1/2-1/(4n)} \exp\left\{\int_{x_0}^{x} p_j(t)dt\right\} \quad x\to\infty, \quad j = 1,\dots,2n,$$

$$y_0(x) = [q_1(x)]^{-1/2+o(1)}, \quad x \to \infty.$$

3.5 Third-Order Equations. We consider the equation

$$ly \equiv (ib_2y'')' + \left[\left(\frac{i}{2}b_2' + a_1\right)y'\right]' + ib_1y' + \left[\frac{i}{2}b_1' + a_0\right]y = i\mu y \tag{46}$$

on the half-line $x \geqslant 0$, where $a_j(x)$ and $b_j(x)$ are real-valued functions, and $b_2(x) \neq 0$. Then the operator l is formally self-adjoint. Suppose that $\mu \neq 0$ is a real constant. In [Gilbert] the asymptotic behaviour of the F.S.S. of equation (46) as $x \to \infty$ was found and the deficiency indices of the symmetric operator L in $L_2(\mathbb{R}^+)$ corresponding to l were calculated under the following assumptions:

1) $b_1(x), b_2(x) \in C^2(\mathbb{R}^+), \quad a_0(x), a_1(x) \in C^3(\mathbb{R}^+),$
$$\lim_{x\to\infty} b_2(x) = 1, \quad \lim_{x\to\infty} |a_0(x)| = \infty,$$
$$\lim_{x\to\infty} a_1(x)a_0^{-1/3}(x) = d \neq (3/2)^{2/3},$$

and $a_0'(x) \neq 0$ for $x \gg 1$.

2) The functions

$$b_1/a_0^{2/3}, \ b_1'/a_0, \ b_2'/a_0^{1/3}, \ b_2''/a_0^{2/3}, \ a_1'/a_0^{2/3}, \ b_1''/a_0^{4/3}, \ a_0'/a_0^{4/3}$$

tend to zero as $x \to \infty$.

The characteristic equation has the form

$$l(x,p) \equiv -p^3b_2(x) + \left(ia_1(x) - \frac{1}{2}b_2'(x)\right)p^2 - b_1(x)p + \left[ia_0(x) - \frac{1}{2}b_1'(x) + \mu\right] = 0.$$

Let us consider the equation

$$\eta^3 - d\eta^2 + 1 = 0 .$$

The roots of this equation are distinct for $d \neq (3/2)^{2/3}$ and non-zero. If $d < (3/2)^{2/3}$ then there is a real root η_1 and two complex conjugate roots $\eta_2 = \bar{\eta}_3$, where Im $\eta_2 > 0$. If $d > (3/2)^{2/3}$ then all three roots are real: $\eta_1 < \eta_2 < \eta_3$. It follows from conditions 1) and 2) that the roots of equation (46) have asymptotic behaviour

$$p_k(x) = ia_0^{1/3}(x)[\eta_k + o(1)] , \quad x \to \infty . \tag{47}$$

The following conditions are restrictions on the regularity of the behaviour of the coefficients of equation (46) at infinity.

3) The functions

$$|b_2'|, \quad |b_1'/a_0^{2/3}|, \quad |b_2'^2/a_0^{1/3}|, \quad |b_2''^2/a_0|, \quad |a_1'^2/a_0|,$$

$$|b_1'^2/a_0^{5/3}|, \quad |b_1''^2/a_0^{7/3}|, \quad |a_0'/a_0^{7/3}|, \quad |b_2''/a_0^{1/3}|,$$

$$|b_2'''/a_0^{2/3}|, \quad |a_1''/a_0^{2/3}|, \quad |b_1'''/a_0^{4/3}|, \quad |a_0''/a_0^{4/3}|$$

lie in the space $L_1[r, \infty)$ for $r \gg 1$.

4) For any j, k and $x \gg 1$ one of the conditions is satisfied:

a) Re $[p_j(x) - p_k(x)] \geqslant 0$;

b) Re $[p_j(x) - p_k(x)] \leqslant 0$, $\int^\infty$ Re $[p_j(x) - p_k(x)]dx = \infty$;

c) $|\int^\infty$ Re $[p_j(x) - p_k(x)]dx| < \infty$.

Then equation (46) has F.S.S. of the form

$$y_k(x) = [1 + o(1)]a_0^{-1/3}(x) \exp\left\{\int_{x_0}^{x} p_k(t)dt\right\} , \quad x \to \infty . \tag{48}$$

Example. Let

$$b_2(x) \equiv 1 , \quad b_1(x) = ax^\alpha , \quad a_1(x) = bx^{\gamma/3} , \quad a_0(x) = cx^\gamma ,$$

where $c \neq 0$, $\gamma > 0$, $\alpha < 2\gamma/3$, and $(b/c)^{1/3} \neq (3/2)^{2/3}$. Then conditions 1) – 3) are satisfied. If $(b/c)^{1/3} < (3/2)^{2/3}$ then condition 4) is satisfied. In this case we can obtain an asymptotic expansion for the roots $p_j(x)$ [Gilbert].

§ 4. Systems of Equations on an Infinite Interval

1. Systems of First-Order Equations. In Chap. 2, § 5 we considered systems close to diagonal ones and constructed the asymptotic behaviour of their fundamental matrices. Here we consider a more general class of system.

1.1 Systems with Asymptotically Simple Roots. We consider the system of n equations

$$y' = A(x)y \tag{1}$$

on the half-line $\mathbb{R}^+ = [0, \infty)$, where $A(x) \in C^2(\mathbb{R}^+)$. Suppose that $q(x)$ is a complex-valued function, $q(x) \neq 0$ for $x \geqslant 1$ and $Q(x) = \operatorname{diag}\{[q(x)]^{\alpha_1}, \ldots, [q(x)]^{\alpha_n}\}$. We formulate conditions on the matrix $A(x)$:

1) $A(x) = q(x)Q(x)B(x)Q^{-1}(x)$, $x \geqslant 0$.

2) The limit $\lim_{x\to\infty} B(x) = B(\infty)$ exists and is finite, and $B(\infty)$ is non-singular and has no repeated eigenvalues.

3) $\lim_{x\to\infty} r_1(x) = 0$, where $r_1 = |q'|\,|q|^{-2} + |q|^{-1}\|B'(x)\|$.

4) $r_2(x) \in L_1[0, \infty)$, where

$$\begin{aligned} r_2 = {} & |q'|^2|q|^{-3} + |q''|\,|q|^{-2} + |q'|\,|q|^{-2}\|B'(x)\| \\ & + |q|^{-1}(\|B'(x)\|^2 + \|B''(x)\|)\,. \end{aligned}$$

Here $\|B(x)\| = \max_{1\leqslant j,k\leqslant n} |b_{jk}(x)|$, and $B'(x)$ is the matrix with elements $b'_{ij}(x)$.

It follows from conditions 1) and 2) that the eigenvalues $p_j(x)$ of $A(x)$ have the form

$$p_j(x) = q(x)\eta_j(x)\,,$$

where the $\eta_j(x)$ are the eigenvalues of $B(x)$. The eigenvalues $p_j(x)$ are asymptotically simple since for all $j \neq k$ there are the limits

$$\lim_{x\to\infty} p_j(x)/p_k(x) = c_{jk} \neq 0, 1, \infty\,. \tag{2}$$

Remarks. 1. If we reduce equation (12) of § 3 to a system of n first-order equations in the standard way then this system satisfies conditions 1) and 2), and further

$$Q(x) = \operatorname{diag}(1, q(x), q^2(x), \ldots, q^n(x))\,.$$

2. Systems (1) satisfying conditions 1) and 2) are, in a certain sense, typical systems. We put $d(x) = \det A(x)$. Suppose that $d(x) \to \infty$ as $x \to \infty$, and that the elements of $A(x)$ behave as powers of $d(x)$ as $x \to \infty$:

$$a_{ij}(x) = [c_{ij} + o(1)]d^{\alpha_{ij}}(x)\,.$$

Suppose also that the matrix $A(x)$ is non-singular in the Volevich sense, that is, in the decomposition of the determinant

$$\det A(x) = \sum \pm a_{1j_1}(x) \dots a_{nj_n}(x)$$

not all the principal terms cancel out. Then

$$A(x) = Q_1(x)B(x)Q_2(x),$$

where

$$Q_1(x) = \operatorname{diag}(d^{s_1}, \dots d^{s_n}), \quad Q_2(x) = \operatorname{diag}(d^{t_1}, \dots, d^{t_n}),$$

and further $\sum_{j=1}^{n}(s_j + t_j) = 1$, the matrix $B(\infty)$ exists, and is finite and non-singular. The eigenvalues of the matrix $A(x)$ are the same as the eigenvalues of the matrix $d^{1/n}BQ_2Q_1d^{-1/n}$ and it follows from condition (2) that $s_j + t_j = 1/n$, so that $A(x)$ satisfies condition 1).

A particular case of the form (1) are systems studied in the analytic theory of differential equations with the irregular singular point $x = \infty$:

$$y' = x^r \sum_{k=0}^{\infty} A_k x^{-k} y.$$

Here $r \geqslant 0$ is an integer, A_k is a constant matrix, the matrix A_0 is non-singular with distinct eigenvalues, and the series converges in the domain $|x| > R$.

Conditions 3) and 4) are the standard requirements for the "regularity" of the behaviour of the coefficients of $A(x)$ at infinity. For instance, if $a_{jk}(x) \sim c_{jk}x^\alpha$ as $x \to \infty$ with $\alpha > -1$, then conditions 3) and 4) follow from conditions 1) and 2).

We denote by $e_j(x)$ and $e_j^*(x)$ the right and left eigenvectors of $A(x)$ corresponding to the eigenvalue $p_j(x)$. Recall that $e_j(x)$ is a column vector and $e_j^*(x)$ is a row vector. Let

$$p_j^{(1)}(x) = -\frac{e_j^*(x)e_j'(x)}{e_j^*(x)e_j(x)}, \quad \tilde{y}_j(x) = \exp\left\{\int_{x_0}^{x} [p_j(t) + p_j^{(1)}(t)]dt\right\}. \tag{3}$$

Since the matrix $B(\infty)$ has no repeated eigenvalues there exists $x_0 > 0$ such that for $x \geqslant x_0$ there are bases $\{f_1(x), \dots, f_n(x)\}$ and $\{f_1^*(x), \dots, f_n^*(x)\}$ of class C^2 consisting of the right and left eigenvectors of $B(x)$. These vectors can be chosen so that they all have finite non-zero limits as $x \to \infty$. We put

$$e_j(x) = Q(x)f_j(x), \quad e_j^*(x) = f_j^*(x)Q^{-1}(x).$$

As in § 3, $p_j^{(1)}(x) = o(p_j(x))$ as $x \to \infty$.

Suppose that for some j and for all k the functions

$$\phi_{jk}(x) = \operatorname{Re}\,[p_j(x) - p_k(x) + p_j^{(1)}(x) - p_k^{(1)}(x)]$$

do not change sign for $x \gg 1$. Then the system (1) has a solution of the form

$$y_j(x) = \tilde{y}_j(x)[e_j(x) + \sum_{k=1}^{n} u_{jk}(x)e_k(x)], \quad \lim_{x\to\infty} u_{jk}(x) = 0. \tag{4}$$

If the condition on the functions $\phi_{jk}(x)$ is satisfied for all j then system (1) has fundamental matrix of the form

$$Y(x) = T_0(x)[I + U(x)] \exp\left\{\int_{x_0}^{x} [\Lambda(t) + \Lambda^{(1)}(t)dt]\right\}, \tag{5}$$

where $T_0(x) = (e_1(x), \dots, e_n(x))$, and Λ and $\Lambda^{(1)}$ are the diagonal matrices with the diagonal elements $p_j(x)$ and $p_j^{(1)}(x)$, $1 \leqslant j \leqslant n$. Also $\lim_{x\to\infty} \|U(x)\| = 0$.

Condition 2) can be weakened; the matrix $B(\infty)$ can have one simple non-zero eigenvalue (this case is similar to that considered in § 3, paragraph 2.2).

Remark. The vector-function

$$z(x) = \exp\left\{\int_{x_0}^{x} p_j^{(1)} dt\right\} e_j(x)$$

is invariant in the sense that it does not depend on the choice of eigenvectors $e_j(x)$, $e_j^*(x)$ up to a constant multiplier.

1.2 Dual Asymptotic Behaviour. We consider the system

$$y' = \lambda A(x)y, \tag{6}$$

for which conditions 1) – 4) are satisfied, and put

$$\tilde{y}_j(x, \lambda) = \exp\left\{\lambda \int_{x_0}^{x} p_j(t) + \int_{x_0}^{x} p_j^{(1)}(t)dt\right\}. \tag{7}$$

We select the eigenvectors $e_j(x)$ and $e_j^*(x)$ in the same way as in paragraph 1.1. Let

$$\int_0^\infty |q(x)|dx = \infty, \tag{8}$$

$$|\mathrm{Re}\,(p_j(x) - p_k(x))| \geqslant c|q(x)| > 0, \quad x \gg 1, \tag{9}$$

for some j and for all $k \neq j$. Then the system (6) has a solution of the form

$$y_j(x, \lambda) = \tilde{y}_j(x, \lambda)\left[e_j(x) + \lambda^{-1} \sum_{k=1}^{n} u_{jk}(x)e_k(x)\right] \tag{10}$$

and for any $\lambda_0 > 0$ there exists $x(\lambda_0) < \infty$ such that

$$|u_{jk}(x, \lambda)| \leqslant k_j(x), \quad \lim_{x\to\infty} k_j(x) = 0 \tag{11}$$

for $x \geqslant x(\lambda_0)$ and $\lambda > \lambda_0$. Thus the asymptotic behaviour (10) is dual.

If condition (9) is also satisfied for all j and k, $j \neq k$, then system (6) has a fundamental matrix of the form

$$Y(x, \lambda) = T_0(x)(I + \lambda^{-1}U)\exp\left\{\int_{x_0}^{x} [\lambda\Lambda(t) + \Lambda^{(1)}(t)]dt\right\}, \tag{12}$$

$$\|U(x, \lambda)\| \leqslant k(x), \quad \lim_{x\to\infty} k(x) = 0. \tag{13}$$

The matrices T_0, Λ, and $\Lambda^{(1)}$ are the same as in formula (5), and the bound for $\|U(x)\|$ holds for $\lambda \geqslant \lambda_0$ and $x \geqslant x(\lambda_0)$.

Under the conditions given, formulae (10) – (12) remain true when λ lies in some sector $S : \alpha \leqslant \arg \lambda \leqslant \beta$, $0 < \beta - \alpha < 2\pi$, of the complex λ-plane. Suppose that there exists

$$\lim_{x\to\infty} \arg q(x) = \phi_0, \quad 0 \leqslant \phi_0 < 2\pi.$$

Then we can specify the sector S more precisely. Suppose that η_j^0 are eigenvalues of $B(\infty)$. We put $\eta_j^* = \eta_j e^{i\phi_0}$, fix j and draw the lines

$$\text{Re}\,[e^{i\phi}(\eta_j^* - \eta_k^*)] = 0; \quad \phi = \arg \lambda,$$

in the complex λ-plane. These lines divide the λ-plane into open sectors $S_1, \ldots, S_N$. Let $\tilde{S}_j$ be a closed sector lying inside the sector S_j. If condition (8) is satisfied then formulae (10) and (11) are true for $\lambda \in \tilde{S}_j$ since condition (9) is satisfied.

We state other sufficient conditions under which formula (12) is true:

(a) all the eigenvalues $p_j(x)$ are purely imaginary;

(b) all the integrals $\int^{\infty} |\text{Re}\, p_j^{(1)}(x)|dx$ converge.

1.3 Bounds for the Remainder Terms. Some of the bounds have been derived in Chap. 2, § 5. Suppose that conditions 1) – 4), (8) and (9) are satisfied for some j. Suppose that for $x \gg 1$ there are the bounds

$$|q'| \leqslant |q|^{\gamma_1}, \quad |q|' \geqslant c|q|^{\gamma_2}, \quad |q''| \leqslant c|q|^{\gamma_3},$$
$$\|B'\| \leqslant c|q'|\,|q|^{-1}, \quad \|B''\| \leqslant c(|q''| + |q'|^2)|q|^{-1},$$

where $c > 0$ is constant and

$$\gamma < 2, \quad \gamma_2 < 2, \quad \gamma_2 < 1 - \gamma, \quad \gamma = \max(2\gamma_1 - 2, \gamma_3 - 1) - \gamma_2.$$

Then in formula (11) we can put

$$k_j(x) = c_j\|q\|^{\gamma_0}, \quad \gamma_0 = \max(\gamma, \gamma_1 - 2).$$

The bounds for $\|B'\|$ and $\|B''\|$ hold when, for instance,

$$B(x) = B_0 + q^{-\alpha}(x)B_1 + \dots,$$

where $\alpha > 0$ and $B_0, B_1, \dots$ are constant matrices.

Let $q(x) = ax^\alpha$, where $\alpha > 0$ and $a \neq 0$. Then we can put

$$\gamma_1 = \gamma_2 = 1 - 1/\alpha, \quad \gamma_3 = 1 - 2/\alpha, \quad \gamma = -1 - 1/\alpha = \gamma_0,$$

so that $k_j(x) = O(x^{-1-1/\alpha})$. Again, let $q(x) = \exp\{ax^\alpha\}$, where $a > 0$ and $\alpha > 0$. Then we can put

$$\gamma_1 = \gamma_2 = \gamma_3 = 1 + \varepsilon,$$

where $\varepsilon > 0$ is arbitrarily small, so that

$$k_j(x) = O(\exp\{(-1+\varepsilon)x^\alpha\}).$$

1.4 Almost Diagonal Systems. We consider the system of n equations

$$y' = [\Lambda(x) + V(x)]y, \quad \Lambda(x) = \operatorname{diag}(p_1(x), \dots, p_n(x)), \tag{14}$$

on the half-line $x \geqslant 0$, where $V(x)$ is continuous. In Chap. 2, § 5 sufficient conditions were given in Levinson's theorem under which we can find the asymptotic behaviour of the fundamental matrix of system (13) as $x \to \infty$. The following theorem gives alternative sufficient conditions.

Theorem (Hartman-Wintner). *Let $\|V(x)\| \in L_2(0, \infty)$ and for all j and k, $j \neq k$, let*

$$|\operatorname{Re}(p_j(x) - p_k(x))| \geqslant c > 0, \quad x \geqslant x_0.$$

Then system (13) has a fundamental matrix of the form

$$Y(x) = [I + o(1)] \exp\left\{\int_{x_0}^{x} [\Lambda(t) + \operatorname{diag} V(t)]dt\right\}.$$

Let us consider the system

$$y' = [A + V(x) + R(x)]y, \tag{15}$$

where A is a constant matrix with distinct eigenvalues $p_1, \dots, p_n$, and the matrix functions $V(x)$ and $R(x)$ are continuous for $x \geqslant 0$. We state sufficient conditions under which system (15) has a fundamental matrix of the form

$$Y(x) = [T + o(1)] \exp\left\{\int_{x_0}^{x} \Lambda(t)dt\right\}, \quad x \to \infty. \tag{16}$$

Here $\Lambda(x)$ has the form (14) where the $p_j(x)$ are the eigenvalues of the matrix $A + V(x)$, and the matrix T reduces A to diagonal form: $T^{-1}AT = \operatorname{diag}(p_1, \dots, p_n) \equiv \Lambda$.

Let

$$S_{kj}(t,x) = \operatorname{Re} \int_t^x [p_k(s) - p_j(s)]ds \,.$$

We say that S_{kj} lies in the class H_1 if

$$S_{kj}(x_0, \infty) = \infty \,, \quad S_{kj}(t,x) \geqslant -c \,, \quad x_0 \leqslant t \leqslant x \,,$$

and S_{kj} lies in the class H_2 if

$$S_{kj}(t,x) \leqslant c \,, \quad x_0 \leqslant t \leqslant x \,.$$

Here c is a constant and $x_0 \geqslant 0$ is a sufficiently large number. Then the conditions to be imposed on (15) in order that (16) holds are as follows.

1) All functions S_{kj} lie in either H_1 or H_2;

$$\lim_{x \to \infty} V(x) = 0 \,, \quad \int^{\infty} (\|V'(x)\| + \|R(x)\|)dx < \infty \,.$$

Examples. 1. Let us consider system (15) where

$$A = \begin{bmatrix} 0 & 0 \\ 0 & 2 \end{bmatrix} , \quad V(x) = x^{-\alpha} \begin{bmatrix} 0 & 1 \\ 1 & 0 \end{bmatrix} , \quad R(x) = 0 \,, \quad 0 < \alpha \leqslant 1 \,.$$

The eigenvalues of the matrix $A + V(x)$ are

$$p_1(x) = 1 - \sqrt{1 + x^{-2\alpha}} = -\frac{1}{2} x^{-2\alpha} + O(x^{-4\alpha}) \,,$$
$$p_2(x) = 1 + \sqrt{1 + x^{-2\alpha}} = 2 + O(x^{-2\alpha}) \,,$$

so that $\operatorname{Re}[p_1(x) - p_2(x)] = 2 + O(x^{-2\alpha})$. The Hartman-Wintner theorem is applicable for $\alpha > 1/2$; in this case

$$Y(x) = [I + o(1)] \begin{bmatrix} 1 & 0 \\ 0 & e^{2x} \end{bmatrix} , \quad x \to \infty \,.$$

Since $\|V'(x)\| = O(x^{-\alpha-1})$ as $x \to \infty$, condition 1) is satisfied and, for $0 < \alpha \leqslant 1$, we have

$$Y(x) = [I + o(1)] \begin{bmatrix} \exp\left\{\int^x p_1(t)dt\right\} & 0 \\ 0 & \exp\left\{\int^x p_2(t)dt\right\} \end{bmatrix} .$$

In particular for $\alpha = 1/2$

$$Y(x) = [I + o(1)] \begin{bmatrix} x^{-1/2} & 0 \\ 0 & x^{1/2} e^{2x} \end{bmatrix} .$$

Another set of conditions which can be imposed on (15) is as follows.

2) There exists an integer $k > 0$ such that $V^{(k)}(x)$ and $R(x)$ are continuous for $x \geqslant 0$, and

$$\lim_{x\to\infty} V^{(j)}(x) = 0, \quad 0 \leqslant j \leqslant k-2;$$

$$\|V^{(j)}(x)\| \in L_2(0,\infty), \quad 1 \leqslant j \leqslant k-1.$$

The functions $\|V(x)\|$, $\|V'(x)\|$, $\|V^{(k)}(x)\|$, $\|R(x)\|$ lie in the space $L_1(0,\infty)$ and all the S_{kj} lie in either H_1 or H_2.

Condition 1) is a special case of condition 2).

2. As an example, we consider the system

$$y' = \begin{bmatrix} i & 0 \\ 0 & -i \end{bmatrix} y + v(x) \begin{bmatrix} 0 & 1 \\ 1 & 1 \end{bmatrix} y, \quad v(x) = x^{-\beta} \sin x^{1-\alpha}.$$

Let $1/2 < \beta < 1$ and $(1-\beta)/(k+1) < \alpha \leqslant (1-\beta)/k$ where $k \geqslant 1$ is an integer. We have

$$p_{1,2}(x) = \pm i\sqrt{1 - v^2(x)}, \quad \operatorname{Re}\,[p_1(x) - p_2(x)] = 0, \quad x \gg 1,$$

and we can check that the other requirements of condition 2) are satisfied. The system has a fundamental matrix of the form

$$Y(x) = [I + o(1)] \begin{bmatrix} \exp\{ix + if(x)\} & 0 \\ 0 & \exp\{ix - if(x)\} \end{bmatrix},$$

where

$$f(x) = \frac{1}{2} \int_{x_0}^{x} t^{-2\beta} \sin^2 t^{1-\alpha} dt.$$

Observe that $v^{(j)}(x) \notin L_1[1,\infty)$ for $0 \leqslant j \leqslant k-1$.

We move on to further sets of conditions on (15).

3) The real parts of the eigenvalues of the matrix A are distinct and the matrix $V'(x)$ is continuous for $x \geqslant 0$ with

$$\lim_{x\to\infty} V(x) = 0, \quad [\|V'(x)\|^2 + \|V(x)\|\,\|V'(x)\|] \in L_1[0,\infty).$$

3. Let us consider the system

$$y' = \begin{bmatrix} 1 & 0 \\ 0 & -1 \end{bmatrix} y + v(x) \begin{bmatrix} 0 & 1 \\ 1 & 0 \end{bmatrix} y,$$

where $v(x)$ is the function in Example 2 with $\beta = 1/2$. In this case $v(x) \notin L_2(1,\infty)$, that is, the conditions of the Hartman-Wintner theorem are not satisfied, but

$$v'(x) \in L_2(1,\infty), \quad v(x)v'(x) \in L_1(1,\infty),$$

and all the requirements of condition 3) are satisfied.

4) $V(x) = V_1(x) + V_2(x)$, where the matrices $V_1(x)$ and $V_1'(x)$ are absolutely continuous on each finite interval $0 \leqslant a < x < b$ and

$$[\|V_1'(x)\| + \|V_2'(x)\|^2 + \|V_2''(x)\| + \|V_2'(x)\|\,\|V(x)\| + \|R(x)\|] \in L_1(0,\infty).$$

For some k and for all $j \neq k$ all the functions S_{kj} lie in either class H_1 or class H_2. Then system (15) has solutions

$$y_k(x) = \exp\left\{\int_{x_0}^{x} p_k(t)dt\right\}[e_k + o(1)], \quad x \to \infty, \tag{17}$$

where e_k is the eigenvector of the matrix A corresponding to the eigenvalue p_k. If the last condition is satisfied for all k then system (15) has a fundamental matrix of the form (16).

4. This example shows that the class of systems satisfying condition 4) is wider than the class of systems satisfying the conditions of Levinsons's theorem (Chap. 2, § 5). Let us consider the function

$$v(x) = -\int_{x^\alpha}^{\infty} \frac{\sin t}{t}dt, \quad \frac{1}{2} < \alpha < 1.$$

Clearly $v(x) \to 0$ as $x \to \infty$, but

$$v(x) = -x^{-\alpha}\cos x^\alpha + O(x^{-2\alpha}),$$

so that $v(x) \notin L_1(1,\infty)$. Further

$$v'(x) = \alpha x^{-1}\sin x^\alpha \notin L_2(1,\infty),$$

$$|v'(x)|^2 \in L_1(1,\infty), \quad v'(x)v(x) \in L_1(1,\infty), \quad v''(x) \in L_1(1,\infty).$$

Let us now give a generalization of the Hartman-Wintner theorem. We consider situations where there exists a matrix $Q(x)$ with zero diagonal elements such that

$$[I + Q(x)]^{-1}T^{-1}[A + V(x)]T[1 + Q(x)] = \Lambda(x),$$

where $\Lambda(x)$ has the form (14), and such a matrix $Q(x)$ is unique. Denote

$$\tilde{\Lambda}(x) = \Lambda(x) - \text{diag}\,[(I + Q(x))^{-1}Q'(x)].$$

5) There is an integer $k > 0$ such that the matrices $V^{(k)}(x)$ and $R(x)$ are continuous for $x \geqslant 0$, all the $S_{jk}(x)$ belong to $H_1 \cup H_2$, and

$$\lim_{x\to\infty} V^{(j)}(x) = 0, \quad 0 \leqslant j \leqslant k-2;$$

$$\|V^{(j)}(x)\| \in L_2(0,\infty), \quad 1 \leqslant j \leqslant k-1;$$

$$[\|V^{(k)}(x)\| + \|R(x)\|] \in L_1(0,\infty).$$

Then system (15) has a fundamental matrix of the form

$$Y(x) = [T + o(1)] \exp\left\{\int_{x_0}^{x} \tilde{\Lambda}(t)dt\right\}, \quad x \to \infty.$$

5. As an example, we consider the system

$$y' = \begin{bmatrix} i & 0 \\ 0 & -i \end{bmatrix} y + 2\begin{bmatrix} 0 & -x^{-1/4}\sin x^{1/2} \\ x^{-1/4}\cos x^{1/2} & 0 \end{bmatrix} y.$$

The eigenvalues $p_j(x)$ are

$$p_{1,2}(x) = \pm i\left(1 + x^{-1/2}\sin 2x^{1/2} - \frac{1}{4}x^{-1}\sin^2 2x^{1/2}\right) + O(x^{-3/2}).$$

In this case $\|V(x)\| \notin L_1(1,\infty)$ and $\|V(x)\|\,\|V'(x)\| \notin L_1(1,\infty)$, but $\|V''(x)\| \in L_1(1,\infty)$ and $\|V'(x)\| \in L_2(1,\infty)$. Further

$$\begin{aligned} \tilde{p}_1(x) &= i\left(1 + x^{-1/2}\sin 2x^{1/2} - \frac{1}{4}x^{-1}\sin^2 2x^{1/2}\right) \\ &\quad + \frac{1}{8x}\sin^2 x^{1/2} + O(x^{-3/2}), \\ \tilde{p}_2(x) &= -i\left(1 + x^{-1/2}\sin 2x^{1/2} - \frac{1}{4}x^{-1}\sin^2 2x^{1/2}\right) \\ &\quad - \frac{1}{8x}\cos^2 x^{1/2} + O(x^{-3/2}), \end{aligned}$$

so that $\operatorname{Re}[p_1(\tilde{x}) - \tilde{p}_2(x)] = 4x^{-1} + O(x^{-3/2})$, and condition 5) is satisfied.

1.5 Systems with Asymptotically Repeated Roots. We consider system (15) on the half-line $x \geqslant 0$. Suppose that the degree of the minimum polynomial $f(A)$ of the matrix A is n, so that it has the form

$$f(p) = \prod_{k=1}^{m}(p - p_k)^{n_k}, \quad \sum_{k=1}^{m} n_k = n,$$

and $p_j \neq p_k$ for $j \neq k$. In this case the Jordan normal form of the matrix A consists of m Jordan blocks of sizes $n_1, \ldots, n_m$ and the matrix A has precisely m linearly independent eigenvectors $e_1, \ldots, e_m$: $Ae_j = p_j e_j$.

Put $r + 1 = \max n_k$. Suppose that the matrix $V(x)$ is absolutely continuous on each finite interval $I \subset \mathbb{R}^+$ and that the matrix $R(x)$ is Lebesgue measurable. Let

$$\lim_{x\to\infty} V(x) = 0, \quad [x^r\|V'(x)\| + x^r\|R(x)\|] \in L_1(0,\infty).$$

Let $p_{jk}(x)$ be the eigenvalues of the matrix $A + V(x)$, $1 \leqslant j \leqslant m$, $1 \leqslant k \leqslant n_j$. Then we introduce the conditions:

1) The functions $p_{jk}(x)$ are absolutely continuous for $x \gg 1$ and can be chosen so that

$$\lim_{x\to\infty} p_{jk}(x) = p_j\,, \quad 1 \leqslant j \leqslant m\,, \quad 1 \leqslant k \leqslant n_j\,.$$

2) $\int^{\infty} x^r |p'_{jk}(x)| dx < \infty$ for all admissible j and k.

Condition 1) is highly restrictive. It is equivalent to the requirement that there exists a matrix $Q(x)$ with $\lim_{x\to\infty} Q(x) = 0$ such that

$$(I + Q(x))^{-1}(A + V(x))(I + Q(x)) = \operatorname{diag}\,(\Lambda_1(x), \ldots, \Lambda_m(x))\,,$$

where the $\Lambda_j(x)$ are the Jordan blocks

$$\Lambda_j(x) = \begin{bmatrix} p_j(x) & 1 & & 0 \\ & \ddots & \ddots & \\ & & \ddots & 1 \\ 0 & & & p_j(x) \end{bmatrix}.$$

We put $\mu_{jp}(x) = p_{j1}(x) + (q-1)/x$ and suppose that all the functions $S_{\alpha\beta}$ lie in one of the classes H_1, H_2. Then the system (15) has a F.S.S. of the form

$$y_{jk}(x) = x^{j-1} \exp\left\{\int_x^x p_{jk}(t)dt\right\} [e_j + o(1)]\,, \qquad (18)$$
$$1 \leqslant j \leqslant m\,, \quad 1 \leqslant k \leqslant n_j\,.$$

The results of paragraphs 1.4 and 1.5 were given in the papers [Devinatz 1–4, Harris 1 and 2].

1.6 Canonical and Hamiltonian Systems. We next consider systems of the form

$$Jy' = H(x)y \qquad (19)$$

on the real line, where $H(x) \in C^2(\mathbb{R})$. The system (19) is called *canonical* if $H(x)$ is a real symmetric matrix and J is a constant non-singular skew-symmetric matrix:

$$H^T(x) = H(x)\,, \quad J^T = -J\,.$$

The number of equations in a canonical system is even, and we denote it by $2n$. With a substitution $y = Tz$, where T is a constant non-singular real matrix, the canonical system can be reduced to a form in which

$$J = J_{2n} = \begin{bmatrix} 0 & I_n \\ -I_n & 0 \end{bmatrix}.$$

Here 0 and I_n are the zero and unit $n \times n$ matrices, and $J^2 = -I_{2n}$. In what follows we take $J = J_{2n}$ in the system (19).

A system of the form (19) is called *Hamiltonian* if $H(x)$ is hermitian and J is constant and skew-hermitian:

$$H^\star(x) = H(x), \quad J^\star = -J.$$

Here the symbol * denotes the operation of hermitian conjugation: if $H(x) = (h_{jk}(x))$ then $H^\star(x) = (\overline{h_{kj}(x)})$. A Hamiltonian system contains an even number of equations (denoted by $2n$) and we can reduce it to the form

$$-iGy' = H(x)y, \tag{20}$$

where $H(x)$ is hermitian and

$$G = \begin{bmatrix} I_p & 0 \\ 0 & -I_q \end{bmatrix}, \quad p + q = 2n.$$

We state some well-known properties of canonical and Hamiltonian systems.

1. Suppose that $Y(x)$ is a fundamental matrix of a Hamiltonian system. Then

$$Y^\star(x)JY(x) \equiv \text{const}.$$

A real square matrix A of order $2n$ is called *symplectic* if

$$A^T J_{2n} A = J_{2n}.$$

2. Suppose that $Y(x)$ is a real fundamental matrix of a canonical system. Then the matrix $Y(x)Y^{-1}(x_0)$ is symplectic.

As in § 2 the scalar product of the vectors $y = (y_1, \dots, y_k)^T$ and $z = (z_1, \dots, z_k)^T$ is defined by the formula

$$(y, z) = \sum_{j=1}^{k} z_j y_j.$$

3. If $y(x)$ and $z(x)$ are solutions of a Hamiltonian system then

$$(Jy(x), \overline{z(x)}) \equiv \text{const}.$$

Properties 2 and 3 are corollaries of property 1. If $J = J_{2n}$ and $y(x)$, $z(x)$ are real solutions of a canonical system then the identity of property 3 becomes

$$\sum_{k=1}^{n} [y_k(x)z_{n+k}(x) - y_{n+k}(x)z_k(x)] \equiv \text{const}.$$

Let us consider the matrix bundles $L(x,p) = H(x) - pJ$ associated with canonical and Hamiltonian systems where $J = J_{2n}$ in the case of the canonical system and $J = -iG$ in the case of a Hamiltonian system. Fix $x_0 \geqslant 0$.

4. If p is an eigenvalue of the canonical bundle L then $-p$, $\overline{p}$, and $-\overline{p}$ are also eigenvalues. If p is an eigenvalue of the Hamiltonian bundle L then $\overline{p}$ is also an eigenvalue.

The right and left eigenvectors of a bundle L are connected by the following relationship. We assume that e is a right eigenvector of the vector bundle L, that is $He = pJe$.

5. If L is a canonical bundle then $e^T H = -pe^T J$. If L is a Hamiltonian bundle then $\overline{e^T} H = -\overline{p}\overline{e^T} J$.

Eigenvectors of the bundles are skew-orthogonal. Assume that $He = pJe$, $Hf = qJf$.

6. If L is a canonical bundle then

$$(Je,\, f) = 0\,, \quad p+q \neq 0\,.$$

If L is a Hamiltonian bundle then

$$(Je,\, \overline{f}) = 0\,, \quad p+q \neq 0\,.$$

We will consider a Hamiltonian bundle L under the following assumptions:

1) The eigenvalues of L are distinct and non-vanishing for $x \geqslant 0$.

For $x = 0$ all the roots of the bundle lying in the first quadrant $\operatorname{Re} p \geqslant 0$, $\operatorname{Im} p \geqslant 0$ can be divided into three groups:

I. Purely imaginary roots.

II. Real roots.

III. Complex roots ($\operatorname{Re} p > 0$, $\operatorname{Im} p > 0$).

For $n = 2$ the roots lie in either group I or group II. If l_j is the number of roots in the jth-group then $2l_1 + 2l_2 + 4l_3 = 2n$. It follows from property 4 and the conditions on the bundle L that, for all $x \geqslant 0$, the roots are in the same group (that is, if $\operatorname{Re} p(0) = 0$, $\operatorname{Im} p(0) > 0$ then $\operatorname{Re} p(x) = 0$, $\operatorname{Im} p(x) > 0$ for $x \geqslant 0$ for the roots of the first group, and so on).

The orthogonality relations in property 6 are also true for all $x > 0$. We consider one of the most important cases which arises in applications.

2) The eigenvalues of the bundle L are purely imaginary. We can then represent them in the form $iq_1(x)$, $-iq_1(x), \dots, -iq_n(x)$, where $q_j(x) > 0$ for $x \geqslant 0$. We denote these eigenvalues by $p_1(x), \dots, p_m(x)$ so that $p_{2k}(x) = \overline{p_{2k-1}(x)}$. The right and left eigenvectors of the bundle L can also be grouped in pairs:

$$\{e_1(x), \overline{e_1(x)}, \dots, e_n(x), \overline{e_n(x)}\}\,,$$

$$\{\overline{e_1^T(x)}, e_1^T(x), \dots, \overline{e_n^T(x)}, e_n^T(x)\}\,.$$

They are indexed in the same way as the roots $p_j(x)$. Let us introduce the matrices

$$\Lambda(x) = \operatorname{diag}\,(p_1(x), \dots, p_{2n}(x)),$$
$$T_0(x) = (e_1(x), \overline{e_1(x)}, \dots, e_n(x), \overline{e_n(x)});$$

then $T_0^{-1}(x)JT_0(x) = -\Lambda(x)$. We normalize the vectors $e_j(x)$ so that

$$e_j^\star(x)Je_j(x) = 1, \quad x \geqslant 0. \tag{21}$$

Then

$$T_0^\star(x) = (\overline{e_1^T(x)}, e_1^T(x), \dots, \overline{e_n^T(x)}, e_n^T(x)),$$
$$T_0^\star(x)JT_0(x) = I_{2n}.$$

We consider first the simplest conditions on the behaviour of $H(x)$ as $x \to \infty$.

3) The limit $H(\infty)$ exists and is finite,

$$\det H(\infty) \neq 0, \quad \int_0^\infty \|H'(x)\|dx < \infty.$$

Then the matrix $T_0(x)$ can be chosen so that there is a finite limit $T_0(\infty)$, and

$$\det T_0(\infty) \neq 0.$$

If conditions 1) – 3) are satisfied then the canonical system (19) has a fundamental matrix of the form (16) where $T = T_0(\infty)$. The solutions forming the F.S.S. have the form

$$y_{2k-1}(x) = \exp\left\{i\int_0^x q_k(t)dt\right\}[e_k + o(1)], \quad x \to \infty,$$
$$y_{2k}(x) = \overline{y_{2k-1}(x)}.$$

For the fundamental matrix of the canonical system

$$Jy' = \lambda H(x)y$$

under conditions 1) – 3) there is an asymptotic formula of the form (12), with dual asymptotic behaviour.

Suppose that conditions 1) and 2) are satisfied, and let the matrix $H(x)$ satisfy conditions 1) – 4) from paragraph 1.1. We introduce the functions

$$p_j^{(1)}(x) = -e_j^\star(x)Je_j'(x). \tag{22}$$

These functions are real-valued and $p_{2k}^{(1)}(x) = p_{2k-1}^{(1)}(x)$ by the normalization (21).

If none of the functions Re $[p_j^{(1)}(x) - p_k^{(1)}(x)]$ changes sign for $x \gg 1$ then the canonical system (19) has a fundamental matrix of the form (5).

2. Systems of Second-Order Equations. We consider the system of n equations

$$y'' = \lambda^2 A(x)y \tag{23}$$

on the half-line $\mathbb{R}^+ : x \geqslant 0$, where $\lambda > 0$ is a parameter and $A(x) \in C^2(\mathbb{R}^+)$. Let $p_1(x), \dots, p_n(x)$ be the eigenvalues of $A(x)$. System (23) corresponds to the matrix bundle $L(x,p) = p^2 I - A(x)$, the eigenvalues of which are $\pm\sqrt{p_j(x)}$. We introduce the following conditions 1) and 2).

1) The functions $p_1(x), \dots, p_n(x)$ are distinct and non-zero for $x \geqslant 0$.

We fix the branches of the functions $\sqrt{p_1(x)}, \dots, \sqrt{p_n(x)}$. Because of condition 1) these functions lie in $C^2(\mathbb{R}^+)$. The next condition means that the eigenvalues of $A(x)$ are asymptotically simple.

2) For all j and k the limits

$$\lim_{x\to\infty} p_j(x)/p_k(x) = c_{jk}$$

exist and are finite, and $c_{jk} \neq 1$ for $j \neq k$.

It follows from conditions 1) and 2) that there is a function $q(x) \in C^2(\mathbb{R}^+)$, with $q(x) \neq 0$ for $x \geqslant 0$, and constants $c_j \neq 0$ such that

$$p_j(x) = [c_j + o(1)]q(x), \quad x \to \infty, \tag{24}$$

where $c_j \neq c_k$ for $j \neq k$. On the strength of condition 1) there are bases $\{e_1(x), \dots, e_n(x)\}$ and $\{e_1^\star(x), \dots, e_n^\star(x)\}$ of class $C^2(\mathbb{R}^+)$ of right and left eigenvectors of $A(x)$ respectively. We normalize them by the condition

$$e_j^\star(x)e_j(x) = 1. \tag{25}$$

Then the matrix $T(x) = (e_1(x), \dots, e_n(x))$ reduces $A(x)$ to the diagonal form

$$T^{-1}(x)A(x)T(x) = \Lambda(x) = \operatorname{diag}\,(p_1(x), \dots, p_n(x)),$$

and $T^{-1}(x)$ is the matrix with rows $e_i^\star(x), \dots, e_n^\star(x)$. We put

$$p_j^{(1)}(x) = -e_j^\star(x)e_j'(x), \quad \Lambda^{(1)}(x) = \operatorname{diag}\,(p_1^{(1)}(x), \dots, p_n^{(1)}(x)). \tag{26}$$

We bring in conditions that are analogous to conditions 1), 3), 4) from paragraph 1.1:

3) $\lim_{x\to\infty}(|q'|\,|q|^{-3/2} + |q|^{-1/2}\|T^{-1}(x)T'(x)\|) = 0$.

4) The function

$$b(x) = |q'|^2|q|^{-5/2} + |q''|\,|q|^{-3/2} + |q'|\,|q|^{-3/2}\|T^{-1}T'\| + |q|^{-1/2}\|T^{-1}T'\|^2 + |q|^{-2}\|(T^{-1}T')'\|$$

lies in $L_1(0,\infty)$.

It follows from condition 3) that $p_j^{(1)}(x) = o(p_j(x))$ as $x \to \infty$. Put

$$\tilde{y}_k^{\pm}(x,\lambda) = \frac{1}{\sqrt[4]{p_k(x)}} \exp\left\{\pm\lambda \int_{x_0}^{x} \sqrt{p_j(t)}dt + \int_{x_0}^{x} p_j^{(1)}(t)dt\right\}, \tag{27}$$

$$y_k^{\pm}(x,\lambda) = \tilde{y}_k^{\pm}(x,\lambda)[e_k(x) + \lambda^{-1} f_k^{\pm}(x,\lambda)]. \tag{28}$$

Suppose now that conditions 1) – 4) are satisfied. Then we can state a further condition under which the solutions of (23) have dual asymptotic behaviour (both as $x \to \infty$ and as $\lambda \to \infty$). This condition can be formulated in two ways. Denote

$$\mu_j(x) = \sqrt{p_j(x)}, \quad \mu_{n+j}(x) = -\sqrt{p_j(x)}, \quad 1 \leqslant j \leqslant n,$$

$$S_{jk}(t,x) = \int_t^x [\mu_j(\tau) - \mu_k(\tau)]d\tau, \tag{29}$$

$$S_{jk}^{(1)}(t,x) = \int_t^x [p_j^{(1)}(\tau) - p_k^{(1)}(\tau)]d\tau.$$

Then the condition is as follows.

5a) $\operatorname{Re} S_{jk}(0,\infty) = \infty, \quad j \neq k,$

$\operatorname{Re}\,[p_j^{(1)}(x) - p_k^{(1)}(x)] = o(\operatorname{Re}\,(\mu_j(x) - \mu_k(x))), \quad x \to \infty.$

5b) $|\operatorname{Re} S_{jk}(0,\infty)| < \infty, \quad \sup_{t,x\geqslant 0} |\operatorname{Re} S_{jk}^{(1)}(t,x)| < \infty.$

If one of 5a) and 5b) holds for some k and for all $j \neq k$ then system (23) has two solutions $y_k^{\pm}$ of the form (28). Moreover for all $\lambda_0 > 0$ there exists $x(\lambda_0) < \infty$ such that for $\lambda \geqslant \lambda_0$ and $x(\lambda) \geqslant x(\lambda_0)$ we have

$$\|f_j^{\pm}(x,\lambda)\| \leqslant \phi(x), \quad \lim_{x\to\infty} \phi(x) = 0. \tag{30}$$

If these conditions hold for all k then system (23) has a F.S.S. consisting of solutions $y_1^+, \ldots, y_n^+, y_1^-, \ldots \overline{y}_n$ of the form (28). Further

$$\frac{d}{dx} y_k^{\pm}(x,\lambda) = \pm\lambda\sqrt{p_k(x)}\tilde{y}_k^{\pm}(x,\lambda)[e_k(x) + \lambda^{-1} f_{1k}^{\pm}(x,\lambda)],$$

where the vector functions $f_{1k}^{\pm}$ have the same properties as $f_k^{\pm}$.

Remark. Fix $\lambda > 0$ and suppose that for some k and for all $j \neq k$ the functions

$$\operatorname{Re}\,[\mu_k(x) - \mu_j(x) + \mu_k^{(1)}(x) - \mu_j^{(1)}(x)]$$

do not change sign for $x \gg 1$. Then system (23) has solutions $y_k^{\pm}$ of the form (28), (30).

Let us consider the matrix equation of the type (23):

$$Y'' - \lambda^2 A(x) Y = 0\,, \tag{31}$$

where $Y(x)$ is an $n \times n$ matrix, and suppose that the above conditions are satisfied. Then any solution of equation (31) can be represented as

$$Y(x,\lambda) = Y^{+}(x,\lambda)C^{+}(\lambda) + Y^{-}(x,\lambda)C^{-}(\lambda)\,,$$

where $C^{\pm}(\lambda)$ are matrices of order $n \times n$, and

$$\begin{aligned} Y^{\pm}(x,\lambda) &= T(x)[I + \lambda^{-1}U^{\pm}(x,\lambda)]\Lambda^{-1/4}(x) \\ &\quad \times \exp\left\{ \pm\lambda \int_{x_0}^{x} \Lambda(t)dt - \int_{x_0}^{x} \operatorname{diag}\,[T^{-1}(t)T'(t)dt] \right\}, \end{aligned} \tag{32}$$

where $\|U^{\pm}(x,\lambda)\|$ has the same properties as $\|f_j^{\pm}(x,\lambda)\|$ (see (30)).

Similar formulae are true for the derivatives $dY^{\pm}(x,\lambda)/dx$; in formula (32) the matrix $\Lambda^{-1/4}(x)$ must be replaced by the matrix $\pm\lambda\Lambda^{1/4}(x)$ and the matrices $U^{\pm}$ by $U_1^{\pm}$ with the same properties.

3. Systems of Equations of Arbitrary Order. We consider the system of n equations

$$y^{(m)} - \sum_{k=0}^{m-1} \lambda^{m-1} A_k(x) y^{(k)} = 0 \tag{33}$$

on the half-line $x \geqslant 0$ where $\lambda > 0$ is a parameter and $A_k(x) \in C^2(\mathbb{R}^+)$ for all k. System (33) corresponds to the matrix bundle

$$L(x,p) = p^m I - \sum_{k=0}^{m-1} p^k A_k(x)\,.$$

Let $p_j(x)$ be the eigenvalues of the bundle L, that is, solutions of the equation $\det L(x,p) = 0$. We will assume that all the roots $p_j(x)$ are simple and take bases of right and left eigenvectors $\{e_1(x), \dots, e_{nm}(x)\}$ and $\{e_1^\star(x), \dots, e_{nm}^\star(x)\}$, normalized by the condition

$$e_j^\star(x) L_p(x, p_j(x)) e_j(x) = 1\,. \tag{34}$$

We bring in the function

$$p_j^{(1)}(x) = -\left(e_j^\star L_p e_j' + \frac{p_j'}{2} L_{pp} e_j \right), \tag{35}$$

where the eigenvectors are taken at the point x and the values of the derivatives of L at the point $(x, p_j(x))$. Suppose that the function $q(x)$ and the matrix function $Q(x)$ are the same as in paragraph 1.1. We bring in the conditions:

1) $A_k(x) = q^{m-k}(x)Q(x)B_k(x)Q^{-1}(x)$, and the limits $\lim_{x\to\infty} B_k(x)$ exist and are finite.

2) The roots η_j of the equation

$$\det\left[\eta^m I - \sum_{k=0}^{m-1} \eta^k B_k(\infty)\right] = 0$$

are distinct and non-zero.

3) conditions 3) and 4) from paragraph 1.1 are satisfied if we replace $\|B(x)\|$ by $\sum_{k=0}^{m-1} \|B_k(x)\|$ in them.

For $m = 1$ those conditions coincide with the conditions of paragraph 1.1. Suppose that conditions (8) and (9) are satisfied for some j and all $k \neq j$, and let $\tilde{y}_j(x)$ be determined by (3). Then system (31) has a solution $y_j(x, \lambda)$ such that

$$y_j^{(s)}(x,\lambda) = \lambda^s p_j^s(x)\tilde{y}_j(x,\lambda)\left[e_j(x) + \lambda^{-1}\sum_{l=1}^{nm} u_{jsl}(x,\lambda)e_l(x)\right], \qquad (36)$$

$$0 \leqslant s \leqslant m.$$

where the functions u_{jsl} have the same properties as the functions u_{jk} in (10).

4. The Fundamental Methods of Proof of Asymptotic Formulae for Solutions of Differential Equations and Systems. Asymptotic formulae are first of all necessarily conjectured and up to this point it is difficult to formulate general principles. After a formula has been suggested its proof in the majority of cases breaks into two stages.

1. Using some suitable transformation (a change of variable and unknown functions) the equation or system is reduced to the form

$$(l_0 + l_1)y = 0,$$

where the equation $l_0 y = 0$ is solvable exactly, and the operator l_1 can be considered as a small perturbation.

2. The equation $l_0 y = -l_1 y$ is then solved as an inhomogeneous equation with right-hand side $-l_1 y$ and one studies the resulting system of integral equations.

Different transformations of equations and systems have been met already (Chap. 2, § 1; Chap. 4, § 2, Chap. 5, §§ 1, 2, etc.). Let us mention the fundamental integral equations. Suppose that we have a system of n equations

$$y' = [A_0(x) + A_1(x)]y, \qquad (37)$$

where the system

$$z' = A_0(x)z$$

is integrable. We denote by $Y(x)$ and $Z(x)$ the fundamental matrices of these systems. Applying the method of variation of constants we arrive at the integral equation

$$Y(x) = Z(x) + \int_{\Gamma(x)} Z(x)Z^{-1}(t)A_1(t)Y(t)dt$$

for the fundamental matrix $Y(x)$. Here $\Gamma(x)$ is an $n \times n$ matrix, the elements $\gamma_{jk}(x)$ of which are intervals of the form (x_{jk}, x) (or curves in the complex x-plane connecting the fixed points x_{jk} and x). Putting

$$Y(x) = Z(x)W(x),$$

we obtain the fundamental integral equation

$$W = W_0 + KW, \tag{38}$$

where

$$(KW)(x) = \int_{\Gamma(x)} Z^{-1}(t)A_1(t)Z(t)W(t)dt.$$

The choice of the matrix of paths $\Gamma(x)$ is still undefined. It is desirable to make the choice so that a fixed-point principle can be applied to equation (38). In this case we can obtain some information on the fundamental matrix $Y(x)$.

In many of the examples which have been considered in the previous chapters, we cannot construct the asymptotic behaviour of the fundamental matrix but only of some of the solutions of (37) and, in place of the matrix integral equation (38), one studies a vector integral equation. We now obtain this equation when

$$A_0(x) = \Lambda(x) = \operatorname{diag}(p_1(x), \ldots, p_n(x)).$$

We bring in the notation

$$S_j(t,x) = \int_t^x p_j(\tau)d\tau, \quad S_{kj}(t,x) = S_k(t,x) - S_j(t,x),$$

$$\tilde{y}_j(x_0,x) = \exp\{S_j(x_0,x)\}, \quad y = \tilde{y}_j(x_0,x)w,$$

$$z = z_j(x_0,x)f_j,$$

where f_j is the column vector with components δ_{jk}. Then for w we obtain the integral equation

$$w = f_j + K_j(w), \tag{39}$$

where K_j is the integral operator

$$(K_j w)_k = \int_{x_{kj}}^{x} \exp\{S_{kj}(t,x)\}[A_1(t)w(t)]_k dt .$$

The choice of paths $\gamma_{jk} = (x_{jk}, x)$ remains undefined. It is desirable to take these paths so that in some Banach space B the norm of the operator K_j is small: $\|K_j\|_B \ll 1$. Then for the solutions of equation (39) we obtain a bound of the type

$$\|w - f_j\|_B \leqslant c\|K_j\|_B .$$

Let us consider two simple cases.

1) Let $I = [a, b]$ be a finite interval, $a < b$, let $\lambda > 0$ be a large parameter, and let

$$A_1(x) = \lambda^{-1}B(x), \quad A(x) = \lambda \operatorname{diag}(p_1(x), \dots, p_n(x)),$$

where the functions $p_j(x)$ and the elements of $B(x)$ are of class $C(I)$. For B we take the space of vector functions $w(x)$ with components of class $C(I)$ and with norm $\|w(x)\|_B = \max_{x\in I}|w(x)|$. Suppose that for each given j none of the differences $\phi_{kj}(x) = \operatorname{Re}[p_j(x) - p_k(x)]$ changes sign. We will put

$$x_{kj} = b, \quad \text{if} \quad \phi_{kj}(x) \leqslant 0,$$

where

$$x_{kj} = a, \quad \text{if} \quad \phi_{kj}(x) \geqslant 0.$$

Then for $t \in (x_{kj}, x)$ we obtain

$$\operatorname{Re}[\lambda S_{kj}(t,x)] \leqslant 0. \tag{40}$$

Consequently

$$|\exp\{\lambda S_{kj}(t,x)\}| \leqslant 1, \quad t \in \gamma_{kj}(x), \tag{41}$$

and therefore $\|K\|_B \leqslant c\lambda^{-1} \ll 1$ for $\lambda \gg 1$. In Chap. 2, § 7 we showed that if the functions $\phi_{kj}(x)$ can change sign then there does not exist an asymptotic formula for the solutions of system (37) that is applicable as $\lambda \to \infty$ on the whole interval I.

2) Suppose that $I = [0, \infty)$, and the matrices $A(x)$ and $A_1(x)$ satisfy the same conditions as in case 1). Also, let $\lambda = 1$ and $\|B(x)\| \in L_1(0,\infty)$ (this case almost completely coincides with the conditions of Levinson's theorem (Chap. 2, § 5)). We require additionally that $\lim_{x\to\infty}\|B(x)\| = 0$ and that

$$\int_0^{\infty} \phi_{kj}(x)dx = +\infty, \text{ if } \phi_{kj}(x) \geqslant 0, \quad k \neq j .$$

We put

$$x_{kj} = a\,, \quad \text{if } \phi_{kj}(x) \leqslant 0\,, \tag{42a}$$

where

$$x_{kj} = \infty\,, \quad \text{if } \phi_{kj}(x) \geqslant 0\,, \tag{42b}$$

where $a > 0$ will be indicated below. Then for $t \in (x_{kj}, x)$ inequality (40) is satisfied. For B we take the space of vector functions of class $C(I)$ that are bounded for $x \in I$ and with the same norm as in case 1). Then

$$|(K_j w_k)(x)| \leqslant \|w\|_B \int_a^x \|B(t)\| dt\,,$$

$$|(K_j w)_k(x)| \leqslant \|w\|_B \int_x^\infty \|B(t)\| dt$$

for $x \geqslant a$ respectively in cases (42a) and (42b). If $a > 0$ is such that $\int_a^\infty \|B(t)\| dt < 1$ then $\|K_j\|_B < 1$ and we can apply the method of successive approximations to equation (37).

However this is not sufficient for the proof of Levinson's theorem: one must still prove that $(K_j w)(x) \to 0$ as $x \to \infty$ if $w(x) \in B$. This follows from the condition $\|B\| \in L_1(0, \infty)$ for the components $(K_j w)_k$ in (42b). For the case (42a) we must make use of the more exact bound

$$|(K_j w)_k(x)| \leqslant \|w\|_B I_{jk}(x)\,,$$

where

$$I_{jk}(x) = \int_a^x \exp\{\operatorname{Re} S_{kj}(t, x)\} \|B(t)\| dt\,.$$

Applying l' Hôpital's rule we obtain

$$\lim_{x\to\infty} I_{jk}(x) = \lim_{x\to\infty} \frac{\int_a^x \exp\{\operatorname{Re} S_{kj}(t,a)\} \|B(t)\| dt}{\exp\{\operatorname{Re} S_{kj}(t,a)\}} = \lim_{x\to\infty} \|B(x)\| = 0\,.$$

Levinson's theorem follows from these relations.

Remark. The solution w_j of the integral equation (39) satisfies the boundary conditions:

$$w_{jk}(a) = \delta_{jk} \qquad \text{for (42a)}$$

and

$$w_{jk}(\infty) = 0 \qquad \text{for (42b)}\,.$$

Solutions of systems with a large parameter λ satisfy similar boundary conditions at the points $x = a$ and $x = b$. The asymptotic behaviour of these solutions involves functions of boundary layer type (Chap. 2, § 3, paragraph 3).

§ 5. Equations and Systems in the Complex Plane

1. Statement of the Problem. We consider the nth-order equation

$$lw \equiv w^{(n)} + \lambda q_1(z)w^{(n-1)} + \ldots + \lambda^n q_n(z)w = 0 \tag{1}$$

and the system of n equations

$$w' = \lambda A(z)w\,, \tag{2}$$

where $\lambda > 0$ is a parameter. To equation (1) and system (2) there correspond the characteristic equations

$$\begin{aligned} l(z,p) &\equiv p^n + q_1(z)p^{n-1} + \ldots + q_n(z) = 0\,,\\ l(z,p) &= \det(A(z) - pI) = 0\,, \end{aligned} \tag{3}$$

the roots of which we will denote $p_1(z), \ldots, p_n(z)$.

Suppose that D is a simply-connected domain in the complex z-plane, possibly unbounded. Throughout this paragraph we will assume that

1) the coefficients of equation (1) and the system (2) are holomorphic in D;

2) the roots $p_1(z), \ldots, p_n(z)$ are distinct for all $z \in D$.

In this way (1) and (2) have no turning points in D. Equations and systems with turning points will be considered in § 6. In this paragraph we will study the asymptotic behaviour of the solutions of (1) and (2) as $\lambda \to \infty$ and also as $z \to \infty$ if D is unbounded. All the asymptotic formulae are almost the same as those in §§ 1–3, only the conditions for their applicability being changed.

2. Equations and Systems in Bounded Domains Without Turning Points.

2.1 Asymptotic Diagonalization of a System. Let D be a simply-connected bounded domain with piecewise smooth boundary D, and let conditions 1) and 2) be satisfied in the closure $[D]$. Then all the roots $p_1(z), \ldots, p_n(z)$ are holomorphic in $[D]$. There exists a matrix function $T(z)$, holomorphic and non-singular in $[D]$, which reduces $A(z)$ to diagonal form

$$T^{-1}(z)A(z)T(z) = \operatorname{diag}\,(p_1(z),\ \ldots,\ p_n(z))\,.$$

In fact the rank of the matrix $B(z) = A(z) - p_1(z)I$ is identically equal to $n-1$ for $z \in [D]$. Let $B_{11}(z), \ldots, B_{1n}(z)$ be the minors corresponding to the elements $b_{11}(z), \ldots, b_{1n}(z)$ of the matrix $B(z)$. Without loss of generality we can assume that at some point $z_0 \in D$, one of these minors is non-zero and therefore in some neighbourhood of z_0 the vector function $\tilde{e}_1(z) = (B_{11}(z), \ldots, B_{1n}(z))$ is holomorphic and non-zero. The vector function $\tilde{e}_1(z)$ can have only a finite number of zeros $z_1, \ldots, z_N$ in $[D]$, and we denote by m_k the least order of the zero of its components at the point z_k. Then the vector function $e_1(z) = (z-z_1)^{k_1}(z-z_2)^{k_2}\dots(z-z_N)^{k_N}\tilde{e}_1(z)$ is holomorphic and non-zero in $[D]$. In precisely the same way we construct the other right eigenvectors $e_2(z),\dots,e_n(z)$ of $A(z)$. The matrix functions $T(z)$, $Z^{-1}(z)$ and all their derivatives are bounded in $[D]$.

Remark. If conditions 1) and 2) are satisfied in an open domain D then as before we can construct a matrix function $T(z)$ that is holomorphic and nondegenerate in D, and which reduces $A(z)$ to diagonal form. However, the derivatives of $T(z)$ and $T^{-1}(z)$ may be unbounded in D.

From the reasoning given in paragraph 1.3, § 1, it follows that there exist matrix functions $T_1(z)$, $T_2(z),\dots,T_{N-1}(z)$, and diagonal matrix functions $\Lambda_0(z),\dots,\Lambda_{N-1}(z)$, such that the transformation

$$w = T(z)\left[I + \sum_{k=0}^{N-1} \lambda^{-k} T_k(z)\right] u$$

reduces system (2) to the diagonal form

$$u' = [\lambda\Lambda_0(z) + \Lambda_1(z) + \dots + \lambda^{-N+1}\Lambda_{N-1}(z) + \lambda^{-N}B_N(z,\lambda)]u$$

to within $O(\lambda^{-N})$. Here $N \geqslant 1$ is arbitrary, all the matrix functions shown are holomorphic in $[D]$,

$$\Lambda_0(z) = \Lambda(z), \quad \|B_N(z,\lambda)\| \leqslant c_N, \quad z \in [D], \quad \lambda \geqslant 1,$$

and

$$\Lambda_1(z) = -\mathrm{diag}\left(T^{-1}(z)\frac{d}{dz}T(z)\right).$$

These formulae also arise for first-order systems equivalent to equation (1).

2.2 Canonical Paths and Admissible Domains. In this section we will use the following notation:

$$S_j(z_0,z) = \int_{z_0}^{z} p_j(t)dt, \quad S_{jk}(z_0,z) = \int_{z_0}^{z} [p_j(t) - p_k(t)]dt. \tag{4}$$

All the integrals are taken along paths lying in $[D]$. Suppose that the curve $\gamma_{jk}(a,b)$ joins the points a and b, and the function Re $S_{jk}(a,z)$ is non-decreasing as z moves from a to b along this curve. Then the curve γ_{jk} is called a *canonical path* (see Chap. 3, § 1). A *canonical vector path* $\gamma_j(b_j,z_0)$, $b_j = (b_{j1},\dots,b_{jn})$, with end-point at z_0 is the vector whose components are the canonical paths $\gamma_{jk}(b_{jk},z_0)$, $1 \leqslant k \leqslant n$. A *matrix* (or *star*) of canonical paths $\Gamma(b,z_0)$ with end-point at z_0 is the matrix whose columns are the canonical vector-paths $\gamma_1(b_1,z_0),\dots,\gamma_n(b_n,z_0)$, where $b = (b_1,\dots,b_n)$. We take two types of admissible domains:

1) a j-admissible domain D: there exist points $b_1, \dots, b_n \in \partial D$ such that for each $z \in [D]$ there is a canonical vector-path $\gamma_j(b_j, z)$;

2) an admissible domain D: for each $j = 1, \dots, n$ the domain D is a j-admissible domain.

2.3 Asymptotic Behaviour of the Solutions. Suppose that D is a j-admissible domain. Then for each integer $N \geqslant 1$ there are solutions w_j of equation (1) and of system (2) with the form

$$w_j(z, \lambda) = \exp\{\lambda S_j(z_0, z) + S_j^{(1)}(z_0, z)\} \times \left[1 + \sum_{k=1}^{N-1} \lambda^{-k} a_{jk}(z) + O(\lambda^{-N})\right], \tag{5}$$

$$w_j(z, \lambda) = \exp\{\lambda S_j(z_0, z) + S_j^{(1)}(z_0, z)\} \times \left[e_j(z) \sum_{k=1}^{N-1} \lambda^{-k} e_{jk}(z) + O(\lambda^{-N})\right], \quad \lambda \to \infty. \tag{6}$$

The coefficients of these asymptotic series coincide respectively with the coefficients of the asymptotic series (7) of § 1 and (11) of § 2 and are holomorphic in $[D]$, where $z_0 \in [D]$ is an arbitrary point. The remainder term in (5) is the function $\lambda^{-N}\phi(z, \lambda)$ of class C^∞ for $z \in [D]$, $\lambda \geqslant 1$, and it is holomorphic in $[D]$ for each fixed $\lambda \geqslant 1$. Also, $|\phi(z, \lambda)| \leqslant c$ for $z \in [D]$, $\lambda \geqslant 1$. The remainder term in (6) has a similar form. The asymptotic expansions (5) and (6) can be differentiated in z and λ an arbitrary number of times, preserving the uniformity in $z \in [D]$ of the bound for the remainder term.

The proof of the asymptotic expansions (5) and (6) follows directly from what was said in § 4, paragraph 4, since

$$|\exp\{\lambda S_{kj}(t, z)\}| \leqslant 1,$$

if the point t lies on the canonical path $\gamma_{jk}(b_{jk}, z)$.

If D is a canonical domain then equation (1) and system (2) have F.S.S. consisting of solutions of the form (5), (6). Further, the fundamental matrix of system (2) has the form

$$W(z, \lambda) = T(z)[I + O(\lambda^{-1})] \exp\left\{\lambda \int_{z_0}^{z} A(t)dt + \int_{z_0}^{z} A_1(t)dt\right\}. \tag{7}$$

Let us mention one important special case. Suppose the matrix function $A(z)$ is holomorphic at the point z_0 and the eigenvalues $p_1(z_0), \dots, p_n(z_0)$ are distinct. Then there is a neighbourhood D of z_0 which is an admissible domain.

Remark. We have not imposed any restrictions on the function Re $(p_j(z) - p_k(z))$ as opposed to the non-analytic case (see Chap. 2, § 7).

The above results carry over in an obvious way to systems of the form

$$w' = \lambda A(z, \lambda^{-1})w\,,$$

where $A(z,\varepsilon)$ ($\varepsilon = \lambda^{-1}$) has the following properties.

1) $A(z,\varepsilon) \in C^\infty([D] \times I)$, where $I = [0, \varepsilon_0]$, $\varepsilon_0 > 0$.
2) For each fixed $\varepsilon \in I$, the matrix $A(z,\varepsilon)$ is holomorphic in $[D]$.
3) As $\varepsilon \to +0$ there is the asymptotic expansion

$$A(z,\varepsilon) = \sum_{k=0}^{\infty} \varepsilon^k A_k(z)\,,$$

uniform in $z \in [D]$.

2.4 Connection Formulae. Suppose that $W(z,\lambda)$ is the fundamental matrix of system (2) and that $w_j(z,\lambda)$ is some solution of this system. Then

$$w_j(z,\lambda) = W(z,\lambda)\omega_j(\lambda)\,,$$

where the n-vector $\omega_j(\lambda)$ does not depend on z. Let M be a connected compact set in the complex z-plane such that the complement of M does not contain bounded domains. A typical example is as follows: $M = [D] \cup l_1 \cup l_2$ where D is a bounded simply-connected domain with piecewise-smooth boundary, and l_1, l_2 are finite simple piecewise smooth curves. The ends of these curves lie on ∂D, and the curves do not intersect and have no other common points with $[D]$.

Suppose that the matrix function $A(z)$ is holomorphic and has no turning points for $z \in M$, and as $\lambda \to \infty$ the solutions w_j and fundamental matrix W have the asymptotic expansions (5) and (7), uniform in $z \in M$. Then

$$\omega_{jj}(\lambda) = 1 + O(\lambda^{-1})\,, \quad \omega_{jk}(\lambda) = O(\lambda^{-1} e^{\lambda a_{jk}})\,, \quad j \neq k\,, \tag{8}$$

where

$$a_{jk} = \min_{z \in M} \operatorname{Re} S_{jk}(z_0, z) < 0\,.$$

Let $W_1(z,\lambda)$, $W_2(z,\lambda)$ be fundamental matrices of system (2). Then

$$W_2(z,\lambda) = W_1(z,\lambda)\Omega(\lambda)\,,$$

where the transition matrix $\Omega(\lambda)$ does not depend on z. If asymptotic expansions of the form (7) hold for both W_1 and W_2 then

$$\Omega(\lambda) = I + \operatorname{diag} O(\lambda^{-1}) + O(e^{-c\lambda})\,, \quad c > 0\,, \quad \lambda \to \infty\,. \tag{9}$$

3. Structure of Admissible Domains

3.1 Local Structure of Turning Points and Stokes Lines. Let z_0 be a turning point, and let U be the disk $|z-z_0| \leqslant r$ for sufficiently small $r > 0$. Any root $p_j(z)$ can be expanded in a series

$$p_j(z) = p_j(z_0) + \sum_{k=1}^{\infty} p_{jk}(z-z_0)^{kp_j/q_j},$$

converging in U, where $p_j \geqslant 1$ and $q_j \geqslant 1$ are mutually coprime integers. If $q_j = 1$ then $p_j(z)$ is holomorphic in U; if $q_j \geqslant 2$ then z_0 is a branch point of order q_j for $p_j(z)$. In U the characteristic polynomial splits into irreducible factors $l(z,p) = l_1^{m_1}(z,p)\dots l_N^{m_N}(z,p)$, $l_j(z,p) \not\equiv \text{const}$, where $m_1, \dots, m_N$ are positive integers, and $l_1(z,p), \dots, l_N(z,p)$ are mutually distinct polynomials of p with functions that are holomorphic in U for their coefficients. Each of the polynomials $l_j(z,p)$ is irreducible, and

$$l_j(z,p) = p^{n_j} + a_{j1}(z)p^{n_j-1} + \dots$$

If $n_j = 1$ then the root $p_j(z)$ of the equation $l_j(z,p) = 0$ is holomorphic in U. If $n_j \geqslant 2$ then the roots of the equation $l_j(z,p) = 0$ can be expanded as a Puiseux series

$$p(z) = p(z_0) + \sum_{k=1}^{\infty} p_{jk}(z-z_0)^{kq_j/n_j}, \tag{10}$$

converging in U, where $q_j \geqslant 1$ is an integer coprime to n_j. Therefore z_0 is a branch point of order n_j for each of the roots.

For second-order equations there are only two possibilities.

1) The polynomial $l(z,p) = p^2 + a(z)p + b(z)$ is irreducible. Then z_0 is a second-order branch point for the roots $p_1(z)$ and $p_2(z)$.

2) The polynomial $l(z,p)$ is reducible. Then $l(z,p) = (p-p_1(z))(p-p_2(z))$, and the functions $p_1(z)$ and $p_2(z)$ are holomorphic in U.

For n^{th}-order equations the roots can coincide at the turning point z_0, for which z_0 is either an analytic point or a branch point of arbitrary integer order.

Suppose that $p_j(z_0) = p_k(z_0)$ for $j \neq k$. The level curve

$$\operatorname{Re} \int_{z_0}^{z} [p_j(t) - p_k(t)]dt = 0,$$

emanating from z_0 is called a *Stokes line*. In view of (10) there is an integer $r \geqslant 1$ such that z_0 is a branch point of order r for the function $p_j(z) - p_k(z)$. Then

$$p_j(z) - p_k(z) \sim a(z-z_0)^{q/r}, \quad z \to z_0,$$

where $a \neq 0$ and $q \geqslant 1$ is an integer, and from z_0 there emerge no more than $2(p+q)$ Stokes lines.

Remark. It is more correct to consider the Stokes lines not on the z-plane, but on the Riemann surface $\tilde{U}$ of the algebroidal function $\tilde{p}(z)$ defined by the equation

$$\tilde{l}(z,\tilde{p}) \equiv \prod_{j\neq k}[\tilde{p} - (p_j(z) - p_k(z))] = 0 . \tag{11}$$

The polynomial $\tilde{l}$ is constructed in the following way. Fix a point $z_1 \in U$ with $z_1 \neq z_0$; then in a small simply-connected neighbourhood V of z_1 equation (3) has n distinct roots $p_1(z), \dots, p_n(z)$ which are holomorphic in V. Thus, the polynomial $\tilde{l}$ is defined in V, and

$$\tilde{l}(z,p) = p^m + b_1(z)p^{m-1} + \dots + b_m(z) ;$$

where the functions $b_1(z), \dots, b_m(z)$ are holomorphic in V. Let γ be the circle $|z - z_0| = |z_1 - z_0|$ starting and finishing at z_1; then for z going around z_0 along γ we have $p_j(z) \to p_{s_j}(z)$ where $\{s_1, \dots, s_n\}$ coincides with $\{1, \dots, n\}$. It follows from this that the coefficients of the polynomial $\tilde{l}$ are holomorphic in U and that (11) determines the multi-valued algebroidal function $\tilde{p}(z)$. The Stokes line is the curve γ: $\operatorname{Re} \int_{\zeta_0}^{z} \tilde{p}(t)dt = 0$, where $\zeta_0 \in \tilde{U}$ and is projected onto z_0, and γ lies on $\tilde{U}$. The Stokes lines lying on $\tilde{U}$ are distinct but their projections onto U can coincide.

Examples. 1. Let $l(z,p) = p^n - i^n z^m$ where m and n are mutually coprime numbers. Then the polynomial l is irreducible and the Stokes lines are the rays

$$l_k : \ \arg = \frac{nk\pi}{m+n}, \qquad k = 0, 1, \dots$$

2. Let $l(z,p) = (p^2 - z)(p^3 - z)$; then

$$\tilde{l}(z,p) = (p^2 - z)(p^3 - 3\sqrt{3}z^2)(p^3 + 3\sqrt{3}z^2) \times [(p^2 - z)^3 - z^2][(p^2 - z)^3 + z^2] .$$

The Riemann surface $\tilde{U}$ consists of one two-sheeted, two three-sheeted and two six-sheeted surfaces.

3.2 Admissible Domains. Let D be a bounded simply-connected domain in the complex z-plane, the boundary of which consists of a finite number of simple analytical curves, and let $f(z)$ be holomorphic in D and continuous in $[D]$. The domain D is called ***admissible*** (for $f(z)$) if there is a point $a \in \partial D$ such that any point $z_0 \in [D]$ can be joined to a by a curve γ along which the function $u = \operatorname{Re} f(z)$ is non-decreasing (travelling from a to z_0 along γ).

In [Kelly] necessary and sufficient conditions were obtained for a domain D to be admissible. We will restrict ourselves to giving the conditions in the simplest case where there are only isolated minima and saddle points of $u(z)$ on ∂D. The point z_0 is called a ***saddle-point*** if the following conditions hold:

(1) D has two or more level curves $\gamma_1, \ldots, \gamma_k$ on which $u(z) = u(z_0)$ with end-points at z_0;

(2) There are not less than two sectors with vertices at z_0, which are bounded by the γ_j and by the arcs of the boundary ∂D, in which $u(z) < u(z_0)$.

In this definition, the domains of the larger values $u(z) > u(z_0)$ and of the smaller values $u(z) < u(z_0)$ are disjoint.

Either of the conditions below is necessary and sufficient for a domain D to be admissible.

A. There are no saddle-points on the boundary of D.

B. There is exactly one minimum point on the boundary of D.

These conditions are equivalent. This result is proved in the general case in [Kelly]. There is a significant difference between 2nd order equations and equations of order greater than 2. For simplicity let the coefficients of these equations be polynomials in z. The domain D is admissible for the equation

$$w'' - \lambda^2 q(z) w = 0\,, \tag{12}$$

if it is admissible for the single function $f(z) = \int^z \sqrt{q(t)}dt$ with suitable choice of the branch of the root. This domain is admissible for an nth-order equation if it is admissible for the collection of $n(n-1)$ functions

$$f_{jk}(z) = \int^z [p_j(t) - p_k(t)]dt\,, \quad 1 \leqslant j, k \leqslant n\,, \quad j \neq k\,.$$

In Chap. 3, § 3, paragraph 2 it was shown that, if there are no multiple turning points and no finite Stokes lines, then an admissible domain for equation (12) is the whole complex z-plane with a finite number of cuts along some Stokes lines. It is evident that the analogous statement is false for third-order equations, even with asymptotic simplicity of the roots at infinity. This is related to the fact that the Stokes lines can intersect on the z-plane.

Example. The equation

$$w''' - 3\lambda w' + \lambda^3 z w = 0$$

has two turning points $z_1 = -2$ and $z_2 = 2$. The Stokes lines intersect in this case (Fig. 24). Moreover the point $z = \infty$ is an irregular singular point for equation (12), while for equation (2) it can be a superposition of regular and irregular singular points, non-singular points and turning points.

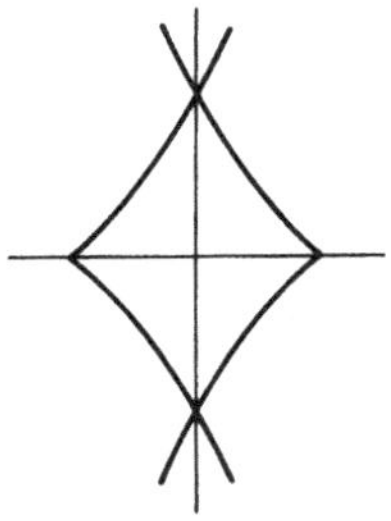

Fig. 24

4. Equations and Systems in Unbounded Domains. Let D be an unbounded simply-connected domain in the complex z-plane. The definitions of canonical paths, j-admissible and admissible domains (paragraph 2.2) carry over completely to the case of unbounded domains, with the only difference being that one of the ends of a canonical path can be the point $z = \infty$.

4.1 Equations and Systems Without a Parameter. We will assume for simplicity that ∂D consists of a single connected component. We consider the system of n equations

$$w' = A(z)w\,, \tag{13}$$

where the matrix function $A(z)$ is holomorphic in $[D] = D \cup \partial D$ and satisfies conditions similar to those in § 4, paragraph 1.1:

1) $$A(z) = q(z)Q^{-1}(z)B(z)Q(z)\,,$$
$$Q(z) = \operatorname{diag}\,(q^{\alpha_1}(z), \dots, q^{\alpha_n}(z))\,,$$

where $q(z)$ and $B(z)$ are holomorphic in $[D]$ and $q(z) \neq 0$.

2) There is a finite limit $\lim_{z\to\infty} B(z) = B(\infty)$ and the eigenvalues η_j of $B(\infty)$ are distinct and non-zero.

3) $$\lim_{z\to\infty,\, z\in[D]} r_1(z) = 0\,,$$

where

$$r_1(z) = |q'(z)||q(z)|^{-2} + |q(z)|^{-1}\|B'(z)\|\,.$$

The eigenvalues of $A(z)$ have the form

$$p_j(z) = [\eta_j + o(1)]q(z)\,, \quad z \to \infty\,, \quad z \in [D]\,.$$

4) $[D]$ is j-admissible.

Let $r_2(z)$ be the function from Chap. 4, paragraph 1.1. In condition 4) of the same paragraph we require that the integral of $r_2(x)$ converges over the half-line $[0, \infty)$. For system (13) in the complex plane we require convergence of the integral of $r_2(z)$ along the canonical path $\gamma_{jk}(z)$. Let us introduce the functions

$$\rho_{jk}(z) = \inf \int_{\gamma_{jk}(z)} |r_2(t)||dt|,$$

where the lower bound is taken over canonical paths joining the points z_{jk} and z. Let us introduce the condition:

5) $\sup_{z\in[D]} \rho_{jk}(z) < \infty$ for all $k = 1, \ldots, n$.

Then the system (13) has a solution of the form

$$w_j(z) = \exp\left\{\int_{z_0}^{z} [p_j(t) + p_j^{(1)}(t)]dt\right\}\left[e_j(z) + \sum_{k=1}^{n} u_{jk}(z)e_k(z)\right],$$

where

$$|u_{jk}(z)| \leqslant k_j(z), \qquad \lim_{z\to\infty,\ z\in[D]} k_j(z) = 0.$$

If D is an admissible domain and condition 5) is satisfied for all j then (13) has a F.S.S. of the form (5) of § 4, where

$$\|U(z)\| \leqslant k(z), \qquad \lim_{z\to\infty,\ z\in[D]} k(z) = 0.$$

The asymptotic formulae (15), (36) and (45) from § 3 and (28) and (36) from § 4 are valid as $z \to \infty$, $z \in D$ under precisely the same conditions.

4.2 Dual Asymptotic Behaviour. We will consider system (2) under the conditions 1) – 5) from paragraph 2.1 together with the following:

$$6) \qquad \lim_{z\to\infty, z\in[D]} \operatorname{Re} \int_{z_0}^{z} [p_j(t) - p_k(t)]dt = \infty$$

for all $k \neq j$.

$$7) \quad |\operatorname{Re} \int_{z_0}^{z} [p_j(t) - p_k(t)]dt| \geqslant c|\operatorname{Re} \int_{z_0}^{z} [p_j^{(1)}(t) - p_k^{(1)}]dt|,$$
$$z \in [D], \quad |z| \gg 1,$$

for all $k \neq j$. Then system (2) has solutions of the form

$$w_j(z,\lambda) = \exp\left\{\lambda \int_{z_0}^{z} p_j(t)dt + \int_{z_0}^{z} p_j^{(1)}(t)dt\right\}$$
$$\times \left[e_j(z) + \lambda^{-1} \sum_{k=1}^{n} u_{jk}(z,\lambda)e_k(z)\right],$$

where for $\lambda \geqslant \lambda_0 > 0$

$$|u_{jk}(z,\lambda)| \leqslant k_j(z)\,, \qquad \lim_{z\to\infty, z\in[D]} k_j(z) = 0\,.$$

The dual asymptotic behaviours given in §§ 3,4 for the half-line $x \geqslant 0$ hold under the same conditions.

4.3 Equation (1) with Polynomial Coefficients. In this case the roots of the characteristic equation can be expanded in Puiseux series which converge for $|z| \leqslant R \ll 1$:

$$p_j(z) = z^{q_j/r_j} \sum_{k=0}^{\infty} a_{jk} z^{-kq_j/r_j}\,,$$

where the q_j and r_j are mutually coprime integers and $r_j \geqslant 1$. In [Leung 3] the following case is considered:

1)The degree $m \geqslant 1$ of $q_n(z)$ is not less than the degrees of the polynomials $q_1(z), \ldots, q_{n-1}(z)$.

2) The roots $p_j(z)$ are asymptotically non-repeated (§ 3). It follows from condition 1) that $a_{j0} \neq 0$ and $q_j \geqslant 0$ for all j, and we can index them so that

$$0 \leqslant q_1/r_1 \leqslant q_2/r_2 \leqslant \ldots \leqslant q_n/r_n\,.$$

Suppose that l is the ray $\arg z = \phi_0$ in the complex z-plane such that none of the differences Re $[p_j(z) - p_k(z)]$, $1 \leqslant j,\, k \leqslant n$, changes sign for $z \in l$, $|z| \gg 1$. Then there exists a sector S: $\phi_0 - \delta \leqslant \arg z \leqslant \phi_0 + \delta$ such that equation (1) has a solution of the form

$$w_j(z,\lambda) = z^{\nu} \exp\left\{\lambda \int_{z_0}^{z} p_j(t)dt\right\} \left[\sum_{k=0}^{N} \lambda^{-k} u_{jk}(z) + \lambda^{-N-1} R_{jN}(z,\lambda)\right],$$

where ν is a rational number and $N \geqslant 1$ is arbitrary. The functions $u_{jk}(z)$ are holomorphic in S, and if we choose $\lambda_0 > 0$ and $R > 0$ sufficiently large, then

$$|R_{jN}(z,\lambda)| \leqslant c_N |z|^{-(N+1)(1+q_1/r_1)}\,, \quad \lambda \geqslant \lambda_0\,, \quad |z| \geqslant R\,.$$

The $u_{jk}(z)$ have order $O(|z|^{-j(1+q_1/r_1)})$ as $z \to \infty$, $z \in S$. The size of S is indicated more precisely in [Leung 3].

Remark. The principal asymptotic term is given by (24) of § 3. In [Leung 3] more complicated formulae for ν and $u_{j0}(z)$ are given.

§ 6. Turning Points

1. Statement of the Problem. We consider the system of n equations

$$\varepsilon y' = A(x, \varepsilon)y \tag{1}$$

and the nth-order equation

$$ly \equiv \varepsilon^n y^{(n)} + \sum_{j=1}^{n} \varepsilon^{n-j} q_j(x, \varepsilon) y^{(n-j)} = 0\,, \tag{2}$$

where ε is a small parameter. We introduce the notation

$$\begin{aligned} l(x, p; \varepsilon) &= \det[A(x, \varepsilon) - pI]\,, \\ l(x, p; \varepsilon) &= p^n + \sum_{j=1}^{n} q_j(x, \varepsilon) p^{n-j} \end{aligned} \tag{3}$$

for sytem (1) and equation (2) respectively. There are two definitions of a turning point.

1. The point x_0 is called a *turning point* of system (1) or equation (2) if the equation $l(x_0, p; 0) = 0$ has a repeated root.

In this case $A(x_0, 0)$ has a repeated eigenvalue. The turning points are found by eliminating p from the system

$$l(x, p; 0) = 0\,, \quad l_p(x, p; 0) = 0\,.$$

2. The point $x_0 = x_0(\varepsilon)$ is called a *turning point* of system (1) or equation (2) if the equation $l(x_0(\varepsilon), p; \varepsilon) = 0$ has a repeated root.

It is clear that the definitions are not equivalent. For instance the equation

$$\varepsilon^2 y'' - (x^n - \varepsilon)y = 0\,,$$

where $n \geqslant 2$ is an integer, has the unique turning point $x_0 = 0$ according to definition 1 and n turning points $x_j(\varepsilon) = \sqrt[n]{\varepsilon}$ according to definition 2. As a rule we will use the first definition.

One of the main problems in the asymptotic theory of ordinary linear differential equations is the construction of a F.S.S. for system (1) and equation (2) in the neighbourhood of a turning point. This problem is highly complex. Even in the case where the coefficients are analytic in x and in ε it is in general unknown how to construct the formal asymptotic solutions, not to speak of rigorous proofs.

In this paragraph we consider the simplest types of turning point. Everywhere in what follows we will assume that one of the following conditions is satisfied.

C. The matrix function $A(x, \varepsilon)$ is infinitely differentiable for $|x - x_0| \leqslant a$, $0 \leqslant \varepsilon \leqslant \varepsilon_0$ where $a > 0$ and $\varepsilon_0 > 0$.

A. The matrix function $A(x,\varepsilon)$ is holomorphic in both the variables x and ε, for $x \in [D]$ and $|\varepsilon| \leqslant \varepsilon_0$, where $\varepsilon_0 > 0$, and D is a bounded simply-connected domain in the complex x-plane with piecewise-smooth boundary.

2. A Simple Turning Point. This type involves turning points in a neighbourhood of which the F.S.S. of the system or equation is expressed in terms of Airy functions.

2.1 Structure of the Roots. We will consider the scalar equation (2) under condition C. A turning point is called *simple* if the following conditions are satisfied:

(1) The equation $l(x_0,p;0)$ has one double root p_0, the other roots being simple.

(2) $l_x(x_0,p_0;0) \neq 0$.

We will assume for definiteness that $p_1(x_0,0) = p_2(x,0) = p_0$; then the values $p_3(x_0,0) = p_2^0, \dots, p_n(x_0,0) = p_n^0$ are different and not equal to p_0. Let $I = [x_0 - a,\ x_0 + a]$, $J = [0, \varepsilon_0]$, where $a > 0$, $\varepsilon_0 > 0$ and $\varepsilon_0 \ll 1$. If there are no other turning points on I then $p_j(x,\varepsilon) \in C^\infty(I \times J)$ for $3 \leqslant j \leqslant n$. The symbol l has a representation as

$$\begin{aligned} l(x,p;\varepsilon) = (p^2 - 2\alpha(x,\ \varepsilon)p + \beta(x,\ \varepsilon)) \\ \times (p - p_3(x,\ \varepsilon)) \dots (p - p_n(x,\ \varepsilon)), \end{aligned} \tag{4}$$

where $\alpha, \beta \in C^\infty(I \times J)$, so that

$$\begin{aligned} &p_{1,2}(x,\ \varepsilon) = \alpha(x,\ \varepsilon) \pm \sqrt{D(x,\ \varepsilon)}, \quad D = \alpha^2 - \beta, \\ &D(x_0,0) = 0, \quad D'_x(x_0,0) \neq 0. \end{aligned} \tag{5}$$

Under condition A the functions $\alpha(x,\varepsilon)$, $\beta(x,\varepsilon)$, $p_3(x,\varepsilon), \dots, p_n(x,\varepsilon)$ are holomorphic in a complex neighbourhood of the point $(x_0,0)$, and $x = x_0$ is a second-order branch point for the roots $p_1(x,0)$ and $p_2(x,0)$.

We consider system (1). Without loss of generality we can assume that $A(x_0,0)$ is reduced to Jordan normal form. The turning point x_0 of (1) is called *simple* if conditions 1) and 2) are satisfied and if its normal form contains one second-order Jordan block:

$$A(x_0,0) = \begin{bmatrix} p_0 & 1 & & & 0 \\ & p_0 & & & \\ & & p_3^0 & & \\ & & & \ddots & \\ 0 & & & & p_n^0 \end{bmatrix}, \quad p_j^0 = p_j(x_0,0).$$

If x_0 is a simple turning point then the following are satisfied:

$$\begin{aligned} &l(x_0,p_j^0;0) = 0, \quad l_p(x_0,p_j^0;0) \neq 0, \quad j = 3, \dots, n, \\ &l(x_0,p_0;0) = l_p(x_0,p_0;0) = 0, \\ &l_{pp}(x_0,p_0;0) \neq 0, \quad l_x(x_0,p_0;0) \neq 0. \end{aligned}$$

Let x_0 be a simple turning point for system (1) or equation (2) in the sense of definition 1. Then for small ε there exists a unique turning point $x_0 = x_0(\varepsilon)$ in the sense of definition 2, that is, the equation

$$l(x_0(\varepsilon), p; \varepsilon) = 0$$

has a repeated root $p_0(\varepsilon)$ for any sufficiently small ε. Further $x_0(\varepsilon)$, $p_0(\varepsilon) \in C^\infty(J)$, $x_0(0) = x_0$, and $p_0(0) = p_0$. A simple turning point is stable under small perturbations of the coefficients.

2.2 Formal Asymptotic Solutions of Equation (2). Equation (2) has F.A.S. $y_3(x,\varepsilon), \ldots, y_n(x,\varepsilon)$ of the form (5) of § 1. Suppose that $x_0 = 0$, and that the function $D(x)$ is real with $D'(0) > 0$ for definiteness; then $D(x) > 0$ for $x > 0$ and $D(x) < 0$ for $x < 0$. We consider first the case where the coefficients of equation (2) do not depend on ε. We look for the missing F.A.S. of equation (2) in the form

$$\begin{aligned} y_0(x,\varepsilon) &= \varepsilon^{-1/6}[Av(-\varepsilon^{-2/3}\xi) + Bi\varepsilon^{1/3}v'(-\varepsilon^{-2/3}\xi)] \\ &\quad \times \exp\left\{\frac{1}{\varepsilon}\int_0^x a(t)dt\right\}, \\ a(x) &= \frac{1}{2}(p_1(x) + p_2(x)), \\ A &= \sum_{k=0}^{\infty} A_k(x)\varepsilon^k, \quad B = \sum_{k=0}^{\infty} B_k(x)\varepsilon^k, \end{aligned} \tag{6}$$

where the v is the Airy-Fock function (Chap. 4, § 1).

We can proceed in the same way as in Chap. 4: substitute the F.A.S (6) into equation (2) and so obtain recurrence relations for the functions $A_j(x)$ and $B_j(x)$. However it is simpler to proceed in another way, using the following facts:

1. A F.A.S. of the form (6) exists.
2. If x is such that $\varepsilon^{2/3}|\xi(x)| \gg 1$ then equation (2) has a F.A.S. $\tilde{y}_{1,2}(x,\varepsilon)$ of the form (5) of § 1.

For $x \leqslant -\delta < 0$ one of these F.A.S. has the form

$$\begin{aligned} \tilde{y}_1(x,\varepsilon) = \exp\Bigg\{\varepsilon^{-1}\int_0^x a(t)dt + \varepsilon^{-1}\int_0^x \sqrt{D(t)}dt \\ + \int^x \phi(t)dt\Bigg\}\left[1 + \sum_{k=0}^{\infty} a_k(x)\varepsilon^k\right], \end{aligned}$$

$$\phi(x) = \sum_{k=2}^{n} \frac{p_1'(x)}{p_k(x) - p_1(x)}.$$

The solution $\tilde{y}_2$ is obtained from $\tilde{y}_1$ on replacing $\sqrt{D}$ by $-\sqrt{D}$ and p_1 by p_2 in the formula for ϕ. We transform ϕ:

$$\phi(x) = -\frac{D'(x)}{4D(x)} + \phi'(x) + \frac{\phi_2(x)}{\sqrt{D(x)}},$$
$$\phi_1(x) = \frac{1}{2}\sum_{k=3}^{n} \frac{2a'(x)p_k(x) + D'(x)}{(p_k(x) - a(x))^2 - D(x)}, \tag{7}$$
$$\phi_2(x) = -\frac{a'(x)}{2} + \frac{1}{2}\sum_{k=3}^{n} \frac{D'(x)(p_k(x) - a(x)) + 2a'(x)D(x)}{(p_k(x) - a(x))^2 - D(x)},$$

where $\phi_1(x), \phi_2(x) \in C^\infty(I)$, and put

$$\exp\left\{\int^x \phi(t)dt\right\} = [D(x)]^{-1/4} \exp\left\{\int_0^x \left[\phi_1(t) + \frac{\phi_2(t)}{\sqrt{D(t)}}\right] dt\right\}.$$

The branches of the roots are chosen so that

$$\sqrt{D(x)} > 0, \quad \sqrt[4]{D(x)} > 0, \quad x > 0,$$
$$\sqrt{D(x)} = -i|\sqrt{D(x)}|, \quad \sqrt[4]{D(x)} = e^{-i\pi/4}|\sqrt[4]{D(x)}|, \quad x < 0.$$

For analytic functions this means that in going from the half-line $x > 0$ to the half-line $x < 0$ we go below the branch point $x = 0$. Let us calculate A_0 and B_0. Replace the function v by its asymptotic expansion as $\varepsilon \to +0$, $\xi(x) \geqslant \delta_1 > 0$; then the principal term of the F.A.S. for the solution y_0 will have the same form as for the F.A.S. of the linear combination

$$\tilde{y}_0 = (-i + O(\varepsilon))\tilde{y}_1 + (1 + O(\varepsilon))\tilde{y}_2 .$$

Comparing the F.A.S.'s for y_0 and $\tilde{y}_0$ we obtain

$$\xi(x) = \left[-\frac{3i}{2}\int_0^x \sqrt{D(t)}dt\right]^{2/3},$$

or, because of the choice of branch for the root,

$$\xi(x) = \left[\frac{3}{2}\int_0^x |\sqrt{D(t)}|dt\right]^{2/3}, \qquad x < 0,$$
$$\xi(x) = -\left[\frac{3}{2}\int_0^x |\sqrt{D(t)}dt|\right]^{2/3}, \quad x > 0.$$

Taking into account that $\xi(x) \in C^\infty(I)$ and that $\operatorname{sgn} \xi(x) = -\operatorname{sgn} x$, we obtain the system of equation for A_0, B_0

$$\frac{1}{2}(A_0\xi^{-1/4} \pm B_0\xi^{1/4}) = \frac{1}{\sqrt[4]{D(x)}} \exp\left\{\int_0^x \left[\phi_1(t) \pm \frac{\phi_2(t)}{\sqrt{D(t)}}\right] dt\right\},$$

from which we find

$$A_0(x) = \frac{\xi^{1/4}(x)}{D^{1/4}(x)} e^{\psi_1(x)} \cosh \psi_2(x),$$

and

$$B_0(x) = \frac{1}{[\xi(x)D(x)]^{1/4}} e^{\psi_1(x)} \sinh \psi_2(x) , \tag{8}$$

where

$$\psi_1(x) = \int_0^x \phi_1(t)dt , \quad \psi_2(x) = \int_0^x \frac{\phi_2(t)}{\sqrt{D(t)}} dt .$$

Then that A_0, $B_0 \in C^\infty$ for $x \leqslant 0$. If we choose the branch $\xi^{1/4}(x) = e^{i\pi/4}|\xi(x)|^{1/4}$, $x \geqslant 0$, then A_0, $B_0 \in C^\infty(I)$. Equation (2) also has solutions y_1, y_2 of the form (6) – (8) in which instead of the function v we have the functions w_1 and w_2 (Chap. 4, § 1).

The asymptotic formula can be simplified considerably for the equation of self-adjoint form

$$ly = \sum_{k=0}^{n} (-1)^n \varepsilon^{2k} [q_{n-k}(x) y^{(k)}]^{(k)} = 0 . \tag{9}$$

Since the characteristic equation (§ 1) contains only even powers of p, its roots form pairs $\{p_j(x, \varepsilon), -p_j(x, \varepsilon)\}$. A turning point x_0 is simple if and only if $p_j = 0$ for some j and the other roots p_k are distinct and non-zero at this point.

Let $x_0 = 0$, let the coefficients of equation (9) be independent of ε and let $p_1(0) = 0$; then $p_1(x) = \sqrt{D(x)}$ and $p_2(x) = -\sqrt{D(x)}$. For $\tilde{y}_1$ and $\tilde{y}_2$ we take solutions with asymptotic behaviour

$$\tilde{y}_1(x, \varepsilon) = \frac{f_j(x)}{\sqrt[4]{D(x)}} [1 + O(\varepsilon)] , \quad f_j(x) = \left[\prod_{k=3}^{n} (p_j(x) - p_k(x)) \right]^{-1/2} ,$$

$$j = 1, 2 ,$$

whose existence was proved in § 1. Then we obtain for A_0 and B_0 the formulae

$$A_0(x) = \sqrt[4]{\frac{\xi(x)}{D(x)}} [f_1(x) + f_2(x)] ,$$

and

$$B_0(x) = \frac{1}{\sqrt[4]{\xi(x)D(x)}} [f_1(x) - f_2(x)] .$$

Suppose that the coefficients of equation (2) depend on ε; then for small $|\varepsilon|$ there is a unique turning point $x = x_0(\varepsilon)$, where $x_0(0) = 0$ and $x_0(\varepsilon) \in C^\infty$, which is found from the system

$$l(x, p; \varepsilon) = 0 , \quad l_p(x, p; \varepsilon) = 0 .$$

We let us change the variable $x - x_0(\varepsilon) = \tilde{x}$; then identity $p_1(\tilde{x}, \varepsilon) = p_2(\tilde{x}, \varepsilon)$ is possible only for $\tilde{x} = 0$, so the discriminant D (see(5)) has the form

$$D(\tilde{x}, \varepsilon) = \tilde{x} D_1(\tilde{x}, \varepsilon), \quad D_1(0,0) \neq 0 .$$

Suppose that the function $D(\tilde{x}, \varepsilon)$ is real with $D_1(0,0) > 0$. Then the above formulae (6) – (9) are true. The difference is that we must replace x by $\tilde{x}$ in all the formulae, so that the functions $\xi_0, \xi_1, p_1, \ldots, p_n$ still depend on ε as well.

2.3 Asymptotic Behaviour of Solutions. Let x_0 be a simple turning point and suppose the following condition is satisfied:

(3) The functions Re $(p_1(x) - p_j(x))$ and Re $(p_2(x) - p_j(x))$ do not change sign for $x \in I$.

Then equation (2) has solutions y_0, y_1, y_2 of the form (6) – (9) (for the solutions $y_{1,2}$, v must be replaced by $w_{1,2}$), in which both the series $A(x, \varepsilon)$ and $B(x, \varepsilon)$ are asymptotic as $\varepsilon \to +0$ uniformly in $x \in I$.

If the coefficients of equation (2) satisfy condition A of paragraph 1, then there exists a complex neighbourhood U of x_0 in which the asymptotic expansions indicated are valid as $x \to +0$ uniformly in $x \in U$. All these asymptotic expansions can be differentiated in x and ε an arbitrary number of times, preserving the uniformity in x of the bound for the remainder term.

2.4 Asymptotic Behaviour of the Solutions of System (1). Without loss of generality we can assume that $x_0(\varepsilon) = 0$ (see paragraph 2.2). All the scalars, vectors and matrix functions considered below lie in $C^\infty(I)$ if $A(x, \varepsilon)$ satisfies condition C of paragraph 1, and they are holomorphic in the domain D if $A(x, \varepsilon)$ satisfies condition A. We will assume that the segment $I \ni 0$ and $D \ni 0$ are small. Denote $D(x) = D(x, 0)$ (see (5)), $A_0(x) = A(x, 0)$ and $p_j(x) = p_j(x, 0)$. There exists a matrix $T_0(x)$ reducing $A_0(x)$ for $x \in I (x \in D)$ to the block-diagonal form

$$T_0^{-1}(x) A_0(x) T_0(x) = \begin{bmatrix} B(x) & 0 \\ 0 & \Lambda(x) \end{bmatrix},$$

where

$$B(x) = \begin{bmatrix} a(x) & 1 \\ D(x) & a(x) \end{bmatrix}, \quad a(x) = \frac{1}{2}[p_1(x) + p_2(x)], \tag{10}$$

$$\Lambda(x) = \text{diag}\,(p_3(x), \ldots, p_n(x)) .$$

The transformation

$$y = \exp\left\{ \frac{1}{2\varepsilon} \int_0^x [p_1(t) + p_2(t)]dt \right\} T_0(x) u \tag{11}$$

reduces system (1) to the form

$$\varepsilon u' = B_0(x)u + \varepsilon B(x,\varepsilon)u\,,$$
$$B(x,\ \varepsilon) = \varepsilon^{-1}T_0^{-1}(x)[A(x,\ \varepsilon) - A(x,\ 0)]T_0(x) - T^{-1}(x)T'(x)\,, \tag{12}$$

where $B_0(x)$ is the block-diagonal matrix

$$B_0(x) = \operatorname{diag}\,(\tilde{B}(x),\ \Lambda(x) - a(x)I_{n-2})\,,$$
$$\tilde{B}(x) = \begin{bmatrix} 0 & 1 \\ D(x) & 0 \end{bmatrix}.$$

We have $B(x,\varepsilon) = B_1(x) + \varepsilon B_2(x) + \dots$. We look for a F.A.S. to system (12) in the form

$$u(x,\varepsilon) = w(\varepsilon^{-2/3}\xi(x))f(x,\varepsilon) + \varepsilon^{1/3}w'(\varepsilon^{-2/3}\xi(x))g(x,\varepsilon)\,,$$
$$f(x,\varepsilon) = \sum_{k=0}^{\infty} f_k(x)\varepsilon^k\,, \quad g(x,\varepsilon) = \sum_{k=0}^{\infty} g_k(x)\varepsilon^k\,, \tag{13}$$

where $w(t)$ is the solution of the Airy equation $w'' - tw = 0$. Substituting (13) into (12) and equating the coefficients of w and w' we arrive at the system

$$\xi'\xi g + \varepsilon f' = (B_0 + \varepsilon B)f\,,$$
$$\xi' f + \varepsilon g' = (B_0 + \varepsilon B)g\,.$$

We expand f, g, and B, as power series in ε and equate coefficients of these powers; then we obtain a recurrence system of equations for f_k and g_k. The first pair of equations has the form

$$B_0 f_0 = \xi'\xi g_0\,, \quad B_0 g_0 = \xi' f_0\,,$$

so that

$$B_0^2 f_0 = \xi'^2\xi f_0\,, \quad B_0^2 g_0 = \xi'^2\xi g_0\,.$$

We note that $B_0^2 = \operatorname{diag}\,(D(x)I_2,\ (\Lambda(x) - Ia(x))^2)$ and put

$$\xi(x) = \left(\frac{3}{2}\int_0^x \sqrt{D(t)}dt\right)^{2/3}. \tag{14}$$

Let $x \in I$, $D(x) > 0$ for $x > 0$ for definiteness, the branch of $\sqrt{D(x)}$ being chosen as in paragraph 2.2. Then

$$\xi(x) \in C^\infty(I)\,, \quad \operatorname{sgn}\,\xi(x) = \operatorname{sgn}\,x\,,$$

and we have

$$g_0(x) = \alpha_1(x)e_1 + \alpha_2(x)e_2\,,$$
$$f_0(x) = \alpha_1(x)\frac{D(x)}{\xi'(x)}e_2 + \alpha_2(x)\frac{e_1}{\xi'(x)}\,, \tag{15}$$
$$e_1 = (1,0,0,\ldots,0)^T\,, \quad e_2 = (0,1,0,\ldots,0)^T\,.$$

The unknown functions $\alpha_1(x)$ and $\alpha_2(x)$ are determined from the system of equations for the second approximation:

$$\begin{bmatrix} B_0 & -\xi'\xi I_n \\ -\xi' I_n & B_0 \end{bmatrix} \begin{bmatrix} f_1 \\ g_1 \end{bmatrix} = \begin{bmatrix} f_0' & -B_1 f_0 \\ g_0' & -B_1 g_0 \end{bmatrix}. \tag{16}$$

The corresponding homogeneous system has non-trivial solutions, so that a necessary condition for solvability is that the right-hand side of system (16) is orthogonal to the solutions of the adjoint homogeneous system. A basis in the space of solutions of the latter system is formed by the $2n$-vectors

$$(e_1^T, e_2^T/\xi')\,, \quad (e_2^T, (D/\xi')e_1^T)\,,$$

where the vectors e_1 and e_2 are as in (15). The conditions for solvability lead to a system of equations for the coefficients α_1 and α_2:

$$(\alpha_1\xi'\xi)' + \alpha_1'\xi'\xi = A\xi'\xi\alpha_1 + \frac{B}{\xi'}\alpha_2\,,$$
$$\left(\frac{\alpha_2}{\xi'}\right)' + \frac{\alpha_2}{\xi'} = \frac{B}{\xi'}\alpha_1 + \frac{A}{\xi'}\alpha_2\,, \tag{17}$$
$$A = b_{11} + b_{22}\,, \quad B = Db_{12} + b_{21}\,.$$

This system is of the same form as the system in § 1 of Chap. 4, so that

$$\alpha_1(x) = \frac{1}{\sqrt{\xi'(x)\xi(x)}}\exp\left\{\frac{1}{2}\int_0^x A(t)dt\right\}\sinh\left\{\frac{1}{2}\int_0^x \frac{B(t)dt}{\sqrt{D(t)}}\right\}\,,$$
$$\alpha_2(x) = \sqrt{\xi(x)}\exp\left\{\frac{1}{2}\int_0^x A(t)dt\right\}\cosh\left\{\frac{1}{2}\int_0^x \frac{B(t)dt}{\sqrt{D(t)}}\right\}\,. \tag{18}$$

Formulae (10), (11), (13)–(15), (17) and (18) determine fully the principal asymptotic term. The vector functions f_k and g_k, $k \geqslant 1$, are determined from the linear system of equations, the matrix of which is the same as that in (16), and the right-hand side is a vector function which is calculated in terms of $f_0, f_1, \ldots, f_{k-1}$, $g_0, g_1, \ldots, g_{k-1}$. The conditions for the existence of the asymptotic formulae for the solutions are formulated in the same way as in paragraph 2.3.

2.5 Superposition of Simple Turning Points. We will restrict ourselves to the case where two simple turning points are superimposed. Suppose that the

Jordan normal form of the matrix $A_0(x_0)$ contains two second-order Jordan blocks

$$A_0(x_0) = \left[\begin{array}{cc|cc|ccc} p_0 & 1 & & & & & \\ 0 & p_0 & & & & 0 & \\ \hline & & q_0 & 1 & & & \\ & & 0 & q_0 & & & \\ \hline & & & & p_5^0 & & \\ & 0 & & & & \ddots & \\ & & & & & & p_n^0 \end{array}\right]$$

where $p_0 \neq q_0$, $p_0 \neq p_j^0$, $q_0 \neq p_j^0$ for $j = 5, \dots, n$ and $l_x(x_0, p_0; 0)$, $l_x(x_0, q_0; 0) \neq 0$. In this case there exists a matrix $T_0(x)$ reducing $A_0(x)$ to the block-diagonal form

$$T_0^{-1}(x)A_0(x)T_0(x) = \operatorname{diag}\left(B_1(x),\, B_2(x),\, \Lambda(x)\right),$$

where

$$B_j(x) = \begin{bmatrix} a_j(x) & 1 \\ D_j(x) & a_j(x) \end{bmatrix}, \quad j = 1, 2, \quad \Lambda(x) = \operatorname{diag}\left(p_5(x), \dots, p_n(x)\right).$$

The properties of all these functions are the same as in paragraph 2.4. We have

$$p_{1,2}(x) = a_1(x) \pm \sqrt{D_1(x)}, \quad p_{3,4}(x) = a_2(x) \pm \sqrt{D_2(x)}.$$

In this case system (1) has a F.A.S. of the form (13), corresponding to the functions

$$\xi_j(x) = \left(\frac{3}{2}\int_0^x \sqrt{D_j(t)}dt\right)^{2/3}, \quad j = 1, 2,$$

and the asymptotic formulae for the solutions have the same form as in paragraph 2.4.

We state the asymptotic formulae for the solutions of equation (9) where two simple turning points are superimposed at $x = 0$. Suppose that the coefficients of equation (9) do not depend on ε and that $p_1^0 = p_2^0 \neq 0$ and the remaining roots $p_3^0, \dots, p_n^0$ are distinct and non-zero at $x = 0$. In this case $p_0 = p_1^0$ and $q_0 = -p_1^0$. We denote

$$a(x) = \frac{1}{2}[p_1(x) + p_2(x)],$$

and

$$f_j(x)[p_j(x)]^{-1/4}\left\{\prod_{k=3}^{n}\sqrt{p_j^2(x)-p_k^2(x)}\right\}^{-1}.$$

Then equation (9) has an asymptotic expansion of the form (6) where

$$A_0(x)=\sqrt[4]{\frac{\xi(x)}{a(x)D(x)}}[f_1(x)+f_2(x)],$$

$$B_0(x)=\frac{1}{\sqrt[4]{a(x)\xi(x)D(x)}}[f_1(x)-f_2(x)].$$

3. Asymptotic Simplification of Systems

3.1. Exact Reduction of a System. Let the matrix function $A(x,\varepsilon)$ satisfy condition A of paragraph 1 and suppose that $0\in D$. Then

$$A(x,\varepsilon)=\sum_{k=0}^{\infty}\varepsilon^k A_k(x),$$

where the $A_k(x)$ are holomorphic in D. Suppose that U is a sufficiently small neighbourhood of $x=0$ and $0\leqslant\varepsilon_1\leqslant\varepsilon_0$ where ε_0 is sufficiently small. For any integer $m\geqslant 0$ there exists a matrix function $T(x,\varepsilon)$, holomorphic in x and ε for $x\in U$, $|\varepsilon|\leqslant\varepsilon_1$ such that

$$T(x,0)=I,\quad x\in U;\qquad T(0,\ \varepsilon)=I,\quad |\varepsilon|\leqslant\varepsilon_1,$$

and the transformation

$$y=T(x,\varepsilon)z \tag{19}$$

reduces system (1) to the form

$$\varepsilon z'=\left[\sum_{k=0}^{\infty}\varepsilon^k A_k(x)+\varepsilon^{m+1}B(x)\right]z.$$

Here $x\in U$, $|\varepsilon|\leqslant\varepsilon_1$, and the matrix function $B(x)$ is holomorphic in U. In particular, system (1) with the substitution (19) can be reduced to the form

$$\varepsilon z'=[A_0(x)+\varepsilon B(x)]z.$$

3.2 Asymptotic Block-Diagonalization of Systems. Let $A_0(x)$ be a square matrix of order n and of class $C^\infty(I)$, $I=[-a,a]$, $a>0$, and suppose that the eigenvalues of A can be split into the two groups: $M_1=\{p_1(x),\dots,p_k(x)\}$ and $M_2=\{p_{k+1}(x),\dots,p_n(x)\}$ such that

$$p_j(x)\neq p_l(x),\quad p_j(x)\in M_1,\quad p_l(x)\in M_2, \tag{20}$$

for all $x \in I$. Then there exist both a number b, $0 < b \leqslant a$, and a matrix function $T_0(x)$ of class $C^\infty(\tilde{I})$, non-singular for $x \in \tilde{I} = [-b, b]$, such that

$$T_0^{-1}(x)A_0(x)T_0(x) = \begin{bmatrix} B_1(x) & 0 \\ 0 & B_2(x) \end{bmatrix}, \quad x \in \tilde{I}. \tag{21}$$

Here $B_1(x)$ and $B_2(x)$ are square matrices of orders k and $n-k$ of class $C^\infty(\tilde{I})$, and their eigenvalues consist of the groups M_1 and M_2 respectively.

An analogous result is true for the system (1) if $A(x, \varepsilon)$ satisfies condition C and if $A_0(x) = A(x, 0)$ satisfies the above conditions. Then for any integer $N \geqslant 1$ there is a matrix $T_N(x, \varepsilon)$ of class $C^\infty(\tilde{I} \times J)$ such that transformation (19) reduces system (1) to the form

$$\varepsilon z' = \begin{bmatrix} B_1(x, \varepsilon) & 0 \\ 0 & B_2(x, \varepsilon) \end{bmatrix} z + \varepsilon^{N+1} C(x, \varepsilon) z. \tag{22}$$

Here B_1, B_2, and C are square matrices of orders $k, n-k, n$ of class $C^\infty(\tilde{I} \times J)$, and

$$\begin{aligned} B_j(x, \varepsilon) &= B_j(x) + \sum_{l=1}^{N} \varepsilon^l B_{jl}(x), \quad j = 1, 2, \\ T(x, \varepsilon) &= T_0(x) + \sum_{l=1}^{N} \varepsilon^l T_l(x), \quad B_{jl}(x), \quad T_l(x) \in C^\infty(\tilde{I}). \end{aligned} \tag{23}$$

Suppose that $A_0(x)$ is holomorphic in the closure $[D]$ of a bounded simply-connected domain D in the complex x-plane and that condition (20) is satisfied everywhere in $[D]$. Then there exists a matrix $T_0(x)$, holomorphic and non-degenerate for $x \in [D]$, reducing $A_0(x)$ to the form (21). If $A(x, \varepsilon)$ is holomorphic in x and ε for $x \in [D]$, $|\varepsilon| \leqslant \varepsilon_0$, then all the above statements concerning system (1) remain in force for $x \in [D]$, $|\varepsilon| \leqslant \varepsilon_1 \leqslant \varepsilon_0$, if $\varepsilon_1 > 0$ is sufficiently small, where the conditions of infinite differentiability of all the matrices are replaced by the condition of being holomorphic.

There are other possible conditions on system (1). Suppose for example that $A(x, \varepsilon)$ is holomorphic in x and ε in the domain

$$D_0 : |x| \leqslant x_0, \quad 0 < |\varepsilon| \leqslant \varepsilon_0, \quad |\arg \varepsilon| \leqslant \theta_0,$$

and has an asymptotic series expansion

$$A(x, \varepsilon) = \sum_{r=0}^{\infty} A_r(x)\varepsilon^r, \quad \varepsilon \to 0, \quad |\arg \varepsilon| \leqslant \theta_0,$$

uniform in $|x| \leqslant x_0$. Then the block-diagonalization of (1) is possible in the domain

$$D_1 : |x| \leqslant x_1, \quad 0 < |\varepsilon| \leqslant \varepsilon_1, \quad |\arg \varepsilon| \leqslant \theta_1,$$

if $x_1, \varepsilon_1, \theta_1$ are sufficiently small. The matrices T, B, B_2, C have the same properties as $A(x, \varepsilon)$ has, but in the smaller domain D_1.

If system (1) is self-conjugate, that is, $A(x, \varepsilon)$ is skew-hermitian:

$$A(x,\ \varepsilon) = -\overline{A^T(\overline{x}, \overline{\varepsilon})}\,,$$

then B_1, B_2 will also be skew-hermitian with a suitable choice of the transformation (19).

In all the cases considered we can reduce the system to the form

$$\varepsilon z' = \begin{bmatrix} B_1(x,\ \varepsilon) & 0 \\ 0 & B_2(x, \varepsilon) \end{bmatrix} z + C(x,\ \varepsilon) z\,,$$

where the $B_j(x, \varepsilon)$ are asymptotic series in powers of ε and $C(x, \varepsilon) = O(\varepsilon^\infty)$, $\varepsilon \to 0$ (for remarks on this see Chap. 1, § 3, paragraph 3).

3.3 Canonical Systems. Let 0 and I_n be the zero and unit matrices of order n,

$$J_{2n} = \begin{bmatrix} 0 & I_n \\ -I_n & 0 \end{bmatrix}, \quad \tilde{J}_{2n} = \begin{bmatrix} J_{2p} & 0 \\ 0 & J_{2(n-p)} \end{bmatrix},$$

and suppose that system (1) is canonical (§ 4) for $x \in \mathbb{R}$ and $0 < \varepsilon \leqslant \varepsilon_0$. The matrix function $A(x, \varepsilon)$ is real and *Hamiltonian*, i.e.

$$J_{2n} A(x,\ \varepsilon) + A^T(x,\ \varepsilon) J_{2n} = 0\,.$$

The eigenvalues of A occur in pairs $(p, -p)$. A complex function is called *tame* if $f(x) \in C^\infty(\mathbb{R})$ and $f^{(k)}(x) \in L_1(\mathbb{R})$ for all $k \geqslant 0$.

Suppose that the following conditions are satisfied:

(1) The elements of $A'(x, \varepsilon)$ and $A(x, \varepsilon) - A(x, 0)$ are tame functions.

(2) There is the asymptotic expansion

$$A(x,\ \varepsilon) = \sum_{r=0}^{\infty} A_r(x) \varepsilon^r \tag{24}$$

as $\varepsilon \to +0$ uniformly for $x \in \mathbb{R}$.

(3) For all $x \in \mathbb{R}$ the characteristic polynomial $d(p^2, x)$ of $A_0(x)$ has representation

$$d(p^2, x) = d_1(p^2, x) d_2(p^2, x)\,,$$

where d_1 and d_2 are non-constant polynomials of powers $2p$ and $2(n-p)$ respectively, having no common roots for each $x \in \mathbb{R}$.

A matrix M of order $(2n \times 2n)$ is called *block-Hamiltonian* if

$$\tilde{J}_{2n} M + M^T \tilde{J}_{2n} = 0\,.$$

Under these conditions there is a real matrix function $T_0(x)$ which reduces $A_0(x)$ to block-diagonal form

$$T_0^{-1}(x)A_0(x)T_0(x) = \operatorname{diag}\,(B_0^{11}(x), B_0^{22}(x)) \tag{25}$$

for all $x \in \mathbb{R}$. The matrices $B_0^{jj}(x)$, $j = 1, 2$, are Hamiltonian matrices of orders $2p$ and $2(n-p)$ respectively, $d_j(p^2, x)$ are their characteristic polynomials, and for all $x \in \mathbb{R}$ we have

$$T_0^T(x)J_{2n}T_0(x) = \tilde{J}_{2n}\,. \tag{26}$$

Elements of the derivatives of the matrices $T_0(x)$, $B_0^{jj}(x)$, $j = 1, 2$ are tame functions.

There is a matrix $T(x, \varepsilon)$ such that transformation (19) reduces (1) to the form

$$\begin{aligned} \varepsilon z' &= \left[\sum_{r=0}^{\infty} B_r(x)\varepsilon^r + C(x,\varepsilon)\right] z\,, \\ B_r(x) &= \begin{bmatrix} B_r^{11}(x) & 0 \\ 0 & B_r^{22}(x) \end{bmatrix}, \end{aligned} \tag{27}$$

where B_r^{jj}, $j = 1, 2$, are Hamiltonian matrices of orders $2p$ and $2(n-p)$ respectively, and for $r = 0$ they coincide with the matrices from (25). As $\varepsilon \to +0$, and for any integer $N \geqslant 1$, we have $C(x, \varepsilon) = O(\varepsilon^N)$ uniformly in $x \in \mathbb{R}$. Also, T satisfies the identity (26) for all $x \in \mathbb{R}$ and has asymptotic series expansion

$$T(x,\varepsilon) = T_0(x) + \sum_{r=1}^{\infty} \varepsilon^r T_r(x)\,, \quad \varepsilon \to +0\,, \tag{28}$$

uniformly in $x \in \mathbb{R}$, $T(x, \varepsilon) - T_0(x)$ vanishes for $x = \pm\infty$, and the elements of the matrices $B_r^{jj}(x)$, $j = 1, 2$, $T_r(x)$ for $r \geqslant 1$ and $T'(x, \varepsilon)$ are tame. The results in paragraphs 3.2 and 3.3 are due to W. Wasow and Y. Sibuya [Sibuya 2 and 3, Wasow 3 and 4].

3.4 The Arnol'd Normal Form of the Matrix and Reduction of Systems. We will consider the system (1) where $A(x, \varepsilon)$ is holomorphic for $|x| \leqslant r \ll 1$, $0 < \varepsilon \leqslant \varepsilon_0 \leqslant 1$. Denote $A_0(x) = A(x, 0)$. Suppose that $A_0(0)$ has Jordan normal form and a single eigenvalue. We can assume that this last condition holds without loss of generality (see paragraph 3.2). Let $m_1 \geqslant m_2 \geqslant \ldots \geqslant m_p$ be the powers of the elementary divisors of $A_0(0)$, and the Jordan blocks are also placed in decreasing order of size. We shall divide the matrix $A_0(x)$ up into blocks A^{jk}. Let us introduce the matrices $\Gamma_1, \ldots, \Gamma_d$, $d = \sum_{k=1}^{p}(2k-1)m_k$, of order $n \times n$. The matrix Γ_s has a single element, equal to 1, all the other elements are zero. If Γ_s^{jk} is a non-zero block then the unit element lies in the last row for $j \leqslant k$ and in the first column for $j = k$. Let the matrices Γ_s be ordered so that the unit element is in the last row of the block Γ_1^{11}

and so on. We give a result due to V.I. Arnol'd [Arnol'd] concerning matrix functions of one variable. There are scalar functions $\rho_1(x), \ldots, \rho_d(x)$ and a matrix function $T_0(x)$, holomorphic at $x = 0$, such that

$$T_0^{-1}(x)A_0(x)T_0(x) = A_0(0) + \sum_{s=1}^{d} \rho_s(x)\Gamma_s \tag{29}$$

in some neighbourhood of $x = 0$. Further

$$\rho_1(0) = \ldots = \rho_d(0) = 0\,, \quad \det T_0(0) \neq 0\,.$$

Without loss of generality we can assume that all the eigenvalues of $A_0(x)$ are zero and that $A_0(x)$ has the canonical form (29). Then there are matrix functions $T_r(x)$ with $T_0(x) = I$ and scalars $\rho_{sr}(x)$ such that transformation (19) reduces (1) to the form [Wasow 1]

$$\varepsilon z' = [A_0(0) + \sum_{s=1}^{d} \rho_s(x, \varepsilon)\Gamma_s]z\,, \quad \rho_s(x, \varepsilon) = \sum_{r=0}^{\infty} \rho_{sr}(x)\varepsilon^r\,. \tag{30}$$

Here $T(x, \varepsilon)$ has the form (28) where $T, \rho_1, \ldots, \rho_d$ are formal power series. It follows from this that there exists a matrix T and a function ρ_s of class C^∞ for $|x| \leqslant r$, $0 \leqslant \varepsilon \leqslant \varepsilon_0$, which are holomorphic for each fixed ε and for which the formal series are asymptotic as $\varepsilon \to +0$ uniformly in x. Transformation (19) reduces (1) to the form

$$\varepsilon z' = [A_0(0) + \sum_{s=1}^{d} \rho_s(x, \varepsilon)\Gamma_s + B(x, \varepsilon)]\,,$$

$$B(x,\ \varepsilon) = O(\varepsilon^N)\,, \quad \varepsilon \to +0\,,$$

for an arbitrary integer $N \geqslant 1$, uniformly in x.

Nothing is known about the asymptotic behaviour of the solutions of those systems in the general case.

3.5 Systems with Moving Singularities. We consider the matrix equation

$$\varepsilon^h \phi(x,\ \varepsilon)Y' = A(x,\ \varepsilon)Y\,, \tag{31}$$

where ϕ is a scalar function, h is a constant, and Y and A are square matrices of order n. System (31) has singularities at points where $\phi(x, \varepsilon) = 0$. A typical example is the system

$$\varepsilon^h(x + \varepsilon)Y' = A(x)Y\,. \tag{32}$$

The asymptotic behaviour of the solutions of the system for $h = 0$, for non-analytic $A(x)$, are investigated in [Lomov].

Let us introduce the following notation: U and V are the disks $|x| \leqslant x_0$ and $|\varepsilon| \leqslant \varepsilon_0$ in the complex x- and ε-planes; $S(x_0, \alpha, \beta)$ and $S(\varepsilon_0, \alpha, \beta)$ are the intersections of $U\backslash\{0\}$ and $V\backslash\{0\}$ respectively with the sectors $\alpha < \arg x < \beta$ and $\alpha < \arg \varepsilon < \beta$; $U_0 \subset U$ and $V_0 \subset V$ are open domains such that $0 \in [U_0]$ and $0 \in [V_0]$. We will assume that the numbers $x_0 > 0$ and $\varepsilon_0 > 0$ and the domains U_0 and V_0 are sufficiently small. Suppose that the following conditions are satisfied:

1) $\phi(x,\varepsilon)$ is holomorphic in $U \times V$, $\phi(x,0) \neq 0$, $\phi(x,\varepsilon) \neq 0$ in $U_0 \times V_0$.

2) The matrix $A(x,\varepsilon)$ is holomorphic in $U \times S(\varepsilon_0,\alpha,\beta)$ and has an asympotic series expansion of the form (24) for $\varepsilon \to 0$, $\varepsilon \in S(\varepsilon_0,\alpha,\beta)$.

3) The eigenvalues $p_1,\ldots,p_n$ of $A(0,0)$ split into two groups $M_1 = \{p_1, \ldots, p_k\}$ and $M_2 = \{p_{k+1},\ldots,p_n\}$ such that $p_j \neq p_l$ for $p_j \in M_1$, $p_l \in M_2$.

4) There is a point $a \in [U_0]$ and a family of simple smooth curves

$$\gamma_x : \ t = t(s,x), \quad t(0,x) = 0, \quad t(1,x) = a,$$

lying in $[U_0]$ that are continuous in $x \in [U_0]$ and such that $t'_s(s_x,x) = x$ at some point on γ_x.

Because of condition 3) we can assume that $A(x,0)$ has been reduced to block-diagonal form (21). Denote

$$p_{jl} = p_j - p_l, \quad p_j \in M_1, \quad p_l \in M_2,$$

$$\psi = \arg\,(p_{jl}t'_s) - \arg\,(\varepsilon^h\phi(t,\varepsilon)).$$

We give two possibilities for the reduction of system (31).

I. $h \geqslant 1$ is an integer. Suppose that there is a δ, $0 < \delta < \pi/2$, such that ψ lies on one of the line segments of the form

$$I_m^+ = [-\pi/2 + \delta + 2\pi m,\ \pi/2 - \delta + 2\pi m],$$
$$I_m^- = [\pi/2 + \delta + 2\pi m,\ 3\pi/2 - \delta + 2\pi m],$$

where $m = 0, \pm 1, \ldots$, when $t \in \gamma_x$, $\varepsilon \in V_0$.

Then there is a transformation of the form (19), reducing system (31) to block-diagonal form

$$\varepsilon z' = B(x,\varepsilon)z, \quad B = \begin{bmatrix} B_1(x,\varepsilon) & 0 \\ 0 & B_2(x,\varepsilon) \end{bmatrix}, \tag{33}$$

where $B(x,0) = A(x,0)$. The matrices $T(x,\varepsilon)$ and $B(x,\varepsilon)$ have the following properties:

1) $T(x,\varepsilon)$ and $B(x,\varepsilon)$ are holomorphic in $U_0 \times V_0$ and are continuous for in $[U_0 \times V_0]$.

2) As $\varepsilon \to 0$, $\varepsilon \in V_0$, the matrices $T(x,\varepsilon)$ and $B(x,\varepsilon)$ have asymptotic series expansions of the form (28) that are uniform in $x \in [U_0]$.

3) For $x \in [U_0]$, $\det T_0(x) \neq 0$, the matrices $B_r(x)$ and $T_r(x)$, $r = 0, 1, \ldots$, are holomorphic in $S(x_0,\alpha,\beta)$ and can be expanded in this sector as an asymptotic series in powers of x as $x \to 0$.

II. Suppose that $h = \phi(0,0) = \phi'_x(0,0) = 0$, m is the multiplicity of the zero ($x = 0$) of $\phi(x,0)$ and $0 \leqslant \beta - \alpha \leqslant \pi/(m-1)$. Suppose that there is an s_0, $0 < s_0 < 1$, such that the value of ψ lies on one of the line segments I_m^+, I_m^-, when $t = t(s,x)$, $0 < s \leqslant s_0$.

Then the transformation (19) exists, reducing (31) to the form (33), and $T(x,\varepsilon)$, $B(x,\varepsilon)$ have properties enumerated above.

The simplest example of this case is the system

$$(x^2 - \varepsilon)Y' = AY\,, \quad A \neq 0\,.$$

For $\varepsilon \neq 0$ the system has the two regular singular points $x = \pm\sqrt{\varepsilon}$ which coalesce as $\varepsilon \to 0$ to the one irregular point $x = 0$.

Remark. The conditions on the function ϕ mean that for any point $x \in [U_1]$ there is a matrix of canonical paths with initial point $x = 0$ or $x = a$.

The results of paragraph 3.5 were obtained in [Zhdanova 1].

4. The Canonical Maslov Operator. The results given in this section are due to V.P. Maslov and are a very special application of his development of the method of a canonical operator [Maslov 1 and 2]. We will consider the equation

$$\hat{L}y \equiv L\left(\overset{2}{x}, \frac{1}{i\lambda}\overset{1}{D}; \frac{1}{i\lambda}\right) y = 0\,, \tag{34}$$

where $\lambda > 0$ is a large parameter, and

$$L(x,p;\varepsilon) = p^n + \sum_{k=0}^{n-1} q_{n-k}(x,\varepsilon)p^k\,, \quad \varepsilon = \frac{1}{i\lambda}\,.$$

The function L is called the ***λ-symbol*** of the operator $\hat{L}$. The function $L^0(x,p) = L(x,p;0)$ is called ***the principal λ-symbol.***

The coefficients of (31) satisfy condition A of paragraph 1. Here the notation used is that due to R. Feynman: the numbers 1, 2 indicate the order of operation of the multiplication operator on the function of x and the differentiation. For example, if $L = xp$ then

$$L\left(\overset{2}{x}, \overset{1}{\frac{d}{dx}}\right) y = x\frac{d}{dx}y\,, \quad L\left(\overset{1}{x}, \overset{2}{\frac{d}{dx}}\right) y = \frac{d}{dx}(xy)\,.$$

4.1 Transition to the p-Representation. We introduce the Fourier λ-transform

$$\{F_{\lambda,x\to p}f(x)\}(p) = \sqrt{\frac{\lambda}{2\pi i}} \int_{-\infty}^{\infty} f(x)\exp\{-i\lambda px\}dx\,,$$

$$\{F^{-1}_{\lambda,p\to x}g(p)\}(x) = \sqrt{\frac{\lambda}{-2\pi i}} \int_{-\infty}^{\infty} g(p)\exp\{i\lambda xp\}dp\,,$$

where $\sqrt{i} = e^{i\pi/4}$ and $\sqrt{-i} = e^{-i\pi/4}$. Then

$$F_{\lambda, x\to p} \hat{L}\left(\overset{2}{x}, \frac{1}{i\lambda}\overset{1}{\frac{d}{dx}}; \varepsilon\right) F^{-1}_{\lambda, p\to x} = \hat{\hat{L}} = L\left(-\frac{1}{i\lambda}\overset{1}{\frac{d}{dp}}, \overset{2}{p}; \varepsilon\right).$$

If the coefficients of $\hat{L}$ are polynomials in x then $\hat{\hat{L}}$ is a differential operator. In the general case it is a pseudo-differential operator.

We seek a F.A.S. to equation (31) in the form of a Fourier transform of an exponent, multiplied by an asymptotic series in powers of ε:

$$y = F^{-1}_{\lambda, p\to x}\tilde{y}, \quad \tilde{y} = \exp\{i\lambda S(p)\}\phi(p, \varepsilon),$$

so that $\hat{\hat{L}}\tilde{y} = 0$. If S is real and $\phi(p)$ is a compactly supported function then there is the asymptotic expansion [Harris 1]

$$\hat{\hat{L}}[\exp\{i\lambda S(p)\}\phi(p)] = \exp\{i\lambda S(p)\}\left\{L\phi + \frac{1}{i\lambda}\right.$$
$$\left.\times\left[-L_x\phi' + \left(\frac{1}{2}L_{xx}S'' - L_{xp}\right)\phi + L\varepsilon\phi\right] + \ldots\right\},$$

where the values of the symbol L and its derivatives are taken at the point $(-S'(p), p; 0)$. We obtain the equation for S

$$L^0(-S'(p), p) = 0.$$

4.2 Structure of the Roots. Suppose that the principal symbol L^0 of the operator L is real and the equation

$$L^0(x, p) = 0 \tag{35}$$

determines a smooth curve Γ on the plane (x, p) such that grad $L^0 \neq 0$ on Γ. If the point $r_0 = (x_0, p_0) \in \Gamma$ then x_0 is a turning point if and only if the tangent to Γ at r_0 is parallel to the p-axis. Since $L^0_x(x_0, p_0) \neq 0$ there is an integer k, $1 \leqslant k \leqslant n$, such that

$$\frac{\partial L^0}{\partial p} = \ldots = \left(\frac{\partial}{\partial p}\right)^{k-1} L^0 = 0, \quad \left(\frac{\partial}{\partial p}\right)^k L^0 \neq 0 \tag{36}$$

at the point r_0. If $k = 2$ then the turning point x_0 is simple.

In a small neighbourhood of the point r_0 the curve Γ is given by the equation $x = x(p)$ where $x(p)$ is a function of class C^∞ for small $|p - p_0|$ and

$$x(p) - x_0 \sim a(p - p_0)^k, \quad p \to p_0, \quad a \neq 0.$$

We point out that at the turning point x_0 the k roots $p_j(x)$ of the characteristic equation (35) coalesce.

Remark. In a small neighbourhood of r_0, using the change of variables $\tilde{p} = p_0 + \psi(x,p)$, $\tilde{x} = x - x_0$, the principal symbol can be reduced to the form

$$L^0(x,p) = (\tilde{p}^k \pm \tilde{x})L_1(\tilde{x},\tilde{p}), \quad L_1(0,0) \neq 0.$$

The function $\psi(x,p)$ is in C^∞ for small $|x - x_0|$ and $|p - p_0|$, is real for real x and p, and is holomorphic in p for fixed x.

Asymptotic Behaviour of the Solutions. Suppose that $x_0 = 0$ and $L = L(x,p)$. The function $\eta(p)$ belongs to $C_0^\infty(\mathbb{R})$ and is defined by $\eta(p) \equiv 1$ in a neighbourhood of the point $p_0 = 0$ and the support of $\eta(p)$ is sufficiently small. We put $S(p) = -\int_0^p x(p')dp'$. Equation (34) has an F.A.S. y of the form

$$\begin{aligned} y(x,\lambda) = \sqrt{\frac{\lambda}{-2\pi i}} \int_0^\infty & \exp\left\{ i\lambda\left[px - \int_0^p x(p')dp'\right] \right. \\ & \left. - \frac{1}{2}\int_0^p \frac{L^0_{px}(x(p'),p')}{l^0_x(x(p'),p')}dp' \right\} [L_x(x(p'),p')]^{-1/2} \\ & \times \left[1 + \sum_{k=1}^\infty \left(\frac{1}{i\lambda}\right)^k \phi_k(p)\right] \eta(p)dp. \end{aligned} \tag{37}$$

The functions $\phi_1, \dots, \phi_N$ can be found from a sequence of recurrence equations of the form

$$-L_x\phi_k' + \left(-\frac{1}{2}L_{xx}S'' + L_{xp}\right)\phi_k = f(\phi_0, \dots, \phi_{k-1}).$$

Formula (37) is the analogue of (7) of § 1 in the p-representation. The existence of a solution having an asymptotic expansion of the form (37) is rigorously proved for $k = n$ (see paragraph 6).

The behaviour of the solution y depends crucially on the evenness of the number k. Suppose that $L_x < 0$ and $(\partial/\partial p)^k L > 0$ at $(0, p_0)$ for definiteness, and let $\delta > 0$ be sufficiently small. If k is odd, then for small $|x|$ the integral in (37) has a unique real stationary phase point determined from the equation $x = x(p)$ which defines the curve Γ, and for $|x| \geqslant \delta$ and $\lambda \to \infty$

$$y(x,\lambda) = A(x)\exp\left\{i\lambda\int_0^x p(t)dt + c\right\}[1 + O(\lambda^{-1})],$$

where c is a constant. Thus the solution oscillates on both sides of the turning point. At the turning point itself

$$y(0,\ \lambda) \sim a\lambda^{-1/2-1/(k+1)}, \quad a \neq 0,$$

so that $y(0,\lambda) \to \infty$ as $\lambda \to \infty$. There is also a solution $\overline{y(x,\lambda)}$.

If k is even, then for $x \geqslant \delta$ there are two real stationary points $p = p_1(x)$ and $p = p_2(x)$, and

$$y(x,\lambda) = \sum_{j=1}^{2} A_j(x)\exp\left\{i\lambda \int_0^{p_j(x)} p(t)dt\right\} + O(\lambda^{-1}).$$

For $x < 0$ there are no real stationary points, so that $y(x,\lambda) = O(\lambda^{-\infty})$, $\lambda \to \infty$, $x \leqslant -\delta$. The solution $y(x,\lambda)$ can be taken to be real. The asymptotic behaviour of $y(0,\lambda)$ has the same form as that for odd k. A typical example is the equation

$$\left(\frac{1}{i\lambda}\frac{d}{dx}\right)^k y - xy = 0$$

(see the remark in paragraph 4.2 and also paragraph 6.2).

5. Turning Points of Self-Adjoint Systems. The method given here is due to V.V. Kucherenko [Kucherenko].

5.1 Systems of First-Order Equations. We will consider the system (1) on the interval $[a,b]$. The matrix $A(x,\varepsilon)$ satisfies condition C from paragraph 1 for $x \in I$ and $\varepsilon \in J = [0,\varepsilon_0]$. We will assume that the matrix $A_0(x)$ is *skew-hermitian* for all x, that is, $A_0^*(x) = -A_0(x)$ where $A_0^*(x) = \bar{A}^T(x)$, and system (1) has precisely one turning point on I, $a < x_0 < b$. Suppose that precisely two eigenvalues of $A_0(x)$ coalesce at this point:

$$p_1(x_0) = p_2(x_0) = p_0\,, \quad p_j(x_0) \neq p_k(x_0)\,, \tag{38}$$

$$p_j(x_0) \neq p_0\,, \quad 3 \leqslant j\,, \quad k \leqslant n\,, \quad j \neq k\,.$$

The eigenvalues of $A_0(x)$ are purely imaginary, and we denote them $p_j(x) = iq_j(x)$. The functions $q_j(x)$ are real-valued and $q_j(x) \in C^\infty(I)$. There exists a basis $\{e_1(x),\ldots,e_n(x)\}$ which consists of orthonormal eigenvectors of the matrix $A_0(x)$ such that $e_j \in C^\infty(I)$. The matrix $T_0(x) = (e_1(x),\ldots,e_n(x))$ is unitary and reduces $A_0(x)$ to diagonal form. The transformation $y = T_0(x)z$ reduces system (1) to the form

$$\varepsilon z' = (\Lambda(x) + \varepsilon B(x,\,\varepsilon))z\,, \tag{39}$$

$$B(x) = \varepsilon^{-1}T_0^{-1}(x)[A_0(x,\,\varepsilon) - A_0(x)]T_0(x) - T_0^{-1}(x)T_0'(x)\,.$$

We will consider the simplest case

$$p_1'(x_0) \neq p_2'(x_0)\,. \tag{40}$$

This condition is stable under small perturbations of the coefficients of system (1) if the perturbation of $A_0(x)$ remains skew-hermitian. Suppose that $x_0 = 0$ and $I = [-a,a]$, $a > 0$.

Remark. Suppose that $n = 2$; then the curve Γ: $\det(pI - A_0(x)) = 0$ breaks down into two smooth curves $p = p_1(x)$, $p = p_2(x)$ which intersect or are tangential at x_0. Therefore Γ is not a smooth curve and the results of paragraph 4 are not applicable to such turning points.

5.2 Systems of Two Equations. Suppose that the system (1) has been reduced to the form (39) so that it has the form

$$\begin{aligned} y_1' &= \lambda p_1(x) y_1 + b_{11}(x, \varepsilon) y_1 + b_{12}(x, \varepsilon) y_2 , \\ y_2' &= \lambda p_2(x) y_2 + b_{21}(x, \varepsilon) y_1 + b_{22}(x, \varepsilon) y_2 , \end{aligned} \tag{41}$$

Since $p_1(0) = p_2(0)$ this system does not admit an asymptotic diagonalization (§ 1, paragraph 1.3). We make the substitution

$$y_j = \exp\left\{ \lambda \int_0^x p_j(t)dt + \int_0^x b_{jj}(t,\varepsilon)dt \right\} z_j , \quad j = 1, 2 ,$$

and replace the system so obtained by the system of integral equations

$$z = c + Kz , \quad (Kz)(x) = \int_0^x K(x,t,\lambda) z(t) dt , \tag{42}$$

$$K(x,t,\lambda) = \begin{bmatrix} 0 & b_{12}(t,\varepsilon)\exp\{H\} \\ b_{21}(t,\varepsilon)\exp\{-H\} & 0 \end{bmatrix} , \tag{43}$$

$$S(t,x) = i\lambda \int_t^x [q_2(\tau) - q_1(\tau)] d\tau , \quad S_1(t,x) = \int_t^x [b_{22}(\tau) - b_{11}(\tau)] d\tau ,$$

where $H = i\lambda S(t,x) + S_1(t,x)$, $z = (z_1, z_2)^T$, $c = (c_1, c_2)^T$, c_j are constants.

The norm $\|K(x,t,\lambda)\|$ of the kernel of the operator K has order $O(1)$ for arbitrary $\lambda > 0$ since $q_1(x)$ and $q_2(x)$ are real. However we can obtain an equation with small kernel from equation (42) by writing it in the form

$$z = c + Kc + K^2 z . \tag{44}$$

Here K^2 is the integral operator:

$$(K^2 z)(x) = \int_t^x K_1(x,t,\lambda) z(t) dt ,$$

where K_1 is a diagonal matrix with diagonal elements

$$\begin{aligned} \phi_1 &= b_{21}(t,\varepsilon) \int_t^x b_{12}(\tau,\varepsilon) \exp\{i\lambda[S(\tau,t) + S(\tau,x)] + S_1(\tau,t)\} d\tau , \\ \phi_2 &= b_{12}(t,\varepsilon) \int_t^x b_{21}(\tau,\varepsilon) \exp\{-i\lambda[S(\tau,t) + S(\tau,x)] - S_1(\tau,t)\} d\tau , \end{aligned} \tag{45}$$

Since $p_1(x) \neq p_2(x)$ in I, $x \neq 0$, we have $\lim_{\lambda\to\infty} \phi_j = 0$ because the exponent is rapidly oscillating. If condition (40) is satisfied then the method of stationary phase allows us to obtain the bound

$$|\phi_j(t,x,\lambda)| \leqslant c\lambda^{-1/2}, \quad \lambda \gg 1,$$

where the constant c does not depend on t and $x \in I$.

We give the asymptotic formulae for the solutions. We put $c = (1,0)^T$; then $z = c + Kc + K^2c + O(\lambda^{-1})$. A more extensive analysis shows that $K^2c = O(\lambda^{-1/2})$, so that

$$\begin{aligned} z^{(1)} &= \left(0, \int_0^x \exp\{-i\lambda S(t,x) - S_1(t,x)\} b_{21}(t,\varepsilon)dt\right)^T + O(\lambda^{-1}), \\ z^{(2)} &= \left(\int_0^x \exp\{i\lambda S(t,x) + S_1(t,x)\} b_{12}(t,\varepsilon)dt,\ 0\right)^T + O(\lambda^{-1}), \end{aligned} \tag{46}$$

where $|O(\lambda^{-1})| \leqslant c\lambda^{-1}$ for $\lambda \geqslant \lambda_0$ and $x \in I$. The integrals that appear in formula (46) do not admit further simplification in practice; we can only calculate them in terms of the Frenel integral

$$\Phi(x) = \int_0^x e^{it^2}\, dt$$

(see [Fedoryuk 11]). However if we move away from the turning point then the method of stationary phase allows us to claculate the asymptotic behaviour of the integrals. Suppose that $\alpha > 0$ is fixed; we can choose α sufficiently small but not depending on λ. Then system (41) has F.S.S. such that as $\lambda \to \infty$, $x \in I$, $|x| \geqslant \alpha$

$$y_j(x,\lambda) = \exp\left\{\lambda \int_0^x p_j(t)dt + \int_0^x b_{jj}(t,0)dt\right\}[f_j + O(\lambda^{-1})], \quad j = 1,2,$$

$$\begin{aligned} f_1 &= \left(1, \sqrt{\frac{\pi}{2\lambda|p_1'(0) - p_2'(0)|}}\, b_{21}(0,0)e^{i\pi\delta/4} \operatorname{sgn} x\right)^T, \\ f_2 &= \left(\sqrt{\frac{\pi}{2\lambda|p_1'(0) - p_2'(0)|}}\, b_{12}(0,0)e^{-i\pi\delta/4} \operatorname{sgn} x,\ 1\right)^T, \\ \delta &= \operatorname{sgn}\, [q_2'(0) - q_1'(0)]. \end{aligned} \tag{47}$$

Recall that the roots $p_j(x)$ are purely imaginary. Since $T_0(x)$ is unitary, $T_0^{-1}(x)T'(x)$ is skew-hermitian. If A does not depend on ε then the elements $b_{11}(x)$ and $b_{22}(x)$ are purely imaginary, $b_{21}(x) = \overline{-b_{12}(x)}$ and $(f_1)_2 = -(f_2)_1$.

Applying the method of successive approximations to equation (44) and the method of stationary phase, we can compute the asymptotic expansion of the solutions of system (1). This method is also applicable when the function $p_1(x) - p_2(x)$ has a zero of arbitrary finite order at the turning point x_0. This case, however, is unstable under small perturbations of the coefficients of the system.

5.8 Systems of n Equations. Using the transformation (19), where $T(x,\varepsilon)$ has the form (23), system (1) can be reduced to the block-diagonal form (22) for $x \in I$, where

$$B_1(x,\varepsilon) = \begin{bmatrix} p_1(x) & 0 \\ 0 & p_2(x) \end{bmatrix} + \sum_{k=1}^{N} \varepsilon^k B_k(x)\,,$$

$$B_2(x,\varepsilon) = \Lambda(x) + \sum_{k=1}^{N} \varepsilon^k \Lambda_k(x)\,,$$

and the $\Lambda_k(x)$ are diagonal matrices. System (1) has solutions $y_3(x,\varepsilon),\ldots,$ $y_n(x,\varepsilon)$ of the form (8) of § 2, and the asymptotic behaviour of the solutions $y_{1,2}(x,\varepsilon)$ is determined by the system $\varepsilon y' = B_1(x,\varepsilon)$.

Suppose that several of the roots coincide at the point x_0, that is $p_1(x_0) = \ldots = p_k(x_0) = p_0$, $p_j'(x_0) \neq p_l'(x_0)$, $j \neq l$, $1 \leqslant j, l \leqslant k$, and that the other roots are distinct and different from p_0. The method given above is applicable in this case since the kernel of the operator K^2 has order $O(\lambda^{-1/2})$ for $\lambda \gg 1$.

Remark. Suppose that all the eigenvalues of $A_0(x)$ are purely imaginary for $x \in I$ and that there is a basis $\{e_1(x),\ldots,e_n(x)\}$ of class $C^\infty(I)$ of eigenvalues of this matrix. Then the above results are valid for system (1).

5.4 Systems of Second-Order Equations. We consider the system of n equations

$$y'' + \lambda^2 A(x,\lambda^{-1})y = 0 \tag{48}$$

with hermitian matrix $A_0(x) = A(x,0)$. Suppose that the matrix $A(x,\varepsilon)$, with $\varepsilon = \lambda^{-1}$, satisfies condition C from paragraph 1 and let the eigenvalues $p_j(x)$ of $A_0(x)$ be positive:

$$p_j(x) > 0\,, \quad x \in I\,, \quad 1 \leqslant j \leqslant n\,.$$

Suppose that on the interval I there is precisely one turning point x_0, $a < x_0 < b$, and that conditions (38) and (40) are satisfied. We take $x_0 = 0$ and $I = [-a,a]$. System (48) is equivalent to the system of $2n$ equations

$$\varepsilon \begin{bmatrix} z \\ \varepsilon z' \end{bmatrix}' + \begin{bmatrix} 0 & I \\ A(x,\varepsilon) & 0 \end{bmatrix} \begin{bmatrix} z \\ \varepsilon z' \end{bmatrix} = 0\,,$$

which satisfies the conditions of the remark in paragraph 4.2. Here 0 and I are respectively the null and unit matrices of order n.

Let the matrix $T_0(x)$ reduce $A_0(x)$ to diagonal form $\Lambda(x)$. Putting

$$\begin{bmatrix} z \\ \varepsilon z' \end{bmatrix} = Tw\,, \quad T(x) = \begin{bmatrix} T_0(x) & T_0(x) \\ T_0(x)\sqrt{\Lambda(x)} & -T_0(x)\sqrt{\Lambda(x)} \end{bmatrix}, \tag{49}$$

we obtain the system

$$\varepsilon w' + \Lambda_0(x)w + \varepsilon B(x,\ \varepsilon)w = 0\,, \tag{50}$$

$$\Lambda_0(x) = \begin{bmatrix} \sqrt{\Lambda(x)} & 0 \\ 0 & -\sqrt{\Lambda(x)} \end{bmatrix}, \quad B(x,\varepsilon) = B_1(x) + B_2(x,\varepsilon),$$

$$\begin{aligned} B_1(x) &= \frac{1}{4}\Lambda^{-1}(x)\Lambda'(x)\begin{bmatrix} I & -I \\ -I & I \end{bmatrix} + \frac{1}{2}T_0^{-1}(x)T_0'(x)\begin{bmatrix} I & I \\ I & I \end{bmatrix} \\ &\quad + \frac{1}{2}\Lambda^{-1/2}(x)T_0(x)T_0'(x)\Lambda^{1/2}(x)\begin{bmatrix} I & -I \\ -I & I \end{bmatrix}, \end{aligned} \tag{51}$$

$$B_2(x,\varepsilon) = \varepsilon^{-1}T^{-1}(x)\begin{bmatrix} 0 & I \\ A(x,\varepsilon) - A_0(x) & 0 \end{bmatrix} T(x).$$

Here the notation used is: if C, C_j are square matrices of order n then

$$C\begin{bmatrix} C_1 & C_2 \\ C_3 & C_4 \end{bmatrix} = \begin{bmatrix} CC_1 & CC_2 \\ CC_3 & CC_4 \end{bmatrix},$$

$$\sqrt{\Lambda(x)} = \operatorname{diag}\left(\sqrt{p_1(x)}, \ldots, \sqrt{p_n(x)}\right),$$

all the roots being positive functions. Observe that

$$\operatorname{diag} B_1(x) = \operatorname{diag}\left[T_0^{-1}(x)T'(x)\right].$$

Because of conditions (38) it is sufficient to consider the system (50) of two equations. Let

$$\tilde{w}_j(x,\lambda) = \frac{1}{\sqrt[4]{\mu_j(x)}} \exp\left\{ i\lambda \int_0^x \mu_j(t)dt - \int_0^x \mu_j^{(1)}(t)dt \right\}, \tag{52}$$

$$\mu_1(x) = -\mu_3(x) = \sqrt{p_1(x)}, \quad \mu_2(x) = -\mu_4(x) = \sqrt{p_2(x)},$$

$$\mu_j^{(1)}(x) = [B_1(x) + B_2(x,\ 0)]_{jj}.$$

System (50) has F.S.S. of the form

$$w_j(x,\ \lambda) = \tilde{w}_j(x,\ \lambda)[f_j + K_j f_j + O(\lambda^{-1})], \tag{53}$$

where f_j has components δ_{jk} and K_j is the integral operator:

$$\begin{aligned} (K_j w)_l(x) = \int_0^x \exp\Bigg\{ i\lambda \int_t^x [\mu_l(\tau) - \mu_j(\tau)]d\tau \\ + \int_t^x [\mu_l^{(1)}(\tau) - \mu_j^{(1)}(\tau)]d\tau \Bigg\} [(\tilde{B}_1(t) + \tilde{B}_2(t,0))w]_l dt, \end{aligned}$$

where

$$\tilde{B}_1(t) = B_1(t) - \operatorname{diag} B_1(t)$$

and $\tilde{B}_2(t,0)$ is defined similarly.

These formulae can be simplified outside some neighbourhood of the turning point. Let $\delta > 0$ be fixed and chosen sufficiently small but not depending on ε. We will restrict ourselves, for simplicity, to the case where $A(x, \varepsilon)$ does not depend on ε and put

$$\delta = \operatorname{sgn}\,[p_1'(0) - p_2'(0)]\,,$$

$$a = -\sqrt[4]{p_1(0)}\sqrt{\frac{\pi}{|p_1'(0) - p_2'(0)|}}\, e^{i\pi\delta/4} e_2^\star(0) e_1'(0)\ \operatorname{sgn}\, x\,. \tag{54}$$

As $\lambda \to \infty$ and $|x| \geqslant \delta$, system (48) has F.S.S. of the form

$$\begin{aligned} y_1(x, \lambda) &= \frac{1}{\sqrt[4]{p(x)}} \exp\left\{ i\lambda \int_0^x \sqrt{p_1(t)}dt \right. \\ &\quad \left. - \int_0^x e_1^\star(t) e_1'(t) dt \right\} \left[e_1(x) + \frac{a}{\sqrt{\lambda}} e_2(x) + O(\lambda^{-1}) \right], \\ y_2(x, \lambda) &= \frac{1}{\sqrt[4]{p_2(x)}} \exp\left\{ i\lambda \int_0^x \sqrt{p_2(t)}dt \right. \\ &\quad \left. - \int_0^x e_1^\star(t) e_1'(t) dt \right\} \left[e_2(x) - \frac{\overline{a}}{\sqrt{\lambda}} e_1(x) + O(\lambda^{-1}) \right], \end{aligned} \tag{55}$$

$$y_3(x, \lambda) = \overline{y_1(x, \lambda)}\,, \quad y_4(x, \lambda) = \overline{y_2(x, \lambda)}\,. \tag{56}$$

The bounds for the remainder terms are uniform in $x \in I$, $|x| \geqslant \delta$, and the asymptotic expansion (55) can be differentiated in x and λ.

6. Turning Points for a Perturbation of the Generalized Airy Equation. The equation

$$y^{(n)} - xy = 0 \tag{57}$$

for $n \geqslant 3$ is called the *generalized Airy equation*, and its solutions *generalized Airy functions*. At this point we mention the results of [Wasov 9] on the asymptotic behaviour of the F.S.S. for system (1) and equation (2) in the whole complex neighbourhood of a turning point in the case when the principal symbol is close to the symbol $l^0 = p^n - x$ for small $|x|$.

6.1 Structure of the Roots. We will consider system (1) where $A(x, \varepsilon)$ is holomorphic in x and ε for $|x| \leqslant r$, $0 < \varepsilon \leqslant \varepsilon_0$ and has asymptotic series expansion

$$A(x,\ \varepsilon) = \sum_{r=0}^{\infty} A_r(x)\varepsilon^r \quad \varepsilon \to +0\,,$$

which is uniform in x for $|x| \leqslant r$. The numbers $r > 0$ and $\varepsilon_0 > 0$ are assumed to be sufficiently small.

6.2 Structure of the Roots. Let the matrix $A_0(0)$ have Jordan normal form

$$A_0(0) = \begin{bmatrix} 0 & 1 & & 0 \\ & 0 & \ddots & \\ & & \ddots & 1 \\ 0 & & & 0 \end{bmatrix}, \quad \frac{d}{dx} \det A_0(x)|_{x=0} = (-1)^{n+1}. \tag{58}$$

Then the principal symbol of system (1) has the form

$$l^0(x,p) = (-1)^n [p^n + xa_{n-1}(x)p^{n-1} + \ldots + xa_1(x)p - x + O(x^2)],$$

where the functions $a_j(x)$ are holomorphic for $|x| \leqslant r$, so that

$$l^0(x,\ p) = p^n - x + O(|x|^2 + |xp|)$$

for small $|x|$ and $|p|$. The equation $l^0(x,p) = 0$, with small $|x|$ and $|p|$, defines a one-dimensional complex analytic manifold Γ in the two-dimensional complex space $C_x \times C_p$.

The eigenvalues $p_j(x)$ of $A_0(x)$ have the form

$$p_j(x) = \omega^{j-1} x^{1/n} + x^{2/n} \tilde{p}_j(x^{1/n}), \quad 1 \leqslant j \leqslant n,$$

where $\omega = e^{2\pi i/n}$ and the functions $\tilde{p}_j(t)$ are holomorphic for $|t| \leqslant r^{1/n}$.

It was shown in paragraph 4 that the above conditions on the turning point $x = 0$ represent a typical situation.

6.3 The Exterior Expansion. The asymptotic expansion of the solutions of system (1) is constructed using the *method of matched asymptotic expansions*, the essence of which is as follows. Two different types of asymptotic expansions are constructed, the *exterior* and the *interior*. The exterior asymptotic expansion is applicable in the annulus $K_1 : 0 \leqslant \delta_1(\varepsilon) \leqslant |x| \leqslant r$ which does not contain the turning point, and the interior expansion is in the neighbourhood of the turning point $K_2 : |x| \leqslant \delta_2(\varepsilon)$. For $0 < \varepsilon \ll 1$ the domains K_1 and K_2 intersect, which allows us to establish the connection between the exterior and the interior asymptotic expansions (that is, to match them). The method of matched asymptotic expansions is widely applicable in the asymptotic theory of linear and non-linear ordinary differential equations and partial differential equations.

The exterior expansion has the same form as the asymptotic expansion in § 2. Let $S(\theta_1, \theta_2)$ be the sector

$$|x| \leqslant r, \quad \theta_1 \leqslant \arg x \leqslant \theta_2, \quad \theta_2 - \theta_1 < \pi n/(n+1).$$

Let

$$\begin{aligned} \Lambda(x) &= \operatorname{diag}(p_1(x), \ldots, p_n(x)), \\ \Omega(t) &= \operatorname{diag}(1, t, \ldots, t^{n-1}), \quad \omega = e^{2\pi i/n}. \end{aligned} \tag{59}$$

System (1) has fundamental matrix of the form

$$Y(x,\varepsilon)=\Omega(x^{1/n})[\varepsilon^{-1/(n+1)}x^{1/n}]^{(1-n)/2}\tilde{Y}(x,\varepsilon)\exp\left\{\frac{1}{\varepsilon}\int_0^x \Lambda(t)dt\right\}, \tag{60}$$

and there is the bound

$$\|\tilde{Y}(x,\ \varepsilon)\|\leqslant c,\quad |\varepsilon x^{-(n+1)/n}|\leqslant c_1,\quad x\in S(\theta_1,\ \theta_2),\quad 0<\varepsilon\leqslant\varepsilon_0. \tag{61}$$

Further

$$\tilde{Y}(x,\ \varepsilon)=U(x,\ \varepsilon)e^{B(\varepsilon)},\quad \varepsilon\to+0,$$

where $B(\varepsilon)$ is a diagonal matrix, and there are the asymptotic expansions

$$B(\varepsilon)=\sum_{r=1}^{\infty}B_r\varepsilon^r,\quad U(x,\varepsilon)=\sum_{r=0}^{\infty}U_r(x^{1/n})[\varepsilon x^{-(n+1)/n}]^r \tag{62}$$

as $\varepsilon\to+0$, $\varepsilon x^{-(n+1)/n}\to 0$. The matrices $U_r(t)$ are holomorphic for $|t|\leqslant r^{1/n}$, and

$$U_0(t)=\begin{bmatrix}1 & 1 & \dots & 1\\ 1 & \omega & \dots & \omega^{n-1}\\ \dots & \dots & \dots & \dots\\ 1 & \omega^{n-1} & \dots & \omega^{(n-1)(n-1)}\end{bmatrix}.$$

6.4 The Interior Expansion. We bring in the interior variable

$$s=x\varepsilon^{-n/(n+1)},$$

and let Σ be the sector

$$\begin{aligned}|\arg s|\leqslant n\pi/2(n+1),\quad & n\ \text{odd},\\ -\pi/2\leqslant\arg s\leqslant(n-1)\pi/2(n+1),\quad & n\ \text{even}.\end{aligned} \tag{63}$$

Also

$$Q(s)=\frac{n}{n+1}\Omega(\omega)s^{(n+1)/n},$$

where Ω is defined by the formula (59).

System (1) has fundamental matrix of the form

$$\begin{aligned}Z(x,\ \varepsilon)&=\Omega(\varepsilon^{1/(n+1)}s^{1/n})^{(1-n)/(2n)}\tilde{Z}(x,\ \varepsilon)e^{Q(s)},\quad |s|\geqslant s_0>0,\\ Z(x,\ \varepsilon)&=\tilde{Z}(x,\ \varepsilon)e^{Q(s)},\quad |s|\leqslant s_0,\end{aligned} \tag{64}$$

and there are the bounds

$$\|Z(x,\ \varepsilon)\|\leqslant c,\ |s|\geqslant s_0,\ |\varepsilon^{1/(n+1)}s^{(n+2)/n}\|\leqslant c_1,\ 0<\varepsilon\leqslant\varepsilon_0,\quad s\in\Sigma.$$

Further

$$\tilde{Z}(x,\ \varepsilon) = V(s,\ \varepsilon) S^{D(\varepsilon)}\,, \tag{65}$$

where $D(\varepsilon)$ is a diagonal matrix and there are the asymptotic expansions

$$D(\varepsilon) = \sum_{r=1}^{\infty} D_r \varepsilon^{r/(n+1)}\,, \tag{66}$$

$$V(s,\ \varepsilon) = \sum_{r=0}^{\infty} V_r(s) [\varepsilon^{1/(n+1)} s^{(n+2)/n}]^r$$

as $\varepsilon \to 0$, $|s| \geqslant s_0$, $s \in \Sigma$, and $\varepsilon^{1/(n+1)} s^{(n+2)/n} \to 0$. The asymptotic expansion of $V_r(s)$ is

$$V_r(s) = \sum_{\nu=0}^{\infty} V_{r\nu} s^{-\nu/n}\,, \quad s \to \infty\,, \quad s \in \Sigma\,, \quad V_{00} = U_0(0)\,.$$

6.5 Matching Asymptotic Series. We choose the sector S such that its bounding rays lie inside Σ. We can put

$$\theta_1 = -\frac{n\pi}{2(n+1)} + \delta = -\theta_2\,, \qquad n \text{ odd}\,,$$

$$\theta_1 = -\frac{\pi}{2} + \delta\,,\ \ \theta_2 = \frac{(n-1)\pi}{2(n+1)} - \delta\,, \quad n \text{ even}\,,$$

where $\delta > 0$ is sufficiently small, and we let $\theta_1 \leqslant \arg x \leqslant \theta_2$.

Since Y and Z are fundamental matrices for system (1),

$$Y(x,\ \varepsilon) = Z(x,\ \varepsilon) \Gamma(\varepsilon)\,,$$

where the matrix $\Gamma(\varepsilon)$ does not depend on x. The domains of application for the exterior and interior series intersect since both asymptotic expansions are applicable for

$$\varepsilon^{n/(n+1)} \ll |x| \ll \varepsilon^{n/(n+2)}\,.$$

We can match these asymptotic series in this zone. We take the interior domain $S_i : \varepsilon^{1/(n+1)} |s|^{\chi(s)(n+2)/n} \leqslant \eta_0$, the exterior domain $S_e : \varepsilon |x|^{-(n+1)/n} \leqslant \xi_0$ and the intermediate domain $S_m = S_i \cap S_e$, where

$$s_0 = \xi_0^{-n/(n+1)}\,, \quad \varepsilon_0 < \xi_0^{n+2} \eta_0^{n+1}\,,$$

$$\chi(s) = 1\,, \quad |s| \geqslant s_0\,, \quad \chi(s) = 0\,, \quad |s| \leqslant s_0\,.$$

If the numbers $\xi_0 > 0$ and $\varepsilon_0 > 0$ are sufficiently small then there are the connection formulae for $j,\ k = 1, \ldots, n$:

$$y_{jk}(x,\varepsilon) = \begin{cases} z_{jk}(x,\varepsilon)[\gamma_{kk}(x,\varepsilon) + \mu_{jk}(x,\varepsilon)], & (x,\varepsilon) \in S_m, \\ z_{jk}(x,\varepsilon) + \mu_{jk}(x,\varepsilon), & (x,\varepsilon) \in (S_i \cap S_m), \end{cases}$$

where $\mu_{jk}(x\varepsilon) = O(\varepsilon^N)$ as $\varepsilon \to +0$, $(x,\varepsilon) \in S$ and N is arbitrary. The asymptotic expansions for the elements $\gamma_{kk}(\varepsilon)$ are

$$\gamma_{kk}(\varepsilon) = \varepsilon^{nb_{kk}(\varepsilon)/(n+1)} \sum_{r=0}^{\infty} \gamma_{kk,r}\varepsilon^{r/(n+1)}, \quad \varepsilon \to +0, \tag{67}$$

where $b_{kk}(\varepsilon)$ are the diagonal elements of $B(\varepsilon)$ from (62).

6.6 Reduction of System (1) for Even n. Let S_0 be the sector

$$|x| \leqslant x_0, \quad \pi + \delta \leqslant \arg x \leqslant \pi + \frac{2\pi}{n+1}\delta,$$

where $\delta > 0$ can be chosen sufficiently small. Then there is a transformation of the form (19) which reduces system (1) to the form

$$\varepsilon z' = A_0(x)z, \quad A_0(x) = \begin{bmatrix} 0 & 1 & 0 & \dots & 0 \\ 0 & 0 & 1 & \dots & 0 \\ \cdot & \cdot & \cdot & \cdot & \cdot \\ 0 & 0 & 0 & \dots & 1 \\ x & 0 & 0 & \dots & 0 \end{bmatrix}.$$

The matrix $T(x,\varepsilon)$ has asymptotic expansion of the form (28) as $\varepsilon \to +0$ uniformly in $x \in S_0$ and is holomorphic in x and ε for $x \in S$ and $0 < \varepsilon \leqslant \varepsilon_0$. The coefficients of the asymptotic series are determined from the formal identity

$$\varepsilon \frac{dT}{dx} = A_0(x)T - TA(x,\varepsilon)$$

for the formal series (28).

7. Almost Diagonal Systems. We now consider system (1) where the matrix $A(x,\varepsilon)$ is holomorphic in x and ε for

$$|x| \leqslant x_0, \ 0 < |\varepsilon| \leqslant \varepsilon_0, \ |\arg \varepsilon| \leqslant \theta_0, \ \varepsilon_0 > 0, \ x_0 > 0, \ \theta_0 > 0 \tag{68}$$

under the conditions:

1. There is a holomorphic matrix $T(x)$, non-singular for $|x| \leqslant x_0$, such that

$$T^{-1}(x)A_0(x)T(x) = \operatorname{diag}(p_1(x), \dots, p_n(x)).$$

2. If $p_1(x), \dots, p_k(x)$ are all the distinct eigenvalues of the matrix $A_0(x)$ then $p_j(x) = x^q\tilde{p}_j(x)$, $1 \leqslant j \leqslant k$, where $q \geqslant 1$ is an integer, the functions $\tilde{p}_j(x)$ are holomorphic for $|x| \leqslant x_0$, and all the $\tilde{p}_j(0)$ are distinct. Here $A_0(x) = A(x,0)$.

Let S be the sector $|\arg x - \alpha| \leqslant \beta$, $0 < \beta < \pi/2(q+1))$, where α is real and the numbers x_0, ε_0, θ_0 are sufficiently small. In [Lee] it was proved that system (1) has a fundamental matrix of the form

$$Y(x,\varepsilon) = U(x,\varepsilon)\exp\left\{\frac{1}{\varepsilon}\int_0^x A_0(t)dt\right\}$$

and there is the asymptotic expansion

$$U(x,\ \varepsilon) = \sum_{r=0}^{\infty} U_r(x,\ \varepsilon)\varepsilon^{r/(q+1)}\,, \quad \varepsilon \to +0\,,$$

uniform in x, ε if $x \in S$ and x, ε lie in the domain (68). Moreover

$$U(x,\varepsilon) - \sum_{r=0}^{N} U_r(x,\varepsilon)\varepsilon^{r/(q+1)} = O(\gamma(\varepsilon)^{N+1})\,,$$

$$\gamma(\varepsilon) = \varepsilon^{1/(q+1)}(\ln\varepsilon)^{\delta_{1q}/2}\,,$$

where δ_{1q} is the Kronecker delta symbol.

Both conditions are satisfied, in particular, if $A_0(x)$ is skew-hermitian for real x (see paragraph 5).

8. The Turrittin Equation. The equation

$$x^n y^{(n)} - x^m y = 0\,, \tag{69}$$

where m is an arbitrary complex number is called the *Turrittin equation*. For $m = 0$ it is the *Euler Equation*; we take $m \neq 0$. For $n = 2$ the solutions of equation (69) are calculated in terms of Bessel functions; in particular for $m = 3$ its solutions are Airy functions. For $m = n + 1$ equation (69) is a generalized Airy equation.

The solutions of equation (69) are calculated in terms of Meijer G-functions (or the generalized hypergeometric series) whose asymptotic behaviour has been little studied. Below we follow [Braaksma].

8.1 Integral Representations and Series. Let C be a contour in the complex t-plane, consisting of the half-line $(+\infty - ia,\ w - ia]$, the segment $[w - ia, w + ia]$ and the half-line $[w + ia,\ +\infty + ia)$, where

$$a > (n-1)\left|\operatorname{Im}\frac{1}{m}\right|\,, \quad w < \operatorname{Re}\left(\frac{n-1}{m}\right)\,, \quad w < 0\,.$$

We put

$$\phi_j(t) = \left\{\prod_{k=0}^{n-1} \Gamma\left(1+t-\frac{k}{m}\right)\right\}^{-1} \left\{\sin\pi\left(t-\frac{i}{m}\right)\right\}^{-1} m^{-nt}e^{-i\pi t},$$

$$j = 0, 1, \ldots, n-1,$$

$$\tilde{\phi}(t) = \prod_{k=0}^{n-1} \Gamma\left(\frac{k}{m}-t\right) m^{-nt}e^{-\pi int}.$$

Then the functions

$$y_j(x) = \frac{1}{2\pi i}\int_C \phi_j(t)x^{mt}\,dt, \quad \tilde{y}(t) = \frac{1}{2\pi i}\int_C \tilde{\phi}(t)x^{mt}\,dt$$

are solutions of equation (69) and have series expansions

$$y_j(x) = -\frac{1}{\pi}e^{-\pi ij/m}m^{-nj/m}x^j \sum_{\nu=0}^{\infty}\left\{\prod_{k=0}^{n-1}\Gamma\left(1+\nu+\frac{j-k}{m}\right)\right\}^{-1} \times m^{-n\nu}x^{m\nu}, \tag{70}$$

which converge for $x^m \neq 0$. The branch

$$x^{mt} = \exp\{mt[\ln|x| + i \arg x]\}$$

depends on the choice of $\arg x$. Formula (70) determines the asymptotic behaviour of the solutions as $x^m \to 0$.

The solutions $y_0(x), \ldots, y_{n-1}(x)$ are linearly independent if $m \neq 0, \pm 1, \pm 2, \ldots, \pm(n-1)$. Let

$$m = \pm 1, \pm 2, \ldots, \pm(n-1). \tag{71}$$

Let us introduce the functions

$$\phi_{jh}(t) = \left\{\prod_{k=0}^{n-1} \Gamma\left(1+t-\frac{k}{m}\right)\right\}^{-1} \left\{\sin\pi\left(t-\frac{i}{m}\right)\right\}^{-1} m^{-nt}e^{-\pi i(h+1)t},$$

where j and h are integers such that

$$0 \leqslant h|m| + j \leqslant n-1, \quad 0 \leqslant h \leqslant n-1, \quad 0 \leqslant j \leqslant |m|-1. \tag{72}$$

Then the functions

$$y_{jh}(x) = \frac{1}{2\pi i}\int_C \phi_{jh}(t)x^{mt}\,dt \tag{73}$$

form the F.S.S. of equation (69) and have series expansions converging for $x^m \neq 0$:

$$y_{jh}(x) = -\sum_{l=0}^{h} [\ln(m^{-n}x^m)]^{h-l} \sum_{\nu=0}^{\infty} a_{l\nu}(j,h) x^{m\nu+j},$$

$$a_{l\nu}(j,h) = \frac{1}{(h-l)!} \pi^{-k-1} m^{-n\nu-nj/m} e^{-j(k+1)\pi i/m} \sum_{\mu=0}^{l} \frac{(2\pi i)^\mu}{\mu!(l-\mu)!} \times B_\mu^{(h+1)} f_j^{(l-j)}(\nu).$$

Here $B_\mu^{(h+1)}$ are the Bernoulli numbers of order $h+1$ and

$$f_j(t) = \left\{ \prod_{k=0}^{n-1} \Gamma\left(1 + t + \frac{j-h}{m}\right) \right\}^{-1}.$$

In particular $y_{j0} = y_j(x)$.

We now state some properties of the solutions of equation (69). Put

$$\omega = e^{2\pi i/m}, \quad \omega^\alpha = e^{2\pi i \alpha/m}. \tag{74}$$

1. If $y(x)$ is a solution then $y(\omega x)$ is also a solution.
2. If m does not have the form (71) then

$$y_j(x) = \omega^p y_j(x\omega^{-p}), \quad p = 0, \pm 1, \pm 2, \ldots, \quad j = 0, 1, \ldots, n-1. \tag{75}$$

3. If m has the form (71) then

$$y_{jh}(x) = \sum_{q=0}^{h} C_p^q (2i)^q \omega^{(2p-q)j} y_{j,h-q}(x\omega^{-p}). \tag{76}$$

4. All the solutions $y_{jh}(x)$ can be expressed in terms of the solutions $\tilde{y}(x)$:

$$y_{jh}(x) = \sum_{k=0}^{n-h-1} c_k(j,h) \tilde{y}(x\omega^k),$$

the constants $c_k(j,h)$ being determined from the relations

$$\sum_{k=0}^{n-h-1} c_k(j,h) x^k = c_0(j,h) \prod_{k=0}^{n-1\,\prime} (1 - x\omega^{-k}),$$

$$c_0(j,h) = (2\pi i)^{-n}(-2i)^{h+1} \exp\left\{ \left[\frac{n(n-1)}{2} - j(h+1) \right] \frac{\pi i}{m} \right\},$$

$$c_{n-h-1}(h) = (-2\pi i)^{-n}(2i)^{h+1} \exp\left\{ \left[j(h+1) - \frac{n(n-1)}{2} \right] \frac{\pi i}{m} \right\}.$$

The prime indicates that the terms with indices $k = j, j+|m|, \ldots, j+h|m|$ are omitted from the product.

5. The solutions $\tilde{y}(x), \tilde{y}(x\omega^{-1}), \ldots, \tilde{y}(x\omega^{-n+1})$ form the F.S.S.

8.2 Asymptotic Behaviour of the Solution $\tilde{y}(x)$. All the asymptotic expansions mentioned in this section are uniform in x as $|x^m| \to \infty$ and ε is a fixed number, $0 < \varepsilon < \pi$. As

$$|x^m| \to \infty, \quad -\pi \leqslant \arg(m^{-n}x^m) \leqslant (2n+1)\pi,$$

there is the asymptotic series expansion

$$\tilde{y}(x) = E(x) \equiv \exp\left\{\frac{n}{m}x^{m/n}\right\}(m^{-n}e^{-n\pi i}x^m)^{\alpha_{mn}} \times \left[n^{-1/2}(2\pi)^{(n-1)/2} + \sum_{k=1}^{\infty} d_k x^{-mk/n}\right],$$

where

$$\alpha_{mn} = \frac{1}{2}(n-1)\left(\frac{1}{m} - \frac{1}{n}\right), \tag{77}$$

uniformly in $\arg x$. The coefficients $d_1, d_2, \ldots$ are chosen so that the formal series $E(x)$ satisfies equation (69).

The identities in paragraph 8.1 allow us to calculate the asymptotic behaviour of the F.S.S. as $|x| \to \infty$ in terms of the asymptotic behaviour of the solution $\tilde{y}(x)$. The solutions $\tilde{y}(x), \tilde{y}(x\omega^{-1}), \ldots, \tilde{y}(x\omega^{-n+1})$ form the F.S.S. which is uniquely determined by the condition that the solution with index j is asymptotically $E(x\omega^{-m})$ for

$$|x^m| \to \infty, \quad -\pi + \varepsilon \leqslant \arg(m^{-n}x^m\omega^{-j}) \leqslant (2n+1)\pi - \varepsilon.$$

The constants $c_k(j,h)$ are the Stokes multipliers of the F.S.S. $\{y_{jh}(x)\}$ with respect to the F.S.S. $\tilde{y}_j(x\omega^{-j})$, $j = 0, 1, \ldots, n-1$. In [Braaksma] another F.S.S. was proposed, analogous to that given by Turrittin earlier, but the Stokes multipliers of the F.S.S. $\{y_{jh}(x)\}$ are more cumbersome.

8.3 Asymptotic Behaviour of the Fundamental System of Solutions. We put

$$c = -\pi^{-n}(2i)^{1-n}\omega^{n(n-1)/4}.$$

We give the asymptotic expansions of the solutions as $|x^m| \to \infty$ in the sectors that are given below. If m does not have the form (71) then $h = 0$ and $j = 0, 1, \ldots, n-1$; if m has the form (71) then j and h are integers satisfying the inequalities (72). We put

$$\alpha = \arg(m^{-n}x^m). \tag{78}$$

1. If $-\pi + \varepsilon \leqslant \alpha \leqslant (2h+3)\pi - \varepsilon$, then

$$y_{jh}(x) = c(-2i)^h\omega^{-j(h+1)/2}E(x) + c(2i)^h\omega^{j(h+1)/2}E(x\,\omega^{-h-1}),$$

excluding the cases $m = \pm 1$, $j = 0$, $h = n-1$.

2. If $-\pi + \varepsilon \leqslant \alpha \leqslant (h+1)\pi - \varepsilon$, then

$$y_{jh}(x) = c(-2i)^h \omega^{-j(h+1)/2} E(x).$$

3. If $(h+1)\pi + \varepsilon \leqslant \alpha \leqslant (2h+3)\pi - \varepsilon$, then

$$y_{jh}(x) = c(2i)^h \omega^{j(h+1)/2} E(x\omega^{-h-1}).$$

4. If $(2p-1)\pi + \varepsilon \leqslant \alpha \leqslant (2p+3)\pi - \varepsilon$, then

$$y_{jh}(x) = x(2i)^h \omega^{j(2p-h-1)/2} \binom{p-1}{h} E(x\omega^{-p}) + \omega^j \binom{p}{h} E(x\omega^{-p-1}),$$

where p is an integer with $p \leqslant -1$ or $p \geqslant h+1$.

5. If $(2p-1)\pi + \varepsilon \leqslant \alpha \leqslant (2p+1)\pi - \varepsilon$, then

$$y_{jh}(x) = x(2i)^h \binom{p-1}{h} \omega^{j(2p-h-1)/2} E(x\omega^{-p}),$$

where p is an integer with $p < 0$ or $p > h+1$.

The constants $c_k(j,h)$ are the Stokes multipliers of the F.S.S. $\{y_{jh}(x)\}$ with respect to the F.S.S. $\{\tilde{y}(x\omega^{-j})\}$. In [Braaksma], together with the first of these F.S.S., there is given a further one where the Stokes multipliers are expressed by more awkward formulae.

8.4 Generalized Airy Functions. We consider the equation

$$y^{(n)} - xy = 0,$$

so that $m = n+1$ in the Turrittin equation (69). We put

$$\phi_p = \frac{2p-1}{m+1}\pi, \quad p = 0, \pm 1, \pm 2,$$

and let $\varepsilon > 0$ be a small fixed number. Let us introduce the function

$$\tilde{E}(x) = -\frac{1}{(2\pi)^{(n-1)/2} n^{1/2} \pi} (n+1)^{(n-1)/2(n+1)} x^{(1-n)/(2n)} \times \exp\left\{\frac{n}{n-1} x^{(n+1)/n}\right\} \left[1 + \sum_{k=1}^{\infty} \tilde{d}_k x^{-(n+1)k/n}\right], \tag{79}$$

which differs from $E(x)$ by a constant multiplier. Then as $|x| \to \infty$

$$y_j(x) = (p-1) \exp\{2pj\pi i/(n+1)\} \tilde{E}(x \exp\{-2p\pi i/m\}) \tag{80}$$

in the sectors

$$\phi_p + \varepsilon < \arg x < \phi_{p+1} - \varepsilon,$$

so that these are exponential solutions. In the remaining sectors, that is, in small neighbourhoods of the rays $\arg x = \phi_p$, the solutions $y_j(x)$ are sums of exponentials. We have

$$y_j(x) = \exp\{j(2p-1)\pi i/(n+1)\}\tilde{E}(x\exp\{-2p\pi i/(n+1)\})$$
$$+ \exp\{j(2p+1)\pi i/(n+1)\}\tilde{E}(x\exp\{-2(p+1)\pi i/(n+1)\}) \quad (81)$$

as $|x| \to \infty$, $|\phi - \phi_{p+1}| \leqslant \varepsilon$. On the ray $\phi = \phi_{p+1}$ we have

$$|\exp\{(xe^{-2\pi ip/(n+1)})^{(n+1)/n}\}| = |\exp\{(xe^{-2\pi i(p+2)/(n+1)})^{(n+1)/n}\}|,$$

so that both terms in (81) have the same order of growth as $|x| \to \infty$, and the solution $y_j(x)$ can have infinitely many zeros near the rays $\arg x = \phi_p$.

8.5 Generalized Airy Functions of the First Kind. Let $n = 2N$ be even. M. Kohno [Kohno 4] introduced the functions $\mathrm{Ai}\,(z)$ which for $n = 2$ are Airy functions (Chap. 4, § 1):

$$\mathrm{Ai}\,(z) = \gamma \sum_{j=1}^{n} (-1)^j (n+1)^{n(j-1)/(n+1)} \eta_{n-j} w_j(z),$$

$$w_j(z) = z^{n-j} \sum_{m=0}^{\infty} \left\{ \prod_{k=1}^{n} \Gamma\left(m + 1 + \frac{k-j}{n+1}\right) \right\}^{-1} [z^{n+1}(n+1)^{-n}]^m, \quad (82)$$

$$j = 1, 2, \ldots, n.$$

Here γ and η_j are real constants defined by

$$\gamma = \left(\frac{\pi}{2}\right)^{\frac{n-1}{2}} n^{\frac{1}{2}} (n+1)^{\frac{(2n+1)(1-n)}{2(n+1)}} \left\{ \prod_{k=1}^{n} \sin\frac{\pi k}{n+1} \right\}^{-1},$$

and

$$\eta_0 x^{n-1} + \eta_1 x^{n-2} + \ldots + \eta_{n-1} = (x+1) \prod_{k=1}^{N-1} \left(x^2 + 2x\cos\frac{\pi k}{n+1} + 1 \right).$$

These functions have the following properties:

1) Suppose that the integers $k_1, \ldots, k_{n+1}$ are unequal mod $(n+1)$, and let $\omega = \exp\{2\pi i/(n+1)\}$. Then the functions $\mathrm{Ai}\,(\omega^{k_j} z)$, $j = 1, \ldots, n$, form the F.S.S. and their Wronskian is

$$W = c_1 \ldots c_n \prod_{1 \leqslant i \leqslant j \leqslant n} \left[\sin\left(\frac{k_i - k_j}{n+1}\pi\right) \right] \exp\left\{ \left(\frac{k_i + k_j}{n+1} + \frac{1}{2} \right) \pi i \right\},$$

where

$$c_j = \gamma(-1)^j (n+1)^{n(n-j)/(n+1)} \eta_{n-j} = \mathrm{Ai}^{(n-j)}(0).$$

2) There is the identity

$$\sum_{k=0}^{n} \beta_k \,\mathrm{Ai}\,(\omega^{-k} z) = 0,$$

where

$$\beta_0 = 1\,, \quad \beta_k = -\omega^{-k} \prod_{j=1}^{k-1} \left|\sin\left(\frac{j-k}{n+1}\pi\right)\right| \left|\sin\frac{j}{n+1}\pi\right|^{-1}\,, \qquad k \geqslant 1\,.$$

3) As $|z| \to \infty$, $\mathrm{Ai}\,(z)$ becomes

$$\bar{d} z^{-(n-1)/(2n)} \exp\left\{-\frac{n}{n+1} z^{(n+1)/n}\omega_n\right\}(1+\varepsilon_1)\,, \qquad -2\pi \leqslant \arg z < -\pi\,,$$

$$z^{-(n-1)/(2n)} \exp\left\{-\frac{n}{n+1} z^{(n+1)/n}\right\}(1+\varepsilon_2)\,, \qquad -\pi \leqslant \arg z < \pi\,,$$

$$d z^{-(n-1)/(2n)} \exp\left\{-\frac{n}{n+1} z^{(n+1)/n}\omega_n^{-1}\right\}(1+\varepsilon_3)\,, \qquad \pi \leqslant \arg z < 2\pi\,,$$

where

$$\omega_n = \exp\left\{\frac{2\pi i}{n}\right\}, \quad d = \exp\left\{\frac{n-1}{n}\pi i\right\}, \quad \varepsilon_j = O(z^{-(n+1)/n})\,.$$

§ 7. A Problem on Scattering, Adiabatic Invariants and a Problem on Eigenvalues

1. The Scattering Matrix. We consider the system of n equations

$$y'' + \lambda^2 A(x) y = 0\,, \tag{1}$$

where $\lambda > 0$ is a parameter. Suppose that the following conditions are satisfied:

1) There exist finite limits $\lim_{x\to\pm\infty} A(x) = A_\pm$ and the eigenvalues p_j^+ and p_j^- of A_+ and A_- are positive and distinct.

2) $A(x) \in C^2(\mathbb{R})$, $A(x) - A_+(x) \in L_1(0,\infty)$, $A(x) - A_-(x) \in L_1(-\infty,0)$.

Then for fixed $\lambda > 0$ any solution of system (1) can be represented in the form

$$\begin{aligned} y(x,\ \lambda) &= A_-^{-1/4} \exp\{i\lambda x\sqrt{A_-}\}[b_+ + \lambda^{-1}\varepsilon_1^-] \\ &\quad + A_-^{-1/4} \exp\{-i\lambda x\sqrt{A_-}\}[b_- + \lambda^{-1}\varepsilon_2^-] \\ &= A_+^{-1/4} \exp\{i\lambda x\sqrt{A_+}\}[c_+ + \lambda^{-1}\varepsilon_1^+] \\ &\quad + A_+^{-1/4} \exp\{-i\lambda x\sqrt{A_+}\}[c_- + \lambda^{-1}\varepsilon_2^+]\,. \end{aligned}$$

The matrix $\sqrt{A_-}$ is similar to the diagonal matrix with elements $\sqrt{p_1^-} > 0$, $\ldots, \sqrt{p_n^-} > 0$ and the other roots of the matrices are defined similarly. The vectors $b_\pm$, $c_\pm$ depend only on λ, while $|\varepsilon_j^-(x,\lambda)| \to 0$ as $x \to -\infty$, and $|\varepsilon_j^+(x,\lambda)| \to 0$ as $x \to +\infty$. The $2n \times 2n$ matrix S such that

$$S = \begin{bmatrix} S_{11} & S_{12} \\ S_{22} & S_{22} \end{bmatrix}, \quad \begin{bmatrix} c_+ \\ b_- \end{bmatrix} = S \begin{bmatrix} b_+ \\ c_- \end{bmatrix}, \tag{2}$$

is called the *scattering matrix* (or *S-matrix*), where the S_{jk} are square matrices of order n. The scattering matrix does not depend on the choice of the solution Y and has the following properties:

1. If $A(x)$ is hermitian for all x, then the S-matrix is unitary.
2. If $A(x)$ is real for all x, then

$$S^{-1} = \begin{bmatrix} \overline{S_{22}} & \overline{S_{12}} \\ \overline{S_{21}} & \overline{S_{11}} \end{bmatrix}.$$

Suppose also that the following are satisfied:

3. $A(x)$ is real and symmetric, and its eigenvalues are strictly positive and distinct for $x \in \mathbb{R}$.
4. The elements of the matrices $A(x) - A_+$ and $A(x) - A_-$ are tame functions (§ 6, paragraph 3.3) on the half-lines $(0, +\infty)$ and $(-\infty, 0)$ respectively.

Then as $\lambda \to \infty$ there are formulae of the type (9) of § 11, Chap. 2:

$$s_{jj}(\lambda) = \exp\left\{ i\lambda \int_0^\infty (\sqrt{p_j(x)} - \sqrt{p_j^+})dx \right.$$
$$\left. + \int_{-\infty}^0 (\sqrt{p_j(x)} - \sqrt{p_j^-})dx + \sum_{k=1}^\infty \alpha_k \lambda^{-k} \right\},$$

$$s_{jl} = O(\lambda^{-\infty}), \quad j \neq l.$$

If the matrix function $A(z)$ is holomorphic in a band of the form Π: $|\mathrm{Im}\, z| < a$, $a > 0$, and conditions 1), 2) and 4) are satisfied in Π, then the non-diagonal elements of the S-matrix have order $O(e^{-c\lambda})$, $c > 0$, $\lambda \to \infty$.

2. Adiabatic Invariants of System (1). With the conditions 1) – 4) and for $\lambda \gg 1$ the equation (1) describes the vibrations of a system of weakly connected linear oscillators, whose frequencies of vibration are slowly changing with time (Chap. 2, § 11, paragraph 3). Let $\{e_1(x), \ldots, e_n(x)\}$ be an orthonormal basis of real eigenvectors of class $C^\infty(\mathbb{R})$ and let y be a solution of (1). Then the functions

$$J_k(x,\varepsilon) = \frac{1}{2\sqrt{p_l(x)}}[p_k(x)(e_k(x), y(x))^2 + (e_k(x), y'(x))^2], \; 1 \leqslant k \leqslant n, \tag{3}$$

are the adiabatic invariants of the system (1). The value

$$J_k(\lambda) = J_k(+\infty, \lambda) - J_k(-\infty, \lambda)$$

is called the *total variation of the adiabatic invariant.*

Let y_j^+ be solutions of (1) such that

$$y_j^+ = (p_j^+)^{-1/4} \exp\{i\lambda\sqrt{p_j^+}x\}[e_j(+\infty) + o(1)]\,, \quad x \to +\infty\,,$$

and let y_j^- be solutions with similar asymptotic behaviour as $x \to -\infty$. The solutions $\{y_j^+, \overline{y_j^+}\}$ and $\{y_j^- \overline{y_j^-}\}$, $1 \leqslant j \leqslant n$, form two F.S.S. and any solution has the form

$$y = Y^+a^+ + \overline{Y^+}b^+ = Y^-a^- + \overline{Y}^-b^-\,.$$

Here $a^\pm$ and $b^\pm$ are n-vectors, Y^+ is the $n \times n$ matrix with columns $y_1^+(x), \ldots, y_n^+(x)$, and the other matrices are defined similarly. Then

$$J_k(\lambda) = 2(a_k^+ b_k^+ - a_k^- b_k^-)\,. \tag{4}$$

If conditions 1) – 4) are satisfied then $J_k(\lambda) = O(\lambda^{-\infty})$ and $J_k(\lambda) = O(e^{-c\lambda})$, $c > 0$, $\lambda \to +\infty$, if $A(x)$ satisfies the conditions given at the end of paragraph 1.

3. Adiabatic Invariants of Canonical Systems. We consider the system of $2n$ equations

$$\varepsilon y' = A(x)y\,, \tag{5}$$

where $J_{2n}A(x)$ is a real symmetric matrix and J_{2n} is as in § 4. We introduce the conditions:

1) The eigenvalues of $A(x)$ are distinct and purely imaginary for all x, $-\infty \leqslant x \leqslant +\infty$.

2) The limits $\lim_{x\to\pm\infty} A(x) = A_\pm$ exist and are finite, and the elements of $A'(x)$ are smooth functions.

Let $p_j(x)$ be the eigenvalues and let $e_j(x)$ be the right eigenvectors of $A(x)$. We index them in such a way that

$$p_{j+n}(x) = -p_j(x) = \overline{p_j(x)}\,, \quad e_{j+n}(x) = \overline{e_j(x)}\,, \quad 1 \leqslant j \leqslant n\,,$$

and take the left eigenvectors as $e_j^\star(x) = e_j^T(x)$. System (5) has n independent adiabatic invariants

$$J_s(x,\varepsilon) = i\frac{(e_s^\star y)\overline{(e_s^\star y)}}{e_s^\star J\overline{e_s}}\,, \quad s = 1, \ldots, n\,, \tag{6}$$

which are real for real solutions y. These invariants are involutive, that is, all the Poisson brackets $\{J_r, J_s\}$, $r \neq s$, are zero. Under conditions 1)and 2), the total variations of the adiabatic invariants,

$$J_s(\varepsilon) = J_s(+\infty, \varepsilon) - J_s(-\infty, \varepsilon)\,,$$

have order $O(\varepsilon^\infty)$ as $\varepsilon \to +0$. Observe that formula (3) is a special case of (6). There is a survey of the theory of adiabatic invariants in [Wasow 8].

4. Eigenvalue Problems

4.1 The Problem on the Half-Line. We consider the system of n equations

$$y' = \lambda A(x) y \tag{7}$$

on the half-line $\mathbb{R}^+ : x \geqslant 0$, and impose the boundary condition

$$Uy(0, \lambda) = 0\,, \tag{8}$$

where U is a constant matrix of order $n \times n$ and of rank m. The point λ_0 is called an *eigenvalue* of the problem (7), (8) on the half-line $\mathbb{R}^+$ if there is a non-trivial solution $y(x, \lambda_0) \in L_2(\mathbb{R}^+)$ of (7) (an *eigenfunction*) satisfying (8). Below we consider the case where the eigenvalues form a discrete set and calculate their asymptotic behaviours for some classes of systems.

We will assume that $A(x)$ satisfies conditions 1) – 4) of § 4, paragraph 1.1 and that

$$q(x) > 0\,, \quad \int_0^\infty q(x)dx = \infty\,. \tag{9}$$

Let η_j and $p_j(x)$ be eigenvalues of $B(+\infty)$ and $A(x)$ repsectively. Then

$$p_j(x) = [\eta_j + o(1)]q(x)\,, \quad x \to \infty\,.$$

Let the numbers η_j be such that

$$\text{Re}\ \eta_1 < \text{Re}\ \eta_2 < \ldots < \text{Re}\ \eta_m < 0 < \text{Re}\ \eta_{m+1} < \ldots < \text{Re}\ \eta_n\,,$$

where $m = \text{rank}\ U$. Then (7) has F.S.S. $\{y_1, \ldots, y_n\}$ of the form (10) of § 4:

$$y_j(x, \lambda) = \exp\left\{\lambda \int_{x_0}^x p_j(t)dt + \int_{x_0}^x p_j^{(1)}(t)dt\right\}[e_j(x) + \lambda^{-1}u_j(x, \lambda)]\,, \tag{10}$$

where for $\lambda \geqslant \lambda_0$ and $x \geqslant x(\lambda_0)$ there are the bounds

$$|u_j(x, \lambda)| \leqslant k(x)\,, \quad \lim_{x \to \infty} k(x) = 0\,, \quad 1 \leqslant j \leqslant n\,.$$

Moreover there is a sector S: $|\lambda| \geqslant \lambda_0 > 0$, $|\arg \lambda| \leqslant \delta$ in the complex λ-plane such that

1) the solution $y_j(x, \lambda)$ is holomorphic in S for each fixed $x \geqslant 0, 1 \leqslant j \leqslant n$;

2) the solutions $y_1, \ldots, y_m$ belong to $L_2(\mathbb{R}^+)$, the solutions $y_{m+1}, \ldots, y_n$ and any non-trivial linear combinations of them do not belong to $L_2(\mathbb{R}^+)$ for each fixed $\lambda \in S$.

We denote $\tilde{Y}(x, \lambda) = (y_1(x, \lambda), \ldots, y_m(x, \lambda))$; then any eigenfunction is of the form

$$y(x, \lambda) = \tilde{Y}(x, \lambda)\, c(\lambda)\,, \tag{11}$$

where $c(\lambda)$ is a vector with m components, and the eigenvalues are given by the equation

$$\det \Omega_0(\lambda) = 0\,, \quad \Omega_0(\lambda) = U\tilde{Y}(0,\lambda)\,. \tag{12}$$

It follows from properties of the F.S.S. that for $\lambda \in S$ either the eigenvalues form a discrete set or $\det \Omega_0(\lambda) \equiv 0$ so that any point of S is a point of the spectrum. The latter case is excluded.

We make the following assumptions:

1) On the half-line $x \geqslant 0$ there is precisely one, and moreover simple, turning point $x_0 > 0$ and $p_m(x_0) = p_{m+1}(x_0)$.

2) $\operatorname{Re} p_m(x) = \operatorname{Re} p_{m+1}(x)$ for $0 \leqslant x \leqslant x_0$, and $\operatorname{Re} p_j(x) < \operatorname{Re} p_{j+1}(x)$ for $x \geqslant 0$ if $j \neq m, n$ and for $x \geqslant x_0$ if $j = m$.

3) The matrix $A(x)$ is holomorphic in a neighbourhood V of the line segment $I = [0, x_0]$.

Here I is a Stokes line and the local structure of the Stokes lines going out from x_0 is the same as that for the Airy equation $y'' - \lambda^2(x - x_0)y = 0$. Another two Stokes lines go out from x_0, l_1 and $l_2 = l_1^*$, and $\operatorname{Im} x > 0$ on l_1.

4.2 Asymptotic Behaviour of the Solutions. In order to find the asymptotic behaviour of the eigenvalues it is necessary to find the asymptotic behaviour of the solutions $y_1(x,\lambda), \ldots, y_m(x,\lambda)$ for $x = 0$ and $\lambda \to +\infty$. Since there is the turning point x_0 on the half-line $x \geqslant 0$ the asymptotic expansions (10) are inapplicable for $x = 0$ and therefore we must consider the asymptotic behaviour of the solutions in different domains of the complex x-plane. Let G, G_m, G^+ be the domains represented in Fig. 25 and sufficiently narrow, and G^- is symmetric to G^+ about the real axis. All these domains contain a point $x_1 > x_0$ and their boundaries contain a point ia, $a > 0$. We denote $\tilde{G} = G \cup [a, +\infty)$ and define $\tilde{G}_m$ and $\tilde{G}^\pm$ similarly.

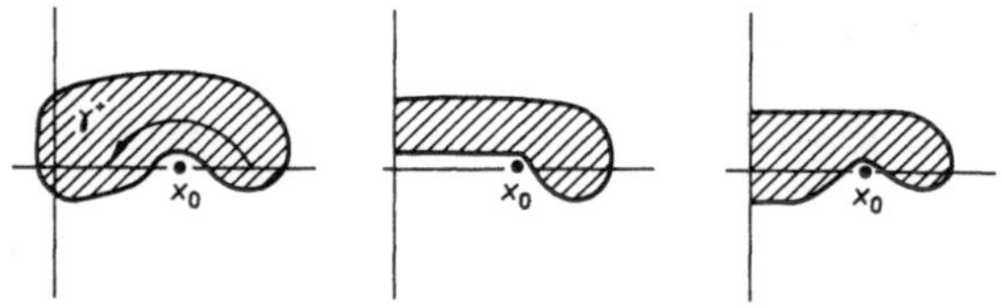

Fig. 25

The following discussion is similar to that given in § 5 of Chap. 3, paragraph 2.1. Let $1 \leqslant j \leqslant m-1$, $1 \leqslant k \leqslant n$, $j \neq k$, and $j \neq m$, $m+1$; then for each $x \in [\tilde{G}]$ there is a canonical path γ_{jk} joining x_{jk} to x and lying in $[\tilde{G}]$. Here $x_{jk} = +\infty$ for $k > j$ and $x_{jk} = ia$, $k < j$, so that there is an asymptotic expansion of the form (10) for the solutions $y_1, \ldots, y_{m-1}$ as $\lambda \to +\infty$, $x \in G$. There are analogous asymptotic expansions for y_m and $y_{m+1}^\pm$ in $[G_m]$ and $[G^\pm]$ respectively.

The solutions $y_1, \dots, y_m, y^+_{m+1}, y^-_{m+1}$ are linearly dependent, so that

$$y_m(x,\lambda) = a_+(\lambda)y^+_{m+1}(x,\lambda) + a_-(\lambda)y^-_{m+1}(x,\lambda) + \sum_{j=1}^{m-1} a_j(\lambda)y_j(x,\lambda)\,. \quad (13)$$

The solutions $y^\pm_{m+1}$ have the greatest and equal orders of growth as $x \to +\infty$, and because the asymptotic behaviour (1) is dual, we have $a_1^+(\lambda) + a_2^+(\lambda) = 0$, $\lambda \gg 1$. We will make more exact the normalization of y_m and $y^\pm_{m+1}$. Since x_0 is a simple turning point, we have

$$p_{m+1}(x) = q(x) + \sqrt{D(x)}\,, \quad p_m(x) = q(x) - \sqrt{D(x)}\,,$$

where $D(x)$ has a simple zero at $x = x_0$, $D(x) > 0$ for $x > x_0$, and $q(x)$ is a smooth function. Further (§ 5),

$$p_j^{(1)}(x) = -\frac{1}{4(x-x_0)} + p_j^{(2)}(x)\,, \quad j = m,\, m+1\,,$$

where $p_j^{(2)}(x) = O((x-x_0)^{-1/2})$ as $x \to x_0$. For $x > x_0$ and $j = m, m+1$ we put

$$\tilde{y}_j(x,\lambda) = (x-x_0)^{-1/4}\exp\{\lambda S_j(x) + S_j^{(2)}(x)\}\,,$$

$$S_j^{(2)}(x) = \int_{x_0}^{x} p_j^{(2)}(t)dt\,,$$

where $\sqrt[4]{x-x_0} > 0$. We put $x = ia$ in (13). The values of all the functions and vector functions for $x = ia$ are obtained as a result of their analytic continuation along paths γ, lying in the corresponding domains. Further, we can put $\gamma = \gamma^+$ in G, G_m, $G^\pm$, and $\gamma = \gamma^-$ in G^- (Fig. 25). We denote the corresponding values of $\tilde{y}_{m+1}$ and e_{m+1} by $\tilde{y}^\pm_{m+1}$ and $e^\pm_{m+1}$. Then

$$(e_{m+1}(x)e^{S^{(2)}_{m+1}(x)})_- = e_m(x)e^{S^{(2)}_m(x)}\,,$$

$$(x-x_0)_+^{1/4} = i(x-x_0)_-^{1/4}\,, \quad (S_{m+1}(x))_- = S_m(x)\,.$$

Thus, for $x = ia$ all the ratios $\tilde{y}_j/y_m$, $\tilde{y}_{m+1}/y^-_m$ are exponentially small and, after cancelling by $\tilde{y}_m$, (13) becomes

$$e_m(x) = ia_-(\lambda)[e_m(x) + O(\lambda^{-1})] + \sum_{j=1}^{m-1} a_j(\lambda)\frac{y_j(x,\lambda)}{\tilde{y}_m(x,\lambda)}[e_j(x) + O(\lambda^{-1})]\,.$$

Multiplying this identity on the left by $e_1^*(x), \dots, e_m^*(x)$, we obtain a system of equations from which we obtain $a_+(\lambda) = i + O(\lambda^{-1})$, so that

$$a_+(\lambda) = i + O(\lambda^{-1})\,, \quad a_-(\lambda) = -i + O(\lambda^{-1})\,.$$

It follows that

$$y_m(0,\lambda) = [i + O(\lambda^{-1})]y^+_{m+1}(0,\lambda) + [-i + O(\lambda^{-1})]y^-_{m+1}(0,\lambda) \\ + O(|y^{\pm}_{m+2}(0,\lambda)|e^{-c\lambda}), \quad c > 0. \tag{14}$$

To achieve this it is sufficient to put $x = x_1 > x_0$ in the system above and to make use of the fact that

$$\operatorname{Re}\int_{x_0}^{x} [p_j(t) - p_k(t)]dt \to -\infty, \quad x \to +\infty,$$

for $1 \leqslant j \leqslant m+1$ and $k = m, m+1$. The latter bound is completely analogous to the connection formula (8) of § 5.

4.3 Asymptotic Behaviour of the Eigenvalues. By construction the asymptotic expansions of $y_1, \ldots, y_{m-1}$, $y^{\pm}_{m+1}$ are applicable at $x = 0$, and y_m is expressed in terms of these solutions. Replacing the values of all those solutions by the asymptotic expansions for $x = 0$ and $\lambda \to \infty$, and considering (13) and (14), we obtain the equation for the eigenvalues from (12). Let

$$\xi_0 = \frac{1}{2}\int_0^{x_0} [p_{m+1}(t) - p_m(t)]dt, \\ \xi_1^{\pm} = [S^{(2)}_{m+1}(x_0, 0)]_{\pm}, \quad \Delta_{\pm} = \det B_{\pm}, \tag{15}$$

where $B_{\pm}$ are the matrices

$$B_{\pm} = U\{e_1(0), \ldots, e_{m-1}(0), e^{\pm}_{m+1}(0)\}.$$

Suppose that at least one of the numbers Δ_+, Δ_- is non-zero. We have

$$B_+e^{-i\pi/4}\exp\{-i\lambda\xi_0 + \xi_1^+\} + B_-e^{i\pi/4}\exp\{i\lambda\xi_0 + \xi_1^-\} = O(\lambda^{-1}),$$

from which we find the asymptotic behaviour of the eigenvalues to be

$$\lambda_k = \left\{k\pi + \frac{\pi}{4} - \frac{i}{2}[(S^{(2)}_{m+1}(x_0,0))_+ - (S^{(2)}_{m+1}(x_0,0))_-] + \frac{i}{2}\ln\frac{A_+}{A_-}\right\}\xi_0^{-1} \\ + O(k^{-1}), \quad k \to +\infty. \tag{16}$$

We can find the asymptotic behaviour of the eigenfunctions on the half-line $x \geqslant 0$ from formulae (12) – (16) (excluding a small neighbourhood of the turning point x_0). The asymptotic expansions of the solutions near to x_0 can be found using the results of § 4.

4.4 Equations of nth-Order. We consider the equation

$$y^{(n)} + \lambda q_1(x)y^{(n-1)} + \ldots + \lambda^n q_n(x)y = 0$$

with the same conditions on the roots of the characteristic equation as those given above, and consider the problem of the eigenvalues on the half-line $x \geqslant 0$ with the simplest boundary conditions

$$y(0) = 0\,, \quad y'(0) = 0, \ldots, y^{(m-1)}(0) = 0\,.$$

Here $\Delta_\pm$ is the Vandermonde determinant of the functions $p_1(x), \ldots, p_{m-1}(x)$, $p^\pm_{m+1}(x)$ for $x = 0$, so that

$$\frac{\Delta_+}{\Delta_-} = \prod_{j=1}^{m-1} \frac{p^+_{m+1}(0) - p_j(0)}{p^-_{m+1}(0) - p_j(0)}\,.$$

Using the explicit expression for $p_j^{(1)}(x)$ we arrive at the asymptotic formula for the eigenvalues

$$\lambda_k = \left[k\pi + \frac{\pi}{4} - \frac{i}{2}\sum_{j=1}^{m-1} \ln \frac{p^+_{m+1}(0) - p_j(0)}{p^-_{m+1}(0) - p_j(0)} \right.$$
$$\left. - \frac{i}{4}\int_0^{x_0} \sum_{j=1}^{n} \left\{ \frac{(p^+_{m+1}(t))'}{p^+_{m+1}(t) - p_j(t)} - \frac{(p^-_{m+1}(t))'}{p^-_{m+1}(t) - p_j(t)} \right\} dt \right]$$
$$\times\, \xi_0^{-1} + O(k^{-1}).$$

For the equation $y^n - \lambda^2 q(x) = 0$ this formula is the same as that in Chap. 3, § 5, paragraph 2.1.

4.5 The Problem on the Whole Line. In this case λ_0 is called an *eigenvalue* if the system (7) has a non-trivial solution $y(x, \lambda_0) \in L_2(-\infty, \infty)$. We assume that the conditions of § 4, paragraph 4.1, are satisfied on each of the half-lines $x \geqslant 0$ and $x \leqslant 0$, so that

$$A(x) = q_+(x) Q_+^{-1}(x) B_+(x) Q_+(x)\,, \quad x \geqslant 0\,,$$
$$A(x) = q_-(x) Q_-^{-1}(x) B_-(x) Q_-(x)\,, \quad x \leqslant 0\,,$$

where each of the matrices $Q_\pm(x)$, $B_\pm(x)$ and functions $q_\pm(x)$ satisfies conditions 3) and 4) of § 4. Suppose that

$$q_\pm(x) > 0\,, \quad \int_0^{+\infty} q_+(x)dx = \infty\,, \quad \int_{-\infty}^0 q_-(x)dx = \infty$$

and the eigenvalues $\eta_j^\pm$ of $B_\pm(\pm\infty)$ are such that

$$\operatorname{Re} \eta_1^+ < \ldots < \operatorname{Re} \eta_m^+ < 0 < \operatorname{Re} \eta_{m+1}^+ < \ldots < \operatorname{Re} \eta_n^+\,,$$
$$\operatorname{Re} \eta_1^- < \ldots < \operatorname{Re} \eta_m^- < 0 < \operatorname{Re} \eta_{m+1}^- < \ldots < \operatorname{Re} \eta_n^-\,.$$

Then (7) has the F.S.S. $\{y_1^+, \ldots, y_n^+\}$, $\{y_1^-, \ldots, y_n^-\}$ for which the asymptotic behaviours of the form (10) are dual respectively as $x \to +\infty$ and $x \to -\infty$. For each fixed λ, $\lambda \geqslant \lambda_0 \gg 1$, we have:

1) the solutions $y_1^+(x, \lambda), \ldots, y_m^+(x, \lambda)$ belong to $L_2(\mathbb{R}^+)$, none of $y_{m+1}^+(x, \lambda)$, $\ldots, y_n^+(x, \lambda)$, and no non-trivial linear combination of them, belong to $L_2(\mathbb{R}^+)$;

2) the solutions $y^-_{m+1}(x,\lambda),\dots,y^-_n(x,\lambda)$ belong to $L_2(\mathbb{R}^-)$, none of $y_1(x,\lambda),\dots,y^-_m(x,\lambda)$, and no non-trivial linear combination of them, belong to $L_2(\mathbb{R}^-)$;

3) there exist $\rho>0$, $\delta>0$, $a>0$ such that the solutions $y^{\pm}_j(x,\lambda)$, $1\leqslant j\leqslant n$, are holomorphic in λ for $|\lambda|\geqslant\rho$ and $|\arg\lambda|\leqslant\delta$ for each fixed $x\geqslant a$ (respectively $x\leqslant a$).

For $\lambda\in S$ and $|\lambda|\gg 1$ any eigenfunction has the form

$$\begin{aligned} y(x,\lambda) &= c^+_1(\lambda)y^+_1(x,\lambda)+\ldots+c^+_m(\lambda)y^+_m(x,\lambda)\\ &= c^-_{m+1}(\lambda)y^-_{m+1}(x,\lambda)+\ldots+c^-_n(\lambda)y^-_n(x,\lambda)\,. \end{aligned}$$

The eigenvalues are determined by the equation $W(\lambda)=0$, where W is the Wronskian of $y^+_1,\dots,y^+_m,\ y^-_{m+1},\dots,y^-_n$. Suppose that the following conditions, similar to 1) – 3) of paragraph 4.1, are satisfied.

1) There are precisely two turning points x_1 and x_2 on the real axis, $x_1<x_2$, both simple, and $p_m(x_j)=p_{m+1}(x_j)$, $j=1,2$.

2) Re $p_m(x)=$ Re $p_{m+1}(x)$, $x_1\leqslant x\leqslant x_2$, and Re $p_j(x)<$ Re $p_{j+1}(x)$ for all x if $j\neq m,n$ and for $x\notin(x_1,x_2)$ if $j=m$.

3) The matrix $A(x)$ is holomorphic in a neighbourhood of $[x_1,x_2]$.

Let C be a simple closed contour in the complex x-plane, enclosing $[x_1,x_2]$ and positively oriented. We denote

$$\begin{aligned} \xi_0 &= \frac{1}{2}\oint_C [p_{m+1}(x)-p_m(x)]dx\,,\\ \xi_1 &= \frac{1}{2}\oint_C [p^{(1)}_{m+1}(x)-p^{(1)}_m(x)]dx\,. \end{aligned}$$

System (7) has an infinite series of eigenvalues of the form

$$\lambda_k=\xi_0^{-1}[\pi(k+1/2)-i\xi_1]+O(k^{-1})\,,k\to\infty\,.$$

Similar results are true when (7) has several simple real turning points [Fedoryuk 3].

§ 8. Examples

1. The Stueckelberg System. This system arose in the study of inelastic collisions between two atoms and has the form

$$h^2u_1''+\phi_1u_1=\alpha u_2\,,\quad h^2u_2''+\phi_2u_2=\alpha u_1\,, \tag{1}$$

where $\phi_1(x)$, $\phi_2(x)$, $\alpha(x)$ are smooth real-valued functions on $I=[a,b]$, and $h>0$ is a small parameter. We put

$$\phi(x)=\frac{1}{2}[\phi_1(x)+\phi_2(x)]\,,\quad \psi(x)=\frac{1}{2}[\phi_1(x)-\phi_2(x)]\,. \tag{2}$$

The eigenvalues and eigenvectors of the matrix of the system are

$$\begin{aligned} p_{1,2}(x) &= \phi(x) \pm \sqrt{D(x)}\,, \quad D = \psi^2 + \alpha^2\,, \\ e_{1,2}(x) &= (\alpha, \psi \mp \sqrt{\psi^2+\alpha^2})^T\,, \quad e_j^*(x) = e_j^T(x)\,, \quad j = 1,2\,. \end{aligned} \tag{3}$$

System (1) has two types of turning points.

A. $\phi^2(x) = \psi^2(x) + \alpha^2(x)$. Here either $p_1 = 0$ or $p_2 = 0$.

B. $\psi^2(x) + \alpha^2(x) = 0$. Here $p_1 = p_2$.

Suppose that there are no turning points on I. Since

$$\exp\left\{-\int^x \frac{e_j^*(t)e_j'(t)}{e_j^*(t)e_j(t)}dt\right\} = \frac{c}{\sqrt{e_j^T(x)e_j(x)}}\,,$$

system (1) has F.S.S. of the form

$$\begin{aligned} u_j^{\pm}(x,h) = {} & \frac{1}{\sqrt[4]{p_j(x)}}\frac{1}{\sqrt{e_j^T(x)e_j(x)}}\exp\left\{\pm\frac{i}{h}\int_{x_0}^x \sqrt{p_j(t)}dt\right\} \\ & \times [e_j(x) + O(h)]\,, \quad j = 1,2\,. \end{aligned} \tag{4}$$

These formulae can be written as

$$\begin{aligned} u_1^+(x,h) = {} & \frac{1}{\sqrt[4]{p_1 D}}\frac{1}{\sqrt{\sqrt{D}-\psi}}\exp\left\{\frac{i}{h}\int_{x_0}^x \sqrt{p_1(t)}dt\right\} \\ & \times \left[\begin{bmatrix} a \\ \psi - \sqrt{D}\end{bmatrix} + O(h)\right]\,, \\ u_2^+(x,h) = {} & \frac{1}{\sqrt[4]{p_2 D}}\frac{1}{\sqrt{\sqrt{D}+\psi}}\exp\left\{\frac{i}{h}\int_{x_0}^x \sqrt{p_2(t)}dt\right\} \\ & \times \left[\begin{bmatrix} a \\ \psi + \sqrt{D}\end{bmatrix} + O(h)\right]\,, \end{aligned} \tag{5}$$

$$u_j^-(x,h) = \overline{u_j^+(x,h)}\,, \quad j = 1,2\,.$$

2. Waves in Plasmas. In the study of electromagnetic wave propagation in a non-uniform planar-layered medium and, in particular, in a magneto-active plasma, there arises the system of equations

$$\begin{aligned} \frac{d^2E_x}{dz^2} + \frac{\omega^2}{c^2}[AE_x + iCE_y] &= 0 \\ \frac{d^2E_y}{dz^2} + \frac{\omega^2}{c^2}[-iCE_x + BE_y] &= 0\,, \end{aligned} \tag{6}$$

where $\omega > 0$ is the frequency, c is the speed of light in a vacuum and A, B, C are functions of z. This case corresponds to the normal incidence of a wave onto a plasma layer. The eigenvalues and eigenvectors of the matrix of the system are

$$p_{1,2}(z) = \frac{1}{2}(A + B \pm \sqrt{D}), \quad D = (A - B)^2 + 4C^2,$$
$$e_{1,2}(z) = \begin{bmatrix} -2iC \\ A - B \mp \sqrt{D} \end{bmatrix}, \quad e^*_{1,2}(z) = [2iC,\ A - B \mp \sqrt{D}].$$

The turning points for the system are determined by the equations $D = 0$ and $AB = C^2$.

Suppose that $\omega/c \gg 1$ and that there are no turning points on I, and let $\text{Re}\sqrt{p_1(z)}$, $\text{Re}\sqrt{p_2(z)}$, $\text{Re}\,(\sqrt{p_1(z)} - \sqrt{p_2(z)})$ retain their signs for $z \in I$. If A, B, C are real then this condition is satisfied. In this case the left and right eigenvectors are such that $e_j = (\alpha_j, \beta_j)^T$ and $e^*_j = (-\alpha_j, \beta_j)$, and therefore

$$\exp\left\{-\int^x \frac{e^*_j(t)e'_j(t)}{e^*_j(t)e_j(t)}dt\right\} = \frac{c}{\sqrt{\beta_j^2(x) - \alpha_j^2(x)}}.$$

The system (6) has an F.S.S. of the form

$$E_1^{\pm} = \frac{1}{\sqrt[4]{D}} \frac{1}{\sqrt{\sqrt{D} - A + B}} \exp\left\{\pm i\frac{\omega}{c}\int_{z_0}^{z} \sqrt{p_1(t)}dt\right\}[e_1(z) + O(h)],$$
$$E_2^{\pm} = \frac{1}{\sqrt[4]{D}} \frac{1}{\sqrt{\sqrt{D} + A - B}} \exp\left\{\pm i\frac{\omega}{c}\int_{z_0}^{z} \sqrt{p_2(t)}dt\right\}[e_2(z) + O(h)],$$

where $E_j = (E_{xj}, E_{yj})^T$.

3. A System for Elasticity Theory. Small vibrations of an elastic isotropic medium are described by the system

$$\begin{aligned} &\frac{\partial \sigma_{11}}{\partial x_1} + \frac{\partial \sigma_{12}}{\partial x_2} + \rho\omega^2 u_1 = 0, \\ &\frac{\partial \sigma_{21}}{\partial x_1} + \frac{\partial \sigma_{22}}{\partial x_2} + \rho\omega^2 u_2 = 0, \end{aligned} \tag{7}$$

Here $u = (u_1, u_2)^T$ is the displacement vector, $\sigma = (\sigma_{jk})$ is the deformation tensor

$$\sigma_{jj} = \lambda \operatorname{div} u + 2\mu\frac{\partial u_j}{\partial x_j}, \quad j = 1, 2,$$
$$\sigma_{12} = \sigma_{21} = \mu\left(\frac{\partial u_1}{\partial x_2} + \frac{\partial u_2}{\partial x_1}\right),$$

λ and μ are the Lame parameters, ρ is the density of the medium, and ω is the frequency of vibration. We consider a laminar non-uniform medium, for which λ, μ, ρ depend only on $x_1 = x$. Then (7) has solutions of the form

$u(x_1, x_2) = v(x_1)\exp(ikx_2)$ where the vector function v is the solution of the system

$$Av'' + (ikB + A')v' + (-k^2C + ikD + \rho\omega^2\dot{I})v = 0\,. \tag{8}$$

Here A, B, C and D are matrices depending on x:

$$A = \begin{bmatrix} \lambda + 2\mu & 0 \\ 0 & \mu \end{bmatrix}, \quad B = \begin{bmatrix} 0 & \lambda + \mu \\ \lambda + \mu & 0 \end{bmatrix},$$
$$C = \begin{bmatrix} \mu & 0 \\ 0 & \lambda + 2\mu \end{bmatrix}, \quad D = \begin{bmatrix} 0 & \lambda' \\ \mu' & 0 \end{bmatrix}.$$

System (8) corresponds to the bundle

$$L(x,p) = Ap^2 + (ikB + A')p - k^2C + ikD + \rho\omega^2 I\,,$$

whose eigenvalues are

$$\begin{aligned} p_{1,2}(x) &= -\frac{1}{2}[\ln(\lambda + 2\mu)]' \pm \sqrt{k^2 + \frac{1}{2}[\ln(\lambda + 2\mu)]' - \frac{\rho\omega^2}{\lambda + 2\mu}}\,, \\ p_{3,4}(x) &= -\frac{1}{2}(\ln\mu)' \pm \sqrt{k^2 + \frac{1}{2}(\ln\mu)' - \frac{\rho\omega^2}{\mu}}\,. \end{aligned} \tag{9}$$

We will consider the two possibilities.

3.1 The Frequency ω does not depend on k. In this case the roots of the bundle are asymptotically repeated:

$$\begin{aligned} p_{1,2}(x) &= \pm k - \frac{1}{2}[\ln(\lambda + 2\mu)]' + O(k^{-1})\,, \\ p_{3,4}(x) &= \pm k - \frac{1}{2}(\ln\mu)' + O(k^{-1})\,. \end{aligned}$$

This case is similar to that considered in Chap. 2, § 8, paragraph 2, A. The principal asymptotic term cannot be found by quadratures but can be expressed in terms of the solutions of the second order differential equation

$$lw \equiv (\mu w)'' + \mu w' + 2\mu\left[\frac{(\mu w)'}{\lambda + \mu}\right]' = 0\,. \tag{10}$$

Let I be a finite closed interval, with $\lambda(x)$, $\mu(x)$, $\rho(x) \in C^\infty(I)$. We look for the F.A.S. to (8) in the form

$$v(x,k) = e^{kx}\left[f_0(x) + \frac{1}{k}f_1(x) + \dots\right].$$

We then arrive at the recurrence system of equations for $f_0(x)$, $f_1(x), \dots$

$$\begin{aligned}
&(A+B-C)f_0 = 0\,,\\
&(A+B-C)f_1 = -(2A+B)f_0' - (A'+D)f_0\,,\\
&(A+B-C)f_j = -(2A+B)f_{j-1}' - (A'+D)f_{j-1}\\
&\qquad\qquad - Af_{j-2}'' + \rho\omega^2 f_{j-2}\,, \quad j = 2,3,\dots\,, \qquad (11)
\end{aligned}$$

in which

$$A+B-C = (\lambda+\mu)\begin{bmatrix} 1 & i \\ -i & 1 \end{bmatrix}.$$

All solutions of the first equation in (11) have the form $f_0(x) = w_0(x)\,(1,i)^T$, where $w_0(x)$ is an arbitrary function. The matrix $A+B-C$ is singular; nevertheless the second equation in (11) is solvable for any function $w_0(x)$ and all solutions of this equation have the form

$$f_1(x) = w_1(x)\begin{bmatrix} 1 \\ i \end{bmatrix} + \frac{1}{\lambda+\mu}\begin{bmatrix} 0 \\ (\lambda+3\mu)w_0'(x) + 2\mu' w_0(x) \end{bmatrix},$$

where $w_1(x)$ is an arbitrary function. The function $w_0(x)$ is still undefined and is determined from the condition for solvability of the third equation in (11). This condition is that $w_0(x)$ must satisfy equation (10). Finally we obtain the expression for the F.A.S. to (9)

$$v^+(x,k) = e^{kx}\begin{bmatrix} w_0(x) + \frac{1}{k}w_1(x) + \frac{1}{k^2}w_2(x) + \dots \\ iw_0(x) + \frac{i}{k}q_1 + \frac{i}{k^2}q_2 + \dots \end{bmatrix},$$

$$\begin{aligned}
a_j(x) &= (\lambda+3\mu)w_{j-1}' + 2\mu' w_{j-1} - (\lambda+\mu)\left(\frac{a_{j-1}}{\lambda+\mu}\right)'\\
&\quad - \frac{\lambda'}{\mu}a_{j-1} + [(\lambda+2\mu)w_{j-2}']' + \rho\omega^2 w_{j-2}\,, \quad j = 1,2,\dots\,, \qquad (12)
\end{aligned}$$

where $q_n(x) = w_n(x) + a_n(x)/(\lambda(x)+\mu(x))$, $n = 1,2$, $a_{-1} = a_0 = 0$, and $w_j(x)$ is a solution of the inhomogeneous equation

$$[(\lambda+\mu)w_j' - a_{j+1}] - 2\mu\left(\frac{a_j+1}{\lambda+\mu}\right)' - \left[\mu\left(\frac{a_j}{\lambda+\mu}\right)'\right]' + \frac{\rho\omega^2 a_j}{\lambda+\mu} = 0\,. \qquad (13)$$

This equation has the form $lw_j = q_j(w_0, w_1, \dots, w_{j-1})$, where l is the operator in (10).

Formula (12) determines two linearly independent F.A.S. $v_1^+(x,h)$ and $v_2^+(x,h)$ to system (8). Namely, let w_{01}, w_{02} be a F.S.S. to equation (10) and let w_j be an arbitrary particular solution of (13). Putting $w_0 = w_{0j}$ into (12) we obtain the F.A.S v_j^+, $j = 1,2$.

System (8) also has a F.A.S. of the form

$$v^-_{1,2}(x) = e^{-kx}\begin{bmatrix} w_0(x) - \frac{1}{k}w_1(x) + \frac{1}{k^2}w_2(x) + \dots \\ -iw_0(x) + \frac{i}{k}q_1 - \frac{i}{k^2}q_2 + \dots \end{bmatrix}, \tag{14}$$

where $q_n = w_n(x) + a_n(x)/(\lambda(x) + \mu(x))$, $n = 1, 2$.

Suppose that the following conditions are satisfied

$$\lambda(x) \neq 0\,, \quad \mu(x) \neq 0\,, \quad \lambda(x) + 2\mu(x) \neq 0\,,$$

$$\lambda'(x)\mu(x) - \lambda(x)\mu'(x) \neq 0\,, \quad x \in I\,.$$

Then (8) has a F.S.S. $\{v_1^+(x,k),\, v_2^+(x,k),\, v_1^-(x,k),\, v_2^-(x,k)\}$, and the asymptotic expansions (12) and (14) are valid for these solutions for Re $k \geqslant 0$, $|k| \to \infty$, uniformly in $x \in I$.

Remark. If $\mu(x) \equiv$ const., then equation (11) is integrable and one of the solutions is $w_0(x) = 1$.

8.2 The Frequency ω Depends on k: $\omega = k\sigma$, $\sigma > 0$. In this case we look for the F.A.S. to (8) in the form

$$v(x,k) = \exp\left\{ik\int_{x_0}^{x} p(t)dt\right\}\left[e_0(x) + \frac{1}{k}e_1(x) + \dots\right].$$

Then we obtain a recurrence system of equations for $p(x)$ and $e_0(x), e_1(x), \dots$. The first two equations are

$$L(x,p)e_0 = 0\,,$$
$$L(x,p)e_1 = i(2pAe_0' + p'Ae_0 + Be_0' + pA'e_0 + De_0)\,,$$

where

$$L(x,p) = A(x)p^2 + B(x)p + C(x) - \rho(x)\sigma^2 I\,.$$

The characteristic equation is $\det L(x,p) = 0$ and its roots are

$$p_{1,2}^2 = -1 + \frac{\rho\sigma^2}{\mu}\,, \quad p_{3,4}^2 = -1 + \frac{\rho\sigma^2}{\lambda + 2\mu}\,. \tag{15}$$

The eigenvectors of $L(x,p)$ are

$$f = (1, -p)^T\,, \quad f^\star = (1, -p)\,,$$

where $p(x)$ is one of the roots of the characteristic equations. The condition for solvability for the second equation is

$$e_0^\star(2pAe_0' + p'Ae_0 + Be_0' + pA'e_0 + De_0) = 0\,.$$

Putting $e_0(x) = \alpha(x)f(x)$ we find

$$\alpha(x) = \frac{c}{\sqrt{\mu p(p^2+1)}}.$$

System (8) has F.A.S. of the form

$$\begin{aligned} v_j(x,k) &= \frac{1}{\sqrt{p_j(x)}} \exp\left\{\pm k \int_{x_0}^{x} \sqrt{p_j(t)}dt\right\} \\ &\quad \times [(1, -p_j(x))^T + O(k^{-1})], \quad j = 1,2, \\ v_j(x,k) &= \sqrt{\frac{\lambda(x)+2\mu(x)}{\mu(x)p_j(x)}} \exp\left\{\pm k \int_{x_0}^{x} \sqrt{p_j(t)}dt\right\} \\ &\quad \times [(1, -p_j(x))^T + O(k^{-1})], \quad j = 3,4. \end{aligned} \tag{16}$$

Let $\lambda(x) > 0$, $\mu(x) > 0$, $\rho(x) > 0$ for $x \in I$, and suppose that one of the following conditions is satisfied:

$$\rho(x)\sigma^2 < \mu(x), \quad \rho(x)\sigma^2 > \lambda(x) + 2\mu(x), \quad x \in I.$$

Then (8) has no turning points in I, and there is a F.S.S. $\{v_1, v_2, v_3, v_4\}$ for which there are asymptotic expansions of the form (16) as $k \to +\infty$, uniformly in $x \in I$. The asymptotic behaviour of the solutions of (8) with turning points is studied in [Alenitsyn 2].

4. The Stueckelberg System with Turning Points. System (1) arises in the study of ineleastic collisions of two atoms [Stueckelberg] with masses M_1, M_2 and is considered on the half-line $0 < r < \infty$, where r is the distance between the atoms. The coefficients in the system have the form

$$\phi_j = [\varepsilon - V_j(r)]2m - \frac{h^2 l(l+1)}{r^2}, \quad \alpha = 2mV_{12}(r), \quad M = \frac{M_1 M_2}{M_1 + M_2},$$

where $\varepsilon > 0$, $V_j(x)$ are the energy terms of the electron states, and $V_{12}(r)$ is the matrix element of interaction of electron states

$$V_1(\infty) = 0, \quad V_2(\infty) = \Delta\varepsilon, \quad 0 < \Delta\varepsilon < \varepsilon, \quad V_{12}(\infty) = 0.$$

The boundary conditions are

$$u_1(0) = 0, \quad u_2(0) = 0, \quad u_2(r) = \eta_l e^{ik_2 r} + o(1),$$

$$u_1(r) = \alpha_l e^{ik_1 r} - e^{-ik_1 r} + o(1), \quad r \to \infty,$$

$$k_1 = \frac{\sqrt{2m\varepsilon}}{h}, \quad k_2 = \frac{\sqrt{2m(\varepsilon - \Delta\varepsilon)}}{h}.$$

Suppose that $V_1(r)$ and $V_2(r)$ are as shown in Fig. 26. Then (1) has two real turning points $r_1 < r_2$ and $\phi_j(r_j) = 0$. If $V_{12}(r)$ is small and the coefficients of

(1) are holomorphic in a neighbourhood of the half-line $r > 0$ then there are two complex turning points near to r_0. The connection formulae for this case were obtained in [Fedoryuk 12] and subsequently repeatedly used without proof. A rigorous mathematical basis for these formulae is still lacking. Below we shall state methods which allow us to obtain some of the connection formulae.

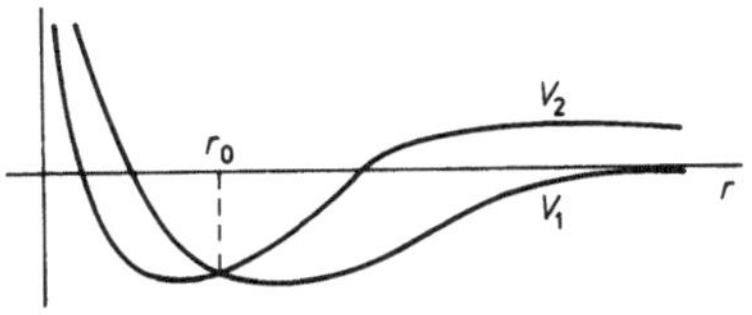

Fig. 26

Let $I = [a, b]$, let $\phi_1(z)$, $\phi_2(z)$ and $\alpha(z)$ be holomorphic in a domain U of the complex z-plane containing I, and suppose that

$$\phi_1(x) + \phi_2(x) > 0\,, \quad 4\phi_1(x)\phi_2(x) > \alpha^2(x)\,, \quad \alpha(x) > 0\,, \quad x \in I\,.$$

Then $p_1(x) > p_2(x) > 0$ for $x \in I$ and (1) has no real turning points. We take the branches of $\sqrt{p_1(x)}$, $\sqrt{p_2(x)}$, $\sqrt{D(x)}$ to be positive for $x \in I$. Suppose that there is a unique point $x_0 \in I$, $a < x_0 < b$, such that $\phi_1(x_0) = \phi_2(x_0)$ and suppose that

$$\phi_1'(x_0) \neq \phi_2'(x_0)\,, \quad \phi_1(x) > \phi_2(x)\,, \quad x \in I\,, \tag{17}$$

$$\alpha(x) = \delta\tilde{\alpha}(x)\,, \quad \delta > 0\,,$$

where δ is a small parameter not depending on h. For $\delta = 0$, x_0 is a turning point of (1) and, for small $\delta > 0$, there are two close complex turning points z_0 and $\overline{z_0}$, given by

$$z_0(\delta) = \frac{\delta i \psi(x_0)}{\phi_1'(x_0) - \phi_2'(x_0)} + O(\delta^2)\,.$$

Suppose that $\varepsilon > 0$ is fixed and does not depend on h or δ. Then (1) has F.S.S. $\{u_1^\pm, u_2^\pm\}$ for which there is the asymptotic expansion (4) as $h \to 0$, $a \leqslant x \leqslant x_0 - \varepsilon$, and F.S.S. $\{v_1^\pm, v_2^\pm\}$ for which (4) is valid as $h \to 0$, $x_0 + \varepsilon \leqslant x \leqslant b$. The asymptotic expansion (4) is not generally applicable close to x_0. It is required to express one of the F.S.S. in terms of the other. This can be effected by making explicit the domain of applicability of (4) for the different solutions. The problem as given is similar to that considered in Chap. 3, § 8, paragraph 7.

4.1 The Solution u_1^+. Let $\delta > 0$ be fixed. Then z_0 is the superposition of two simple turning points, since $p_1(z_0) = p_2(z_0) \neq 0$. From z_0 there arise three Stokes lines l_1, l_2, l_3 which are determined by the equation

$$\operatorname{Im} \int_{z_0}^{z} (\sqrt{p_1(t)} - \sqrt{p_2(t)})dt = 0\,,$$

and from $\overline{z_0}$ there arise the Stokes lines $l_1^\star$, $l_2^\star$, $l_3^\star$.

For small $|z - z_0|$ we have

$$\sqrt{p_1(z)} - \sqrt{p_2(z)} = \frac{2\sqrt{D(z)}}{\sqrt{\phi(z) + \sqrt{D(z)}} + \sqrt{\phi(z) - \sqrt{D(z)}}}$$
$$\sim \sqrt{\frac{2i\alpha(x_0)(\phi_1'(x_0) - \phi_2'(x_0))}{\phi(x_0)}}\sqrt{z - z_0}\,, \tag{18}$$

so that one of the Stokes lines (we will call it l_2) touches the imaginary axis at z_0, and Im $z >$ Im z_0 for $z \in l_2$. We denote by l_1 the Stokes line on which Re $z <$ Re z_0. Further, there is a line l joining the points z_0 and $\overline{z_0}$, with $l = l^\star$, on which

$$\operatorname{Re} \int_{z_0}^{z} (\sqrt{p_1(t)} - \sqrt{p_2(t)})dt = 0\,.$$

Locally the Stokes lines are the same as those for $h^2 w'' + D(z)w = 0$.

We will find the 1-admissable domain D_1^+ (§ 5): for each $z \in D_1^+$ there are canonical paths γ_{1k}, $1 \leqslant k \leqslant 4$, joining z_{1k} and z. The path γ_{11} is arbitrary, and the functions

$$\operatorname{Re}\left[i \int_{z_{12}}^{\zeta} (\sqrt{p_2(t)} - \sqrt{p_1(t)})dt\right]\,, \quad \operatorname{Re}\left[-i \int_{z_{13}}^{\zeta} \sqrt{p_1(t)}dt\right]\,,$$
$$\operatorname{Re}\left[-i \int_{z_{14}}^{\zeta} (\sqrt{p_1(t)} + \sqrt{p_2(t)})dt\right]$$

are non-increasing as ζ moves along the canonical path γ_{1k} from z_{1k} to z.

Let D be a rectangle $a_1 \leqslant$ Re $z < b_1$, $|\text{Im } z| \leqslant c$, containing the interval I and the points z_0, $\overline{z_0}$, where $c > 0$ is sufficiently small so that the coefficients of (1) are holomorphic in D, and put $\tilde{D}_1 = D \backslash (l_3 \cup l_2^\star)$. Choose a_1, b_1 close to a, b and put $z_{12} = a_1 + ic$. Then for any $z \in \tilde{D}_1$ there is a canonical path γ_{12} joining z_{12} and z and lying in $\tilde{D}_1$ (Chap. 3, § 8, paragraph 8).

Let us construct the canonical path $\gamma_{13}(z)$. Since $p_1(x) > 0$ for $x \in I$ and $\delta > 0$ is small, the level curves Re $S_1(z) =$ const., lying in D and not passing through z_0 and $\overline{z}_0$, are smooth curves close to the intervals Re $z =$ const. The function $-iS_1(0, z)$ maps $\tilde{D}$ one-to-one onto a domain U in the complex z-plane which is close to the rectangle Π with axes parallel to the coordinate axes and with two cuts L_3, $L_2^\star$. Here L_3, $L_2^\star$ are the images of the cuts l_3, $l_2^\star$ which differ only slightly from the segments Im $S =$ const., Re $S =$ const. respectively. We put $z_{13} = \overline{z_{12}}$ and $A = -iS_1(0, z_{13})$. If we remove from U the points for which Re $S_1 <$ Re A and, if possible, the right-hand half-neighbourhood of L_3, then any point z in the pre-image $\tilde{D}_1$ of this domain can be joined to z_3 by a canonical path $\gamma_{13}(z)$. In precisely the same way we can construct the canonical path $\gamma_{14}(z)$ with initial point z_{13}.

In this way there is a domain $D_1^+ \supset (I \cup z_0 \cup \overline{z_0})$ in which the asymptotic expansion (4) is valid, and this domain differs only slightly from $\tilde{D}_1$.

4.2 The Solutions u_1^-, u_2^+, u_2^-. Let us construct the canonical paths $\gamma_{2j}(z)$, $j = 1, 3, 4$, along which the functions

$$\mathrm{Re}\left[i \int_{z_{21}}^{\zeta} (\sqrt{p_1(t)} - \sqrt{p_2(t)})dt\right], \quad \mathrm{Re}\left[-i \int_{z_{23}}^{\zeta} (\sqrt{p_1(t)} + \sqrt{p_2(t)})dt\right],$$

$$\mathrm{Re}\left[-i \int_{z_{24}}^{\zeta} \sqrt{p_1(t)}dt\right]$$

are non-increasing. Let D be the same domain as in paragraph 4.1 and $D_2 = D \backslash (l_2 \cup l_3^*)$. The same reasoning as in paragraph 2.1 shows that (4) is valid for u_2^+ in the domain $D_2^+ \supset (I \cup z_0 \cup \overline{z_0})$ which only slightly differs from D_2. We put

$$u_1^-(z,h) = \overline{u_1^+(\overline{z},h)}, \quad u_2^-(z,h) = \overline{u_2^+(\overline{z},h)}.$$

Then the asymptotic expansions for u_1^- and u_2^- are valid in the domains $(D_1^+)^\star$ and $(D_2^+)^\star$ respectively. In particular the asymptotic expansions for $u_{1,2}^\pm$ are valid on $[a,\ x_0 - \varepsilon]$.

4.3 The Solutions $v_j^\pm$. Let D be the same domain as in paragraph 4.1, $\tilde{D} = D \backslash (l_1 \cup l_2^*)$, and $z_2 = b_1 + ic$. Then for any $z \in \tilde{D}$ there is a canonical path $\gamma_{12}(z)$ with initial point z_2, along which Re $[i(S_2(0,z) - S_1(0,z))]$ is non-increasing. In the same way as in paragraph 4.1 one shows that (4) is applicable for v_1^+ in the domain D_1^+, which differs only slightly from $\tilde{D}_1$. Let $\tilde{D}_3 = D \backslash (l_2 \cup l_1^*)$; then (4) is applicable for v_2^+ in D_2^+, close to $\tilde{D}_2$. We put

$$v_1^-(z,h) = \overline{v_1^+(\overline{z},h)}, \quad v_2^-(z,h) = \overline{v_2^+(\overline{z},h)}.$$

4.4 Connection Formulae. We have

$$\begin{aligned} v_1^+ &= A_1^+ u_1^+ + A_2^+ u_2^+ + A_1^- u_1^- + A_2^- u_2^- , \\ v_2^+ &= B_1^+ u_1^+ + B_2^+ u_2^+ + B_1^- u_1^- + B_2^- u_2^- , \end{aligned} \tag{19}$$

where $A_j^\pm$ and $B_j^\pm$ depend only on h. Because of the choice of the solutions

$$\begin{aligned} v_1^- &= \overline{A_1^-} u_1^+ + \overline{A_2^-} u_2^+ + \overline{A_1^+} u_1^- + \overline{A_2^+} u_2^- , \\ v_2^- &= \overline{B_1^-} u_1^+ + \overline{B_2^-} u_2^+ + \overline{B_1^+} u_1^- + \overline{B_2^+} u_2^- . \end{aligned} \tag{20}$$

It is required to find the asymptotic behaviour of the coefficients $A_j^\pm$ and $B_j^\pm$ as $h \to 0$, $0 \leqslant \delta \leqslant \delta_0$, where $\delta_0 > 0$ is sufficiently small, but not depending on h.

From paragraphs 4.1–4.3 it follows that there is an asymptotic expansion of the form (4) for v_1^+ and $u_{1,2}^{\pm}$ at points lying in a domain bounded by l_1 and l_2. We fix such a point $\tilde{z}$ and put $z = \tilde{z}$ in the identities

$$\begin{aligned} v_1^+ &= A_1^+ u_1^+ + A_2^+ u_2^+ + A_1^- u_1^- + A_2^- u_2^- , \\ \frac{d}{dz} v_1^+ &= \frac{d}{dz}(A_1^+ u_1^+ + A_2^+ u_2^+ + A_1^- u_1^- + A_2^- u_2^-) . \end{aligned} \tag{21}$$

Then $A_1^- = \Delta_1^- / \Delta$, where Δ is the determinant of the system and Δ_1^- is the determinant obtained from Δ by replacing the third column by $(v_1^+, (v_1^+)'^T)$. The branches of all the roots appearing in the formulae (4) are positive for $z \in I$ and all the values of these branches at $\tilde{z}$ for $u_j^{\pm}$ are obtained by extending analytically along $\gamma_1 = [x_0, \tilde{z}]$. Therefore $\Delta = 16 + O(h)$.

The branches of the roots appearing in formula (4) for the asymptotic expansion of v_1^+ are obtained by extending analytically along the path γ_2, which joins x_0 with $\tilde{z}$ and bypasses the turning point on the right. Since $\delta > 0$ is small

$$\begin{aligned} &\operatorname{Im} \int_{\gamma_1} \sqrt{p_j(z)} dz > 0 , \quad j = 1, 2 , \\ &\operatorname{Im} \int_{\gamma_2} \sqrt{p_1(z)} dz > 0 , \end{aligned} \tag{22}$$

so the solutions v_1^+, u_1^+, u_2^+ decrease exponentially, while u_1^- and u_2^- increase exponentially at $\tilde{z}$. Also

$$A_1^- = \exp\left\{ \frac{i}{h} \left(\int_{\gamma_2} \sqrt{p_1(z)} dz + \int_{\gamma_1} \sqrt{p_1(z)} dz \right) \right\} O(1)$$

and therefore

$$A_1^- = O(e^{-\xi/h}) , \quad \xi > 0 , \quad \xi = \operatorname{Im} \int_0^z \sqrt{\phi(t)} dt + O(\delta) . \tag{23}$$

In precisely the same way we get $A_2^- = O(\exp(-\xi/h))$. Further, because of the choice of the branch for the roots we have

$$\int_{\gamma_2} \sqrt{p_1(z)} dz - \int_{\gamma_1} \sqrt{p_1(z)} dz = A , \quad \operatorname{Im} A < 0 ,$$

so that $A_1^+ = O(e^{a/h})$. The right-hand side of this formula increases exponentially as $h \to 0$ and therefore the asymptotic behaviour of the coefficient A_1^+ is still undetermined. We will find the asymptotic behaviour of $A_2^+ = \Delta_2^+ / \Delta$, where the determinant Δ_2^+ differs from Δ in that the second column is replaced by the column $(u_2^+, (u_2^+)'^T)$. We make a cut along l, going through z_0 and $\tilde{z}$, up to the boundary of D. Then the branches of $\sqrt{D(z)}$ on the sides of the cut differ by the factor -1, so that $p_1(z)|_{l_+} = p_2(z)|_{l_-}$, where l_+, l_- are the right and left sides of the cut. Therefore

$$\int_0^{\tilde{z}} \sqrt{p_1(t)}dt - \int_0^{\tilde{z}} \sqrt{p_2(t)}dt = \int_0^{z_0} (\sqrt{p_1(t)} - \sqrt{p_2(t)})dt = \eta\,, \quad \operatorname{Im} \eta > 0\,,$$

where the integrals are taken along the paths γ_1 and γ_2, and

$$v_1^+(\tilde{z}, h) = e^{i\eta/h} u_1^+(\tilde{z}, h)[1 + O(h)]\,.$$

Hence we obtain

$$A_2^+ = e^{i\eta/h}[1 + O(h)]\,. \tag{24}$$

Thus the coefficients A_1^- and A_2^- are exponentially small, the asymptotic behaviour of A_2^+ is given by (24), and A_1^+ is still undetermined. Similarly we can prove that

$$B_1^+ = e^{-2i\eta/h}[1 + O(h)]\,, \quad B_{1,2}^- = O(e^{-\xi/h})\,,$$

while the coefficient B_2^+ is still undetermined. Information about A_1^+ and B_2^+ can be found from the identity

$$v^T u' - u^T v' = \text{const.}\,,$$

where u and v are solutions of (1). In particular there is the formula

$$|A_1^+|^2(1 + \varepsilon_1) + |A_2^+|^2(1 + \varepsilon_2) + \varepsilon_3 \operatorname{Im} (\overline{A_1^+ A_2^+}) = 1 + \varepsilon_4\,,$$

$$\varepsilon_j = o(h)\,,$$

and a similar identity is true for B_1^+ and B_2^+.

References*

Abramov 1 — Abramov, A.A.: Boundary conditions at singular points for systems of linear ordinary differential equations. Zh. Vychisl. Mat. Mat. Fiz. *11* (1971), No.1, 275–278; MR *44* #1857, Zbl.244.34011. Engl. transl.: USSR Comput. Math. Math. Phys. *11* (1971), No.1, 363–367 (1973)

Abramov 2 — Abramov, A.A., Balla, K., Konyukhova, N.B.: Moving boundary conditions away from singular points for systems of ordinary differential equations. Soobshch. Vychisl. Mat. Akad. Nauk SSSR, Moscow (1981) 1–64 [Russian]; Zbl.488.34002

Abramovitz — Abramovitz, M., Stegun, I.A. (eds.): Handbook of Mathematical Functions. Nat. Bureau Stand., Appl. Math. Series *55*, 1964. Table errata, Math. Comput. *21*, 747 (1967); Zbl.171.385

Alenitsyn 1 — Alenitsyn, A.G.: Splitting of the spectrum generated by a potential barrier in problems with symmetric potential. Differ. Uravn. *18* (1982), No.11, 1971–1975 [Russian]; MR *84* b#34027, Zbl.522.34020

Alenitsyn 2 — Alenitsyn, A.G.: The Rayleigh's waves in nonhomogeneous elastic band. Prikl. Mat. Mekh. *28* (1964), No.5, 880–888. MR *32*#3361. Engl. transl.: J. Appl. Math. Mech. *28*, 1067–1076 (1965); Zbl. 139,199

Arnol'd — Arnol'd, V.I.: Matrices depending on parameters. Usp. Mat. Nauk *62* (1971), No.2, 101–114; MR *46* 400, Zbl.259.15011. Engl. transl.: Russ. Math. Surv. *62* (1971), No.2, 29–43 (1972)

Babich — Babich, V.M., Buldyrev, V.S.: Asymptotic Methods in Problems of Diffraction of Short Waves. Moscow, Nauka, 1972 [Russian]; Zbl.255.35002

Bagirov — Bagirov, L.A., Myshkis, P.A.: Integral representations of the fundamental system of solutions of one class of equations of high order. Differ. Uravn. *23* (1987), No.6, 1072–1074 [Russian]; Zbl. 641.34057

Bellman — Bellman, R.: Stability Theory of Differential Equations. London, McGraw-Hill, 1953; Zbl.53,247

Berk — Berk, H.L., Roberts, K.V.: New Stokes line in WKB-theory. J. Math. Phys. *23* (1982), No.6, 988–1002; Zbl.488.34050

Birkhoff — Birkhoff, G.D.: Quantum mechanics and asymptotic series. Bull. Am. Math. Soc. *39* (1933), 681–700; Zbl.8,89

* For the convenience of the reader, references to reviews in Zentralblatt für Mathematik (Zbl.), compiled using the MATH database, Mathematical Reviews (MR), and Jahrbuch über die Fortschritte der Mathematik (Jbuch) have, as far as possible, been included in this bibliography.

Bogaevskij — Bogaevskij, V.N., Povzner, A. Ya.: Algebraic Methods in Nonlinear Perturbation Theory. Moscow, Nauka, 1987; Zbl.611.34002

Braaksma — Braaksma, B.L.J.: Asymptotic analysis of a differential equation of Turrittin. SIAM J. Math. Anal. *2* (1971), No.1, 1–16; MR *45*#2271, Zbl. 213,103

Buldyrev — Buldyrev, V.S., Slavyanov, S. Yu.: Uniform asymptotic expansions for solutions of Schrödinger type equations with two transition points. Vestn. Leningr. Univ., Mat. *22* (1968), No.4, 70–84 [Russian]; MR *39*#3775

Butuzov — Butuzov, V.F., Vasil'eva, A.B., Fedoryuk, M.V.: Asymptotic methods in the theory of ordinary differential equations. Itogi Nauki Tekh., Ser. Mat. Anal. (1969), 5–73. Engl. transl.: Prog. Math. *8*, 1–82 (1970); Zbl.246.34055

Cherry — Cherry, T.M.: Uniform asymptotic formulae for functions with transition points. Trans. Am. Math. Soc. *68*, 224–257 (1950); Zbl.36,61

Coddington — Coddington, E.A., Levinson, N.: Theory of Ordinary Differential Equations. New York, McGraw-Hill, 1955; Zbl.64,330

Cole — Cole, J.: Perturbation Methods in Applied Mathematics. Blaisdell Publ. Co. Waltham, Massachusetts Toronto, London, 1968; Zbl.162,126

Devinatz 1 — Devinatz, A.: An asymptotic theorem for systems of linear differential equations. Trans. Am. Math. Soc. *160* (1971), No.10, 353–363; MR *44*#545, Zbl.252.34062

Devinatz 2 — Devinatz, A.: The deficiency index of a certain class of ordinary self-adjoint differential operators. Adv. Math. *8* (1972), 434–473; MR *45*#7154, Zbl.259.34019

Devinatz 3 — Devinatz, A.: The deficiency index of certain fourth-order ordinary self-adjoint differential operators. J. Math. Oxford, II. Ser. *23* (1972), 267–286; MR *47*#5350, Zbl.263.34022

Devinatz 4 — Devinatz, A., Kaplan, J.L.: Asymptotic estimates for solutions of linear systems of ordinary differential equations having multiple characteristic roots. Indiana Univ. Math. J. *22* (1972), No.4, 355–366; Zbl.267.34044

Dnestrovskij — Dnestrovskij, Yu. N., Kostomarov, D.P.: The asymptotic behaviour of eigenvalues of a nonselfadjoint boundary value problem. Zh. Vychisl. Mat. Mat. Fiz. *4* (1964), Vol. 2, 267–277; MR *29*#1379. Engl. transl.: USSR Comput. Math. Math. Phys. *4* (1964), No.2, 77–91 (1966); Zbl.154,91

Dorodnitsyn — Dorodnitsyn, A.A.: Asymptotic laws for the distribution of eigenvalues for some particular forms of second-order differential equations. Usp. Mat. Nauk 7(1952), No.6, 3–96; [Russian] MR *14*#876, Zbl.48,324

Eastham 1 — Eastham, M.S.P.: A repeated transformation in the asymptotic solutions of linear differential systems. Proc. Soc. Edinb., Sect. A 102, (1986), 173–188; Zbl.595.34059

Eastham 2 — Eastham, M.S.P.: Asymptotic formulae of Liouville-Green type for higher-order differential equations. J. Lond. Math. Soc. II. Ser. *28* (1983), No.3, 507–528; Zbl.532.34038

Eastham 3 — Eastham, M.S.P.: The Liouville-Green asymptotic theory for second-order differential equations: a new approach and some extensions. Lect. Notes Math. *1032* (1983), 110–122; Zbl.525.34045

Erdelyi — Erdelyi, A.: Asymptotic Expansions. New York, Dover Publ., 1956; Zbl.70,290

Evgrafov — Evgrafov, M.A., Fedoryuk, M.V.: Asymptotic behaviour of the solutions of the equation $w''(z) - p(z,\lambda)w(z) = 0$ as $\lambda \to +\infty$ in the complex z-plane. Usp. Mat. Nauk. *21* (1966), Vol. 1, 3–50; MR *35*#459. Engl. transl.: Russ. Math. Surv. *21* (1966), 1–48; Zbl.173,338

Fedoryuk 1 — Fedoryuk, M.V.: Adiabatic invariants of a system of linear operators in scattering theory. Differ. Uravn. *12* (1976), No.6, 1012–1018; MR *55*#1961, Zbl.338.34057. Engl. transl.: Differ. Equations *12*, 713–718 (1977)

Fedoryuk 2 — Fedoryuk, M.V.: Analytic proportion of the scattering amplitude in the one-dimensional case. I. Differ. Uravn. *4* (1968), No.10, 1842–1853. II. Differ. Uravn. *5* (1969), No.3, 507–517; MR *39*#7894. Engl. transl.: I. Differ. Equations *4* (1968), No.10, 948–954. II. Differ. Equations *5* (1969), No.3, 402–410; Zbl.202,82

Fedoryuk 3 — Fedoryuk, M.V.: Asymptotic behaviour of eigenvalues and eigenfunctions of one-dimensional singular differential operators. Dokl. Acad. Nauk SSSR *169* (1966), No.2, 288–291; MR *34*#2992. Engl. transl.: Sov. Math. Dokl. *7* (1966), No.4, 929–932; Zbl.154.90

Fedoryuk 4 — Fedoryuk, M.V.: Asymptotic behaviour of eigenvalues and eigenfunctions of the Sturm-Liouville operator with a complex-valued potential-polynominal. I. Differ. Uravn. *8* (1972), No.5, 811–816. II. Differ. Uravn. *10* (1974), No.6, 1068–1073. I; MR *46*#5722, Zbl.247.34021. II; MR *50*#678, Zbl.307.34017. Engl. transl.: I. Differ. Equations *8* (1972), No.5, 616–620. II. Differ. Equations *10* (1974), No.6, 823–828

Fedoryuk 5 — Fedoryuk, M.V.: Asymptotic behaviour of the discrete spectrum of the operator $w''(x) - \lambda^2 p(x)w(x)$. Mat. Sb. *68* (1965), No.1, 81–110; MR *32*#4315, Zbl.238.34032

Fedoryuk 6 — Fedoryuk, M.V.: Asymptotic behaviour of the solutions of ordinary linear nth-order differential equations. Differ. Uravn. *2* (1966), No.4, 492–507; MR *34*#446. Engl. transl.: Differ. Equations *2* (1966), No.4, 250–258; Zbl.176,55

Fedoryuk 7 — Fedoryuk, M.V.: Asymptotic methods in analysis. In: Itogi Nauki Tekh., Sovrem. Probl. Mat. Fund. Napr. *13*, Moscow, VINITI, 1985, 93–210. Engl. transl. in: Encycl. Math. Sc. *13*, Berlin Heidelberg New York, Springer-Verlag, 1989, 83–192; Zbl.655.41034

Fedoryuk 8 — Fedoryuk, M.V.: Asymptotic methods in the theory of homogeneous singular differential operators. Tr. Mosk. Mat. Ob.-va *15* (1966), 296–345; MR *34*#7870. Engl. transl.: Trans. Mosc. Math. Soc. *15* (1966), 333–386; Zbl.163,324

Fedoryuk 9 — Fedoryuk, M.V.: Asymptotic methods in the theory of ordinary linear differential equations. Mat. Sb., Nov. Ser. *79*, 477–516 (1969). Engl. transl.: Math. USSR, Sb. *8* (1969), No.4, 451–491; Zbl.215, 448

Fedoryuk 10 — Fedoryuk, M.V.: Homogeneous problem on scattering in quasi-classical approximation. I. Differ. Uravn. *1* (1965), No.5, 631–646. II. Differ. Uravn. *1* (1965), No.11, 1525–1536 [Russian]; MR *35*#4517

Fedoryuk 11 — Fedoryuk, M.V.: The Saddle-Point Method. Moscow, Nauka, 1977 [Russian]; Zbl.463.41020

Fedoryuk 12 Fedoryuk, M.V.: The Sturm-Liouville problem with regular singular points. I. Differ. Uravn. *18* (1982), No.12, 2166–2173; MR *84*k:34030a, Zbl.531.34016. II. Differ. Uravn. *19* (1983), No.2, 278–286; MR *84*k:34030b, Zbl.532.34010. Engl. transl.: I. Differ. Equations *18* (1982), No.12, 1550–1557. II. Differ. Equations *19* (1983), No.2, 208–215

Fedoryuk 13 Fedoryuk, M.V.: Topology of Stokes lines for second-order equations. Izv. Akad. Nauk SSSR. Ser. Mat. *29* (1965), No.3, 645–656; MR *31*#3663; Zbl.168,64. Engl. transl.: Am. Math. Soc., Transl., II. Ser. *89*, 89–102 (1970)

Feshchenko Feshchenko, S.F., Shkil', N.I., Nikolenko, L.D.: Asymptotic Methods in the Theory of Linear Differential Equations. Kiev, Naukova Dumka, 1966 [Russian]; Zbl.141,280

Fröman Fröman, N., Fröman, P.O.: JWKB-Approximation. Amsterdam, North-Holland Publ. Co., 1965; Zbl.129,419

Gilbert Gilbert, R.C.: The deficiency index of a third-order operator. Pac. J. Math. *68* (1977), No.2, 369–397; MR *57*#3526, Zbl.335.34009

Gingold Gingold, H.: Simplification of linear homogeneous differential equations with moving singularities. Funkc. Ekvacioj, Ser.Int. *19* (1976), 203–225; MR *55*#5698, Zbl.392.34023

Gollwitzer Gollwitzer, H.E., Sibuya, Y.: Stokes multipliers for subdominant solutions of second-order differential equations with polynomial coefficients. Univ. Minnesota, Inst. Technology, School Math., Minneapolis (1967), 1–67; MR *42*#8009; appeared in: J. Reine Angew. Math. *243*, 98–119 (1970); Zbl.215,141

Golubev Golubev, V.V.: Lectures in the Analytic Theory of Differential Equations. Moscow Leningrad, Gostekhizdat, 1950; Zbl.38,242. German transl.: Vorlesung über Differentialgleichungen in Komplexen. Berlin, VEB Deutscher Verlag der Wissenschaften, 1958 (Hochschulbücher für Mathematik *43*); Zbl.82,70

Harris 1 Harris, W.A., Lutz, D.A.: Asymptotic integration of adiabatic oscillators. J. Math. Anal. Appl. *51* (1975), 76–93; MR *51*#6009, Zbl. 315.34070

Harris 2 Harris, W.A., Lutz, D.A.: On the asymptotic integration of linear differential systems. J. Math. Anal. Appl. *48* (1974), 1–16; MR *50*#7698, Zbl.304.34043

Hartman 1 Hartman, P.: Asymptotic integration of ordinary differential equations. SIAM J. Math. Anal. *14* (1982), No.4, 772-779; Zbl.5453.4038

Hartman 2 Hartman, P.: Ordinary Differential Equations. New York, Wiley, 1964; Zbl.125,321

Heading 1 Heading, J.: An Introduction to Phase Integral Methods. Methuen and Co., London, 1962; Zbl.115,71

Heading 2 Heading, J.: Generalized approximate methods for transmission through a barrier governed by a differential equation of order $2n$. Math. Proc. Camb. Philos. Soc. *85* (1979), 361–377; MR *83*e:81035, Zbl.417.34026

Hille Hille, E.: Ordinary Differential Equations in the Complex Domain. New York, Wiley, 1976; Zbl.343.34007

Ince Ince, E.L.: Ordinary Differential Equations. London, Longmans, Green & Co, 1927; Jbuch 53, 399

Iwano Iwano, N.: Asymptotic solutions of a system of linear ordinary differential equations containing a small parameter. Funkc. Ekvacioj, Ser. Int. *5* (1963), 71–134; MR *30*#1277, Zbl.123,49

Jenkins — Jenkins, J.A.: Univalent functions and conformal mapping. Berlin Heidelberg New York, Springer-Verlag, 1958; Zbl.83,296

Kamke — Kamke, E.: Differential Equations. Methods of Solution and Solutions. Vol. 1: Ordinary Differential Equations. 4th ed., Akad. Verlag, Geest and Portig, Leipzig, 1951; Zbl.145,100

Kelly — Kelly, B.J.: Admissible domains for higher-order differential operators. Stud. Appl. Math. *60* (1979), 211–240; MR *80*h:34005, Zbl.407.34043

Kohno 1 — Kohno, M.: A multipotent connection problem. Lect. Notes Math. *1015* (1982), 136–171; Zbl.538.34035

Kohno 2 — Kohno, M.: Derivatives of Stokes multipliers. Hiroshima Math. J. *14* (1984), No.2, 247–256; Zbl.568.34008

Kohno 3 — Kohno. M., Yokoyama, T.: A central connection problem for a normal system of differential equations. Hiroshima Math. J. *14* (1984), No.2, 257–263; Zbl.572.34006

Kohno 4 — Kohno, M., Ohkohchi, S., Kohmoto, T.: On full uniform simplification of even order linear differential equations with a parameter. Hiroshima Math. J. *9* (1979), 747–767; MR *80*j:34071, Zbl.432.34003

Konyukhova — Konyukhova, N.B.: The stable manifolds of solutions for some non-linear systems of ordinary differential equations with singularities. Zh. Vychisl. Mat. Mat. Fiz. *13* (1973), No.3, 609–626; MR *48*#6570; Zbl.289.34026. Engl. transl.: USSR Comput. Math. Math. Phys. *13*, No.3, 91–113 (1974)

Kostyuchenko — Kostyuchenko, A.G., Sargsyan, I.S.: Distribution of Eigenvalues. Moscow, Nauka, 1970 [Russian]; Zbl.478.34022

Kucherenko 1 — Kucherenko, V.V.: Asymptotic behaviour of the solution to the system $A(x, -ih\delta/\gamma x)u = 0$ as $h \to 0$. Izv. Akad. Nauk SSSR, Ser. Mat. *38* (1974), 625–662; MR *51*#3663, Zbl.308.35080. Engl. transl.: Math. USSR, Izv. *8* (1974), 631–666 (1975)

Kucherenko 2 — Kucherenko, V.V., Osipov, Yu. V.: Asymptotic solution of ordinary differential equations with degenerate symbols. Mat.Sb. *118* (1982), No.1, 74–103; MR *83*g:34054, Zbl.514.34046. Engl. transl.: Math. USSR, Sb.. *46*, 75–104 (1983); Zbl.527.34058

Landau — Landau, L.D., Lifshitz, E.M.: Quantum Mechanics (Theoretical Physics, Vol. 3). London, Pergamon, 1958; Zbl.81,222

Langer — Langer, R.E.: The asymptotic solutions of certain linear ordinary differential equations of the second order. Trans. Am. Math. Soc. *36* (1934), 90–106; Zbl.8,312

Lee — Lee, R.Y.: Turning point problems of almost diagonal systems. J. Math. Anal. Appl. *24*, 509–526 (1968); MR *38*#2390, Zbl.193,59

Leung 1 — Leung, A.: A doubly asymptotic existence theorem and application to order reduction. Proc. Lond. Math. Soc., III. Ser. *33* (1976), 151–176; MR *54*#666, Zbl.411.34006

Leung 2 — Leung, A.: Lateral connections for asymptotic solutions for higher order turning points in unbounded domains. J. Math. Anal. Appl. *50* (1975), 560–578; MR *51*#8572, Zbl.303.34044

Leung 3 — Leung, A.: Studies of doubly asymptotic series solutions for differential equations in unbounded domains. J. Math. Appl. *44* (1973), 238–263; MR *50*#5118, Zbl.274.34007

Leung 4 — Leung, A., Meyer, K.: Adiabatic invariants of Hamiltonian systems. J. Differ. Equations *17* (1975), 32–43; MR *51*#13374, Zbl. 265.34018

Levitan — Levitan, B.M, Sargsyan, I.S.: Introduction to Spectral Theory. Moscow, Nauka, 1970. Engl. transl.: Transl. Math. Monogr. Vol. *39*, Providence (1975); Zbl.225.47019

Lomov — Lomov, S.A.: Introduction ot the General Theory of Singular Perturbations. Moscow, Nauka, 1981 [Russian]; Zbl.514.34049

Maslov 1 — Maslov, V.P.: Perturbation Theory and Asymptotic Methods. Moscow, Moscow Univ. Press, 1965. French transl.: Paris, Gauthier-Villars, 1972; Zbl.247.47010. New Russian edition: Moscow, Nauka, 1988; Zbl.653.35002

Maslov 2 — Maslov, V.P., Fedoryuk, M.V.: Quasi-Classical Approximation for the Equations of Quantum Mechanics. Moscow, Nauka, 1976. Engl. transl.: Semi-Classical Approximations in Quantum Mechanics. Reidel, Dordrecht, 1981; Zbl.449.58002

Meyer 1 — Meyer, R.E., Painter, J.F.: Connection for wave modulation. SIAM J. Math. Anal. *14* (1983), No.3, 450–462; Zbl.511.34043

Meyer 2 — Meyer, R.E., Painter, J.F.: Irregular points of modulation. Adv. Appl. Math. *4* (1983), No.2, 145–174; Zbl.511.34042

Meyer 3 — Meyer, R.E., Painter, J.F.: On the Schroedinger connection. Bull. Am. Math. Soc., New Ser. *8* (1983), No.1, 73–76; Zbl.511.34044

Moiseev — Moiseev, N.N.: Asymptotitic Methods in Non-Linear Mechanics. 2nd ed. Moscow, Nauka, 1981 [Russian]; Zbl.527.70024, Zbl.193,248

Naimark — Naimark, M.A.: Linear Differential Operators. I, II. Moscow, Nauka, 1974. Engl. transl.: F. Unger Publ. Co., New York, 1967, 1968; Zbl.219,190, Zbl.227.34020

Nakano 1 — Nakano, M.: Second-order linear ordinary differential equations with turning points and singularities. I. Kodai Math. Sem. Reports *29* (1977), 88–102; MR *57*#16879, Zbl.409.34055

Nakano 2 — Nakano, M., Nishimoto, T.: On a secondary turning point problem. Kodai. Math. Sem. Reports *22* (1970), 355–384; MR *43*#2302, Zbl.208,111

Nayfeh — Nayfeh, A.H.: Perturbation Methods. N.Y., London Sydney Toronto, John Wiley and Sons, 1973; Zbl.265.35002

Nishimoto 1 — Nishimoto, T.: On an extension theorem and its application for turning point problems of large order. Kodai Math. Sem. Reports *25* (1973), 458–489; MR *49*#762, Zbl.275.34060

Nishimoto 2 — Nishimoto, T.: On the central connection problem at a turning point. Kodai Math. Sem. Reports *22* (1970), 30–44; MR *42*#605, Zbl.193,52

Nishimoto — Nishimoto, T.: Uniform asymptotic properties of the WKB method. Kodai Math. J. *4* (1981), 71–81; MR *82*m:34058, Zbl.486.34049

Ohkohchi — Ohkohchi, S.: Uniform simplification in a full neighbourhood of a turning point. Hiroshima Math. J. *15* (1985), No.3, 493–580; Zbl.595.34034

Olver 1 — Olver, F.W.J.: Asymptotics and Special Functions. New York London, Academic Press, 1974; Zbl.303.41035

Olver 2 — Olver, F.W.J.: Connection formulae for second-order differential equations having an arbitrary number of turning points of arbitrary multiplicities. SIAM J. Math. Anal. *8* (1977), No.4, 673–700; Zbl.353.34064

Olver 3 — Olver, F.W.J.: Connection formulae for second-order differential equations with multiple turning points. SIAM J. Math. Anal. *8* (1977), No.1, 127–154; MR *55*#793, Zbl.344.34050

Olver 4 Olver, F.W.J.: General connection formulae for Liouville-Green approximations in the complex plane. Philos. Trans. R. Soc. Lond., Ser. A *289* (1978), 501–548; MR *80*a:34083, Zbl.389.34040

Olver 5 Olver, F.W.J.: Improved error bounds for second-order differential equations with two turning points. J. Res. Nat. Bur. Stand. Sect. B *80* (1976), No.4, 437–440 (1977); MR *56*#7212, Zbl.357.34009

Olver 6 Olver, F.W.J.: Introduction to Asymptotics and Special Functions. New York, Academic Press, 1974; Zbl. 308.41023

Olver 7 Olver, F.W.J.: Second-order differential equations with fractional transition points. Trans. Am. Math. Soc. *266* (1977), 227–241; MR *55*#3450, Zbl.355.34004

Olver 8 Olver, F.W.J.: Second-order differential equations with two turning points. Philos. Trans. R. Soc. Lond., Ser. A *278* (1975), 137–174; MR *51*#6073, Zbl.301.34072

Paris Paris, R.B., Wood, A.D.: On the asymptotic expansions of solutions of an nth-order linear differential equation with power coefficients. Proc. R. Ir. Acad., Sect. A *85* (1985), No.2, 201–220; Zbl.605.34049

Povzner Povzner, A. Yu.: Stokes constants for the Schroedinger equation with polynomial coefficients. Teor. Mat. Fiz. *51* (1982), No.1, 54–72 [Russian]; MR *84*m:81055, Zbl.509.34022

Raman Raman, V.M.: On singular ordinary linear differential operators related to Orr-Sommerfeld stability equation. J. Math. Anal. Appl. *94* (1983) No.2, 536–560; Zbl.529.34032

Rapoport Rapoport, I.M.: Some asymptotic methods in the theory of differential equations. Kiev, Izd-vo Akad. Nauk. SSSR, 1954 [Russian]; MR *17*#734

Roos 1 Roos, H.E.: Die asymptotische Lösung einer linearen Differentialgleichung mit dreisegmentigem charakteristischen Polygon. Math. Nachr. *88* (1979), 93–103; MR *80*g:34058, Zbl.429.34055

Roos 2 Roos, H.E.: Die asymptotische Lösung einer linearen Differentialgleichung zweiter Ordnung mit zweisegmentigem charakteristischem Polygon. Beitr. Anal. 7 (1975), 55–63; MR *57*#10138, Zbl.271.34068

Schäfke 1 Schäfke, R.: A connection problem for a regular and irregular singular point of complex ordinary differential equations. SIAM J. Math. Anal. *15* (1984), No.2, 253–271; Zbl.536.34001

Schäfke 2 Schäfke, R.: Über das globale Verhalten der Normallösungen von $x(t) = (B + t^{-1}A)x(t)$ und zweier Arten von assoziierten Funktionen. Math. Nachr. *121* (1985), 123–145; Zbl.563.34003

Shirikyan Shirikyan, R.A.: Asymptotic methods in the theory of homogeneous singular differential operators of odd order. Differ. Uravn. *3* (1967), No.11, 1942–1956; MR *36*#5451. Engl. transl.: Differ. Equations *3* (1967), No.11, 1010–1017; Zbl.173,105

Sibuya 1 Sibuya, Y.: Global Theory of a Second-Order Linear Differential Equation with a Polynomial Coefficient. Amsterdam, North Holland Publishing Company, 1975; Zbl.322.34006

Sibuya 2 Sibuya, Y.: Some global properties of functions of one variable. Math. Ann. *161* (1965), 67–77; MR *33*#2644, Zbl.229.15011

Sibuya 3 Sibuya, Y.: Uniform simplification in a full neighbourhood of a transition point. Mem. Am. Math. Soc. *149* (1974); MR *55*#13020, Zbl.297.34051

Simonyan — Simonyan, S.G.: Asymptotic behaviour of wide gaps in the spectrum of a Sturm-Liouville operator with periodic potential. Differ. Uravn. *6* (1970), No.7, 1265–1272; MR *43*#6495, Zbl.272.34029. Engl. transl.: Differ. Equations *6*, 965–971 (1973)

Slavyanov — Slavyanov, S. Yu.: Asymptotic behaviour of Sturm-Liouville problems with a large parameter with close transition points. Differ. Uravn. *5* (1969), No.2, 313–325; MR *39*#7227. Engl. transl.: Differ. Equations *5* (1969), No.2, 258–267; Zbl.164,390

Smirnov — Smirnov, V.G.: Asymptotic behaviour as $\lambda \to \infty$ of the monodromy matrix of the equation $w''(z) - \lambda^2 q(z)w(z) = 0$. MIEM *5* (1969), 153–191

Stengle — Stengle, G.: Asymptotic estimates for the adiabatic invariance of a simple oscillator. SIAM J. Math. Anal. *8* (1977), No.4, 640–654; MR *56*#12467, Zbl.355.34046

Stueckelberg — Stueckelberg, E.C.: Theorie der unelastischen Stösse zwischen Atomen. Helv. Phys. Acta. *5* (1932), 370–442

Tamarkin — Tamarkin, Ya. D.: Some general problems of the theory of ordinary linear differential equations and the expansion of an arbitrary function in a series of fundamental functions. Petrograd (1917). Engl. transl.: Math. Z. *27* (1927), 1–54; Jbuch 53, 419

Titchmarsh — Titchmarsh, E.C.: Eigenfunction expansions associated with second-order differential equations. I and II. Clarendon Press, 1958–62; Zbl.99,52, Zbl.97,276

Turrittin — Turrittin, H.L.: Asymptotic expansion of the solutions of a system of ordinary linear differential equations in terms of parameter. Ann. Math. Stud. *29*, 81–116 (1952); Zbl.47,86

Vasil'eva — Vasil'eva, A.B., Butuzov, V.F.: Asymptotic Expansions of Solutions of Singular Perturbation Equations. Moscow, Nauka, 1973 [Russian]; Zbl.364.34028

Vishik 1 — Vishik, M.I., Lyusternik, L.A.: Regular degeneracy and boundary layers of linear differential equations with a small parameter. Usp. Mat. Nauk *12* (1957), No.5, 3–122; MR *25*#322. Engl. transl.: Am. Math. Soc., Transl., II. Ser. *20*, 239–364 (1962); Zbl.87,296

Vishik 2 — Vishik, M.I., Lyusternik, L.A.: The solution of some perturbation problems in the case of matrices of both self-adjoint and non-self-adjoint differential equations. Usp. Mat. Nauk *15* (1960). No.3, 3–80; MR *23*#A1920. Engl. transl.: Russ. Math. Surv. *15* (1960), No.3, 1–73; Zbl.96,87

Wasow 1 — Wasow, W.: Arnold's canonical matrices and the asymptotic simplification of ordinary differential equations. Linear Algebra Appl. *18* (1977), 163–170; MR *56*#9014, Zbl.364.34003

Wasow 2 — Wasow, W.: Asymptotic Expansions for Solutions of Ordinary Differential Equations. N.Y. London Sydney, John Wiley and Sons, 1965; Zbl.133,353

Wasow 3 — Wasow, W.: Asymptotic simplification of linear Hamiltonian differential equations with a parameter. J. Differ. Equations *2* (1966), No.4, 378–390; MR *34*#4632, Zbl.148,65

Wasow 5 — Wasow, W.: Calculation of an adiabatic invariant by tuning point theory. SIAM J. Math. Anal. *5* (1974), No.4, 673–700; MR *51*#6075, Zbl.289.34084

Wasow 6 — Wasow, W.: Linear Turning Point Theory. Berlin Heidelberg New York, Springer-Verlag, 1985; Zbl.558.34049

Wasow 7 Wasow, W.: Simple turning point problems in unbounded domains. SIAM J. Math. Anal. *1* (1970), No.2, 153–170; MR *41*#3901, Zbl.211,110

Wasow 8 Wasow, W.: Some recent results in the theory of adiabatic invariants. Int. Conf. Differ. Equat. New York, Acad. Press, 1975, 747–764; MR *53*#11179, Zbl.316.34062

Wasow 9 Wasow, W.: The central connection problem at turning points of linear differential equations. Comment. Math. Helv. *46* (1971), 65–86; MR *44*#2992, Zbl.211,110

Wasow 10 Wasow, W.: Turning point problems for systems of linear equations. I. Commun. Pure Appl. Math. *14* (1961), 657–673, II. Commun. Pure Appl. Math. *15* (1962), 173–187; MR *24*#A2094, Zbl.106,293, Zbl.142,344

Weinstein Weinstein, M.L., Keller, J.B.: Hill's equation with a large potential. SIAM J. Appl. Math. *45* (1985), No.2, 200–214; Zbl.578.34038

Yokoyama Yokoyama, T.: On connection formulae for a fourth-order hypergeometric system. Hiroshima Math. J. *15* (1985), No.2, 297–320; Zbl.585.34002

Zhdanova 1 Zhdanova, G.V.: Asymptotic behaviour of the eigenvalues for a self-conjugate singular operator of order $2n$. Differ. Uravn. *6* (1970), No.5, 639–652. Engl. transl.: Differ. Equations *6*, 639–651 (1973); Zbl.264.34032

Zhdanova 2 Zhdanova, G.V.: Formal asymptotic behaviour of a fundamental system of solutions of an equation in elasticity theory. Differ. Uravn. *18*, No.9, 1820–1821; MR *84*c:34081, Zbl.511.73016

Zhdanova 3 Zhdanova, G.V., Fedoryuk, M.V.: Asymptotic theory of systems of ordinary differential equations of second order and a scattering problem. Tr. Mosk. Mat. O.-va *34* (1977), 213–242; MR *57*#6682. Engl. transl.: Trans. Mosc. Math. Soc. *34* (1978), No.2, 205–236; Zbl.401.34042

Subject Index